# THE SLEEPWALKERS

Arthur Koestler alternated all his life between the man of action and the man of letters. Born in Budapest in 1905, he studied at Vienna University before becoming a journalist. As a foreign correspondent he travelled widely, visting the Middle East, Paris and Moscow. In 1937, while representing the *News Chronicle* in Spain, he was captured by Franco's troops and imprisoned under sentence of death. He was eventually released on the intervention of the British Government and returned to London. During the war he served with the French Foreign Legion and the British Army and in 1945 became Special Correspondent for *The Times* in Palestine. In the 1940s and early 1950s he was perhaps the most widely read political novelist of the time. *Darkness at Noon*, considered to be his masterpiece, was published in 1940, followed by *Arrival and Departure* (1943), *Thieves in the Night* (1946), *The Age of Longing* (1951) and *The Call Girls* (1972). From 1956, he became immersed in questions of science and mysticism, and had a huge following among young people. Explaining his varied interests, Koestler wrote: 'Out of my quarrels with the human condition I made my novels; the other books are attempts to analyse that same condition in scientific terms. In my more optimistic moments it seems to me that the two add up to a whole. At any rate, without both media I would feel only half alive.' *The Sleepwalkers* (1959) is the first book in his classic trilogy on the mind of man, which continued with *The Act of Creation* (1964) and ended with *The Ghost in the Machine* (1967). All three are published by Arkana. Among his other books are *The Yogi and the Commissar* (1945), *The Case of the Midwife Toad* (1971), two volumes of autobiography, *Arrow in the Blue* (1952) and *The Invisible Writing* (1954), and a selection of his writings with comments by the author entitled *Bricks to Babel* (1980).

Koestler recieved the Sonning Prize from the University of Copenhagen in 1968 and was awarded a number of honorary doctorates. He was a Fellow of both the Royal Society of Literature and of the Royal Astronomical Society. He was made a CBE in 1972 and a Companion of Literature in 1974, and on three occasions was nominated for the Nobel Prize. Arthur Koestler died in 1983. *The Times* obituary called him 'a consistently lucid and humane writer . . . His reputation as one of the most versatile and protean writers of our century is thoroughly deserved.'

Arthur Koestler

# The Sleepwalkers

A History of Man's changing vision
of the Universe

With an Introduction by
Herbert Butterfield

ARKANA
PENGUIN BOOKS
by arrangement with Hutchinson of London

ARKANA

Published by the Penguin Group
Penguin Books Ltd, 27 Wrights Lane, London W8 5TZ, England
Penguin Putnam Inc., 375 Hudson Street, New York, New York 10014, USA
Penguin Books Australia Ltd, Ringwood, Victoria, Australia
Penguin Books Canada Ltd, 10 Alcorn Avenue, Toronto, Ontario, Canada M4V 3B2
Penguin Books (NZ) Ltd, 182–190 Wairau Road, Auckland 10, New Zealand

Penguin Books Ltd, Registered Offices: Harmondsworth, Middlesex, England

First published by Hutchinson 1959
First published in the United States of America by Macmillan 1959
Published in Penguin Books 1964
Reissued in Pelican Books 1968
Reprinted in Peregrine Books 1986
Reprinted in Arkana 1989
10 9 8 7



Printed in England by Clays Ltd, St Ives plc
Set in Linotype Times

# Contents

To the memory of Mamaine

# Preface

In the index to the six hundred odd pages of Arnold Toynbee's *A Study of History*, abridged version, the names of Copernicus, Galileo, Descartes, and Newton do not occur.[1]* This one example among many should be sufficient to indicate the gulf that still separates the Humanities from the Philosophy of Nature. I use this outmoded expression because the term 'Science', which has come to replace it in more recent times, does not carry the same rich and universal associations which 'Natural Philosophy' carried in the seventeenth century, in the days when Kepler wrote his *Harmony of the World* and Galileo his *Message from the Stars*. Those men who created the upheaval which we now call the 'Scientific Revolution' called it by a quite different name: the 'New Philosophy'. The revolution in technology which their discoveries triggered off was an unexpected by-product; their aim was not the conquest of Nature, but the understanding of Nature. Yet their cosmic quest destroyed the medieval vision of an immutable social order in a walled-in universe together with its fixed hierarchy of moral values, and transformed the European landscape, society, culture, habits, and general outlook, as thoroughly as if a new species had arisen on this planet.

This mutation of the European mind in the seventeenth century is merely the latest example of the impact of the 'Sciences' on the 'Humanities' – of the inquiry into the nature of Nature on the inquiry into the nature of Man. It also illustrates the wrongheadedness of erecting academic and social barriers between the two; a fact which is at last beginning to gain recognition, nearly half a millennium after the Renaissance discovered the *uomo universale*.

Another result of this fragmentation is that there exist Histories of Science, which tell one at what date the mechanical

* Superior figures refer to notes at the end of the book.

clock or the law of inertia made their first appearance, and Histories of Astronomy which inform one that the precession of the equinoxes was discovered by Hipparchus of Alexandria; but, surprisingly, there exists to my knowledge no modern History of Cosmology, no comprehensive survey of man's changing vision of the universe which encloses him.

The above explains what this book is aiming at, and what it is trying to avoid. It is not a history of astronomy, though astronomy comes in where it is needed to bring the vision into sharper focus; and, though aimed at the general reader, it is not a book of 'popular science' but a personal and speculative account of a controversial subject. It opens with the Babylonians and ends with Newton, because we still live in an essentially Newtonian universe; the cosmology of Einstein is as yet in a fluid state, and it is too early to assess its influence on culture. To keep the vast subject within manageable limits, I could attempt only an outline. It is sketchy in parts, detailed in others, because selection and emphasis of the material was guided by my interest in certain specific questions, which are the *leitmotifs* of the book, and which I must briefly set out here.

Firstly, there are the twin threads of Science and Religion, starting with the undistinguishable unity of the mystic and the savant in the Pythagorean Brotherhood, falling apart and re-uniting again, now tied up in knots, now running on parallel courses, and ending in the polite and deadly 'divided house of faith and reason' of our day, where, on both sides, symbols have hardened into dogmas, and the common source of inspiration is lost from view. A study of the evolution of cosmic awareness in the past may help to find out whether a new departure is at least conceivable, and on what lines.

Secondly, I have been interested, for a long time, in the psychological process of discovery[2] as the most concise manifestation of man's creative faculty – and in that converse process that blinds him towards truths which, once perceived by a seer, become so heartbreakingly obvious. Now this black-out shutter operates not only in the minds of the 'ignorant and superstitious masses' as Galileo called them, but is even more strikingly evident in Galileo's own, and in other geniuses like Aristotle, Ptolemy, or Kepler. It looks as if, while part of their spirit was asking for more light, another part had been crying out for more darkness. The History of Science is a relative newcomer on the scene, and the biographers of its Cromwells and Napoleons are as yet little concerned with psychology; their heroes are mostly represented as reasoning-machines on austere marble pedestals, in a manner long outdated in the mellower branches of historiography – probably on the assumption that in the case of a Philosopher of Nature, unlike that of a statesman or conqueror, character and personality are irrelevant. Yet all cosmological systems, from the Pythagoreans to Copernicus, Descartes, and Eddington, reflect the unconscious prejudices, the philosophical or even political bias of their authors; and from physics to physiology, no branch of Science, ancient or modern, can boast freedom from metaphysical bias of one kind or another. The progress of Science is generally regarded as a kind of clean, rational advance along a straight ascending line; in fact it has followed a zigzag course, at times almost more bewildering than the evolution of political thought. The history of cosmic theories, in particular, may without exaggeration be called a history of collective obsessions and controlled schizophrenias; and the manner in which some of the most important individual discoveries were arrived at reminds one more of a sleepwalker's performance than an electronic brain's.

Thus, in taking down Copernicus or Galileo from the pedestal on which science-mythography has placed them, my motive was not to 'debunk', but to inquire into the obscure workings of the creative mind. Yet I shall not be sorry if, as an accidental by-product, the inquiry helps to counteract the legend that Science is a purely rational pursuit, that the Scientist is a more 'level-headed' and 'dispassionate' type than others (and should therefore be given a leading part in world affairs); or that he is able to provide for himself and his contemporaries, a rational substitute for ethical insights derived from other sources.

It was my ambition to make a difficult subject accessible to the general reader, but students familiar with it will, I hope, nevertheless find some new information in these pages. This refers mainly to Johannes Kepler, whose works, diaries, and correspondence have so far not been accessible to the English reader; nor does a serious English biography exist. Yet Kepler is one of the few geniuses who enables one to follow, step by step, the tortuous path that led him to his discoveries, and to get a really intimate glimpse, as in a slow-motion film, of the creative act. He accordingly occupies a key-position in the narrative.

Copernicus' *magnum opus, On the Revolutions of the Heavenly Spheres*, also had to wait until 1952 for a first English translation – which perhaps explains certain curious misunderstandings about his work, shared by practically all authorities who have written on the subject, and which I have tried to rectify.

The reader with a scientific education is asked to forbear with explanations which might seem an insult to his intelligence. So long as in our educational system a state of cold war is maintained between the Sciences and the Humanities, this predicament cannot be avoided.

One significant step towards ending this cold war was Professor Herbert Butterfield's *Origins of Modern Science*, first published in 1949. Apart from the work's profundity and excellence *per se*, I was much impressed by the fact that the Professor of Modern History in the University of Cambridge should venture into medieval Science and undertake such a gulf-bridging task. Perhaps this age of specialists is in need of creative trespassers. It was this shared conviction which made me ask Professor Butterfield for the favour of a short Introduction to another trespassing venture.

My sincere thanks are due to Professor Max Caspar, Munich, and to Bibliotheksrat Dr Franz Hammer, Stuttgart, for help and advice on Johannes Kepler; to Dr Marjorie Grene for her help on medieval Latin sources and various other problems; to Professor Zdenek Kopal, University of Manchester, for his critical reading of the text; to Professor Alexandre Koyré,

École des Hautes Études, Sorbonne, and Professor Ernst Zinner, Bamberg, for information quoted in the Notes; to Professor Michael Polanyi for his sympathetic interest and encouragement; lastly to Miss Cynthia Jefferies for her unending patient labours on the typescript and galleys.

# Introduction

No field of thought can be properly laid out by men who are merely measuring with a ruler. Sections of history are liable to be transformed – or, even where not transformed, greatly vivified – by an imagination that comes, sweeping like a searchlight, from outside the historical profession itself. Old hunches are then confirmed by fresh applications of the evidence or by unexpected correlations between sources. New matter emerges because things are joined together which it had not occurred to one to see in juxtaposition. New details are elicited, different details become relevant, because of a fresh turn that the argument has taken.

We are constantly finding that we have been reading too much modernity into a man like Copernicus, or have merely been selecting from Kepler (and plucking out of their context) certain things which have a modern ring; or, in a similar manner, we have been anachronistic in our treatment of the mind and life of Galileo. The present author carries this particular process further, picks up many loose ends, and gives the whole subject a number of unexpected ramifications. Looking not only at the scientific achievements but at the working-methods behind them, and at a good deal of private correspondence, he has illuminated great thinkers, putting them back into their age, and yet not making them meaningless – not leaving us with anomalies and odds-and-ends of antiquated thought, but tracing the unity, recovering the texture and showing us the plausibility and the self-consistency of the underlying mind.

It is particularly useful for English readers that Mr Koestler has concentrated on some of the aspects of the story that have been neglected, and has paid great attention to Kepler, who most required exposition and most called for historical imagination. History is not to be judged by negatives; and those of us who differ from Mr Koestler in respect of some of the outer framework of his ideas or who do not follow him in certain details, can hardly fail to catch the light which not only modifies and enlivens the picture but brings out new facts, or makes dead ones dance before our eyes.

It will be surprising if even those who are familiar with this subject do not often feel that here is a shower of rain where every drop has caught a gleam.

HERBERT BUTTERFIELD

# Part One. The Heroic Age

# 1 Dawn

## 1. *Awakening*

We can add to our knowledge, but we cannot subtract from it. When I try to see the Universe as a Babylonian saw it around 3000 B.C., I must grope my way back to my own childhood. At the age of about four I had what I felt to be a satisfactory understanding of God and the world. I remember an occasion when my father pointed his finger at the white ceiling, which was decorated with a frieze of dancing figures, and explained that God was up there, watching me. I immediately became convinced that the dancers were God and henceforth addressed my prayers to them, asking for their protection against the terrors of day and night. Much in the same manner, I like to imagine, did the luminous figures on the dark ceiling of the world appear as living divinities to Babylonians and Egyptians. The Twins, the Bear, the Serpent were as familiar to them as my fluted dancers to me; they were thought to be not very far away, and they held power of life and death, harvest and rain.

The world of the Babylonians, Egyptians, and Hebrews was an oyster, with water underneath, and more water overhead, supported by the solid firmament. It was of moderate dimensions, and as safely closed in on all sides as a cot in the nursery or a babe in the womb. The Babylonians' oyster was round, the earth was a hollow mountain, placed in its centre, floating on the waters of the deep; above it was a solid dome, covered by the upper waters. The upper waters seeped through the dome as rain, and the lower waters rose in fountains and springs. Sun, moon, and stars progressed in a slow dance across the dome, entering the scene through doors in the East and vanishing through doors in the West.

The universe of the Egyptians was a more rectangular oyster or box; the earth was its floor, the sky was either a cow whose feet rested on the four corners of the earth, or a woman supporting herself on her elbows and knees; later, a vaulted metal lid. Around the inner walls of the box, on a kind of elevated gallery, flowed a river on which the sun and moon gods sailed their barques, entering and vanishing through various stage doors. The fixed stars were lamps, suspended from the vault, or carried by other gods. The planets sailed their own boats along canals originating in the Milky Way, the celestial twin of the Nile. Towards the fifteenth of each month, the moon god was attacked by a ferocious sow, and devoured in a fortnight of agony; then he was re-born again. Sometimes the sow swallowed him whole, causing a lunar eclipse; sometimes a serpent swallowed the sun, causing a solar eclipse. But these tragedies were, like those in a dream, both real and not; inside his box or womb, the dreamer felt fairly safe.

This feeling of safety was derived from the discovery that, in spite of the tumultuous private lives of the sun and moon gods, their appearances and movements remained utterly dependable and predictable. They brought night and day, the seasons and the rain, harvest, and sowing time, in regular cycles. The mother leaning over the cradle is an unpredictable goddess; but her feeding breast can be depended on to appear when needed. The dreaming mind may go through wild adventures, it may travel through Olympus and Tartarus, but the pulse of the dreamer has a regular beat that can be counted. The first to learn counting the pulse of the stars were the Babylonians.

Some six thousand years ago, when the human mind was still half asleep, Chaldean priests were standing on watch-towers, scanning the stars, making maps and time-tables of their motions. Clay tablets dating from the reign of Sargon of Akkad, around 3800 B.C., show an already old-established astronomical tradition.[1] The time-tables became calendars which regulated organized activity, from the growing of crops to religious ceremonies. Their observations became amazingly precise: they computed the length of the year with a deviation of less than 0·001 per cent from the correct value,[2] and their figures relating to the motions of sun and moon have only three times the margin of error of nineteenth-century astronomers armed with mammoth telescopes.[3] In this respect, theirs was an

Exact Science; their observations were verifiable, and enabled them to make precise predictions of astronomical events; though based on mythological assumptions. the theory 'worked'. Thus at the very beginning of this long journey, Science emerges in the shape of Janus, the double-faced god, guardian of doors and gates: the face in front alert and observant, while the other, dreamy and glassy-eyed, stares in the opposite direction.

The most fascinating objects in the sky – from both points of view – were the planets, or vagabond stars. Only seven of these existed among the thousands of lights suspended from the firmament. They were the Sun, the Moon, Nebo – Mercury, Ishtar – Venus, Nergal – Mars, Marduk – Jupiter, and Ninib – Saturn. All other stars remained stationary, fixed in the pattern of the firmament, revolving once a day round the earth-mountain, but never changing their places in the pattern. The seven vagabond stars revolved with them, but at the same time they had a motion of their own, like flies wandering over the surface of a spinning globe. Yet they did not wander all across the sky: their movements were confined to a narrow lane, or belt, which was looped around the firmament at an angle of about twenty-three degrees to the equator. This belt – the Zodiac – was divided into twelve sections, and each section was named after a constellation of fixed stars in the neighbourhood. The Zodiac was the lovers' lane in the skies, along which the planets ambled. The passing of a planet through one of the sections had a double significance: it yielded figures for the observer's time-table, and symbolic messages of the mythological drama played out behind the scenes. Astrology and Astronomy remain to this day complementary fields of vision of Janus sapiens.

## 2. *Ionian Fever*

Where Babylon and Egypt left off, Greece took over. At the beginning, Greek cosmology moved much on the same lines – Homer's world is another, more colourful oyster, a floating disc surrounded by Okeanus. But about the time when the texts of the Odyssey and Iliad became consolidated in their final version, a new development started in Ionia on the Aegean coast. The sixth pre-Christian century – the miraculous century of Buddha, Confucius, and Lao-tze, of the Ionian philosophers and Pythagoras – was a turning point for the human species. A March breeze seemed to blow across this planet from China to Samos, stirring man into awareness, like the breath in Adam's nostrils. In the Ionian school of philosophy, rational thought was emerging from the mythological dream-world. It was the beginning of the great adventure: the Promethean quest for natural explanations and rational causes, which, within the next two thousand years, would transform the species more radically than the previous two hundred thousand had done.

Thales of Miletos, who brought abstract geometry to Greece, and predicted an eclipse of the sun, believed, like Homer, that the earth was a circular disc floating on water, but he did not stop there; discarding the explanations of mythology, he asked the revolutionary question out of what basic raw material, and by what process of nature, the universe was formed. His answer was, that the basic stuff or element must be water, because all things are born from moisture, including air, which is water evaporated. Others taught that the prime material was not water, but air or fire; however, their answers were less important than the fact that they were learning to ask a new type of question, which was addressed not to an oracle, but to dumb nature. It was a wildly exhilarating game; to appreciate it, one must again travel back along one's own private time-track to the fantasies of early adolescence when the brain, intoxicated with its newly discovered powers, let speculation run riot. 'A case in point,' Plato reports, 'is that of Thales, who, when he was star-gazing and looking upward, fell into a well, and was rallied (so it is said) by a clever and pretty maidservant from Thrace because he was eager to know what went on in the heaven, but did not notice what was in front of him, nay, at his very feet.' [4]

The second of the Ionian philosophers, Anaximander, displays all the symptoms of the intellectual fever spreading through Greece. His universe is no longer a closed box, but infinite in extension and duration. The raw material is none of the familiar forms of matter, but a substance without definite properties except for being indestructible and everlasting. Out of this stuff all things are developed, and into it they return; before this our world, infinite multitudes of other universes have already existed, and been dissolved again into the amorphous mass. The earth is a cylindrical column, surrounded by air; it floats upright in the centre of the universe without support or anything to stand on, yet it does not fall because, being in the centre, it has no preferred direction towards which to lean; if it did, this would disturb the symmetry and balance of the whole. The spherical heavens enclose the atmosphere 'like the bark of a tree', and there are several layers of this enclosure to accommodate the various stellar objects. But these are not what they seem, and are not 'objects' at all. The sun is merely a hole in the rim of a huge wheel. The rim is filled with fire, and as it turns round the earth, so does the hole in it – a puncture in a gigantic tyre filled with flames. For the moon we are given a similar explanation; its phases are due to recurrent partial stoppages of the puncture, and so are the eclipses. The stars are pin-holes in a dark fabric through which we glimpse the cosmic fire filling the space between two layers of 'bark'.

It is not easy to see how the whole thing works, but it is the first approach to a mechanical model of the universe. The boat of the sun god is replaced by the wheels of a clockwork. Yet the machinery looks as if it had been dreamed up by a surrealist painter; the punctured fire-wheels are certainly closer to Picasso than to Newton. As we move along past other cosmologies, we shall get this impression over and again.

The system of Anaximenes, who was an associate of Anaximander, is less inspired; but he seems to have been the originator of the important idea that the stars are attached 'like nails' to a transparent sphere of crystalline material, which turns round the earth 'like a hat round the head'. It sounded so plausible and convincing, that the crystal spheres were to dominate cosmology until the beginning of modern times.

The Ionian philosophers' home was Miletos in Asia Minor; but there existed rival schools in the Greek towns of Southern Italy, and rival theories within each school. The founder of the Eleatic school, Xenophanes of Kolophon, is a sceptic who wrote poetry to the age of ninety-two, and sounds as if he had served as a model for the author of Ecclesiastes:

From earth are all things and to earth all things return. From earth and water come all of us. ... No man hath certainly known, nor shall certainly know, that which he saith about the gods and about all things; for, be that which he saith ever so perfect, yet does he not know it; all things are matters of opinion. ... Men imagine gods to be born, and to have clothes and voices and shapes like theirs. ... Yea, the gods of the Ethiopians are black and flat-nosed, the gods of the Thracians are red-haired and blue-eyed. ... Yea, if oxen and horses and lions had hands, and could shape with their hands images as men do, horses would fashion their gods as horses, and oxen as oxen. ... Homer and Hesiod have ascribed to the gods all things that are a shame and a disgrace among men, theft, adultery, deceit, and other lawless acts. ...

As against this:

There is one God ... neither in shape nor thought like unto mortals. ... He abideth ever in the same place motionless ... and without effort swayeth all things by his force of mind. ...[5]

The Ionians were optimistic, heathenly materialists; Xenophanes was a pantheist of a sorrowful brand, to whom change was an illusion, and effort vanity. His cosmology reflects his philosophical temper; it is radically different from the Ionians'. His earth is not a floating disc, or column, but is 'rooted in the infinite'. The sun and the stars have neither substance nor permanence, they are merely cloudy exhalations of the earth which have caught fire. The stars are burnt out at dawn, and in the evening a new set of stars is formed from new exhalations. Similarly, a new sun is born every morning from the crowding together of sparks. The moon is a compressed, luminous cloud, which dissolves in a month; then a new cloud starts forming. Over different regions of the earth, there are different suns and moons, all cloudy illusions.

In this manner do the earliest rational theories of the Universe betray the bias and temperament of their makers. It is generally believed that with the progress of scientific method, the theories became more objective and reliable. Whether this belief is justified, we shall see. But *à propos* of Xenophanes we may note that two thousand years later Galileo also insisted on regarding comets as atmospheric illusions – for purely personal reasons, and against the evidence of his telescope.

Neither the cosmology of Anaxagoras, nor of Xenophanes, gained a considerable following. Every philosopher of the period seems to have had his own theory regarding the nature of the universe around him. To quote Professor Burnet, 'no sooner did an Ionian philosopher learn half a dozen geometrical propositions and hear that the phenomena of the heavens recur in cycles than he set to work to look for law everywhere in nature and with an audacity amounting to *hybris* to construct a system of the universe'.[6] But their divers speculations had this one feature in common, that the sun-eating serpents and Olympian string-pullers were discarded; each theory, however strange and bizarre, was concerned with natural causes.

The sixth century scene evokes the image of an orchestra expectantly tuning up, each player absorbed in his own instrument only, deaf to the caterwaulings of the others. Then there is a dramatic silence, the conductor enters the stage, raps three times with his baton, and harmony emerges from the chaos. The maestro is Pythagoras of Samos, whose influence on the ideas, and thereby on the destiny, of the human race was probably greater than that of any single man before or after him.

# 2 The Harmony of the Spheres

## 1. *Pythagoras of Samos*

Pythagoras was born in the early decades of that tremendous century of awakening, the sixth; and may have seen it go out, for he lived at least eighty, and possibly over ninety, years. Into that long life-span he packed, in the words of Empedokles, 'all things that are contained in ten, even in twenty, generations of men'.

It is impossible to decide whether a particular detail of the Pythagorean universe was the work of the master, or filled in by a pupil – a remark which equally applies to Leonardo or Michelangelo. But there can be no doubt that the basic features were conceived by a single mind; that Pythagoras of Samos was both the founder of a new religious philosophy, and the founder of Science, as the world is understood today.

It seems reasonably certain that he was the son of a silversmith and gem engraver named Mnesarchos; that he was a pupil of Anaximander, the atheist, but also of Pherekydes, the mystic who taught the transmigration of souls. He must have travelled extensively in Asia Minor and Egypt, as many educated citizens of the Greek Islands did; and it is said that he was charged with diplomatic missions by Polycrates, the enterprising autocrat of Samos. Polycrates was an enlightened tyrant who favoured commerce, piracy, engineering, and the fine arts; the greatest poet of the time, Anakreon, and the greatest engineer, Eupalinos from Megara, both lived at his court. According to a story by Herodotus, he became so powerful that, to placate the jealousy of the gods, he threw his most precious signet ring into the deep waters. A few days later, his cook cut open a large fish, freshly caught, and found the ring in its stomach. The doomed Polycrates promptly walked into a trap set by a small Persian ruler, and was crucified. But by that time Pythagoras, with his family, had emigrated from

Samos, and around 530 B.C. settled in Kroton – which, next to its rival Sybaris, was the largest Greek town in Southern Italy. The reputation which preceded him must have been tremendous, for the Pythagorean Brotherhood which he founded on his arrival soon ruled the town, and for a time gained supremacy over a considerable part of Magna Grecia. But its secular power was short-lived; Pythagoras, at the end of his life, was banished from Kroton to Metapontion; his disciples were exiled or slain, and their meeting-houses burnt down.

This is the meagre stem of more or less established fact, around which the ivy of legend began to grow even during the master's lifetime. He soon achieved semi-divine status; according to Aristotle, the Krotonians believed him to be a son of the Hyperborean Apollo, and there was a saying that 'among rational creatures there are gods and men and beings like Pythagoras'. He worked miracles, conversed with demons from heaven, descended to Hades, and possessed such power over men, that after his first sermon to the Krotonians, six hundred joined the communal life of the Brotherhood without even going home to bid their families farewell. Among his disciples his authority was absolute – 'the master said so' was their law.

## 2. *The Unifying Vision*

Myths grow like crystals, according to their own, recurrent pattern; but there must be a suitable core to start their growth. Mediocrities or cranks have no myth-generating power; they may create a fashion, but it soon peters out. Yet the Pythagorean vision of the world was so enduring, that it still permeates our thinking, even our vocabulary. The very term 'philosophy' is Pythagorean in origin; so is the word 'harmony' in its broader sense; and when we call numbers 'figures', we talk the jargon of the Brotherhood.[1]

The essence and power of that vision lies in its all-embracing, unifying character; it unites religion and science, mathematics and music, medicine and cosmology, body, mind, and spirit in an inspired and luminous synthesis. In the Pythagorean philosophy all component parts interlock: it presents a homogeneous surface, like a sphere, so that it is difficult to decide from which side to cut into it. But the simplest approach is through music. The Pythagorean discovery that the pitch of a note depends on the length of the string which produces it, and that concordant intervals in the scale are produced by simple numerical ratios (2 : 1 octave, 3 : 2 fifth, 4 : 3 fourth, etc.), was epoch-making: it was the first successful reduction of quality to quantity, the first step towards the mathematization of human experience – and therefore the beginning of Science.

But here an important distinction must be made. The twentieth-century European regards with justified misgivings the 'reduction' of the world around him, of his experiences and emotions, into a set of abstract formulae, deprived of colour, warmth, meaning, and value. To the Pythagoreans, on the contrary, the mathematization of experience meant not an impoverishment, but an enrichment. Numbers were sacred to them as the purest of ideas, disembodied and ethereal; hence the marriage of music to numbers could only ennoble it. The religious and emotional *ekstasis* derived from music was canalized by the adept into intellectual *ekstasis*, the contemplation of the divine dance of numbers. The gross strings of the lyre are recognized to be of subordinate importance; they can be made of different materials, in various thicknesses and lengths, so long as the proportions are preserved: what produces the music are the ratios, the numbers, the pattern of the scale. Numbers are eternal while everything else is perishable; they are of the nature not of matter, but of mind; they permit mental operations of the most surprising and delightful kind without reference to the coarse external world of the senses – which is how the divine mind must be supposed to operate. The ecstatic contemplation of geometrical forms and mathematical laws is therefore the most effective means of purging the soul of earthly passion, and the principal link between man and divinity.

The Ionian philosophers had been materialists in the sense that the chief accent of their inquiry was on the stuff from which the universe was made; the Pythagoreans' chief accent was on form, proportion, and pattern; on the *eidos* and *schema*, on the relation, not on the relata. Pythagoras is to Thales what Gestalt philosophy is to the materialism of the nineteenth century. The pendulum has been set swinging; its ticking will be heard through the entire course of history, as the blob alternates between the extreme positions of 'all is body', 'all is mind'; as the emphasis shifts from 'substance' to 'form', from 'structure' to 'function', from 'atoms' to 'patterns', from 'corpuscles' to 'waves' and back again.

The line connecting music with numbers became the axis of the Pythagorean system. This axis was then extended in both directions: towards the stars on one side, the body and soul of man on the other. The bearings, on which the axis and the whole system turned, were the basic concepts of *armonia*: harmony, and *katharsis*: purge, purification.

The Pythagoreans were, among other things, healers; we are told that 'they used medicine to purge the body, and music to purge the soul'.[2] One of the oldest forms, indeed, of psychotherapy consists in inducing the patient, by wild pipe music or drums, to dance himself into a frenzy followed by exhaustion and a trance-like, curative sleep – the ancestral version of shock-treatment and abreaction therapy. But such violent measures were only needed where the patient's soul-strings were out of tune – overstrung or limp. This is to be taken literally, for the Pythagoreans regarded the body as a kind of musical instrument where each string must have the right tension and the correct balance between opposites such as 'high' and 'low', 'hot' and 'cold', 'wet' and 'dry'. The metaphors borrowed from music which we still apply in medicine – 'tone', 'tonic', 'well-tempered', 'temperance', are also part of our Pythagorean heritage.

However, the concept *armonia* did not have quite the same meaning that we lend to 'harmony'. It is not the pleasing effect of simultaneously-sounded concordant strings – 'harmony' in that sense was absent from classical Greek music – but something more austere: *armonia* is simply the attunement of the strings to the intervals in the scale, and the pattern of the scale itself. It means that balance and order, not sweet pleasure, are the law of the world.

Sweetness does not enter the Pythagorean universe. But it contains one of the most powerful tonics ever administered to the human brain. It lies in the Pythagorean tenets that 'philosophy is the highest music', and that the highest form of philosophy is concerned with numbers: for ultimately 'all things are numbers'. The meaning of this oft-quoted saying may perhaps be paraphrased thus: 'all things have form, all things *are* form; and all forms can be defined by numbers'. Thus the form of the square corresponds to a 'square number', i.e. 16 = 4 + 4, whereas 12 is an oblong number, and 6 a triangular number:

Numbers were regarded by the Pythagoreans as patterns of dots which form characteristic figures, as on the sides of a dice; and though we use arabic symbols, which have no resemblance to these dot-patterns, we still call numbers 'figures', i.e. shapes.

Between these number-shapes unexpected and marvellous relations were found to exist. For instance, the series of 'square numbers' was formed simply by the addition of successive odd numbers: 1 +3= 4 +5= 9 +7= 16 +9= 25 and so forth:

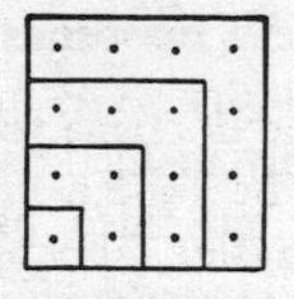

The addition of even numbers formed 'oblong numbers', where the ratio of the sides represented exactly the concordant intervals of the musical octave: 2 (2 : 1, octave) + 4 = 6 (3 : 2, fifth) + 6 = 12 (4 : 3, fourth).

In a similar manner, 'cubic' numbers and 'pyramidal' numbers were obtained. Mnesarchos had been a gem engraver, so Pythagoras in his youth must have been familiar with crystals whose form imitated those of pure number-shapes: quartz the pyramid and double pyramid, beryl the hexagon, garnet the dodocaeder. It all went to show that Reality could be reduced to number-series and number-ratios, if only the rules of the game were known. To discover these was the chief task of the *Philosophos*, the Lover of Wisdom.

An example of the magic of numbers is the famous theorem, by which alone Pythagoras is consciously remembered today – the visible peak of the submerged iceberg.* There is no obvious relationship between the lengths of the sides of a right-angled triangle; but if we build a square over each side, the areas of the two smaller squares will exactly equal the area of the larger. If such wonderfully ordered laws, hitherto hidden from the human eye, could be discovered by the contemplation of number-shapes, was it not legitimate to hope that soon all secrets of the universe would be revealed through them? Numbers were not thrown into the world at random; they arranged themselves into balanced patterns, like the shapes of crystals and the concordant intervals of the scale, according to the universal laws of harmony.

## 3. '*Soft Stillness and the Night*'

Extended to the stars, the doctrine took the form of the 'Harmony of the Spheres'. The Ionian philosophers had begun to prise open the cosmic oyster, and to set the earth adrift; in Anaximander's universe the earth-disc no longer floats in water, but stands in the centre, supported by nothing and surrounded by air. In the Pythagorean universe the disc changes into a spherical ball.[3] Around it, the sun, moon, and planets revolve in concentric circles, each fastened to a sphere or wheel. The swift revolution of each of these bodies causes a swish, or musical

* Ironically, Pythagoras seems to have had no complete proof of the Pythagorean theorem.

hum, in the air. Evidently each planet will hum on a different pitch, depending on the ratios of their respective orbits – just as the tone of a string depends on its length. Thus the orbits in which the planets move form a kind of huge lyre whose strings are curved into circles. It seemed equally evident that the intervals between the orbital cords must be governed by the laws of harmony. According to Pliny,[4] Pythagoras thought that the musical interval formed by earth and moon was that of a tone; moon to Mercury, a semi-tone; Mercury to Venus, a semi-tone; Venus to Sun, a minor third; Sun to Mars, a tone; Mars to Jupiter, a semi-tone; Jupiter to Saturn, a semi-tone; Saturn to the sphere of the fixed stars, a minor third. The resulting 'Pythagorean Scale' is C, D, ♭ E, G, A, ♭ B, B, D – though the accounts of the scale given by different writers vary slightly.

According to tradition, the Master alone had the gift of actually hearing the music of the spheres. Ordinary mortals lack this gift, either because they are from the moment of birth, unknowingly but constantly bathed in the celestial humming; or because – as Lorenzo explains to Jessica – they are too grossly constituted:

> . . . soft stillness and the night
> Become the touches of sweet harmony . . .
> Look how the floor of heaven
> Is thick inlaid with patines of bright gold;
> There's not the smallest orb which thou behold'st
> But in his motion like an angel sings . . .
> Such harmony is in immortal souls;
> But, whilst this muddy vesture of decay
> Doth grossly close it in, we cannot hear it.[5]

The Pythagorean dream of musical harmony governing the motion of the stars never lost its mysterious impact, its power to call forth responses from the depth of the unconscious mind. It reverberates through the centuries, from Kroton to Elizabethan England; I shall quote two more versions of it – with a purpose that will become apparent later. The first is Dryden's well-known:

From harmony, from heavenly harmony,
This universal frame began:
When nature underneath a heap
Of jarring atoms lay
And could not heave her head.
The tuneful voice we heard from high:
Arise, ye more than dead.

The second is from Milton's *Arcades*:

But els in deep of night when drowsiness
Hath lockt up mortal sense, then listen I
To the celestial Sirens harmony . . .
Such sweet compulsion doth in music ly,
To lull the daughters of *Necessity*,
And keep unsteddy Nature to her law,
And the low world in measur'd motion draw
After the heavenly tune, which none can hear
Of human mould with grosse unpurged ear.

But, one might ask, was the 'Harmony of the Spheres' a poetic conceit or a scientific concept? A working hypothesis or a dream dreamt through a mystic's ear? In the light of the data which astronomers collected in the centuries that followed, it certainly appeared as a dream; and even Aristotle laughed 'harmony, heavenly harmony' out of the courts of earnest, exact science. Yet we shall see how, after an immense detour, at the turn of the sixteenth century, one Johannes Kepler became enamoured with the Pythagorean dream, and on this foundation of fantasy, by methods of reasoning equally unsound, built the solid edifice of modern astronomy. It is one of the most astonishing episodes in the history of thought, and an antidote to the pious belief that the Progress of Science is governed by logic.

## 4. *Religion and Science Meet*

If Anaximander's universe reminds one of a Picasso painting, the Pythagorean world resembles a cosmic musical box playing the same Bach prelude from eternity to eternity. It is not surprising, then, that the religious beliefs of the Pythagorean

Brotherhood are closely related to the figure of Orpheus, the divine fiddler, whose music held not only the Prince of Darkness, but also beasts, trees, and rivers under its spell.

Orpheus is a late arrival on the Greek stage, overcrowded with gods and demigods. The little we know about his cult is clouded in conjecture and controversy; but we know, at least in broad outlines, its background. At an unknown date, but probably not much before the sixth century, the cult of Dionysus-Bacchus, the 'raging' goat-god of fertility and wine, spread from barbaric Thracia into Greece. The initial success of Bacchism was probably due to that general sense of frustration which Xenophanes so eloquently expressed. The Olympian Pantheon had come to resemble an assembly of wax-works, whose formalized worship could no more satisfy truly religious needs than the pantheism – this 'polite atheism' as it has been called – of the Ionian sages. A spiritual void tends to creat emotional outbreaks; the Bacchae of Euripides, frenzied worshippers of the horned god, appears as the forerunners of the medieval tarantula dancers, the bright young things of the roaring twenties, the maenads of the Hitler youth. The outbreak seems to have been sporadic and short-lived: the Greeks, being Greeks, soon realized that these excesses led neither to mystic union with God, nor back to nature, but merely to mass-hysteria:

> Theban women leaving
> Their spinning and their weaving
> Stung with the maddening trance
> Of Dionysus! . . .
> Brute with bloody jaws agape
> God-defying, gross and grim,
> Slander of the human shape.[6]

The authorities seemed to have acted with eminent reasonableness: they promoted Bacchus-Dionysus to the official Pantheon with a rank equal to Apollo's. His frenzy was tamed, his wine watered down, his worship regulated, and used as a harmless safety-valve.

But the mystic craving must have persisted, at least in a sensitivized minority, and the pendulum now began to swing in the opposite direction: from carnal ecstasy to other-worldliness. In the most telling variant of the legend, Orpheus appears as a victim of Bacchic fury: when, having finally lost his wife, he decides to turn his back on sex, the women of Thrace tear him to pieces, and his head floats down the Hebrus – still singing. It sounds like a cautionary tale; but the tearing and devouring of the living god, and his subsequent rebirth, is a *leitmotif* that recurs in Orphism on a different level of meaning. In Orphic mythology, Dionysus (or his Thracian version, Zagreus) is the beautiful son of Zeus and Persephone; the evil Titans tear him to pieces and eat him, all but his heart, which is given to Zeus, and he is born a second time. The Titans are slain by Zeus' thunderbolt; but out of their ashes man is born. By devouring the god's flesh, the Titans have acquired a spark of divinity which is transmitted to man; and so is the desperate evil that resided in the Titans. But it is in the power of man to redeem this original sin, to purge himself of the evil portion of his heritage by leading an other-worldly life and performing certain ascetic rites. In this manner he can obtain liberation from the 'wheel of rebirth' – his imprisonment in successive animal and even vegetable bodies, which are like carnal tombs to his immortal soul – and regain his lost divine status.

The Orphic cult was thus in almost every respect a reversal of the Dionysian; it retained the name of the god and some features of his legend, but all with a changed emphasis and different meaning (a process that will repeat itself at other turning points of religious history). The Bacchic technique of obtaining emotional release by furiously clutching at the Now and Here, is replaced by renunciation with an eye on after-life. Physical intoxication is superseded by mental intoxication; the 'juice that streams from the vine-clusters to give us joy and oblivion' now serves only as a sacramental symbol; it will eventually be taken over, together with the symbolic swallowing of the slain god and other basic elements of Orphism, by Christianity. 'I am perishing with thirst, give me to drink of the waters of memory', says a verse on an Orphic gold tablet, alluding to the divine origin of the soul: the aim is no longer oblivion but remembrance of a knowledge which it once possesed. Even words change their meaning: 'orgy' no longer means Bacchic revelry, but religious ecstasy leading to liberation from the wheel of rebirth.[7] A similar development is the transformation of the carnal union between the King and the Shulamite into the mystic union of Christ and his Church; and, in more recent times, the shift of meaning in words like 'rapture' and 'ravishment'.

Orphism was the first universal religion in the sense that it was not regarded as a tribal or national monopoly, but open to all who accepted its tenets; and it profoundly influenced all subsequent religious development. It would nevertheless be a mistake to attribute too much intellectual and spiritual refinement to it; the Orphic purification rites, which are the hub of the whole system, still contain a series of primitive taboos – not to eat meat, or beans, not to touch a white cock, not to look in a mirror beside the light.

But this is precisely the point where Pythagoras gave Orphism a new meaning, the point where religious intuition and rational science were brought together in a synthesis of breathtaking originality. The link is the concept of *katharsis*. It was a central concept in Bacchism, Orphism, in the cult of the Delian Apollo, in Pythagorean medicine and science; but it had different meanings, and entailed different techniques in all of them (as it still does in the various schools of modern psychotherapy). Was there anything in common between the raving Bacchante and the aloof mathematician, the fiddle of Orpheus and a laxative pill? Yes: the same yearning for release from various forms of enslavement, from passions and tensions of body and mind, from death and the void, from the legacy of the Titans in man's estate – the yearning to re-light the divine spark. But the methods of achieving this must differ according to the person. They must be graded according to the disciple's lights and degree of initiation. Pythagoras replaced the soul-purging all-cures of competing sects, by an elaborate hierarchy of kathartic techniques; he purified the very concept of purification, as it were.

At the bottom of the scale are simple taboos, taken over from Orphism, such as the interdiction of eating meat and beans; for the coarse-natured the penance of self-denial is the only effective purge. At the highest level katharsis of the soul is achieved by contemplating the essence of all reality, the harmony of forms, the dance of numbers. 'Pure science' – a strange expression that we still use – is thus both an intellectual delight and a way to spiritual release; the way to the mystic union between the thoughts of the creature and the spirit of its creator. 'The function of geometry,' says Plutarch of the Pythagoreans, 'is to draw us away from the world of the senses and of corruption, to the world of the intellect and the eternal. For the contemplation of the eternal is the end of philosophy as the contemplation of the mysteries is the end of religion.' [8] But to the true Pythagorean, the two have become indistinguishable.

The historical importance of the idea that disinterested science leads to purification of the soul and its ultimate liberation, can hardly be exaggerated. The Egyptians embalmed their corpses so that the soul might return to them and need not be reincarnated again; the Buddhists practised non-attachment to escape the wheel; both attitudes were negative and socially sterile. The Pythagorean concept of harnessing science to the contemplation of the eternal, entered, via Plato and Aristotle. into the spirit of Christianity and became a decisive factor in the making of the Western world.

Earlier in this chapter I have tried to show how, by relating music to astronomy and both to mathematics, emotional experience became enriched and deepened by intellectual insight. Cosmic wonder and aesthetic delight no longer live apart from the exercise of reason; they are all inter-related. Now the final step has been taken: the mystic intuitions of religion have also been integrated into the whole. Again, the process is accompanied by subtle changes in the meaning of certain key-words, such as *theoria* – theory. The word was derived from *theorio* – 'to behold, contemplate' (*thea*: spectacle, *theoris*: spectator, audience). But in orphic usage, *theoria* came to signify 'a state of fervent religious contemplation, in which the spectator is identified with the suffering god, dies in his death, and rises again in his new birth'.[9] As the Pythagoreans canalized religious fervour into intellectual fervour, ritual ecstasy into the ecstasy of discovery, *theoria* gradually changed its meaning into 'theory' in the modern sense. But though the raucous cry of the ritual worshippers was replaced by the Eureka of the new theorizers, they remained aware of the common source from which both sprang. They were aware that the symbols of mythology and the symbols of mathematical science were different aspects of the same, indivisible Reality.* They did not live in a 'divided house of faith and reason'; the two were interlocking, like ground-plan and elevation on an architect's drawing. It is a state of mind very difficult for twentieth-century man to imagine – or even to believe that it could ever have existed. It may help to remember though, that some of the greatest pre-Socratic sages formulated their philosophies in verse; the unitary source of inspiration of prophet, poet, and philosopher was still taken for granted.

It did not last long. Within a few centuries, the unitary awareness faded, religious and rational philosophizing split apart – were partially reunited, then divorced again; with results that will become apparent as the story unfolds.

The Pythagorean synthesis would have been incomplete had it not also included precepts for a way of life.

The Brotherhood was a religious order, but at the same time an academy of science, and a power in Italian politics. The ascetic rules of life seem to have anticipated the Essenes', which in turn served as a model to primitive Christian communities. They shared all property, led a communal existence, and gave equal status to women. They observed rites and abstinences, gave much time to contemplation and examinations of conscience. According to the degree of purification which a Brother achieved, he was gradually initiated into the higher mysteries of musical, mathematical, and astronomical *theoria*. The secrecy surrounding these was partly due to the tradition of the older mystery cults, whose adepts had known that the Bacchic,

* Hence the short-cuts, or short-circuits, between different sets of symbols in Pythagorean mystic number-lore, such as the correlation of odd and even numbers were male and female, right and left; or the magic quality attributed to the pentagram.

and even the Orphic, ecstasies would cause havoc if offered to all and sundry. But the Pythagoreans also realized that similar dangers inhered in the orgies of reasoning. They apparently had an intuition of the *hybris* of science, and recognized it as a potential means both of man's liberation and destruction; hence their insistence that only those purified in body and spirit should be trusted with its secrets. In a word, they believed that scientists ought to be vegetarians, as Catholics believe that priests ought to live in celibacy.

It may be thought that this interpretation of the Pythagorean insistence on secretiveness is far-fetched, or that it implies prophetic foresight on their part. The answer to this is that Pythagoras was, by personal experience, well aware of the immense technological potentialities of geometry. I have mentioned already that Polycrates, and the islanders he ruled, were devoted to engineering. Herodotus, who knew the island well, reports:[10]

> I have written thus at length of the Samians, because they are the makers of the three greatest works to be seen in any Greek land. First of these is the double-mouthed tunnel they pierced for a hundred and fifty fathoms through the base of a high hill ... through which the water, coming from an abundant spring, is carried by its pipes to the city of Samos.

Herodotus is fond of telling tall stories, and his report was not taken very seriously, until, at the beginning of our century, the tunnel was actually found and excavated. It is no less than nine hundred yards long, complete with water-course and inspection-pathway, and its shape shows that it was begun from *both* ends. It further shows that the two digging parties, one working from the north, the other from the south, had met in the centre only a couple of feet apart. Having watched this fantastic feat being performed (by Eupalinos, who also built the second marvel mentioned by Herodotus, a huge mole to protect the Samian war-fleet), even a lesser genius than Pythagoras might have realized that Science may become a hymn to the creator or a Pandora's box, and that it should be trusted only to saints. It is said, incidentally, that Pythagoras, like St Francis, preached to animals, which would seem rather odd behaviour in a modern mathematician; but in the Pythagorean view nothing could be more natural.

### 5. *Tragedy and greatness of the Pythagoreans*

Towards the end of the Master's life, or shortly after his death, two misfortunes befell the Pythagoreans, which would have meant the end of any sect or school with a less universal outlook. They triumphantly survived both.

One blow was the discovery of a type of numbers such as $\sqrt{2}$ – the square root of 2 – which could not be fitted into any dot-diagram. And such numbers were common: they are, for instance, represented by the diagonal of any square. Let the side of the square be called $a$, and the diagonal $d$. It can be proved that if I assign to $a$ any precise numerical value, then it becomes impossible to assign a precise numerical value to $d$. The side and the square are 'incommensurable'; their ratio $a/d$ cannot be represented by any real numbers or fractions thereof; it is an irrational' number; *it is both odd and even at the same time.** I can easily *draw* the diagonal of a square, but I cannot express its length in numbers – I cannot count the number of dots it contains. The point-to-point correspondence between arithmetic and geometry has broken down – and with it the universe of number-shapes.

It is said that the Pythagoreans kept the discovery of irrational numbers – they called them *arrhētos*, unspeakable – a secret, and that Hippasos, the disciple who let the scandal leak out, was put to death. There is also another version, in Proclos:[11]

* The simplest manner of proving this is as follows. Let $d$ be represented by a fraction $\frac{m}{n}$, where $m$ and $n$ are unknown. Let $a=1$, then $d^2=1^2+1^2$ and $d=\sqrt{2}$. Then $\frac{m^2}{n^2}=2$. If $m$ and $n$ have a common factor, divide it out, then either $m$ or $n$ must be odd. Now $m^2=2n^2$, therefore $m^2$ is even, therefore $m$ is even, therefore $n$ is odd. Suppose $m=2p$. Then $4p^2=2n^2$, therefore $n=2p^2$ and therefore $n$ is even, *contra hyp.* Therefore no fraction $\frac{m}{n}$ will measure the diagonal.

It is told that those who first brought out the irrationals from concealment into the open perished in shipwreck, to a man. For the unutterable and the formless must needs be concealed. And those who uncovered and touched this image of life were instantly destroyed and shall remain forever exposed to the play of the eternal waves.

Yet, Pythagoreanism survived. It had the elastic adaptability of all truly great ideological systems which, when some part is knocked out of them, display the self-regenerating powers of a growing crystal or a living organism. The mathematization of the world by means of atom-like dots proved a premature short-cut; but on a higher turn of the spiral, mathematical equations proved once again the most serviceable symbols for representing the physical aspect of reality. We shall meet with further examples of prophetic intuition supported by the wrong reasons; and we shall find that they are rather the rule than the exception.

Nobody before the Pythagoreans had thought that mathematical relations held the secret of the universe. Twenty-five centuries later, Europe is still blessed and cursed with their heritage. To non-European civilizations, the idea that numbers are the key to both wisdom and power, seems never to have occurred.

The second blow was the dissolution of the Brotherhood. We know little of its causes; it probably had something to do with the equalitarian principles and communist practices of the order, the emancipation of women, and its quasi-monotheistic doctrine – the eternal messianic heresy. But persecution remained confined to the Pythagoreans as an organized body – and probably prevented them from degenerating into sectarian orthodoxy. The Master's principal pupils – among them Philolaus and Lysis – who had gone into exile, were soon allowed to return to Southern Italy and to resume teaching. A century later, that teaching became one of the sources of Platonism, and thus entered the mainstream of European thought.

In the words of a modern scholar. 'Pythagoras is the founder of European culture in the Western Mediterranean sphere.' [12] Plato and Aristotle, Euclid and Archimedes, are landmarks on the road; but Pythagoras stands at the point of departure, where it is decided which direction the road will take. Before that decision, the future orientation of Greco-European civilization was still undecided: it may have taken the direction of the Chinese, or Indian, or pre-Columbian cultures, all of which were still equally unshaped and undecided at the time of the great sixth-century dawn. I do not mean to say that if Confucius and Pythagoras had exchanged their places of birth, China would have beaten us to the Scientific Revolution, and Europe become a land of tea-sipping mandarins. The interactions of climate, race, and spirit, the directional influence of outstanding individuals on the course of History, are so obscure that no predictions are possible even in reverse; all 'if' statements about the past are as dubious as prophecies of the future are. It seems fairly plausible that if Alexander or Ghengis Khan had never been born, some other individual would have filled his place and executed the design of the Hellenic or Mongolic expansion; but the Alexanders of philosophy and religion, of science and art, seem less expendable; their impact seems less determined by economic challenges and social pressures; and they seem to have a much wider range of possibilities to influence the direction, shape, and texture of civilizations. If conquerors be regarded as the engine-drivers of History, then the conquerors of thought are perhaps the pointsmen who, less conspicuous to the traveller's eye, determine the direction of the journey.

# 3 The Earth Adrift

I have tried to give a brief general description of Pythagorean philosophy, including aspects of it that are only indirectly related to the subject of this book. In the following sections, some important schools of Greek philosophy and science – Eleatics and Stoics, Atomists and Hippocratics – will hardly be mentioned at all, until we arrive at the next turning point in cosmology, Plato and Aristotle. The development of man's views about the cosmos cannot be treated in isolation from the philosophical background which coloured these views; on the other hand, if the narrative is not to be swallowed up by the background, the latter can only be sketched in at certain turning points of the tale, where the general philosophical climate had a direct impact on cosmology and altered its course. Thus, for instance, the political views of Plato, or of the religious convictions of Cardinal Bellarmine, profoundly influenced astronomical developments for centuries, and must accordingly be discussed; whereas men like Empedokles and Democritus, Socrates, and Zeno, who had a lot to say about the stars, but nothing that is really relevant to our subject, must be passed in silence.

## 1. *Philolaus and the Central Fire*

From the end of the sixth century B.C. onward, the idea that the earth was a sphere, freely floating in air, made steady headway. Herodotus[1] mentions a rumour that there exist people far up in the north who sleep six months of the year – which shows that some of the implications of the earth's roundness (such as the polar night) had already been grasped. The next, revolutionary step was taken by a pupil of Pythagoras, Philolaus, the first philosopher to attribute *motion* to our globe. The earth became air-borne.

The motives which led to this tremendous innovation we can only guess. Perhaps it was the realization that there is something illogical in the apparent movements of the planets. It seemed crazy that the sun and planets should turn round the earth once a day, but at the same time slowly crawl along the Zodiac on their annual revolutions. Everything would be much simpler if one assumed that the *daily* revolution of the entire sky was an illusion caused by the earth's own motion. If the earth existed free and unattached in space, could she not also *move*? Yet the apparently obvious idea of letting the earth rotate on her own axis did not occur to Philolaus. Instead, he made her revolve, in twenty-four hours, round an extraneous point in space, By describing one complete circle a day, the observer on earth would have the illusion, like a traveller on a roundabout, that the whole cosmic fair was turning in the opposite direction.

In the centre of his roundabout, Philolaus placed the 'watch-tower of Zeus', also called 'the hearth of the universe' or the 'central fire'. But this 'central fire' is not to be confused with the sun. It could never be seen; for the inhabited part of the earth – Greece and its neighbours – was always turned away from it, as the dark side of the moon is always turned away from the earth. Moreover, between the earth and the central fire Philolaus inserted an invisible planet: the *antichton* or counter-earth. Its function was, apparently, to protect the anti-podes from being scorched by the central fire. The ancient belief that the far-western regions of the earth, beyond the straits of Gibraltar, are shrouded in eternal twilight[2] was now explained by the shadow which the counter-earth threw on those parts. But it is also possible – as Aristotle contemptuously remarks – that the counter-earth was invented merely to bring the number of moving things in the universe up to ten, the sacred number of the Pythagoreans.[3]

Around the central fire, then, revolved in concentric orbits these nine bodies: the *antichton* innermost, then the earth, the moon, the sun, and the five planets; then came the sphere carrying all the fixed stars. Beyond this outer shell there was a wall of fiery ether, enclosing the world from all sides. This 'outer fire' was the second and main source from which the universe drew its light and breath. The sun served merely as a kind of transparent window or lens, through which the outer light was filtered and distributed. The picture reminds one of Anaximander's holes in the flame-filled tyre. But these fantastic imaginations were perhaps less fantastic than the notion of a ball of fire hurtling across the sky through eternity, without burning out; a preposterous idea at which the mind boggles. Looking at the sky with eyes washed clean of theories, is it not more convincing to regard the sun and stars as holes in the curtain enclosing the world?

The only heavenly object considered to be similar to the earth was the moon. It was supposed to be inhabited by plants and animals fifteen times as strong as ours, because the moon enjoyed daylight for fifteen days in succession. Other Pythagoreans thought that the lights and shadows on the moon were reflections of our oceans. As for eclipses, some were caused by the earth, some by the counter-earth, which also accounted for the faint ashen light on the lunar disc at new moon. Still others seem to have assumed the existence of several counter-earths. It must have been a lively debate.

## 2. *Herakleides and the Sun-Centred Universe*

In spite of its poetic oddities, the system of Philolaus opened up a new cosmic perspective. It broke away from the geocentric tradition – the sturdy conviction that this earth occupies the centre of the Universe, from which, massive and immobile, it never budges an inch.

But it was also a landmark in another direction. It separated neatly two phenomena which had previously been mixed up: the succession of day and night, that is, the *diurnal* rotation of the sky as a whole; and the *annual* motions of the seven wandering planets.

The next improvement of the model concerned the daily motions. The central fire dropped out; the earth, instead of going round it, was now made to spin on her own axis, like a top. The reason was, presumably,[4] that the Greek seafarers' growing contacts with distant regions – from the Ganges to the Tagus, from the island of Thule to Taprobrana – had failed to produce any sign, or even rumour, of the central fire or the *antichton*, both of which should have been visible from the other side of the earth. I have said before that the Pythagoreans' world-view was elastic and adaptable. They did not drop the idea of the central fire as a source of heat and energy; but they transferred it from outer space into the core of the earth, and the counter-earth they simply identified with the moon.[5]

The next great pioneer in the Pythagorean tradition is Herakleides of Pontus. He lived in the fourth century B.C., studied under Plato, and presumably also under Aristotle; hence, by chronological order, he ought to be discussed after these. But I shall first follow the development of the Pythagorean cosmology, the boldest and most hopeful in antiquity, to its end – which came in the generation of Herakleides.

Herakleides took the earth's rotation round its own axis for granted. This explained the daily round of the skies, but left the problem of the annual motion of the planets untouched. By now, these annual motions had become the central problem of astronomy and cosmology. The multitude of fixed stars presented no problem. They never altered their positions relative to each other or to the earth.[6] They were a permanent guarantee of law and order and regularity in the universe, and could be imagined, without much difficulty, as a pattern of pinheads (or pin-holes) in the celestial pin-cushion which either turned, as a unit, around the earth, or appeared to do so owing to the earth's rotation. But the planets, the tramp stars, moved with a shocking irregularity. Their only reassuring feature was that they all moved along the same narrow belt or lane looped around the sky (the Zodiac): which meant that their orbits all lay very nearly in the same plane.

To get an idea of how the Greeks perceived the universe, imagine all transatlantic traffic – submarines, ships, aircraft – to be confined to the same trade-route. The 'orbits' of all craft will then be along concentric circles round the earth's centre, all in the same plane. Let an observer lie on his back in a cavity in the centre of the transparent earth, and watch the traffic. It will appear to him as points moving at different speeds along a single line: his zodiacal lane. If the transparent sphere is set rotating round the observer (who, himself, remains at rest) the traffic-lane will rotate with the sphere, but the traffic will still remain confined to this lane. The traffic consists of: two submarines ploughing the waters at different depths under the lane: they are the 'lower' planets, Mercury and Venus; then a single ship with blazing lights: the sun; then three aeroplanes at different heights: the 'upper' planets, Mars, Jupiter, and Saturn, in that order. Saturn would be very high up in the stratosphere; above it there is only the sphere of the fixed stars. As for the moon, she is so close to the observer in the centre, that she must be considered a ball rolling on the concave wall inside his cavity; but still in the same plane with all the other craft. This, then, in broad outlines, is the antique model of the world (Fig. A).

But model A could never be made to work properly. To our hind-sight, the reason is obvious: the planets were arranged in the wrong order; the sun should be in the centre, and the earth should take the sun's place between the 'lower' and 'upper' planets, taking the moon with her (Fig. D). This basic fault in the model caused incomprehensible irregularities in the apparent motions of the planets.

By the time of Herakleides, these irregularities had become the principal worry of the philosophers concerned with the universe. The sun and moon seemed to move in a more or less regular manner along the traffic lane; but the five planets travelled in a most erratic way. A planet would amble for a while along the lane, in the general direction of the traffic, West to East; but at intervals he would slow down, come to a stop as if he had reached a station in the sky, and retrace his steps; then change his mind again, turn round and resume his wandering in the original direction. Venus behaved even more capriciously. The pronounced periodical changes in her brightness and size seemed to indicate that she alternately approached and receded from us, and this suggested that she

(A) Classical geocentric system

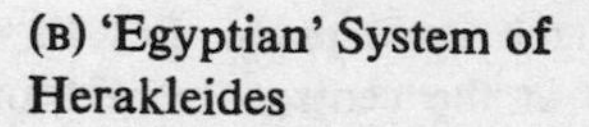
(B) 'Egyptian' System of Herakleides

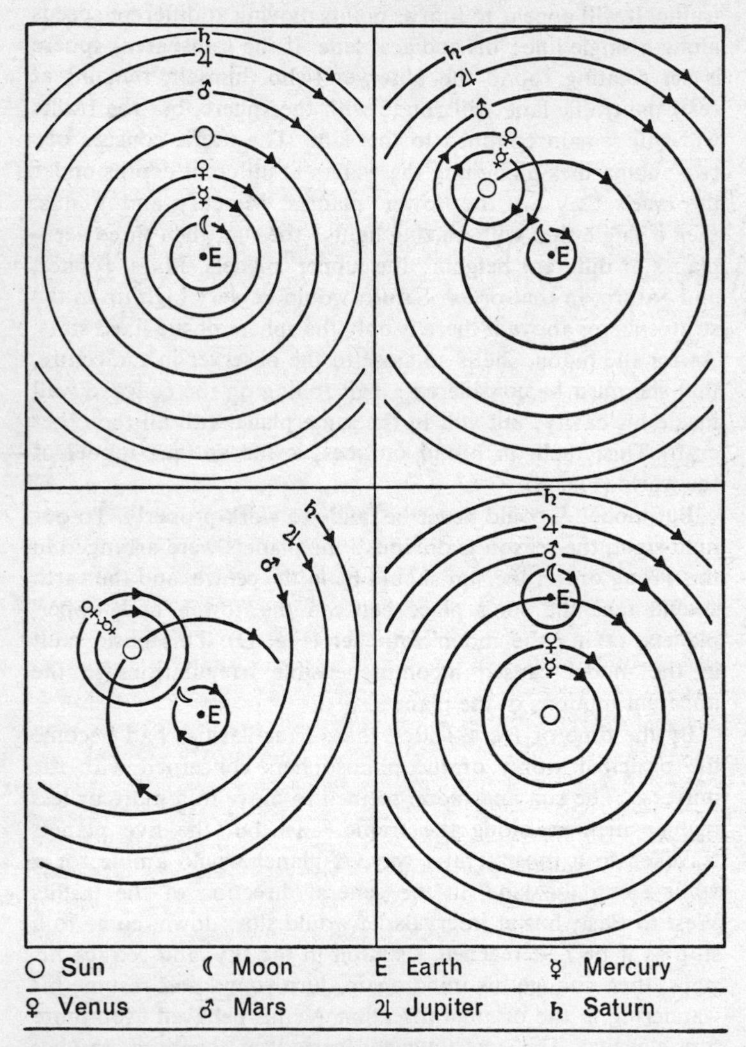


(C) System of Tycho de Brahe (and of Herakleides?)

(D) Aristarchus' heliocentric system

did not really move in a circle round the earth, but along some unthinkable, wavy line. Moreover, both she and Mercury, the second inner planet, now raced ahead of the steadily moving sun, now fell behind, but always stuck close to him, like dolphins playing around a ship. Accordingly, Venus at times apeared as Phosphoros the 'morning star', rising with the sun in her wake, at other times as Hesperos the 'evening star' at the sun's tail; Pythagoras seems to have been the first to recognize that they were one and the same planet.

Once more, in the rear-view mirror, Herakleides' solution of the puzzle seems simple enough. If Venus moved in an irregular manner relative to the earth, the supposed centre of her orbit, yet danced attendance to the sun, then she obviously was attached to the sun, and not to the earth: she was a satellite of the sun. And since Mercury behaved in the same manner, both inner planets must revolve round the sun – and with the sun round the earth, like a wheel turning on a wheel.

Figure B on page 48 explains at a single glance why Venus alternately approaches and recedes from the earth; why she is at times ahead of, at others behind, the sun; and also why she intermittently moves in reverse gear along the Zodiacal lane.[7]

It all looks beautifully obvious – in the rear mirror. But there are situations where it needs great imaginative power, combined with disrespect for the traditional current of thought, to discover the obvious. The scant information we have about the personality of Herakleides shows that he had both: originality, and contempt for academic tradition. He was nicknamed by his acquaintances the *paradoxolog* – a maker of paradoxes; Cicero relates that he was fond of telling 'puerile fables' and 'marvellous stories'; and Proclus tells us that he had the audacity to contradict Plato, who taught the immobility of the earth.[8]

The idea that the two lower planets – and only these two – were satellites of the sun, while the sun itself and the remaining planets still revolved round the earth, became later known by the misnomer 'Egyptian System' and gained great popularity (Fig. B, p. 48). It was evidently a half-way house between the geocentric (earth-centred) and heliocentric (sun-centred) conceptions of the universe. We do not know whether Herakleides stopped there, or whether he took the further step of letting the three outer planets also go around the sun, and the sun, with all his five satellites, go round the earth (Fig. C, p. 48). It would have been a logical step, and some modern scholars believe that Herakleides did reach this three-quarter-way house.[9] Some even believe that he also took the ultimate step of making *all* the planets, including the earth, revolve around the sun.

But whether he went the whole way to the modern conception of the solar system or not, is merely a matter of historic curiosity, for his successor, Aristarchus, certainly did.

## 3. *The Greek Copernicus*

Aristarchus, last of the line of the Pythagorean astronomers, came, like the Master, from Samos; and he is supposed to have been born, symbolically, in the same year, 310 B.C., in which Herakleides died.* Only one short treatise of his survives: *On the Sizes and Distances of the Sun and Moon.* It shows that he had the basic gifts required of a modern scientist: originality of thought and meticulousness in observation. The elegant method he designed for calculating the distance of the sun was followed by astronomers throughout the Middle Ages; if his actual figures were wrong, it was due to the fact that he was born two thousand years before the telescope. But though an equal distance separated him from the invention of the pendulum clock, he improved the estimates of the length of the solar year by adding 1/1623 to the previous estimate of 365¼ days.

The treatise in which Aristarchus proclaimed that the sun, not the earth, was the centre of our world around which all planets revolve – this crowning achievement of Pythagorean cosmology, which Copernicus was to rediscover seventeen

* These dates are rather conjectural. But astronomers have a knack for timing their life-orbits: Galileo died in the year Newton was born; and Newton was born exactly a hundred years after Copernicus died.

centuries later – is lost. But fortunately, we have the testimony of no smaller authorities than Archimedes and Plutarch, among others; and the fact that Aristarchus taught the heliocentric system is unanimously accepted by the ancient sources and modern scholars.

Archimedes, the greatest mathematician, physicist, and inventor of antiquity, was a younger contemporary of Aristarchus. One of his most curious works is a little treatise called *The Sand Reckoner*, dedicated to King Gelon of Syracuse. It contains the crucial phrase: 'For he (Aristarchus of Samos) supposed that the fixed stars and the sun are immovable, but that the earth is carried round the sun in a circle. . . .'[10]

Plutarch's reference to Aristarchus is equally important. In his treatise *On the Face in the Moon Disc*, one of the characters refers to Aristarchus of Samos who thought 'that the heaven is at rest, but that the earth revolves in an oblique orbit, while it also rotates about its own axis'.[11]

Thus Aristarchus of Samos had carried the development which started with Pythagoras and was continued by Philolaus and Herakleides, to its logical conclusion: the sun-centred universe. But here the development comes to an abrupt end. Aristarchus had no disciples and found no followers.[12] For nearly two millennia the heliocentric system was forgotten – or, shall one say, repressed from consciousness? – until an obscure Canon in Varmia, a remote outpost in Christendom, picked up the thread where the Samian had left off.

This paradox would be easier to understand if Aristarchus had been a crank, or a dilettante whose ideas were not taken seriously. But his treatise *On the Sizes and Distances of the Sun and Moon* became a classic of antiquity, and shows him as one of the foremost astronomers of his time; his fame was so great that nearly three centuries later Vitruvius, the Roman architect, starts his list of universal geniuses of the past with: 'Men of this type are rare, men such as were in times past Aristarchus of Samos. . . .'[13]

In spite of all this, his correct hypothesis was rejected in favour of a monstrous system of astronomy, which strikes us today as an affront to human intelligence, and which reigned supreme for fifteen hundred years. The reasons for this benightedness will emerge only gradually, for we are faced here with one of the most astonishing examples of the devious, nay crooked ways of the 'Progress of Science' – which is one of the main topics of this book.

# 4 The Failure of Nerve

## 1. *Plato and Aristotle*

By the end of the third century, B.C., the heroic period of Greek science was over. From Plato and Aristotle onward, natural science begins to fall into disrepute and decay, and the achievements of the Greeks are only rediscovered a millennium and a half later. The Promethean venture which had started around 600 B.C., had within three centuries spent its elan; it was followed by a period of hibernation, which lasted five times as long.

From Aristarchus there is, logically, only one step to Copernicus; from Hippocrates, only a step to Paracelsus; from Archimedes, only a step to Galileo. And yet the continuity was broken for a time-span nearly as long as that from the beginning of the Christian era to our day. Looking back at the road along which human science travelled, one has the image of a destroyed bridge with rafters jutting out from both sides; and in between, nothing.

We know all this happened; if we knew exactly why it happened, we would probably have the remedy to the ills of our own time. For the breakdown of civilization during the Dark Ages is in some respects the reverse of the breakdown that started, though less dramatically, in the Age of Enlightenment. The former can be broadly described as a withdrawal from the material world, contempt for knowledge, science, and technology; rejection of the body and its pleasures in favour of the life of the spirit. It reads like a mirror-writing to the tenets of the age of scientific materialism which begins with Galileo and ends with the totalitarian state and the hydrogen bomb. They have only one factor in common: the divorce of reason from belief.

On the watershed that separates the heroic age of science from the age of its decline, stand the twin peaks, Plato and

Aristotle. Two quotations may illustrate the contrast in philosophical climate on the two sides of the watershed. The first is a passage from a writer belonging to the Hippocratic school, and dates presumably from the fourth century B.C. 'It seems to me,' he says, dealing with that mysterious affliction, epilepsy, 'that the disease is no more "divine" than any other. It has a natural cause, just as other diseases have. Men think it divine merely because they do not understand it. But if they called everything divine which they do not understand, why, there would be no end of divine things!'[1] The second quotation is from Plato's *Republic* and sums up his attitude to astronomy. The stars, he explains, however beautiful, are merely part of the *visible* world which is but a dim and distorted shadow or copy of the *real* world of ideas; the endeavour to determine exactly the motions of these imperfect bodies is therefore absurd. Instead: 'let us concentrate on (abstract) problems, said I, in astronomy as in geometry, and dismiss the heavenly bodies, if we intend truly to apprehend astronomy'.[2]

Plato is equally hostile to the Pythagoreans' first and favourite branch of science. 'The teachers of harmony,' he lets Socrates complain, 'compares the sounds and consonances which are heard only, and their labour, like that of the astronomers, is vain'.[3]

None of this was probably meant to be taken quite literally, but it was – by that extremist school of Neoplatonism which dominated Western philosophy for several centuries, and stifled all progress in science – until, in fact, Aristotle was rediscovered and interest in nature revived. I have called them twin peaks separating two epochs of thought; but insofar as their influence on the future is concerned, Plato and Aristotle should rather be called twin-stars with a single centre of gravity, which circle round each other and alternate in casting their light on the generations that succeed them. Until the end of the twelfth century, as we shall see, Plato reigned supreme; then Aristotle was resurrected and remained for two hundred years *the* philosopher, as he was commonly called; then Plato made a come-back, in an entirely different guise. Professor

Whitehead's famous remark: 'the safest general characterization of the European philosophical tradition is that it consists in a series of footnotes to Plato' could be amended by: 'Science, up to the Renaissance, consisted in a series of footnotes to Aristotle.'

The secret of their extraordinary influence, intermittently stimulating and choking European thought, during such a near-astronomical period, has been the subject of passionate and never-ending controversy. It is, of course, not due to any single reason, but to the confluence of a multitude of causes at a particularly critical point of history. To mention only a few, starting with the most obvious: they are the first philosophers of antiquity whose writings survived not in odd fragments, in second- or third-hand quotations, but in massive bulk (Plato's authenticated dialogues alone make a volume of the length of the Bible), embracing all domains of knowledge and the essence of the teachings of those who came before them; as if after an atomic war, among the torn and charred fragments, a complete *Encyclopaedia Britannica* had been preserved. Apart from bringing together all the relevant items of available knowledge in an individual synthesis, they were of course, in their own right, original thinkers of great creative power in such varied fields as metaphysics, biology, logics, epistemology, and physics. They both founded 'schools' of a new kind: the first Academy and the first Lyceum, which survived for centuries as organized institutions, and transformed the founders' once fluid ideas into rigid ideologies, Aristotle's hypotheses into dogmas, Plato's visions into theology. Then again, they were truly twin-stars, born to complement each other; Plato the mystic, Aristotle the logician; Plato the belittler of natural science, Aristotle the observer of dolphins and whales; Plato, the spinner of allegorical yarns, Aristotle the dialectician and casuist; Plato, vague and ambiguous, Aristotle precise and pedantic. Lastly – for this catalogue could be continued forever – they evolved systems of philosophy which, though different and even opposed in detail, taken jointly seemed to provide a complete answer to the predicament of their time.

The predicament was the political, economic, and moral bankruptcy of classical Greece prior to the Macedonian conquest. A century of constant war and civil strife had bled the country of men and money; venality and corruption were poisoning public life; hordes of political exiles, reduced to the existence of homeless adventurers, were roaming the countryside; legalized abortion and infanticide were further thinning out the rank of citizens. The history of the fourth century, wrote a modern authority,

> is in some of its aspects that of the greatest failure in history. ... Plato and Aristotle ... each in his different way tries (by suggesting forms of constitution other than those under which the race had fallen into political decadence) to rescue that Greek world which was so much to him from the political and social disaster to which it is rushing. But the Greek world was past saving.[4]

The political reforms suggested by them concern us only insofar as they indicate the unconscious bias which permeates their cosmology; but in this context, they are relevant. Plato's *Utopia* is more terrifying than Orwell's *1984* because Plato desires to happen what Orwell fears might happen. 'That Plato's Republic should have been admired, on its political side, by decent people, is perhaps the most astonishing example of literary snobbery in all history,' remarked Bertrand Russell.[5] In Plato's Republic, the aristocracy rules by means of the 'noble lie', that is, by pretending that God has created three kinds of men, made respectively of gold: the rulers, silver: the soldiers, and base metals: the common man. Another pious lie will help to improve the race: when marriage is abolished, people will be made to draw mating-lots, but the lots will be secretly manipulated by the rulers according to the principles of eugenics. There will be rigid censorship; no young person must be allowed to read Homer because he spreads disrespect of the gods, unseemly merriment, and the fear of death, thus discouraging people from dying in battle.

Aristotle's politics move along less extreme, but essentially similar lines. He criticizes some of Plato's most provocative formulations, but not only does he regard slavery as the natural basis of the social order – 'the slave is totally devoid of any faculty of reasoning'[6]; he also deplores the existence of a 'middle' class of free artisans and professional men, because their superficial resemblance to the rulers brings discredit on the latter. Accordingly, all professionals are to be deprived of the rights of citizenship in the Model State. It is important to understand the source of this contempt of Aristotle for artisans, craftsmen, architects, engineers, and the like – by contrast, say, to the high esteem in which an Eupalinos, the tunnel-builder, had been held in Samos. The point is that Aristotle believed them no longer to be necessary, because *applied science and technology had already completed their task*. Nothing further need, or could, be invented to make life more comfortable and enjoyable, because 'nearly all requisites of comfort and social refinement have been secured' and 'everything of these kinds has already been provided'.[7] Pure science and philosophy 'which deal neither with the necessities nor with the enjoyment of life' only arose, in Aristotle's view, after the practical sciences had done all that they can ever do, and material progress had come to a halt.

Even these cursory remarks may indicate the general mood underlying these philosophies: the unconscious yearning for stability and permanence in a crumbling world where 'change' can only be a change for the worse, and 'progress' can only mean progress toward disaster. 'Change' for Plato is virtually synonymous with degeneration; his history of creation is a story of the successive emergence of ever lower and less worthy forms of life – from God who is pure self-contained Goodness, to the World of Reality which consists only of perfect Forms or Ideas, to the World of Appearance, which is a shadow and copy of the former; and so down to man: 'Those of the men first created who led a life of cowardice and injustice were suitably reborn as women in the second generation, and this is why it was at this particular juncture that the gods contrived the lust for copulation.' After the women we come to the animals: 'Beasts who go on all fours came from men who were wholly unconversant with philosophy and had never gazed on the heavens'.[8] It is a tale of the Fall in permanence: a theory of *descent* and *devolution* – as opposed to evolution by ascent.

As so often with Plato, it is impossible to say whether all this is to be taken literally, or allegorically, or as an esoteric leg-pull. But there can be no doubt concerning the basic trend of the whole system.

We shall have to hark back time and again to Plato, to pick up the scent of some particular later development. For the time being, let us retain this essential clue to Plato's cosmology: his fear of change, his contempt and loathing for the concepts of evolution and mutability. It will reverberate all through the Middle Ages, together with its concomitant yearning for a world of eternal, changeless perfection:

> Then agin I think on that which Nature said
> Of that same time when no more change shall be,
> But steadfast rest of all things, firmly stay'd
> Upon the pillars of eternity,
> That is contrary to mutability.[9]

This 'mutation phobia' seems to be mainly responsible for the repellent aspects of Platonism. The Pythagorean synthesis of religion and science, of the mystical and empirical approach is now in shambles. The mysticism of the Pythagoreans is carried to sterile extremes, while empirical science is ridiculed and discouraged. Physics is separated from mathematics and made into a department of theology. The Pythagorean Brotherhood is transformed into the Guides of a totalitarian Utopia; the transmigration of souls on their way to God is debased by old-wife's tales, or edifying lies, about cowards being punished by feminine reincarnations; orphic asceticism curdles into hatred of the body and contempt for the senses. True knowledge cannot be obtained by the study of nature; for 'if we would have true knowledge of anything, we must be quit of the body. ... While in company with the body, the soul cannot have true knowledge'.[10]

All this is not an expression of humility – neither of the humility of the mystic seeker for God, nor the humility of reason acknowledging its limits; it is the half-frightened, half-arrogant philosophy of the genius of a doomed aristocracy and a bankrupt civilization. When reality becomes unbearable, the mind must withdraw from it and create a world of artificial perfection. Plato's world of pure Ideas and Forms, which alone is to be considered as real, whereas the world of nature which we perceive is merely its cheap Woolworth copy, is a flight into delusion. The intuitive truth expressed in the allegory of the Cave is here carried to absurdity by over-concretization – as if the author of the line 'this world is a vale of tears' were to proceed with a factual survey of the distribution of tear-drops in the vale.

Once again one must remember, that in the surrealistic cosmogony of the *Timaeus* it is impossible to draw the line between philosophy and poetry, metaphorical and factual statement; and that long passages in the *Parmenides* virtually destroy the doctrine that the world is a copy of models in heaven. But if some of my previous paragraphs sound like a harsh and one-sided view of what Plato meant, this is essentially *what he came to mean* to a long row of future generations – the one-sided shadow that he threw. We shall also see that the *second* Platonic revival, in the fifteenth century, highlighted a quite different side of Plato, and threw his shadow into the opposite direction. But that turn is still a long way ahead.

## 2. *Rise of the Circular Dogma*

I must now turn to Plato's contribution to astronomy – which insofar as concrete advances are concerned, is nil; for he understood little of astronomy, and was evidently bored by it. The few passages where he feels moved to broach the subject are so muddled, ambiguous, or self-contradictory, that all scholarly efforts have failed to explain their meaning.[11]

However, by a process of metaphysical and *a priori* reasoning, Plato came to certain general conclusions regarding the shape and motions of the universe. These conclusions, of paramount importance for everything which follows, were that *the shape of the world must be a perfect sphere, and that all motion must be in perfect circles at uniform speed.*

> And he gave the universe the figure which is proper and natural. ... Wherefore he turned it, as in a lathe, round and spherical, with its extremities equidistant in all directions from the centre, the figure of all figures most perfect and most like to itself, for he deemed the like more beautiful than the unlike. To the whole he gave, on the outside round about, a surface perfectly finished and smooth, for many reasons. It had no need of eyes, for nothing visible was left outside it; nor of hearing, for there was nothing audible outside it; and there was no breath outside it requiring to be inhaled. ... He allotted to it the motion which was proper to its bodily form, that motion of the seven motions which is most bound up with understanding and intelligence. Wherefore, turning it round in one and the same place upon itself, he made it move with circular rotation; all the other six motions [i.e., straight motion up and down, forward and back, right and left] he took away from it and made it exempt from their wanderings. And since for this revolution it had no need of feet, he created it without legs and without feet. ... Smooth and even and everywhere equidistant from the centre, a body whole and perfect, made up of perfect bodies. ...[12]

Accordingly, the task of the mathematicians was now to design a system which would reduce the apparent irregularities in the motions of the planets to regular motions in perfectly regular circles. This task kept them busy for the next two thousand years. With his poetic and innocent demand, Plato laid a curse on astronomy, whose effects were to last till the beginning of the seventeenth century, when Kepler proved that planets move in oval, and not circular orbits. There is perhaps no other example in the history of thought of such dogged, obsessional persistence in error, as the circular fallacy which bedevilled astronomy for two millennia.

But here again, Plato had merely thrown out, in semi-allegorical language, a suggestion which was quite in keeping with the Pythagorean tradition; it was Aristotle who promoted the idea of circular motion to a dogma of astronomy.

## 3. *The Fear of Change*

In Plato's world the boundaries between the metaphorical and the factual are fluid; all such ambiguity disappears as Aristotle takes over. With pedantic thoroughness the vision is dissected, its poetic tissue is preserved *in vitro*, its volatile spirit condensed and frozen. The result is the Aristotelian model of the universe.

The Ionians had prised the world-oyster open, the Pythagoreans had set the earth-ball adrift in it, the Atomists dissolved its boundaries in the infinite. Aristotle closed the lid again with a bang, shoved the earth back into the world's centre, and deprived it of motion.

I shall describe the model first in its broad outline, and fill in the details later.

The immobile earth is surrounded, as in the earlier cosmologies, by nine concentric, transparent spheres, enclosing each other like the skins of an onion (see Fig. A, p. 48). The innermost skin is the sphere of the moon; the two outermost are the sphere of the fixed stars, and beyond that, the sphere of the Prime Mover, who keeps the whole machinery turning: God.

The God of Aristotle no longer rules the world from the inside, but from the outside. It is the end of the Pythagorean central fire, the hearth of Zeus, as a divine source of cosmic energy; the end of Plato's mystic conception of the *anima mundi*, of the world as a living animal possessed with a divine soul. Aristotle's God, the Unmoved Mover, who spins the world round from outside it, is the God of abstract theology. Goethe's '*Was wär' ein Gott der nur von aussen stiesse*' – seems to be aimed directly at him. The removal of God's home from the centre to the periphery automatically transformed the central region, occupied by earth and moon, into the farthest away from Him: the humblest and lowliest of the whole universe. The space enclosed by the sphere of the moon and containing the earth – the 'sub-lunary region' – is now considered definitely non-U. To this region, and to this region alone, are the horrors of Change, of mutability confined. Beyond the sphere of the moon, the heavens are eternal and unalterable.

This splitting-up of the universe into two regions, the one lowly, the other exalted, the one subject to change, the other not, was to become another basic doctrine of medieval philosophy and cosmology. It brought a serene, cosmic reassurance to a frightened world by asserting its essential stability and permanence, but without going so far as to pretend that all change was mere illusion, without denying the reality of growth and decline, generation and destruction. It was not a reconciliation of the temporal and the eternal, merely a confrontation of the two; but to be able to take in both in one glance, as it were, was something of a comfort.

The division was made intellectually more satisfactory and easier to grasp, by assigning to the two parts of the universe different raw materials and different motions. In the sub-lunary region, all matter consisted of various combinations of the four elements, earth, water, air, and fire, which themselves were combinations of two pairs of opposites, hot and cold, dry and wet. The nature of these elements requires that they move in straight lines: earth downward, fire upward, air and water horizontally. The atmosphere fills the whole sub-lunary sphere, though its upper reaches consist not of proper air, but of a substance which, if set in motion, will burn and produce comets and meteors. The four elements are constantly being transformed one into the other, and therein lies the essence of all change.

But if we go beyond the moon's sphere, nothing changes, and none of the four terrestrial elements is present. The heavenly bodies consist of a different, pure, and immutable 'fifth element', which becomes the purer the farther away from the earth. The natural motion of the fifth element, as opposed to the four earthly elements, is circular, because the sphere is the only perfect form, and circular motion is the only perfect motion. Circular motion has no beginning and no end; it returns into itself and goes on forever: it is motion without change.

The system had yet another advantage. It was a compromise between two opposite trends in philosophy. On the one side there was the 'materialistic' trend, which had started with the Ionians, and was continued by men like Anaxagoras, who believed that *homo sapiens* owed his superiority to the dexterity of his hand; by Heraklitus, who regarded the universe as a product of dynamic forces in eternal flux; and culminated in Leucippus and Democritus, the first atomists. The opposite tendency, which originated with the Eleatics, found its extreme expression in Parmenides, who taught that all apparent change, evolution and decline, were illusions of the senses, because whatever exists cannot arise from anything that does not, or is different from it; and that the Reality behind the illusion is indivisible, unchangeable, and in a state of static perfection. Thus for Heraklitus Reality is a continuous process of Change and Becoming, a world of dynamic stresses, of creative tensions between opposites; whereas for Parmenides Reality is a solid, uncreated, eternal, motionless, changeless, uniform sphere.[13]

The preceding paragraph is, of course, a woeful over-simplification of developments in one of the liveliest periods of philosophic debate; but my purpose is merely to show how neatly the Aristotelian model of the universe solved the basic dilemma by handing over the sub-lunary region to the Materialists, and letting it be governed by Heraklitus' motto 'all is change'; whereas the rest of the universe, eternal and immutable, stood in the sign of the Parmenidian 'nothing ever changes'.

Once again, it was not a reconciliation, merely a juxtaposition, of two world-views, or 'world-feelings', both of which have a profound appeal to the minds of men. This appeal was increased in power when, at a later stage, mere juxtaposition yielded to *gradation* between the opposites; when the original Aristotelian two-storey universe – all basement and loft – was superseded by an elaborately graded, multi-storeyed structure; a cosmic hierarchy where every object and creature had its exact 'place' assigned to it, because its position in the many-layered space between lowly earth and high heaven defined its rank on the Scale of Values, in the Chain of Being. We shall see that this concept of a closed-in cosmos graded like the Civil Service (except that there was no advancement, only demotion) survived for nearly a millennium and a half. It was really a Mandarin Universe. During these long centuries, European thought had more in common with Chinese or Indian philosophy than with its own past and future.

However, even if European philosophy were only a series of footnotes to Plato, and even though Aristotle had a millennial stranglehold on physics and astronomy, their influence, when all is said, depended not so much on the originality of their teaching, as on a process of natural selection in the evolution of ideas. Out of a number of ideological mutations, a given society will select that philosophy which it unconsciously feels to be best suited for its need. Each time, in subsequent centuries, when the cultural climate changed in Europe, the twin stars also changed their aspect and colour: Augustine and Aquinas, Erasmus and Kepler, Descartes and Newton each read a different message in them. Not only did the ambiguities and contradictions in Plato, the dialectical twists in Aristotle, admit a wide range of interpretations and shifts of emphasis; but, by taking the two jointly or in alternation, by combining selected facets of each, the total effect could virtually be reversed; we shall see that the 'New Platonism' of the sixteenth century was in most respects the opposite of the Neoplatonism of the early Middle Ages.

In this context I must briefly return to Plato's loathing for change – for 'generation and decay' – which made the sublunary sphere such a disreputable slum-district of the universe. Aristotle himself did not share this loathing. As a keen biologist, he regarded all change, all movement in nature as purposeful and goal-directed – even the motions of inanimate bodies: a stone will fall to the earth, as a horse will canter to its stable, because that is its 'natural place' in the universal hierarchy. We shall have occasion, later on, to marvel at the disastrous effects of this Aristotelian brainwave on the course of European science; at the moment, I merely wish to point out that Aristotle's attitude to Change, though he rejects evolution and progress, is not quite as defeatist as Plato's.[14] Yet Neoplatonism, in its dominant trend, ignores Aristotle's dissent on this essential point, and manages to make the worst of both worlds. It adopts the Aristotelian scheme of the universe, but makes the sublunary sphere a Platonic vale of shadows; it follows the Platonic doctrine of the natural world as a dim copy of ideal Forms – which Aristotle rejected – yet follows Aristotle in placing the Prime Mover outside the confines of the world. It follows both in their anxious effort to build a walled-in universe, protected against the Barbarian incursions of Change; a nest of spheres-within-spheres, eternally revolving in themselves, yet remaining in the same place; thus hiding its one shameful secret, that centre of infection, safely isolated in the sub-lunary quarantine.

In the immortal parable of the Cave, where men stand in their chains backs to the light, perceiving only the play of shadows on the wall, unaware that these are but shadows, unaware of the luminous reality outside the Cave – in this allegory of the human condition, Plato hit an archetypal chord as pregnant with echoes as Pythagoras' Harmony of the Spheres. But when we think of Neoplatonism and scholasticism as concrete philosophies and precepts of life, we may be tempted to reverse the game, and to paint a picture of the founders of the Academy and the Lyceum as two frightened men standing in the self-same Cave, facing the wall, chained to their places in a catastrophic age, turning their back on the flame of Greece's heroic era, and throwing grotesque shadows which are to haunt mankind for a thousand years and more.

# 5 The Divorce from Reality

## 1. *Spheres Within Spheres (Eudoxus)*

In a closed universe, where the fixed stars offered as yet no specific problems, the challenge to understanding came from the planets; the chief task of cosmology was to devise a system which explained how sun, moon, and the remaining five planets moved.

This task was further narrowed down when Plato's dictum that all heavenly bodies move in perfect circles, became the first Academic dogma in the first institution that bore that solemn name. The task of Academic astronomy was now to prove that the apparently irregular meanderings of the planets were the result of some combination of several simple, circular, uniform motions.

The first serious attempt was made by Plato's pupil Eudoxus, and improved by the latter's pupil, Calippus. It is an ingenious attempt – Eudoxus was a brilliant mathematician, to whom most of Euclid's fifth book is due. In the earlier geocentric models of the universe, each planet, we remember, was attached to a transparent sphere of its own, and all spheres were turning round the earth. But, since this did not account for the irregularities of their motions, such as standing occasionally still and going backward for a while: their 'stations' and 'retrogressions', Eudoxus assigned to each planet not one, but several spheres. The planet was attached to a point on the equator of a sphere, which rotates round its axis, A. The two ends of this axis are let into the inner surface of a concentric larger sphere $S_2$, which rotates round a different axis, $A_2$ and carries A around with it. The axis of $S_2$ is attached to the next larger sphere $S_3$, which rotates again round a different axis $A_3$: and so on. The planet will thus participate in all the independent rotations of the various spheres which form its 'nest'; and by letting each sphere rotate at the appropriate tilt and speed, it was possible to reproduce roughly – though only very roughly – the actual motion of each planet.[1] The sun and moon needed a nest of three spheres each, the other planets four spheres each, which (with the modest single sphere assigned to the multitude of fixed stars) made altogether twenty-seven spheres. Calippus improved the system at the price of adding seven more spheres, making a total of thirty-four. It is at this point that Aristotle came in.

In the previous chapter I concentrated on the broad outlines and the metaphysical implications of Aristotle's universe, without going into astronomical detail. Thus I spoke of the classic *nine* spheres, from the moon's sphere to that of the Prime Mover (which alone were, in fact, remembered during the Middle Ages), without mentioning that each of these nine spheres was actually a nest of spheres-within-spheres. In reality, Aristotle used altogether fifty-four spheres to account for the motions of the seven planets. The reason for this additional investment of twenty spheres is interesting. Eudoxus and Calippus were not concerned with constructing a model that would be physically possible; they were not concerned with the real machinery of the heavens; they constructed a purely geometrical device, which, they knew, could exist only on paper. Aristotle wanted to do better, and transform it into a true physical model. The difficulty about this was that all adjoining spheres must be mechanically connected, yet the individual motion of each planet must not be transmitted to the others. Aristotle tried to solve this problem by inserting a number of 'neutralizing' spheres, which turned in the opposite direction to the 'working spheres', between two successive nests; in this manner, the effect of the motions of, say, Jupiter on his neighbour was eliminated, and the nest of Mars could be started from scratch, as it were. But insofar as the reproduction of the actual planetary motions is concerned, Aristotle's model was no improvement.

Besides, another difficulty remained. While each sphere participated in the motion of the next larger one enclosing it, it needed a special moving force to impart to it its independent rotation on its own axis; which meant, that there had to be no less than fifty-five 'unmoved movers', or spirits, to keep the system going.

It was an extremely ingenious system – and completely mad, even by contemporary standards; as shown by the fact that in spite of Aristotle's enormous prestige, it was quickly forgotten and buried. Yet it was only the first of several equally ingenious and equally mad systems which astronomers created out of their tortured brains, in obedience to Plato's post-hypnotic suggestion that all heavenly motion must be circular motion centred round the earth.

There was also a certain dishonesty about it. The spheres of Eudoxus could account – however imprecisely – for the existence of 'stations' and 'retrogressions' in the progress of a planet; but it could never account for the variations in size and brightness, caused by variations of the planet's distance from the earth. These were particularly evident in the case of Venus and Mars, and most of all, the moon: thus central eclipses of the sun are 'annular' or 'total', according to the moon's momentary distance from the earth. Now all this was known before Eudoxus, and thus to Eudoxus himself as well as to Aristotle;[2] yet their system simply ignores the fact: however complicated the planet's motion is, it is confined to a sphere centred on the earth, and its distance to the earth can therefore never vary.

It was this unsatisfactory state of affairs which gave rise to the unorthodox branch of cosmology developed by Herakleides and Aristarchus (*see* Chapter III). The system of Herakleides eliminated (though merely for the inner planets) *both* the most conspicuous scandals: the 'stations-and-retrogressions', and the varying distances from the earth. Moreover, it explained (as a glance at Fig. B on p. 48 will show) the logical relatedness of the two scandals: why Venus was always brightest when she was moving crabwise, and vice versa. When Herakleides and/or Aristarchus made the remaining planets, including the earth, move round the sun, Greek science was on the straight road to the modern universe; then abandoned it again. Aristarchus' sun-centred model was discarded as a freak; and academic science marched on triumphantly from Plato, via Eudoxus, and Aristotle's fifty-five spheres, to an even more ingenious and improbable artefact: the maze of epicycles devised by Claudius Ptolemy.

## 2. *Wheels Within Wheels (Ptolemy)*

If we call Aristotle's world an onion universe, we might as well call Ptolemy's the Ferris Wheel universe. It was begun by Apollonius of Perga in the third century B.C., developed by Hipparchus of Rhodes in the century that followed, and completed by Ptolemy of Alexandria in the second century A.D. The Ptolemaic system remained, with minor modifications, the last word in astronomy until Copernicus.

Any rhythmic movement, even the dance of a bird, can be imagined as being caused by a clockwork in which a great number of invisible wheels cooperate to produce the motions. Ever since 'uniform circular motion' had become the law governing the heavens, the task of astronomy was reduced to designing, on paper, just such imaginary clockworks which explained the dance of the planets as a result of the gyrations of perfectly circular, ethereal components. Eudoxus had used spheres as components; Ptolemy used wheels.

It is perhaps easiest to visualize the Ptolemaic universe not as an ordinary clockwork, but a system of 'Big Wheels' or 'Ferris Wheels' as one sees them in amusement parks – a huge, upright, slowly revolving wheel with seats or small cabins hanging suspended from its rim. Let us imagine the passenger safely strapped to his seat in the little cabin, and let us further imagine that the machinery has gone crazy – the cabin, instead of hanging down quietly from the rim of the Big Wheel, rotates wildly round the pivot from which it is suspended, while the pivot itself revolves slowly with the Wheel. The unhappy passenger – or planet – is now describing a curve in space which is not a circle, but is nevertheless produced by a combination of circular motions. By varying the size of the Wheel, the length of the arm by which the cabin is suspended, and the speeds of the two rotations, an amazing variety of curves can be produced, such as the one shown on the diagram – but also kidney-shaped curves, garlands, ovals, and even straight lines!

Seen from the earth, which is in the centre of the Big Wheel, the planet-passenger in the cabin will move clockwise until he reaches the 'stationary point' $S_1$, then regress anti-clockwise to $S_2$, then move again clockwise to $S_3$, and so on.* The rim of the Big Wheel is called the *deferent*, and the circle described by

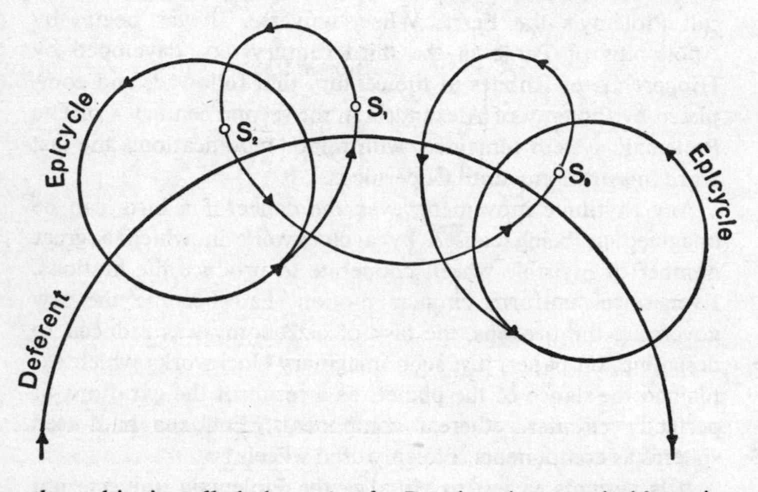


the cabin is called the *epicycle*. By choosing a suitable ratio between the diameters of epicycle and deferent, and suitable velocities for each, it was possible to achieve a fair approximation to the observed motions of the planet, insofar as the 'stations and retrogressions', and its varying distances from the earth were concerned.

These, however, were not the only irregularities in the planetary motions. There was yet another scandal, due (as we today know) to the fact that their orbits are not circular, but elliptic, that is, oval-shaped; they 'bulge'. To do away with this anomaly, another device was brought in, called a 'movable eccentric': the hub of the Big Wheel no longer coincided with

* At this point the reader may think that I am repeating myself – for the diagram on this page seems to express the same idea as Fig. B on p. 48 – the idea of Herakleides. But there is a difference: in Herakleides' scheme the planet's epicycle is centred on the sun. In Ptolemy's, it is centred on nothing. It is a purely geometrical construction.

the earth, but moved on a small circle in the vicinity of the earth; and in this manner a suitably eccentric, i.e. 'bulgy' orbit, was produced.*

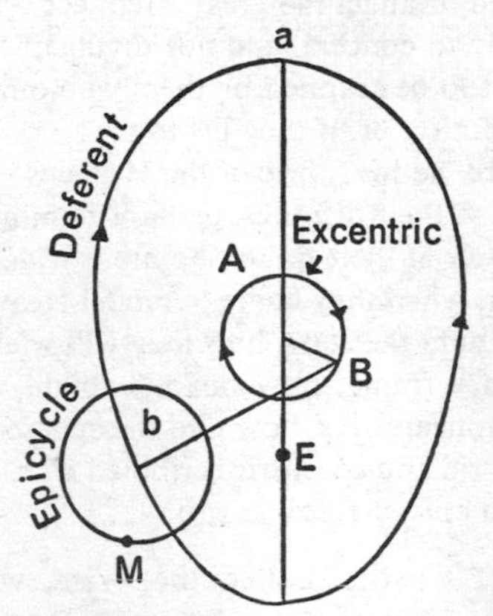


Egg-shaped Orbit of Mercury, according to Ptolemy: E=Earth; M=Mercury.

In the figure above, the hub of the Big Wheel moves clockwise on the small circle, from A to B; the point on the rim from which the cabin is suspended moves anti-clockwise in an egg-shaped curve from a to b; and the cabin spins round the final epicycle. But this was not enough; in the case of some recalcitrant planets it was found necessary to suspend a second cabin from the cabin suspended on the Big Wheel, with a different radius and a different speed; and then a third, fourth, and fifth, until the passenger in the ultimate cabin did indeed describe a trajectory more or less conforming to the one he was meant to describe.

By the time the Ptolemaic system was perfected, the seven passengers, sun, moon, and five planets, needed a machinery of not less than thirty-nine wheels to move through the sky; the outermost wheel, which carried the fixed stars, made the number an even forty. This system was still the only one recognized

* The 'movable eccentric' is in fact merely a kind of epicycle-in-reverse; and since the two are geometrically interchangeable, I shall use the term 'epicycle' for both.

by academic science in Milton's day – and was caricatured by him in a famous passage of *Paradise Lost*:

> From man or angel the great Architect
> Did wisely to conceal, and not divulge,
> His secret to be scanned by them who ought
> Rather admire; or, if they list to try
> Conjecture, he his fabric of the Heavens
> Hath left to their disputes, perhaps to move
> His laughter at their quaint opinions wide
> Hereafter, when they come to model Heaven
> And calculate the stars, how they will wield
> The mighty frame, how build, unbuild, contrive
> To save appearances, how gird the sphere
> With centric and eccentric scribbled o'er,
> Cycle and epicycle, orb in orb.

Alphonso X of Castile, called the Wise, who was a pious man and a great patron of astronomy, put the matter more succinctly. When he was initiated into the Ptolemaic system, he sighed: 'If the Lord Almighty had consulted me before embarking upon the Creation, I should have recommended something simpler.'

## 3. *The Paradox*

There is something profoundly distasteful about Ptolemy's universe; it is the work of a pedant with much patience and little originality, doggedly piling 'orb in orb'. All the basic ideas of the epicyclic universe, and the geometrical tools for it, had been perfected by his predecessor, Hipparchus; but Hipparchus had only applied them to the construction of orbits for the sun and moon. Ptolemy completed the unfinished job, without contributing any idea of great theoretical value.[3]

Hipparchus flourished around 125 B.C., more than a century after Aristarchus; and Ptolemy flourished around A.D. 150, nearly three centuries after Hipparchus. In this span of time, nearly equal to the duration of the Heroic Age, practically no progress was made. The landmarks were thinning out, and were soon to vanish altogether in the desert; Ptolemy was the last great astronomer of the Alexandrian School. He picked up the threads which had been left trailing behind Hipparchus, and completed the pattern of loop entwined in loop. It was a monumental and depressing tapestry, the product of tired philosophy and decadent science. But nothing else turned up to replace it for nearly a millennium and a half. Ptolemy's *Almagest*,[4] remained the Bible of astronomy until the beginning of the seventeenth century.

To get this extraordinary phenomenon into a proper perspective, one must not only be on one's guard against the wisdom of hindsight, but also against the opposite attitude, that kind of benevolent condescension which regards the past follies of Science as the unavoidable consequences of ignorance or superstition: 'our forebears just did not know better'. The point I shall try to make is that they did know better; and that to explain the extraordinary *cul de sac* into which cosmology had manoeuvred itself, we must look for more specific causes.

In the first place, the Alexandrian astronomers can hardly be accused of ignorance. They had more precise instruments for observing the stars than Copernicus had. Copernicus himself, as we shall see, hardly bothered with star-gazing; he relied on the observations of Hipparchus and Ptolemy. He knew no more about the actual motions in the sky than they did. Hipparchus' Catalogue of the fixed stars, and Ptolemy's Tables for calculating planetary motions, were so reliable and precise that they served, with some insignificant corrections, as navigational guides to Columbus and Vasco da Gama. Eratosthenes, another Alexandrian, computed the diameter of the earth as 7,850 miles, with an error of only ½ per cent;[5] Hipparchus calculated the distance of the moon as 30¼ earth diameters – with an error of only 0.3 per cent.[6]

Thus, insofar as factual knowledge is concerned, Copernicus was no better off, and in some respects worse off, than the Greek astronomers of Alexandria who lived at the time of Jesus Christ. They had the same observational data, the same instruments, the same know-how in geometry, as he did. They were giants of 'exact science'. Yet they failed to see what Copernicus saw after, and Herakleides-Aristarchus had seen before them:

that the planets' motions were obviously governed by the sun.

Now I have said before that we must beware of the word 'obvious'; but in this particular case its use is legitimate. For Herakleides and the Pythagoreans had not been led to the heliocentric hypothesis by a lucky guess, but by the observed fact that the inner planets behaved like satellites of the sun, and that the outer planets' retrogressions and changes in earth-distance were equally governed by the sun. Thus, by the end of the second century B.C., the Greeks had all the major elements of the puzzle in their hands,[7] and yet failed to put them together; or rather, having put them together, they took them to pieces again. They knew that the orbits, periods, and velocities of the five planets were connected with, and dependent on, the sun – yet in the system of the universe which they bequeathed to the world, they managed to ignore completely this all-important fact.

This mental snow-blindness is all the more remarkable as, *qua philosophers, they were aware of the dominant part played by the sun which, qua astronomers, they nevertheless denied.*

A few quotations will illustrate this parodox. Cicero, for instance, whose knowledge of astronomy is, of course, entirely based on Greek sources, writes in *The Republic*: 'The sun ... rules, prince, and leader of the other stars, sole and ordering principle of the universe (is) so large that its light brightens and fills the all. ... *The orbits of Mercury and Venus follow him as his companions.*'[8]

Pliny writes a century later: 'The sun is carried around in the midst of the planets, *directing not only the calendar and the earth but also the stars themselves and the sky.*'[9]

Plutarch speaks in a similar vein in *On the Face in the Moon Disc*:

> But in general how can we say: the earth is in the centre – in the centre of what? The universe is infinite; and the infinite, which has neither beginning nor end, has no centre either. ... The universe does not assign any fixed centre to the earth, which drifts homelessly and unsteadily through the infinite emptiness without a proper goal. ...[10]

In the fourth century A.D., when darkness was finally closing in on the world of antiquity, Julian the Apostate wrote about the sun: 'He leads the dance of the stars; his foresight guides all generation in nature. Around him, their King, the planets dance their rounds; they revolve around him in the perfect harmony of their distances which are exactly circumscribed, as the sages affirm, who contemplate the events in the skies. . .'[11]

Lastly Macrobius, who lived around 400 A.D., comments on the passage from Cicero which I have just quoted:

He calls the sun the ruler of the other stars because the sun regulates their progression and retrogression within spatial limits, for there are spatial limits which confine the planets in their advance and regress relative to the sun. Thus the force and power of the sun regulates the course of the other stars within fixed limits.[12]

Here, then, is evidence that to the very end of the antique world, the teaching of Herakleides and Aristarchus was well remembered; that a truth, once found, can be hidden away, buried under the surface, but not undone. And yet the Ptolemaic earth-centred universe, ignoring the specific role of the sun, held the monopoly in scientific thought for fifteen centuries. Is there an explanation for this remarkable paradox?

It has been frequently suggested that the explanation is fear of religious persecution. But all the evidence quoted in support of this view consists of a single, facetious remark by a character in Plutarch's dialogue *On the Face in the Moon Disc*, which I have mentioned before. The character, Lucius, is playfully accused of 'turning the universe upside down' by pretending that the moon consists of solid matter like the earth; he is then invited to explain his views further:

Lucius smiled and said: 'Very well; only do not bring against me a charge of impiety such as Cleanthes used to say that it behoved Greeks to bring against Aristarchus of Samos for moving the Hearth of the Universe, because he tried to save the phenomena by the assumption that the heaven is at rest, but that the earth revolves in an oblique orbit, while also rotating about its own axis.'[13]

However, the charge was never brought; neither Aristarchus, who was held in the highest esteem, nor Herakleides or any other adherent of the earth's motion, was persecuted or indicted. If Cleanthes had really tried to have anybody indicted on the grounds of 'moving the Hearth of the Universe', then the first person charged with impiety would have been the venerated Aristotle; for Aristarchus merely made the Hearth move with the earth through space, whereas Aristotle removed the Hearth to the periphery of the world, deprived the earth altogether of the divine presence, and made it the lowliest place in the world. In reality, the 'Hearth of the Universe' was no more than a poetic allusion to the Pythagorean Central Fire, and it would be absurd to regard it as a religious dogma. Cleanthes himself was a mystically inclined, and rather sour Stoic philosopher, who wrote a hymn to Zeus and despised science. His attitude to Aristarchus, a scientist and a Samian to boot, that island from which no good has ever come, was evidently 'the fellow deserves to be hanged'. Apart from this bit of academic gossip in Plutarch, there is no mention in any of the sources of religious intolerance towards science in the Hellenistic Age.[14]

## 4. *Knowing and Un-Knowing*

Thus neither ignorance, nor the threats of an imaginary Alexandrian inquisition, can serve to explain why the Greek astronomers, after having discovered the heliocentric system, turned their backs on it.[15] However, they never did so entirely; as the passages previously quoted, from Cicero and Plutarch to Macrobius, indicate, they knew that the sun governed the motions of the planets, but at the same time closed their eyes to the fact. But perhaps it is this irrationality itself which provides the clue to the solution, by jolting us out of the habit of treating the history of science in purely rational terms. Why should we allow artists, conquerors, and statesmen to be guided by irrational motives, but not the heroes of science? The post-Aristotelian astronomers denied the rule of the sun over the planets and affirmed it at the same time; while conscious reasoning rejects such a paradox, it is in the nature of the unconscious that it may simultaneously affirm and deny, say yes and no to the same question; to know and to un-know, as it were. Greek science in the age of decline was faced with an insoluble conflict, which resulted in a split of the mind; and this 'controlled schizophrenia' continued throughout the Dark and Middle Ages, until it came to be almost taken for granted as the normal condition of man. It was maintained, not by threats from outside, but by a kind of censor planted inside the mind, who kept it separated into strictly non-communicating compartments.

Their main concern was 'to save the appearances'. The original meaning of this ominous phrase was that a theory must do justice to the observed phenomena or 'appearances'; in plain words, that it must agree with the facts. But gradually, the phrase came to mean something different. The astronomer 'saved' the phenomena if he succeeded in inventing a hypothesis which resolved the irregular motions of the planets along irregularly shaped orbits into regular motions along circular orbits – *regardless whether the hypothesis was true or not*, i.e. whether it was physically possible or not. Astronomy, after Aristotle, becomes an abstract sky-geometry, divorced from physical reality. Its principal task is to explain away the scandal of non-circular motions in the sky. It serves a practical purpose as a method for computing tables of the motions of the sun, moon, and planets; but as to the real nature of the universe, it has nothing to say.

Ptolemy himself is quite explicit about this: 'We believe that the object which the astronomer must strive to achieve is this: to demonstrate that all the phenomena in the sky are produced by uniform and circular motions. ...'[16] And elsewhere: 'Having set ourselves the task to prove that the apparent irregularities of the five planets, the sun and moon can all be represented by means of uniform circular motions, because only such motions are appropriate to their divine nature. ... We are entitled to regard the accomplishment of this task as the ultimate aim of mathematical science based on philosophy.'[17] Ptolemy also makes it clear why astronomy must renounce all attempts to explain the *physical* reality behind it: because the heavenly bodies, being of a divine nature, obey laws different from those to be found on earth. No common link exists between the two; therefore we can know nothing about the physics of the skies.

Ptolemy was a wholehearted Platonist; the effect of the twin-stars on the course of science now makes itself fully felt. The divorce which they effected between the four elements of the sublunary region and the fifth element of the heavens, leads directly to a divorce of sky-geometry from physics, of astronomy from reality. The split world is reflected in the split mind. It knows that *in reality the* sun has a physical influence on the planets; but reality is no longer its concern.[18]

The situation is summed up in a striking passage by Theon of Smyrna, a contemporary of Ptolemy. After expressing his opinion that Mercury and Venus may, after all, be revolving round the sun, he goes on to say that the sun should be called the heart of the universe, which is both 'a world and an animal'.

> But [he reflects] in animated bodies the centre of the animal is different from the centre of its mass. For instance, for us who are both men and animals, the centre of the animated creature is in the heart, always in motion and always warm, and therefore the source of all the faculties of the soul, of desire, imagination, and intelligence; but the centre of our volume is elsewhere, about the navel. . . . Similarly, . . . the mathematical centre of the universe is where the earth is, cold and immovable, but the centre of the world as an animal is in the sun, which is, so to say, the heart of the universe.[19]

The passage is both appealing and appalling; it strikes a note which will reverberate throughout the Dark and Middle Ages. It appeals to the archetypal craving to comprehend the world as a live, pulsating animal; and it appalls by its unholy mix-up of allegorical and physical statements, by its pedantic variations on the inspired Platonic leg-pull. The contrast between navel and heart is witty but unconvincing; it does not explain why two planets should revolve round the heart and the other three round the navel. Did Theon and his readers believe in this sort of thing? The answer is, apparently, that one compartment of their minds did, the other did not; the process of divorcement was nearly completed. Observational astronomy was still progressing; but what a regression in philosophy compared to the

Pythagorean, and even the Ionian, school of seven centuries before!

## 5. *The New Mythology*

It looks as if the wheel had come full circle, back to the early Babylonians. They too had been highly competent observers and calendar-makers, who combined their exact science with a mythological dream-world. In the universe of Ptolemy, interlocking canals of perfect circles have replaced the heavenly waterways, along which the star-gods sail their barges on their precisely charted journeys. The Platonic mythology of the sky was more abstract and less colourful, but as irrational and dream-born as the older one.

The three fundamental conceits of this new mythology were: the dualism of the celestial and sub-lunary worlds; the immobility of the earth in the centre; and the circularity of all heavenly motion. I have tried to show that the common denominator of the three, and the secret of their unconscious appeal, was the fear of change, the craving for stability and permanence in a disintegrating culture. A modicum of split-mindedness and double-think was perhaps not too high a price to pay for allaying the fear of the unknown.

But whether the price was high or low, it had to be paid: the universe was put into the deep freeze, science was paralysed, and the manufacture of artificial moons and nuclear warheads was delayed by a millennium or more. Whether, *sub specie aeternitatis*, this was a Good Thing or a Bad Thing, we shall never know; but as far as our limited topic is concerned, it was clearly a bad thing. The earth-centred, dualistic, circular view of the cosmos excluded all progress and all compromise for fear of endangering its main principle, stability. Thus, it could not even be admitted that the two inner planets circled round the sun, because once you gave way on this apparently harmless minor point, the next logical step would be to extend the idea to the outer planets and to the earth itself – as the development of the Herakleidian deviation had clearly shown. The frightened mind, always on the defensive, is particularly aware of the dangers of yielding an inch to the devil.

The anxiety complex of the late Greek cosmologists becomes almost palpable in a curious passage[20] by Ptolemy himself, in which he defends the immobility of the earth. He starts with the usual commonsense argument that if the earth moved, 'all the animals and all separate weights would be left behind floating on the air' – which sounds plausible enough, though the Pythagoreans and atomists had long before Ptolemy realized its fallacious nature. But then Ptolemy continues to say that if the earth were really moving, it would 'at its great speed, have fallen completely out of the universe itself'. Now this is not plausible even on a naïve level, for the only motion attributed to the earth was a circular motion round the sun, which entailed no risk of falling out of the universe, just as the sun incurred no risk by circling the earth. Ptolemy, of course, knew this quite well – or, more precisely, one compartment of his mind knew it, while the other was hypnotized by the fear that once the earth's stability was shaken, the world would fly to pieces.

The myth of the perfect circle had an equally deep-rooted, spell-binding power. It is, after all, one of the oldest symbols; the ritual of drawing a magic circle around a person protects him against hostile spirits and perils of the soul; it marks off a place as an inviolable sanctuary; it was commonly used in tracing out the *sulcus primigenius*, the first furrow, when founding a new city. Apart from being a symbol of stability and protection, the circle, or wheel, had a technological plausibility, as it were, as a suitable element for any machine. But on the other hand, the planetary orbits were evidently *not* circles; they were eccentric, bulging, oval – or egg-shaped. They could be made *to appear* as the product of a combination of circles by geometrical artifices, but only at the price of renouncing any semblance of physical reality. There exist some fragmentary remains, dating from the first century A.D., of a small-sized Greek planetarium – a mechanical model designed to reproduce the motions of sun, moon, and perhaps also of the planets. But its wheels, or at least some of them, are not circular – they are egg-shaped.[21] A glance a the orbit of Mercury in the Ptolemaic system on p. 71 shows a similar egg-shaped curve staring into one's face. All these pointers were ignored, relegated into limbo as a sacrifice to circle-worship.

And yet there was nothing *a priori* frightening about oval or elliptic curves. They too were 'closed' curves, returning into themselves, and displayed a reassuring symmetry and mathematical harmony. By an ironical coincidence, we owe the first exhaustive study of the geometrical properties of the ellipse to the same man, Apollonius of Perga, who, never realizing that he had the solution in his hands, started the development of the epicyclic monster-universe. We shall see that, two thousand years later, Johannes Kepler, who cured astronomy of the circular obsession, still hesitated to adopt elliptical orbits, because, he wrote, if the answer were as simple as that, 'then the problem would already have been solved by Archimedes and Apollonius'.[22]

## 6. *The Cubist Universe*

Before bidding farewell to the Greek world, an imaginary parallel may help to bring matters into focus.

In 1907, simultaneously with the Cézanne memorial exhibition in Paris, a collection of the master's letters was published. A passage in one of the letters ran:

> Everything in nature is modelled on the sphere, the cone and the cylinder. One must teach oneself to base one's painting on these simple figures – then one can accomplish anything one likes.

And further:

> One must treat nature by reducing its forms to cylinder, sphere, and cone, all put into perspective, meaning that each side of an object, each plane, is directed towards a central plane.[23]

This pronouncement became the gospel of a school of painting known under the misnomer 'Cubism'. Picasso's first 'Cubist' picture was in fact constructed entirely of cylinders, cones, and circles; while other members of the movement saw nature in terms of angular bodies – pyramids, and bricks, and octaeders.*

But whether they painted in terms of cubes, cylinders, or cones, the declared aim of the Cubists was to resolve every object to a configuration of regular geometrical solids. Now the human face is not constructed out of regular solids any more than the orbits of the planets are made of regular circles; but in both cases it is possible to 'save the phenomena': in Picasso's *Femme au miroir*, the reduction of the model's eyes and upper lip to an interplay of spheres, pyramids, and parallelepipedes, displays the same ingenuity and inspired madness as Eudoxus' spheres pivoting within spheres.

It is rather depressing to imagine what would have happened to painting if Cézanne's Cubist pronouncement had been turned into a dogma as Plato's spherist pronouncement was. Picasso would have been condemned to go on painting more elaborate cylindrical bowls to the bitter end; and lesser talents would have found out soon that it is easier to save the phenomena with compass and ruler on graph-paper under a neon lamp, than by facing the scandals of nature. Luckily, Cubism was only a passing phase because painters are free to choose their style; but the astronomers of the past were not. The style in which the cosmos was presented had, as we saw, a direct bearing on the fundamental questions of philosophy; and later, during the Middle Ages it acquired a bearing on theology. The curse of 'spherism' upon man's vision of the universe lasted for two thousand years.

During the last few centuries, from about A.D. 1600 onwards, the progress of science has been continuous and without a break; so we are tempted to extend the curve back into the past and to fall into the mistaken belief that the advance of knowledge has always been a continuous, cumulative process along a road which steadily mounts from the beginnings of civilization to our present dizzy height. This, of course, is not

* The name of the movement derives from a slighting remark by Matisse, who said of a landscape by Braque that it was 'entirely constructed in little cubes'.[24]

the case. In the sixth century B.C., educated men knew that the earth was a sphere; in the sixth century A.D., they again thought it was a disc, or resembling in shape the Holy Tabernacle.

In looking back at the part of the road travelled so far, we may well wonder at the shortness of those stretches where the progress of science was guided by rational thought. There are tunnels on the road, whose length in time is measured in miles, alternating with stretches in full sunlight of no more than a few yards. Up to the sixth century B.C., the tunnel is filled with mythological figures; then for three centuries there is a shrill light; then we plunge into another tunnel, filled with different dreams.

## *Chronological Table to Part One**

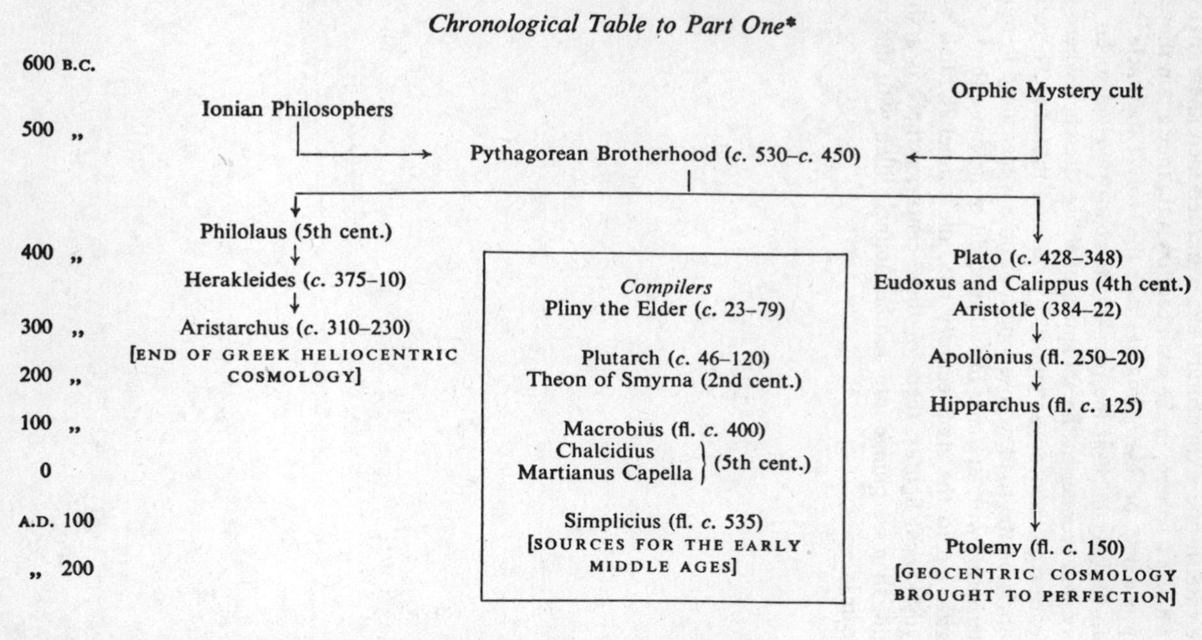


* Only the main lines of development of cosmological systems are represented.

# Part Two. Dark Interlude

# 1 The Rectangular Universe

## 1. *The City of God*

Plato had said that mortal man was prevented from hearing the Harmony of the Spheres by the grossness of his bodily senses, the Christian Platonists said that he lost that faculty with the Fall.

When Plato's images strike an archetypal chord, they continue to reverberate on unexpected levels of meaning, which sometimes reverse the messages originally intended. Thus one might venture to say that it was Plato who caused that Fall of philosophy which made his followers deaf to the harmonies of nature. The sin which led to the Fall was the destruction of the Pythagorean union of natural and religious philosophy, the denial of science as a way of worship, the splitting up of the very texture of the universe into a vile lowland and ethereal highlands, made of different materials, governed by different laws.

'This 'dualism of despair', as one might call it, was carried over into medieval philosophy by the Neoplatonists. It was the legacy of one bankrupt civilization: Greece at the age of the Macedonian conquest, to another bankrupt civilization: the Latin world at the age of its conquest by the Germanic tribes. From the third century A.D to the end of the Empire, Neoplatonism had reigned without a rival at the three main centres of philosophy, Alexandria, Rome, and the Athenian Academy. By that process of natural selection in the realm of ideas which we have already seen at work, the Middle Ages took over precisely those elements in Neoplatonism which appealed to their mystic aspirations toward the Kingdom of Heaven, and which echoed their despair of this world as 'the lowest and vilest element in the scheme of things;[1] while the more optimistic aspects of Neoplatonism were ignored. Of Plato himself only the *Timaeus*, that masterpiece of ambiguity, was available in Latin translation (the knowledge of Greek was dying out); and though Plotinus, the most influential among the Neoplatonists, affirmed that the material world partook to some extent of the goodness and beauty of its Creator, he was best remembered for the saying that he 'blushed because he had a body'. It was in this distorted and extreme form that Neoplatonism was absorbed into Christianity after the collapse of the Roman Empire, and became the main link between antiquity and medieval Europe.

The dramatic symbol of this fusion is the chapter in St Augustine's *Confessions* in which he describes how God 'brought in my way by means of a certain man – an incredibly conceited man – some books of the Platonists translated from Greek into Latin'.[2] Their impact on him was so powerful that, 'being admonished by all this to return to myself, I entered into my own depth'[3] and was thus set on the road to conversion. Although, after his conversion, he complained about the Neoplatonists' failure to realize that the Word was made Flesh in Christ, this proved no obstacle. The mystic union between Platonism and Christianity was consummated in the *Confessions* and the *City of God*.

A modern translator of the *Confessions* wrote about Augustine:

> In him the Western Church produced its first towering intellect – and indeed its last for another six hundred years. . . . What he was to mean for the future can only be indicated. All the men who had to bring Europe through the six or seven centuries that followed fed upon him. We see Pope Gregory the Great at the end of the sixth century reading and rereading the *Confessions*. We see the Emperor Charlemagne at the end of the eighth century using the *City of God* as a kind of Bible.[4]

Now this Bible of the Middle Ages, the *City of God*, was begun in 413, under the impact of the sack of Rome; and Augustine died in 430, while the Vandals were besieging his episcopal city of Hyppo. This goes a long way toward explaining his catastrophic views about humanity as a *massa perditiones*, a heap of depravity, in a state of moral death where even the newborn child carries the hereditary stigma of original sin; where infants who die unbaptized share the fate of eternal damnation with the vast majority of mankind, pagan and Christian. For salvation is only possible through an act of Grace which God extends to individuals predestined to receive it by an apparently arbitrary selection; because 'fallen man cannot do anything well-pleasing to God'.[5] This terrible doctrine of predestination was taken up again in various forms at various ages by Cathars, Albigenses, Calvinists, and Jansenists, and was also to play a curious part in the theological struggles of Kepler and Galileo.

Again, there are countless redeeming aspects, ambiguities, and contradictions in Augustine's writings, such as his passionate pleading against the death penalty and judicial torture; his repeated affirmation that *Omnis natura, inquantum natura est, bonum est*;* it may even be said that 'Augustine was not an Augustinian'.[6] But these brighter elements were ignored by the generations after him, and the shadow he threw was dark and oppressive; it blotted out what little interest in nature, or inclination to science, still remained.

Since, in the Middle Ages, the churchmen became the successors to the philosophers of antiquity, and, in a manner of speaking, the Catholic Church took over from the Academy and the Lyceum, its attitude now determined the whole climate of culture and the course of learning. Hence the importance of Augustine, who was not only the most influential churchman of the earlier Middle Ages, the chief promoter of the Papacy as a supranational authority, and the originator of the rules of monastic life; but above all the living symbol of continuity between the vanished ancient, and the emerging new civilization. A modern Catholic philosopher justifiably said that Augustine was 'to a greater degree than any emperor or barbarian war-lord, a maker of history and a builder of the bridge which was to lead from the old world to the new'.[7]

## 2. *The Bridge to the City*

The tragedy lies in the selective nature of the traffic which passed across the bridge that Augustine built. At the tollgate

* All nature, in as much as it is nature, is good.

of the City of God, all vehicles carrying the treasures of ancient learning, beauty, and hope were turned back, for all pagan virtue is 'prostituted with the influence of obscene and filthy devils. . . .[8] Let Thales depart with his water, Anaximenes with the air, the Stoics with their fire, Epicurus with his atoms.'[9]

And depart they did. Only Plato and his disciples were allowed to pass the bridge and were welcomed, for they knew that knowledge cannot be obtained through the eyes of the body, and provided an allegorical supplement to Genesis, as it were: Adam, expelled from the Garden, was made to proceed straight to Plato's Cave, and to take up the existence of a chained troglodyte.

Most welcome of all was the Neoplatonists' contempt for all branches of science. From them Augustine 'derived the conviction, which he transmitted to the succeeding generations of many centuries, that the only type of knowledge to be desired was knowledge of God and the soul, and that no profit was to be had from investigating the realm of Nature'.[10]

A few quotations from the *Confessions* will illustrate more vividly the mental attitude toward knowledge at the opening of the Christian era. In the Tenth Book, which concludes his personal narrative, Augustine describes his state of mind twelve years after his conversion, and implores the help of God to overcome various forms of temptations which are still assailing him: the lust of the flesh, which he can resist when awake but not in sleep; the temptation to enjoy his food instead of taking it as a necessary medicine 'until the day when Thou wilt destroy both the belly and the meat'; the allurement of sweet scents, to which he is fairly immune; the pleasures of the ear derived from church music at the risk of being 'more moved by the singing than by the thing that is sung'; the lure to the eye of 'diverse forms of beauty, of brilliant and pleasing colours'; and, last but one, the temptation of 'knowing for knowing's sake':

At this point I mention another form of temptation more various and dangerous. For over and above that lust of the flesh which lies in the delight of all our senses and pleasures – whose slaves are wasted unto destruction as they go far from You – there can also be in the mind itself, through those same bodily senses, a certain vain desire and curiosity, not of taking delights in the body, but of making experiments with the body's aid, and cloaked under the name of learning and knowledge. ... Pleasure goes after objects that are beautiful to see, smell, taste, touch; but curiosity for the sake of experiment can go after quite contrary things, not in order to experience their unpleasantness, but through a mere itch to experience and find out. ... Because of this disease of curiosity you have the various freaks shown in the theatres. Thus men proceed to investigate the phenomena of nature – the part of nature external to us – though the knowledge is of no value to them: for they wish to know simply for the sake of knowing. ...

In this immense forest of snares and perils, I have cut off and thrust from my heart many sins, as you have given me to do, O God of my salvation; yet when would I dare to say – with so many things of the sort buzzing about our daily life on every side – when dare I say that no such thing can draw me to look at it or through vain curiosity to desire it? Certainly the theatres no longer attract me, nor do I care to know the course of the stars. ...[11]

But he has not yet succeeded in plucking out of the human heart that sinful desire for knowledge.

He came perilously near to it, though.

## 3. *The Earth as a Tabernacle*

Compared with the other early Fathers, Augustine was still by far the most enlightened. Saint Lactantius, who lived in the century before him, set himself to demolish the notion of the rotundity of the earth, with resounding success. The third volume of his *Divine Institutions* is called 'On the False Wisdom of the Philosophers', and contains all the naïve arguments against the existence of the antipodes – people can't walk with their feet above their heads, rain and snow can't fall upwards – which, seven hundred years earlier, no educated person could have used without making a fool of himself. Saint Jerome, the translator of the Vulgate, fought a life-long battle against the temptation of reading the pagan classics, until he finally defeated 'the stupid wisdom of the philosophers': 'Lord, if ever again I possess worldly books, or if ever again I read such, I have denied Thee'.[12] Not until about the end of the ninth century was the spherical shape of the earth, and the possible existence of the antipodes reinstated, fifteen hundred years after Pythagoras.

The cosmology of this period goes straight back to the Babylonians and Hebrews. Two main ideas dominate it: that the earth is shaped like the Holy Tabernacle, and that the firmament is enclosed by water. The latter idea was based on Genesis I: 6, 7:

> And God said let there be a firmament in the midst of the waters, and let it divide the waters from the waters. And God made the firmament, and divided the waters which were under the firmament from the waters which were above the firmament.

From this, the notion was derived that the super-celestial waters were resting on top of the firmament, and that their purpose was – as Basil the Great* explained[13] – to protect the world against the celestial fire. His contemporary, Severianus, further explained, that the lower heaven consisted of crystalline or 'congealed' water, which prevented it from being set aflame by the sun and stars; and that it was kept cool by the liquid water on top of it, which, on the Last Day, God would use to extinguish all the lights.[14] Augustine, too, believed that Saturn was the coolest planet because it was closest to the upper waters. In answer to those who objected to the presence of heavy water on top of the heavens, he pointed out that there is liquid phlegm present in the heads of men too.[15] The further objection that the spherical surface of the firmament and its motion would cause the waters to slide down or be spilled, was met by several Fathers who explained that the heavenly vault may be round inside but flat on top; or contain grooves and vessels to hold the water in.[16]

At the same time the notion was spreading that the firmament is not round, but a tent or tabernacle. Severianus refers to Isaiah, XL: 22, that God 'stretches out the heavens as a curtain and spreadeth them out as a tent to dwell in',[17] and others follow suit. However, the Fathers and Doctors were not sufficiently interested in these worldly matters to go into detail.

* Fourth century A.D.

The first comprehensive cosmological system of the early Middle Ages, destined to replace the teachings of pagan astronomers from Pythagoras to Ptolemy, was the famous *Topographica Christiana* by the monk Cosmas. He lived in the sixth century, was born in Alexandria, and, as a merchant and seaman, had travelled wide and far through the known world, including Abyssinia, Ceylon, and Western India, which earned him the title *Indicopleustus*, the Indian traveller. He subsequently became a monk, and wrote his great work in a monastery on Sinai.

The first of its twelve books is entitled 'Against those who, while wishing to profess Christianity, think and imagine like the pagans that the heaven is spherical'. The Holy Tabernacle, described in Exodus, was rectangular and twice as long as it was wide; hence the earth has the same shape, placed lengthwise from East to West at the bottom of the universe. It is surrounded by the ocean – as the table of shew-bread is surrounded by its wavy border; and the ocean is surrounded by a second earth which was the seat of Paradise, and the home of man until Noah crossed the ocean, but is now uninhabited. From the edges of this deserted outer earth rise four vertical planes, which are the walls of the universe. Its roof is a half-cylinder which rests on the north and south walls, making the universe look like a Nissen hut or a Victorian travelling trunk with a curved lid.

However, the floor, that is the earth, is not flat but slants from north-west to south-east – for it is written in Ecclesiastes I: 5 that 'the sun goes *down*, and hasteth to his place where he arose'. Accordingly, rivers like the Euphrates and Tigris which flow southward, have a faster current than the Nile which flows 'uphill'; and ships sail faster toward the south and east than those which must 'climb' to the north and west; the latter are therefore called 'lingerers'. The stars are carried round the space under the roof of the universe by angels, and are hidden when they pass behind the uptilted northern part of the earth, which is topped by a huge conical mountain. This mountain also hides the sun at night, the sun being much smaller than the earth.

Cosmas himself was not a high ecclesiastical authority, but his ideas are all derived from the Fathers of the preceding two centuries. There were more enlightened men among them, such as Isidore of Seville (sixth to seventh century) and the Venerable Bede (seventh to eighth century). Yet Cosmas' *Topographica Christiana* is typical of the general view of the universe prevailing during the early Middle Ages. Long after the spherical shape of the earth was reinstated, and indeed up to the fourteenth century, maps were still produced representing the earth either as rectangular, after the shape of the Tabernacle, or as a circular disc with Jerusalem as its centre, because Isaiah had spoken of the 'circuit of the earth' and Ezekiel had stated that 'God had set Jerusalem in the midst of the nations and countries'. A third type of map made the earth oval-shaped, as a compromise between the tabernacle and circuitous view; the Far East was usually occupied by Paradise.

Once again we are impelled to ask ourselves: Did they really believe in all this? And again the answer must be both yes and no – depending on which compartment of the split mind was involved. For the Middle Ages were the era of the split mind *par excellence*; I shall return to the subject at the end of this chapter.

### 4. *The Earth is Round Again*

The first medieval churchman to state unequivocally that the earth is a sphere, was the English monk Bede, who rediscovered Pliny, as it were, and often quoted him verbatim; yet he still clung to the notion of the super-celestial waters and denied that there were people living in the antipodal regions; for those regions being inaccessible on account of the vast ocean, its supposed inhabitants could neither have descended from Adam, nor be redeemed by Christ.

A few years after Bede's death, a curious incident took place. A certain Irish ecclesiastic by name of Fergil or Virgil, who lived as an abbot at Salzburg, became involved in a quarrel with his superior, Boniface, who denounced Virgil to Pope Zacharias on the grounds that the Irishman taught the existence 'of another world and other people under the earth' –

meaning the antipodes. The Pope replied that Boniface should call a council and expel Virgil from the Church for his scandalous teaching. But nothing happened – except that Virgil in due time became Bishop of Salzburg and held that see till his death. The episode reminds one of the futile denunciation of Aristarchus by Cleanthes; it seems to indicate that even in this period of benightedness, orthodoxy in matters of natural philosophy (as distinct from matters theological) was maintained less by threats than by inner compulsion. At least I am not aware of any recorded instance of a cleric or layman being indicted for heresy in this heresy-ridden age because of his cosmological views.

This danger was further diminished when, in A.D 999, Gerbert, the most accomplished classical scholar, geometer, musician and astronomer of his age, ascended the papal throne as Sylvester II. He died four years later, but the impression that the 'magician Pope' made on the world was so powerful that he soon became a legend. Though he was an exceptional individual, far in advance of his age, his papacy, at the symbolical date A.D. 1000, nevertheless marks the end of the darkest period of the Middle Ages, and a gradual change of attitude toward the pagan science of antiquity. From now onward, the spherical shape of the earth, and its position in the centre of space, surrounded by the spheres of the planets, became again respectable. What is more, several manuscripts from approximately the same period show that the 'Aegyptian' system of Herakleides (where Mercury and Venus are satellites of the sun) had been rediscovered, and that elaborate drawings of the planetary orbits were circulating among the initiates. But they did not make any noticeable impression on the dominant philosophy of the age.

Thus by the eleventh century A.D., a view of the universe had been achieved roughly corresponding to that of the fifth century B.C. It had taken the Greeks some two hundred and fifty years to progress from Pythagoras to Aristarchus' heliocentric system; it took Europe more than twice that time to achieve the corresponding progress from Gerbert to Copernicus. The Greeks, once they had recognized that the earth was a ball floating in space, had almost at once set that ball in motion; the Middle Ages hastily froze it into immobility at the centre of a rigid cosmic hierarchy. It was not the logic of science, not rational thought that determined the shape of the next development, but a mythological concept which symbolized the needs of the age: the tabernacular universe was succeeded by the universe of the Golden Chain.

# 2 The Walled-in Universe

## 1. *The Scale of Being*

It was a walled-in universe like a walled-in medieval town. In the centre lies the earth, dark, heavy, and corrupt, surrounded by the concentric spheres of the moon, sun, planets, and stars in an ascending order of perfection, up to the sphere of the *primum mobile*, and beyond that the Empyrean dwelling of God.

But in the hierarchy of values, which is attached to this hierarchy in space, the original simple division into sub-lunary and supra-lunary regions has now yielded to an infinite number of sub-divisions. The original, basic difference between coarse, earthly mutability and ethereal permanence is maintained; but both regions are sub-divided in such a manner that the result is a continuous ladder, or graded scale, which stretches from God down to the lowliest form of existence. In a passage, frequently quoted throughout the Middle Ages, Macrobius sums up the idea:

> Since, from the Supreme God Mind arises, and from Mind, Soul, and since this in turn creates all subsequent things and fills them all with life ... and since all things follow in continuous succession, degenerating in sequence to the very bottom of the series, the attentive observer will discover a connexion of parts, from the Supreme God down to the last dregs of things, mutually linked together and without a break. And this is Homer's golden chain, which God, he says, bade hang down from heaven to earth.[1]

Macrobius echoes the Neoplatonist 'theory of emanations' which goes back to Plato's *Timaeus*. The One, the Most Perfect Being 'cannot remain shut up in itself'; it must 'overflow' and create the World of Ideas, which in turn creates a copy or image of itself in the Universal Soul, which generates 'the sentient and vegetative creatures' – and so on in a descending series, to the 'last dregs of things'. It is still a process of degeneration by descent, the very opposite of the evolutionary idea; but since every created being is ultimately an emanation of God, partaking of His essence in a measure diminishing with distance, the soul will always strive upward, to its source.

The emanation-theory was put into a more specifically Christian shape in *The Celestial Hierarchy* and *The Ecclesiastical Hierarchy* by the second most influential among the Neoplatonists, known as the pseudo-Dionysius. He lived probably in the fifth century, and perpetrated the most successful pious hoax in religious history by pretending that the author of his works was Dionysius Areopagite, the Athenian mentioned in Acts XVII: 34, as a convert of St Paul's. He was translated into Latin in the ninth century by John the Scot, and from then on exerted an immense influence on medieval thought. It was he who provided the upper reaches of the ladder with a fixed hierarchy of angels, which afterwards were attached to the star-spheres to keep them in motion: the Seraphim turning the Primum Mobile,[2] the Cherubim the sphere of the fixed stars, the Thrones the sphere of Saturn; the Dominations, Virtues, and Powers the spheres of Jupiter, Mars, and the sun; the Principalities and Archangels the spheres of Venus and Mercury, while the lower Angels look after the moon.[3]

If the upper half of the ladder was Platonic in origin, the lower rungs were provided by Aristotelian biology, which was rediscovered around A.D. 1200. Particularly important became his 'principle of continuity' between apparently divided realms of nature:

> Nature passes so gradually from the inanimate to the animate that their continuity renders the boundary between them indistinguishable; and there is a middle kind that belongs to both orders. For plants come immediately after inanimate things; and plants differ from one another in the degree in which they appear to participate in life. For the class taken as a whole seems, in comparison with other bodies, to be clearly animate; but compared with animals to be inanimate. And the transition from plants to animals is continuous; for one might question whether some marine forms are

animals or plants, since many of them are attached to the rock and perish if they are separated from it.[4]

The 'principle of continuity' made it not only possible to arrange all living beings into a hierarchy according to criteria such as 'degrees of perfection', 'powers of soul' or 'realization of potentialities' (which, of course, were never exactly defined). It also made it possible to connect the two halves of the chain – the sub-lunary and the celestial – into a single, continuous one, without denying the essential difference between them. The connecting link was found, by St Thomas Aquinas, in the dual nature of man. In the continuity of all that exists, 'the lowest member of the higher genus is always found to border upon the higher member of the lower genus'; this is true of the zoophytes, which are half plant, half animal, and it is equally true of man, who

> has in equal degree the characters of both classes, since he attains to the lowest member of the class above bodies, namely, the human soul, which is at the bottom of the series of intellectual beings – and is said, therefore, to be the horizon and boundary line of things corporeal and incorporeal.[5]

The chain, thus unified, now reached from God's throne down to the meanest worm. It was further extended downward through the hierarchy of the four elements into inanimate nature. Where no obvious clues could be found to determine an object's 'degree of excellence', astrology and alchemy provided the answer by establishing 'correspondences' and 'influences', so that each planet became associated with a day of the week, a metal, a colour, a stone, a plant, defining their rank in the hierarchy. A further downward extension led into the conic cavity in the earth, around whose narrowing slopes the nine hierarchies of devils were arranged in circles, duplicating the nine heavenly spheres; Lucifer, occupying the apex of the cone in the precise centre of the earth, marked the bitter end of the chain.

The medieval universe, as a modern scholar remarked, is thus not really geocentric, but 'diabolocentric'.[6] Its centre, once the Hearth of Zeus, is now occupied by Hell. In spite of the continuous nature of the chain, the earth, compared to the incorruptible heavens, still occupies the lowest place, described by Montaigne as 'the filth and mire of the world, the worst, lowest, most lifeless part of the universe, the bottom storey of the house'.[7] In a similar vein his contemporary, Spenser, bemoans the sway of the Goddess Mutability over the earth, which makes him

> Loathe this state of life so tickle
> And love of things so vain to cast away;
> Whose flow'ring pride, so fading and so fickle, –
> Short time shall soon cut down with his consuming sickle.[8]

The extraordinary power of this medieval vision of the universe is illustrated by the fact that it had the same, undiminished hold on the imagination of the Elizabethan poets at the turn of the sixteenth century as it had on Dante's at the turn of the thirteenth; and it is still echoed, in a famous passage by Pope, in the eighteenth. The concluding half of the quotation provides a clue to the understanding of the great stability of the system:

> Vast chain of being! which from God began,
> Nature's aethereal, human, angel, man,
> Beast, bird, fish, insect . . .
> . . . from Infinite to thee,
> From thee to nothing. – On superior pow'rs
> Were we to press, inferior might on ours;
> Or in the full creation leave a void,
> Where, one step broken, the great scale's destroy'd;
> From Nature's chain whatever link you strike,
> Tenth, or ten thousandth, breaks the chain alike.[9]

The consequence of such a break would be disintegration of the cosmic order. The same moral, the same warning of the catastrophic consequences of any change, however small, in the rigid, graded hierarchy, of any disturbance in the fixed order of things, returns, as a leitmotif, in Ulysses's speech in *Troilus and Cressida* and in countless other places. The secret of the medieval universe is that it is static, immune against change; that every item in the cosmic inventory has its permanent place and rank assigned to it on a rung of the ladder. It reminds one of the pecking hierarchy in a henyard. There is no evolution of biological species, and no social progress; no traffic moves up or down the ladder. Man may aspire to a higher life or condemn himself to an even lower one; but he will only move up or down the ladder after his death; while he is in this world, his preordained rank and place cannot be altered. Thus blessed immutability is made to prevail even in the lowly world of mutability and corruption. The social order is part of the chain, the part which connects the hierarchy of angels with the hierarchy of animal, vegetable, and mineral. To quote another Elizabethan, Raleigh – in straight prose for a change:

Shall we therefore value honour and riches at nothing and neglect them as unnecessary and vain? Certainly not. For that infinite wisdom of God, which hath distinguished his angels by degrees, which hath given greater and less light and beauty to heavenly bodies, which hath made differences between beasts and birds, created the eagle and the fly, the cedar and the shrub, and among stones given the fairest tincture to the ruby and the quickest light to the diamond, hath also ordained kings, dukes or leaders of the people, magistrates, judges, and other degrees among men.[10]

Not only Kings and Barons, Knights and Squires, have their fixed place in the cosmic hierarchy; the Chain of Being runs even through the kitchen.

Who is to take the chief cook's place in case he is absent: the spit-master or the soup-master? Why do the bread-bearers and cup-bearers form the first and second ranks, above carvers and cooks? – Because they are in charge of bread and wine, to which the sanctity of the sacrament gives a holy character.[11]

The Middle Ages had an even greater horror of change, and desire for permanence than the age of Plato, whose philosophy they carried to obsessional extremes. Christianity had saved Europe from a relapse into barbarism; but the catastrophic conditions of the age, its climate of despair, prevented it from evolving a balanced, integrated, evolutionary view of the universe and of man's role in it. The recurrent, panic expectations of the End of the World, the outbreaks of dancing and flagellating manias, were symptoms of mass hysteria,

> brought on by terror and despair, in populations oppressed, famished, and wretched to a degree almost unimaginable today. To the miseries of constant war, political and social disintegration, there was added the dreadful affliction of inescapable, mysterious, and deadly disease. Mankind stood helpless as though trapped in a world of terror and peril against which there was no defence.[12]

It was against this background that the vision of the walled universe was taken over from the Platonists as a protection against the Black Death of Change – rigid, static, hierarchic, petrified. The Babylonian oyster-world, which lay three and four thousand years back, was full of dynamism and imagination compared with this pedantically graded universe, wrapped in cellophane spheres, and kept by God in the deep-freeze locker to hide its eternal shame. Yet the alternative was even worse:

> ... when the planets
> In evil mixture to disorder wander,
> What plagues and what portents, what mutiny,
> What raging of the sea, shaking of earth,
> Commotion in the winds, frights, changes, horrors,
> Divert and crack, rend and deracinate
> The unity and married calm of states
> Quite from their fixture ...
> Take but degree away, untune that string,
> And hark, what discord follows. Each thing meets
> In mere oppugnancy. The bounded waters
> Should lift their bosoms higher than the shores
> And make a sop of all this solid globe.[13]

## 2. *The Age of Double-Think*

I have said that the Herakleidian system, in which the two inner planets circle the sun, and not the earth, had been rediscovered toward the end of the first millennium. But it would be more correct to say that heliocentricism had never been quite forgotten, even at the time of the tabernacular universe. I have already quoted (pp. 71–2), among others, Macrobius to that effect. Now Macrobius, Chalcidius, and Martianus Capella, three encyclopaedic compilers of the period of Roman decadence (all three of the fourth-fifth century A.D.), were, together with Pliny, the main sources on natural science available till the Greek revival; and they all propounded the system of Herakleides.[14] It was again taken up by John the Scot in the ninth century, who made not only the inner planets, but all of them except distant Saturn, satellites of the sun; and from then onward, Herakleides remains firmly established on the medieval scene.[15] In the words of the best authority on the subject: 'the majority of the men who, from the ninth to the twelfth century, have written on astronomy, and whose books are preserved, were acquainted with and adopted the planetary theory designed by Herakleides of Pontus.'[16]

And yet at the same time, cosmology had reverted to a naïve and primitive form of geocentrism – with concentric crystal spheres determining the order of the planets and the accompanying hierarchy of angels. The highly ingenious systems of Aristotle's fifty-five spheres, of Ptolemy's forty epicycles were forgotten, and the complex machinery was reduced to ten revolving spheres – a kind of poor man's Aristotle which had nothing whatever in common with any of the observed motions in the sky. The Alexandrian astronomers had at least tried to save the phenomena; the medieval philosophers disregarded them.

But a complete disregard for reality would make life impossible; and thus the split mind must evolve two different codes of thought for its two separate compartments: one conforming to theory, the other to cope with fact. Up to the end of the first millennium and beyond, the rectangular and oval, tabernacle-inspired maps were piously copied out by the monks; they provided a kind of Sunday idea of the shape of the earth according to the patristic interpretation of Scripture. But coexisting with these was an entirely different kind of map of amazing accuracy, the so-called Portolano charts, for practical use among Mediterranean seamen. The shapes of countries and seas on the two types of maps are as unrelated to each other as the medieval idea of the cosmos and the observed events in the sky.[17]

The same split can be traced through the most heterogeneous fields of medieval thought and behaviour. Since it is against man's nature to go on blushing because he has a body and a brain, a thirst for beauty and an appetite for experience, the frustrated half took its revenge through extremes of coarseness and obscenity. The disembodied, ethereal love of the troubadour or serving knight for his lady, coexist with the brutal publicity given to the wedding bed, which makes marriages resemble public executions. The fair lady is compared to the Goddess of Virtue, but is made to wear a cast-iron chastity belt on her sub-lunary sphere. Nuns must wear shirts even in the privacy of their baths, because, though nobody else, God can see them. When the mind is split, both halves are debased: earthly love sinks to the animal level, the mystic union with God acquires an erotic ambiguity. Confronted with the Old Testament, the theologians save the phenomena in the Song of Songs by declaring that the King is Christ, the Shulamite the Church, and that the praise for various parts of her anatomy refers to corresponding excellences in the edifice that St Peter built.

Medieval historians must also live by double-think. The cosmology of the age explained away the disorder in the skies by ordered motions in perfect circles; the chroniclers, faced with worse disorder, had recourse to the notion of perfect chivalry as the moving force of History. It became to them

> ... a sort of magic key by the aid of which they explained to themselves the motives of politics and of history. ... What they saw about them looked primarily mere violence and confusion. ... Yet they required a form for their political conceptions and here the idea of chivalry came in. ... By this traditional fiction they succeeded in explaining to themselves, as well as they could, the motives and the course of history, which was thus reduced to a spectacle of the honour of princes and the virtue of knights, to a noble game with edifying and heroic rules.[18]

The same dichotomy is carried into social behaviour. A grotesque and rigid etiquette governs every activity, designed to freeze life in the image of the heavenly clockwork, whose crystal spheres turn on themselves yet always remain in the same place. Humble refusals to take precedence in passing through a door take up a quarter of an hour, yet bloody feuds are fought for that same right of precedence. The ladies at Court pass their time poisoning each other with words and philtres, yet etiquette

> not only prescribes which ladies may hold each other by the hand, but also which lady is entitled to encourage others to this mark of intimacy by beckoning them. ... The passionate and violent soul of the age, always vacillating between tearful piety and frigid cruelty, between respect and insolence, between despondency and wantonness, could not dispense with the severest rules and the strictest formalism. All emotions required a rigid system of conventional forms, for without them passion and ferocity would have made havoc of life.[19]

There are mental disorders whose victims feel compelled to walk on the centres of flagstones, avoiding the edges, or to count the matches in the box before going to sleep, as a protective ritual against their fears. The dramatic outbursts of mass-hysteria during the Middle Ages tend to divert our attention from the less spectacular, but chronic and insoluble mental conflicts which underlie them. Medieval life in its typical aspects resembles a compulsive ritual designed to provide protection against the all-pervading potato-blight of sin, guilt, and anguish; yet it was unable to provide it so long as God and Nature, Creator and Creation, Faith and Reason, were split apart. The symbolic prologue to the Middle Ages is Origen cutting off his private parts *ad gioriam dei*; and the epilogue is provided by the parched voices of the schoolmen: Did the first man have a navel? Why did Adam eat an apple and not a pear? What is the sex of the angels, and how many can dance on the point of a pin? If a cannibal and all his ancestors have lived on human flesh so that every part of his body belongs to somebody else and will be claimed by its owner on the day of resurrection, how can the cannibal be resurrected to face his judgement? This last problem was earnestly discussed by Aquinas.

When the mind is split, departments of it which should complete each other, develop autonomously by inbreeding, as it were, insulated from reality. Such is medieval theology, cut off from the balancing influence of the study of nature; such is medieval cosmology, cut off from physics; such is medieval physics, cut off from mathematics. The purpose of the digressions in this chapter, which seem to have led us so far away from our topic, is to show that the cosmology of a given age is not the result of a unilinear, 'scientific' development, but rather the most striking, imaginative symbol of its mentality – the projection of its conflict, prejudices, and specific ways of double-think on to the graceful sky.

# 3 The Universe of the Schoolmen

## 1. *The Thaw*

I have compared Plato and Aristotle to twin stars which alternate in visibility. Broadly speaking, from the fifth to the twelfth century, Neoplatonism in the form in which St Augustine and the pseudo-Dionysius had imported it into Christianity, held the sway. From the twelfth to the sixteenth century, it was the turn of Aristotle.

Except for two of his logical treatises,[1] Aristotle's works had been unknown before the twelfth century – buried and forgotten, together with Archimedes, Euclid, the atomists, and the rest of Greek science. What little knowledge survived had been handed down in sketchy, distorted versions by the Latin compilers and the Neoplatonists. Insofar as science is concerned, the first six hundred years of established Christendom were a glacial period with only the pale moon of Neoplatonism reflected on the icy steppes.

The thaw came not by a sudden rise of the sun, but by ways of a devious Gulf-stream which wended its way from the Arab peninsula through Mesopotamia, Egypt, and Spain: the Moslems. In the seventh and eighth centuries, this stream had picked up the wreckage of Greek science and philosophy in Asia Minor and in Alexandria, and carried it in a circumambient and haphazard fashion into Europe. From the twelfth century onwards, the works, or fragments of works, of Archimedes and Hero of Alexandria, of Euclid, Aristotle, and Ptolemy, came floating into Christendom like pieces of phosphorescent flotsam. How devious this process of Europe's recovery of its own past heritage was, may be gathered from the fact that some of Aristotle's scientific treatises, including his *Physics*, had been translated from the original Greek into Syriac, from Syriac into Arabic, from Arabic into Hebrew, and finally from Hebrew into medieval Latin. Ptolemy's

*Almagest* was known in various Arab translations throughout the Empire of Harun Al Rashid, from the Indus to the Ebro, before Gerardus of Cremona, in 1175, retranslated it from the Arabic into Latin. Euclid's *Elements* were rediscovered for Europe by an English monk, Adelard of Bath, who around 1120, came across an Arabic translation in Cordova. With Euclid, Aristotle, Archimedes, and Ptolemy recovered, science could start again where it had left off a millennium earlier.

But the Arabs had merely been the go-betweens, preservers and transmitters of the heritage. They had little scientific originality and creativeness of their own. During the centuries when they were the sole keepers of the treasure, they did little to put it to use. They improved on calendrical astronomy and made excellent planetary tables; they elaborated both the Aristotelian and the Ptolemaic models of the universe; they imported into Europe the Indian system of numerals based on the symbol zero, the sine function, and the use of algebraic methods; but they did not advance theoretical science. The majority of the scholars who wrote in Arabic were not Arabs but Persians, Jews, and Nestorians; and by the fifteenth century, the scientific heritage of Islam had largely been taken over by the Portuguese Jews. But the Jews, too, were no more than go-betweens, a branch of the devious Gulf-stream which brought back to Europe its Greek and Alexandrine heritage, enriched by Indian and Persian additions.

It is a curious fact that the Arab-Judaic tenure of this vast body of knowledge, which lasted two or three centuries, remained barren; whilst as soon as it was reincorporated into Latin civilization, it bore immediate and abundant fruit. The heritage of Greece was obviously of no benefit to anybody without some specific receptiveness for it. How this readiness to rediscover its own past, and be fertilized by it, as it were, arose in Europe is a question that belongs to the field of general history. The slow increase in security, in trade and communications; the growth of towns and the development of new crafts and techniques; the invention of the magnetic compass and the mechanical clock, which gave man a more concrete feeling of space and time; the utilization of water power, and even the improved harnessing of horses, were some of the material factors which quickened and intensified the pulse of life and led to a gradual change in the intellectual climate, a thaw in the frozen universe, a diminution of apocalyptic fear. As men ceased to blush at the fact of having a body, they also ceased to be frightened of using their brains. It was still a long way to the Cartesian *cogito ergo sum*. But at least the courage was reborn to say: *sum, ergo cogito.*

The dawn of this early, or 'first' Renaissance is intimately connected with the rediscovery of Aristotle – more precisely, of the naturalistic and empirical elements in him, of that side of Aristotle which is averted from his twin star. The alliance, born of catastrophe and despair, between Christianity and Platonism, was replaced by a new alliance between Christianity and Aristotelianism, concluded under the auspices of the Angelic Doctor, Thomas Aquinas. Essentially, this meant a change of fronts from the negation to the affirmation of life, a new, positive attitude to Nature, and to man's striving to understand nature. Perhaps the greatest historical achievement of Albert the Great and Thomas Aquinas lies in their recognition of the 'light of reason' as an independent source of knowledge beside the 'light of grace'. Reason, hitherto regarded as *ancilla fidei*, the handmaid of faith, was now considered the bride of faith. A bride must, of course, obey her spouse in all important matters; nevertheless, she is recognized as a being in her own right.

Aristotle had not only been a philosopher, but also an encyclopaedist in whom a little of everything could be found; by concentrating on his hard-headed, down-to-earth, non-Platonic elements, the great schoolmen brought back to Europe a whiff of the heroic age of Greece. They taught respect for 'irreducible and stubborn facts'; they taught 'the priceless habit of looking for an exact point and of sticking to it when found. Galileo owes more to Aristotle than appears on the surface . . .: he owes him his clear head and his analytic mind'.[2]

By using Aristotle as a mental catalyzer, Albert and Thomas taught men to think again. Plato maintained that true knowledge could only be obtained intuitively, by the eye of the soul, not of the body; Aristotle had stressed the importance of experience – *empiria* – as against intuitive *aperia*:

> It is easy to distinguish those who argue from fact and those who argue from notions. ... The principles of every science are derived from experience: thus it is from astronomical observation that we derive the principles of astronomical science.[3]

The sad truth is that neither Aristotle himself, nor his Thomist disciples, lived up to their lofty precepts, and as a result scholasticism went into decline. But during the honeymoon period of the new alliance all that mattered was that 'the philosopher' (a title for which Aristotle acquired the exclusive monopoly among the schoolmen), had upheld the rationality and intelligibility of Nature; that he made it a duty of man to take an interest in the world around him by observation and reasoning; and that this fresh, naturalistic outlook freed the human mind from its sickly infatuation with the Neoplatonic *Weltschmerz*.

The renaissance of learning in the thirteenth century was full of promise – the stirrings of a patient who emerges from a long, comatose state. It was the century of Robert of Lincoln and Roger Bacon, the first who understood, far ahead of his time, the principles and methods of empirical science; of Peter Peregrine, who wrote the first scientific treatise on the magnetic compass, and of Albert the Great, the first serious naturalist since the Plinys, who studied insects, whales, and polar bears, and gave a fairly complete description of German mammals and birds. The young universities of Salerno and Bologna, of Paris, Oxford and Cambridge, radiated the new fervour of learning which had brought on the thaw.

## 2. *Potency and Act*

And yet after these great and hopeful stirrings, the philosophy of nature gradually froze up again in scholastic rigidity – though not entirely this time. The reason for this short splendour and long decline can be summed up in a single phrase: the rediscovery of Aristotle had changed the intellectual climate of Europe by encouraging the study of nature; the concrete teachings of Aristotelian science, elevated into dogmas, paralysed the study of nature. If the schoolmen had merely listened to the cheerful and encouraging timbre in the Stagyrite's voice, all would have been well; but they made the mistake of taking in what it actually said – and insofar as the physical sciences are concerned, what it said was pure rubbish. Yet for the next three hundred years this rubbish came to be regarded as gospel truth.[4]

I must now say a few words about Aristotelian physics, for it is an essential part of the medieval universe. The Pythagoreans had shown that the pitch of a tone depends on the length of a cord, and had thus pointed the way to the mathematical treatment of physics. Aristotle divorced science from mathematics. To the modern mind, the most striking fact about medieval science is that it ignores numbers, weight, length, speed, duration, quantity. Instead of proceeding by observation and measurement, as the Pythagoreans did, Aristotle constructed, by that method of *a priori* reasoning which he so eloquently condemned, a weird system of physics 'argued from notions and not from facts'. Borrowing his ideas from his favourite science, biology, he attributed to all inanimate objects a purposeful striving toward an end, which is defined by the inherent nature or essence of the thing. A stone, for instance, is of an earthly nature, and while it falls toward the centre of the earth it will increase its speed, because of its impatience to get 'home'; and a flame will strive upward because its home is in the sky. Thus all motion, and all change in general, is the realization of what exists potentially in the nature of the thing: it is a transition from 'potency' to 'act'. But this transition can only be achieved with the help of some other agent which itself is in the 'act';[5] thus wood which is potentially hot, can be made *actually* hot only by fire, which is *actually* hot. Similarly, an object moving from A to B, being 'in a state of potency with respect to B', can only reach B with the help of an *active mover*: 'whatever is moved must be moved by another'. All this terrifying verbal acrobacy can be summed up in the statement that things only move when they are pushed – which is as simple as it is untrue.

Indeed, Aristotle's *omne quod movetur ab alio movetur* – whatever is moved must be moved by another – became the main obstacle to the progress of science in the Middle Ages. The idea that things only move when they are pushed seems, as a modern scholar remarks,[6] to have originated with the painful motion of oxcarts over bad Grecian roads, where friction was so great that it annihilated momentum. But the Greeks also shot arrows, threw the discus and spears – and yet chose to ignore the fact that once the initial impulse had been imparted to the arrow, it continued its motion, without being pushed, until gravity brought it to an end. According to Aristotelian physics, the arrow, the moment it ceased to have contact with its mover, the bowstring, ought to have fallen to the earth. To this the Aristotelians gave the answer that when the arrow started moving while still pushed by the bow, it created a disturbance in the air, a kind of vortex, which kept dragging it along its course. Not before the fourteenth century, not for seventeen hundred years, was the objection raised that the air-commotion caused by the arrow's start could not be strong enough to make it continue its flight against the wind; and furthermore that if a boat, kicked away from the shore, continued to move merely because it was pulled along by the commotion in the water which the boat itself had caused, then the initial kick should be sufficient to make it traverse the ocean.

This blindness to the fact that moving bodies tend to persist in their movement unless stopped or deflected, prevented the emergence of a true science of physics until Galileo.[7] The necessity for every moving body to be constantly accompanied and pushed along by a mover, created 'a universe in which unseen hands had to be in constant operation'.[8] In the sky, a host of fifty-five angels were needed to keep the planetary spheres moving around; on earth, each stone rolling down a slope, and each drop of rain falling from the sky, needed a quasi-sentient purpose functioning as its 'mover', to get from 'potency' to 'act'.

There was also a distinction between 'natural' and 'violent' motion. Heavenly bodies moved in perfect circles, because of their perfect nature; the natural motion of the four elements on earth was along straight lines – earth and fire along vertical, water and air along horizontal lines. Violent motion was everything that departed from the natural. Both types of motion needed movers, spiritual or material; but the heavenly bodies were incapable of violent motion; hence objects in the sky, such as comets, whose motion was not circular, had to be placed in the sub-lunary sphere – a dogma to which even Galileo conformed.

How is it to be explained that a view of the physical world, so fantastic to the modern mind, could survive even the invention of gunpowder, into an age when bullets and cannon-balls were flying about in obvious defiance of the prevailing laws of physics? Part of the answer is contained in the question: the small child, whose world is still closer to the primitive than to the modern mind, is an unrepentant Aristotelian by investing dead objects with a will, a purpose, an animal spirit of their own; and we all revert to Aristotle in moments when we curse an obstinate gadget or a temperamental motor-car. Aristotle regressed from the abstract-mathematical treatment of physical objects to the animistic view, which evokes so much deeper, primordial responses in the mind. But the days of primitive magic were then past; Aristotle's is a highbrow version of animism, with quasi-scientific concepts like 'embryonic potentialities' and 'degrees of perfection' imported from biology, with a highly sophisticated terminology and an impressive logic-chopping apparatus. Aristotelian physics is really a pseudo-science, out of which not a single discovery, invention or new insight has come in two thousand years; nor could it ever come – and that was its second profound attraction. It was a static system, describing a static world, in which the natural state of things was to be at rest, or to come to rest at the place where by nature they belonged, unless pushed or dragged; and this scheme of things was the ideal furnishing for the walled-in universe, with its immutably fixed Scale of Being.

So much so, that Aquinas' celebrated First Proof of the existence of God was entirely based on Aristotelian physics. Everything that moves needs something else that moves it; but this regress cannot go on to infinity; there must be a limit to it, an agency which moves other things without itself being moved; this unmoved mover is God. In the following century, William of Ockham,* the greatest of the Franciscan schoolmen, made mincemeat of the tenets of Aristotelian physics on which Aquinas' First Proof rested. But by that time scholastic theology had completely fallen under the spell of Aristotelianism – and particularly of the most sterile, pedantic, and at the same time ambiguous elements in Aristotle's logical apparatus. Another century later Erasmus cried out:

> They will smother me beneath six hundred dogmas; they will call me heretic and they are nevertheless Folly's servants. They are surrounded with a bodyguard of definitions, conclusions, corollaries, propositions explicit and propositions implicit. Those more fully initiated explain further whether God can become the substance of a woman, of an ass, of a pumpkin, and whether, if so, a pumpkin could work miracles, or be crucified. ... They are looking in utter darkness for that which has no existence whatever.[9]

The union between the Church and the Stagyrite, which had started with so much promise, turned out to be a misalliance, after all.

## 3. *The Weeds*

Before we take leave of the medieval universe, a brief word must be said of astrology, which will crop up again repeatedly in later parts of this book.

In the days of Babylon, science and magic, calendar-making and augury, were an indivisible unity. The Ionians separated the wheat from the chaff; they took over Babylonian astronomy, and rejected astrology. But three centuries later, in the spiritual bankruptcy following the Macedonian conquest, 'astrology fell upon the Hellenistic mind as a new disease falls upon some remote island people'.[10] The phenomenon repeated itself after the collapse of the Roman Empire. The medieval landscape is grown over with the weeds of astrology and alchemy, which invade the ruins of the abandoned sciences.

* 1300–49.

When building started again, they got mixed up in the materials, and it took centuries to clean them out.*

But the medieval addiction to astrology is not merely a sign of 'failure of nerve'. According to Aristotle, everything that happens in the sub-lunary world is caused and governed by the motions of the heavenly spheres. This tenet served as a rationale for the defenders of astrology, both in antiquity and the Middle Ages. But the affinity between astrological reasoning and Aristotelian metaphysics goes deeper. In the absence of quantitative laws and causal relations, the Aristotelian thought in terms of affinities and correspondences between the 'forms' or 'natures' or 'essences' of things; he classified them by categories, and sub-categories: he proceeded by deduction from analogies, which were often metaphorical, or allegorical, or purely verbal. Astrology and alchemy employed the same methods, only more freely and imaginatively, undeterred by academic pedantry. If they were weeds, medieval science itself had become so weedy, that it was difficult to draw the line between the two. We shall see that Kepler, the founder of modern astronomy, was chronically unable to do so. No wonder, then, that 'influences', 'sympathies' and 'correspondences' between planets and minerals, humours and temperaments, played an integral part in the medieval universe, as a semi-official complement to the Great Chain of Being.

## 4. *Summary*

'In the year 1500 Europe knew less than Archimedes who died in the year 212 B.C.,' Whitehead remarks in the opening pages of his classic work.[11]

I shall try to sum up briefly the main obstacles which arrested the progress of science for such an immeasurable time. The first was the splitting up of the world into two spheres, and the mental split which resulted from it. The second was the geocentric dogma, the blind eye turned on the promising

* Even today, when the house-physician diagnoses *influenza*, he unknowingly ascribes its cause to the evil *influence* of the stars, from which all plagues and pestilences are derived.

line of thought which had started with the Pythagoreans and stopped abruptly with Aristarchus of Samos. The third was the dogma of uniform motion in perfect circles. The fourth was the divorcement of science from mathematics. The fifth was the inability to realize that while a body at rest tended to remain at rest, a body in motion tended to remain in motion.

The main achievement of the first part of the scientific revolution was the removal of these five cardinal obstacles. This was done chiefly by three men: Copernicus, Kepler, and Galileo. After that, the road was open to the Newtonian synthesis; from there on the journey led with rapidly gathering speed to the atomic age. It was the most important turning point in man's history; and it caused a more radical change in his mode of existence than the acquisition of a third eye or some other biological mutation could have achieved.

At this point the method and style of this narrative will change. The emphasis will shift from the evolution of cosmic ideas to the individuals who were chiefly responsible for it. At the same time, we plunge into a new landscape under a different climate: the Renaissance of the fifteenth century. The sudden transition will leave certain gaps in continuity; these will be filled in as the occasion arises.

However, the first of the pioneers of the new era did not belong to it, but to the old one. Though born into the Renaissance, he was a man of the Middle Ages: haunted by its anxieties, ridden with its complexes, a timid, conservative cleric, who started the revolution against his will.

## *Chronological Table to Part Two*

| Date | Greco-Roman Heritage | Early Christianity | Cosmos |
|---|---|---|---|
| | Greco-Roman Heritage ↓ | Early Christianity ↓ | |
| | Neoplatonism ↓ | | |
| A.D. 200 | Plotinus (205–70) ↓ | | |
| „ 300 | | Lactantius (*c.* 260–*c.* 340) | |
| „ 400 | Proclus (410–80) | | |
| | Pseudo-Dionysius (5th cent. ?) | Augustine (354–430) | |
| „ 500 | | Cosmas (6th cent.) → | RECTANGULAR COSMOS |
| „ 600 | Influence of Moors ↓ | Influence of Latin compilers | |
| „ 700 | | → Bede (*c.* 672–735) | |
| | | John the Scot (*c.* 815–*c.* 875) | |
| „ 800 | | → 'Pseudo-Bede' (900) → | SPHERICAL COSMOS REINSTATED |
| „ 900 | | | |
| „ 1000 | | Gerbert (Pope 999–1003) → ↓ | SPHERICAL COSMOS REINSTATED |
| „ 1100 | Euclid rediscovered (*c.* 1120) | | |
| | Ptolemy rediscovered (1175) | | |
| „ 1200 | | Albert the Great (*c.* 1206–80) → | ARISTOTELIAN COSMOS – HIERARCHY OF BEING |
| | Aristotelian Revival ↓ | Thomas Aquinas (*c.* 1225–74) | |
| „ 1300 | | Roger Bacon (1214–94) ↓ | |
| „ 1400 | Scholasticism in Decline ↓ | Occam, Buridan, Oresme (14th cent.) | |
| | Platonic Revival | | |
| „ 1500 | | | |

# Part Three. The Timid Canon

# 1 The Life of Copernicus

## 1. *The Mystifier*

On 24 May 1543, Canon Nicolas Koppernigk,[1] by his Latin name Copernicus, was dying of a haemorrhage of the brain. He had reached the age of three score years and ten, and had published only one scientific work, which he knew to be unsound: *On the Revolutions of the Heavenly Spheres*.[2] He had delayed publishing his theory for some thirty years; the first completed copy of it arrived from the printers a few hours before his death. It was placed on his bed, so that he could handle it. But by then the Canon's mind was wandering, and he could not comment on the anonymous preface to the book, which told the reader that its contents need not be regarded as true, or even as probable. Thus posterity never knew for certain whether Canon Koppernigk had authorized that preface, and whether he really believed in his system or not.

The room where the Canon was dying was in the north-west tower of the fortified wall surrounding the Cathedral hill of Frauenburg in East Prussia, on the outskirts of civilized Christendom. He had lived in that tower for thirty years. It was three storeys high: from the second floor a small door led out on to a platform on top of the wall. It was a grim and forbidding place, but it gave Canon Nicolas an open view over the Baltic Sea to the north and west, the fertile plain to the south, and of the stars at night.

Between the town and the sea stretched a fresh-water lagoon, three or four miles wide and some fifty miles long – a famous landmark of the Baltic coast, known as the *Frisches Haff*. But in the *Book of the Revolutions* the Canon insisted on calling it the Vistula. In one of his asides, he remarked enviously that the astronomers of Alexandria 'were favoured by a serene sky, for the Nile, according to their reports, does not exhale such vapours as the Vistula does hereabouts'.[3] Now the Vistula falls

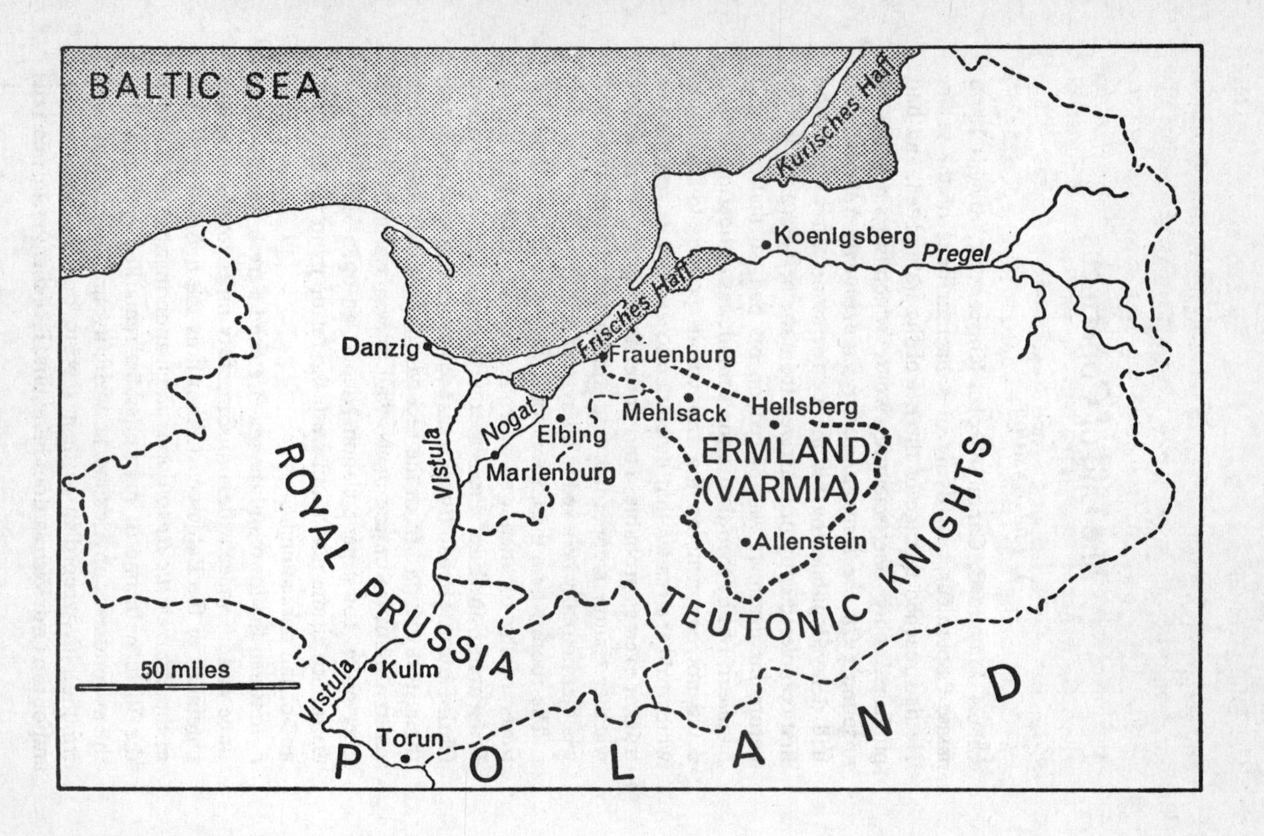
BALTIC SEA
Kurisches Haff
Koenigsberg
Pregel
Frisches Haff
Danzig
Frauenburg
Mehlsack
Heilsberg
Nogat
Elbing
Marienburg
Vistula
ERMLAND
(VARMIA)
Allenstein
ROYAL PRUSSIA
TEUTONIC KNIGHTS
50 miles
Kulm
Vistula
Torun
POLAND

into the sea at Danzig, forty-two miles to the west of Frauenburg; and the Canon, who had lived in these parts nearly all his life, knew perfectly well that the vast expanse of water under his tower was not the Vistula but the *Frisches Haff*, which in German means 'fresh lake'. It was a curious mistake to be made by a man dedicated to scientific precision – and who, incidentally, had been commissioned to make a geographical map of the region. The same mistake is repeated in another passage of the *Book of the Revolutions*: in the chapter 'On the Longitudinal Places and Anomalies of the Moon' it is said that 'all the foregoing observations refer to the meridian of Cracow, since most of these were made from Frauenburg, on the estuary of the Vistula, which lies on the same meridian'.[4] But Frauenburg lies neither on the estuary of the Vistula, nor on the meridian of Cracow.

Posterity had such faith in the precision and trustworthiness of Canon Koppernigk's statements, that a number of scholars blandly transferred Frauenburg down to the Vistula, and as late as 1862, a German encyclopaedia did the same.[5] The foremost of his biographers, Herr Ludwig Prowe, mentioned this puzzle in a single footnote[5A]. Herr Prowe thought the Canon wanted to help readers of his book to locate Frauenburg by displacing it to the shore of a well-known river; and this explanation was taken over by others who wrote after him. But it misses the point. For in the casual remark about the noxious vapours the Canon was clearly not concerned with giving locational clues; and in the second remark, which indeed purports to locate his observatory for other astronomers, a matter which requires utmost precision, the displacement of forty miles was preposterously misleading.

Another of Canon Koppernigk's whims had been to call Frauenburg 'Gynopolis'. Nobody before or after him had thus graecised the German name of the little town; and this might perhaps provide a clue to the apparently senseless mystification of calling the *Haff* the Vistula, and placing both on the meridian of Cracow. Frauenburg, and with it the whole Province of Ermland, lay wedged in between the lands of the Polish King and of the Order of Teutonic Knights. It had frequently served as a battleground before and during the Canon's life-time. The burning, plundering, peasant-slaying Knights, and the vapours of the *Haff* had grievously interfered with the Canon's work; he loathed them both. Ensconced in his tower, he longed for the civilized life of his youth – which was spent on the friendly banks of the Vistula and at Cracow, the brilliant Polish capital. Besides, the Vistula *did* send out a small, half-dry side-branch which trickled into the *Haff* a mere twenty miles from Frauenburg – so that, stretching a few points, one could almost think of himself as living not in Frauenburg on the *Frisches Haff*, but in Gynopolis on the Vistula, and also, more or less, on the meridian of the Polish capital.[6]

This explanation is merely guesswork, but whether true or not, it is in keeping with a curious feature in Canon Koppernigk's character: his inclination to mystify his contemporaries. Half a century of bitter experiences, alternating between the tragic and the sordid, had turned him into a weary and morose old man, given to secretiveness and dissimulation; his sealed-up feelings leaked out only rarely, in roundabout ways. When, two years before his death, he was at last persuaded by his old friend Bishop Giese and the young firebrand Rheticus to publish the *Book of the Revolutions*, he went about it in the same secretive and mystifying manner. Did he really believe, when he looked down from the small window of his tower on the famous lagoon, that his eyes beheld the waters of the distant Vistula – or did he merely wish to believe it? Did he really believe that the forty-eight epicycles of his system were physically present in the sky, or did he merely regard them as a device, more convenient than Ptolemy's, to save the phenomena? It seems that he was torn between the two views; and it was perhaps this doubt about the real value of his theory which broke his spirit.

In the room leading to the platform on the wall lived the Canon's instruments for observing the sky. They were simple, and mostly made by himself according to the instructions given by Ptolemy in the *Almagest*, a thousand and three hundred years before. They were, in fact, cruder and less reliable than the instruments of the ancient Greeks and the Arabs. One was the *triquetrum* or 'cross-bow', about twelve feet high; it consisted of three bars of pine. One bar stood upright; a second bar, with two sights on it, as on the barrel of a gun, was hinged to the top of the first, so that it could be pointed at the moon or a star; the third was a cross-piece, marked with ink like a yardstick, from which the angle of the star above the horizon could be read. The other main instrument was an upright sundial, its base pointing north and south, which indicated the sun's altitude at midday. There was also a 'Jacob's Staff' or *Baculus astronomicus*, which was simply a long staff with a shorter, movable crossbar. Lenses or mirrors were nowhere to be seen; astronomy had not yet discovered the uses of glass.

Nevertheless, better and more precise instruments would have been available to the Canon – quadrants and astrolabes and huge armillary spheres of shining copper and bronze, such as the great Regiomontanus had installed at his observatory in Nuremberg. Canon Koppernigk had always enjoyed a comfortable income, and could well afford to order these instruments from the Nuremberg workshops. His own cross-bow and cross-staff were crude; on an unguarded occasion he had remarked to young Rheticus that if he were able to reduce observational errors to ten minutes arc, he would be as happy as Pythagoras was on discovering his famous theorem.[7] But an error of ten minutes arc amounts to one-third of the apparent width of the full moon in the sky; the Alexandrian astronomers had done better than that. Having made the stars his main business in life, why in heaven's name did the prosperous Canon never order the instruments which would have made him happier than Pythagoras?

Apart from his niggardliness, which had grown worse as the bitter years went by, there existed a deeper, anxious reason for this: Canon Koppernigk was not particularly fond of stargazing. He preferred to rely on the observations of Chaldeans, Greeks, and Arabs – a preference that led to some embarrassing results. The *Book of the Revolutions* contains, altogether, only twenty-seven observations made by the Canon himself; and these were spread over thirty-two years! The first he made as a student in Bologna, aged twenty-four; the last referred to in his Book, an eclipse of Venus, was made no less than fourteen years before he sent the manuscript to the printers; and though during these fourteen years he continued to make occasional observations, he did not bother to enter them into his text. He merely scribbled them on the margin of the book he happened to be reading, in between other marginal jottings, such as recipes against toothaches and kidney stones, for the dyeing of the hair, and for an 'imperial pill' which 'may be taken at any time and has a curative effect on every disease'.[8]

All in all, Canon Koppernigk noted down between sixty and seventy observations in a lifetime. He regarded himself as a philosopher and mathematicus of the skies, who left the work of actual stargazing to others, and relied on the records of the ancients. Even in the position he assumed for his basic star, the Spica, which he used as a landmark, was erroneous by about forty minutes' arc, more than the width of the moon.

As a result of all this, Canon Koppernigk's life-work seemed to be, for all useful purposes, wasted. From the seafarers' and stargazers' point of view, the Copernican planetary tables were only a slight improvement on the earlier Alphonsine tables, and were soon abandoned. And insofar as the theory of the universe is concerned, the Copernican system, bristling with inconsistencies, anomalies, and arbitrary constructions, was equally unsatisfactory, most of all to himself.

In the lucid intervals between the long periods of torpor, the dying Canon must have been painfully aware that he had failed. Before sinking back into the comforting darkness, he probably saw, as dying men do, scenes of his frigid past warmed by the merciful glow of memory. The vineyards of Torun; the golden pomp of the Vatican gardens in the jubilee year 1500; Ferrara entranced by its lovely young duchess, Lucretia Borgia; the precious letter from the most reverend Cardinal Schoenberg; the miraculous arrival of young Rheticus. But if memory could lend some deceptive warmth and colour to Canon Koppernigk's past, its soothing grace does not extend to posterity. Copernicus is perhaps the most colourless figure among those who, by merit or circumstance, shaped mankind's destiny. On the luminous sky of the Renaissance, he appears as one of those dark stars whose existence is only revealed by their powerful radiations.

## 2. *Uncle Lucas*

Nicolas Koppernigk was born in 1473, half-way between the transformation of the old world through Coster of Haarlem's invention of the printing press with movable metal types, and Columbus' discovery of a new world beyond the seas. His life overlapped with Erasmus of Rotterdam's who 'laid the egg of the Reformation', and with Luther's who hatched it; with Henry VIII who broke away from Rome and Charles V who brought the Holy Roman Empire to its climax; with the Borgias and Savonarola, with Michelangelo and Leonardo, Holbein and Dürer; with Machiavelli and Paracelsus, Ariosto and Rabelais.

His birthplace was Torun on the Vistula, formerly an outpost of the Teutonic Knights against the Prussian pagans, later a member of the Hanseatic League and a trading centre between East and West. At the time when Nicolas Koppernigk was born, the town was already in decline, steadily losing its trade to Danzig which lay closer to the river's estuary. Yet he could still watch the merchants' fleets sailing down the broad, muddy waters towards the sea, loaded with timber, and coal from the Hungarian mines, with pitch and tar, and honey and wax from Galiczia; or making their way upstream with textiles from Flanders, and silk from France, and herring and salt and spices: always in convoys, to be safe from pirates and brigands.

It is unlikely, though, that the boy Nicolas spent much time watching life on the river wharfs, for he was born inside the sheltering walls where, protected by moat and drawbridge, the gabled, narrow-chested patrician houses stood hemmed in between church and monastery, town-hall and school. Only the lowly folk lived outside the crenelated walls, among the wharfs and storehouses, in the noise and stench of the suburban artisanate: the wheel and waggon-makers, blacksmiths, coppersmiths and gun-barrel makers, salt refiners and saltpetre boilers, schnapps distillers and hop brewers.

Perhaps Andreas, the elder brother who was something of a

scamp, liked to loiter in the suburbs, hoping to become one day a pirate; but Nicolas remained, through all his life, fearful of venturing, in any sense, outside the walls. He must have had an early awareness of the fact that he was the son of a wealthy magistrate and patrician of Torun: of one of those prosperous merchants whose ships had, only a generation or two earlier, roamed the seas as far as Brugge and the Scandinavian ports. Now, when the fortunes of their town were on the decline, they were becoming all the more self-important, stuffy and ultra-patrician. Nicolas Koppernigk senior had come from Cracow to Torun in the late fourteen-fifties, as a wholesale dealer in copper, the family business from which the Koppernigks derived their name. Or so at least it is presumed, for everything connected with the ancestry of Canon Koppernigk is shrouded in the same secretive and uncertain twilight through which he moved during his life on earth. There lived no historic personality in that epoch of whom less is known by way of documents, letters, or anecdotes.

About the father we know at least where he came from, and that he owned a vineyard in the suburbs, and that he died in 1484, when Nicolas was ten. About the mother, *née* Barbara Watzelrode, nothing is known except her name; neither the date of her birth, nor of her marriage, nor of her death could be found on any record. This is the more remarkable, as Frau Barbara came from a distinguished family: her brother, Lucas Watzelrode, became the Bishop and ruler of Ermland. There are detailed records of the life of Uncle Lucas, and even of Aunt Christina Watzelrode; only Barbara, the mother, is blotted out – eclipsed, as it were, by the persistent shadow thrown by the son.

Of his childhood and adolescence up to the age of eighteen, only one event is known – but an event which became decisive in his life. At the death of Koppernigk senior, Nicolas, his brother and two sisters, became the charges of Uncle Lucas, the future Bishop. Whether at that time their mother was still alive, we do not know; at any rate, she fades out of the picture (not that she had ever been much present in it); and henceforth Lucas Watzelrode plays the part of father and protector, employer and maecenas to Nicolas Koppernigk. It was an intense and intimate relationship which lasted to the end of the Bishop's life, and which one Laurentius Corvinus, town scribe and poetaster of Torun, compared to the attachment between Aeneas and his faithful Achates.

The Bishop, twenty-six years older than Nicolas, was a powerful and irascible, proud and sombre personality; an autocrat and a bully who brooked no contradiction, never listened to others' opinions, never laughed, and was loved by nobody. But he was also a fearless and dedicated man, impervious to slander, and just according to his own lights. His historic merit is the relentless fight he put up against the Teutonic Knights, preparing the way to the eventual dissolution of their Order – that anachronistic survival of the Crusades, which had degenerated into a rapacious, plundering horde. One of the Order's last Grand Masters called Bishop Lucas 'the devil in human shape', and its Chronicler reports that the Knights prayed every day for his death. They had to wait till he was sixty-five; but when death came to the vigorous Bishop, it came through such sudden and suspicious illness that it was assumed they poisoned him.

The only endearing feature of that hard Prussian Prince of the Church was his nepotism – the loving care he took of his numerous nephews, nieces, in-laws, and his bastard son. He procured Nicolas and brother Andreas the fat prebends of the Canonry of Frauenburg; through his influence, the older of the Koppernigk sisters became Mother Superior of the Cistercian convent at Kulm, while the younger was married off to a nobleman. A contemporary chronicler further reports that 'Philip Teschner, by birth a son-of-a-whore, born to Luca the Bishop by a pious virgin when Luca was still a magistrate at Torun, was promoted by the Bishop to the post of mayor of Braunsberg'.[9]

But his favourite, his *fidus Achates*, was young Nicolas. It was evidently a case of attraction by opposites. The Bishop was overbearing, the nephew self-effacing. The Bishop was impetuous and irritable, the nephew meek and submissive. The Uncle was sanguine and unpredictable, the nephew pedestrian and pedantic. Both in their private relationship and in the eyes of their small provincial world, Bishop Lucas was the brilliant star, Canon Nicolas the pale satellite.

### 3. *The Student*

In the winter of 1491–2, at the age of eighteen, Nicolas Koppernigk was sent to the famous University of Cracow. The only record of his four years of study there is an entry according to which 'Nicolas, the son of Nicolas of Torun' was immatriculated and paid his fee in full. Brother Andreas was accepted too, but the record says that he paid only part of the fee down. Also, Andreas was late at the immatriculation: fifteen other names were entered after Nicolas' name in the roster before the older brother turned up. Neither of them took a degree.

At twenty-two, Nicolas returned to Torun on Bishop Lucas' request. One of the Canons of his Cathedral at Frauenburg was dying, and the Bishop was anxious to secure the prebend for his favourite nephew. He had good reason to make haste, for the patricians of Torun were in a state of grave anxiety regarding their economic future. For several months they had been receiving disquieting letters from their business relations and their agents in Lisbon, concerning the alleged opening of a sea route to India by a Genoese captain, and about the endeavours of Portuguese seafarers to achieve the same aim by rounding the south Cape of Africa. Rumour became certainty when the report which Columbus, after his return from the first crossing, had addressed to the Chancellor Raphael Sanches, was printed as a broadsheet first in Rome, then in Milan, and finally in Ulm. There could no longer be any doubt: these new trading routes to the Orient were a grave menace to the prosperity of Torun and the whole Hanseatic League. For a young man of good family and uncertain vocation, the safest thing was to secure a nice, comfortable prebend. It is true that he was only twenty-two; but after all, Giovanni di Medici, the future Leo X, had been made a Cardinal at fourteen.

Unfortunately, the expected death of Canon Matthias de Launau, Precentor of the Frauenburg Cathedral, occurred ten days too early, on 21 September. Had he died in October, Bishop Lucas could have made Nicolas a Canon without further ado; but in all uneven months of the year the privilege of filling vacancies in the Ermland Chapter belonged not to the Bishop, but to the Pope. There were other candidates, and complicated intrigues for the prebend; Nicolas was defeated and complained about his misfortune in several letters which were still extant in the seventeenth century, but have vanished since.

Two years later, however, a new vacancy occurred in the Chapter, this time conveniently in the month of August, and Nicolas Koppernigk was duly appointed a Canon of Frauenburg Cathedral; whereupon he promptly departed for Italy, to continue his studies. He drew his prebend, but he neither took Holy Orders, nor was his physical presence at Frauenburg required for the next fifteen years. During this period, the new Canon's name only appears twice on the Cathedral records: the first time in 1499, when his appointment was officially confirmed, the second time in 1501, when his original leave of absence of three years was extended by another three years. A Canonry in Ermland seems to have been, in the vulgar parlance of our century, a cushy job.

From the age of twenty-two to thirty-two, the young Canon studied at the Universities of Bologna and Padua; added to his four years at Cracow, this makes altogether fourteen years spent at various universities. According to the Renaissance ideal of *l'uomo universale*, he studied a little of everything: Philosophy and Law, Mathematics and Medicine, Astronomy and Greek. He took his degree as a doctor of Canon Law at Ferrara in 1503, aged thirty. Apart from paying his immatriculation fees and taking his degree, he left no trace, either of distinction or scandal, on the records of his various universities.

While the majority of the young men from Torun went for their preliminary studies to the German university of Leipzig, Koppernigk went to Polish Cracow; but at the next stage, in Bologna, he joined the German, not the Polish *natio* or student fraternity, whose list of new members enrolled in 1496 shows the name of 'Nicolaus Kopperlingk de Thorn'. The *natio Germanorum* was the most powerful in Bologna, both in the frequent street brawls and inside the *alma mater*. Its roster contained the names of many illustrious German scholars, among them Nicolas of Cusa. Uncle Lucas, too, had first studied in Cracow, and then joined the German *natio* in Bologna; and young Nicolas can hardly be blamed for following in his footsteps. Besides, nationalism by rigid ethnic divisions was still a plague of the future; thus beside the *natio Germanorum* there existed independent Swabian, Bavarian, etc., *natios*. Yet for the last four hundred years a bitter and silly feud has been raging between Polish and German scholars, both claiming Copernicus as a true son of their nation.[10] All one can say is, in the manner of Solomon, that his forebears came from the proverbially mixed stock of the border provinces between the Germanic and Slavonic peoples; that he lived in a contested territory; that the language he mostly wrote was Latin, the vernacular of his childhood German, while his political sympathies were on the side of the Polish King against the Teutonic Order, and on the side of his German Chapter against the Polish King; lastly, that his cultural background and heritage were neither German, nor Polish, but Latin and Greek.

Another much discussed question was why, having completed his studies in Canon Law at the world-famous university of Padua, Copernicus chose to take his degree at the small and insignificant University at Ferrara, where he had never studied. The puzzle was solved only at the end of the last century, when an Italian scholar[11] unearthed the fact that around A.D. 1500, degrees could be obtained in Ferrara not only more easily, but also considerably cheaper. A newly promoted Doctor at Bologna or Padua was expected to provide lavish entertainment to celebrate the event; by slipping away from his teachers and friends to obscure Ferrara, Canon Nicolas, following the precedent set by some other members of the *natio Germanorum*, successfully avoided the burdens of hospitality.

Copernicus' diploma reveals another interesting detail: that the candidate was not only a Canon of Frauenburg Cathedral, but also enjoyed a second absentee prebend as 'Scholasticus of the Collegiate Church of the Holy Cross in Breslau'. What rights and duties, apart from drawing a steady income, this impressive title entailed, the historians are unable to tell. It is doubtful whether Canon Koppernigk ever visited Breslau; one can only assume that he obtained this additional benefice through some Silesian business relation of his late father's, or the loving care of Uncle Lucas. He kept the matter characteristically secret throughout his life; neither in the records of the Frauenburg Chapter, nor in any other document, is Canon Koppernigk's second ecclesiastical function mentioned; it appears only in his promotion paper. It is not difficult to guess that on this particular occasion the candidate in Canon Law found it expedient to reveal his learned title.

In between his studies in Bologna and Padua, he also spent a year in Rome – the jubilee year 1500. There, according to his disciple, Rheticus, Copernicus 'being twenty-seven years of age, more or less, he lectured on mathematics before a large audience of students and a throng of great men and experts in this branch of knowledge'.[12] This assertion, based on the Canon's scant remarks about his life to Boswell Rheticus, was eagerly taken up by subsequent biographers. Yet neither the records of the University, nor of any college, seminary, or school in Rome mention lectures by Copernicus. It is now assumed that he may have given some casual talks, as travelling scholars and humanists usually did when visiting a centre of learning. The lectures, and his ten years' presence in Italy, left no echo or trace in the countless letters, diaries, chronicles, or memoirs of that hyper-awake, garrulous, and graphomane age, when Italy was like a floodlit stage over which no foreign scholar of any personality could pass without being noticed and recorded in one way or another.

The only treat for the biographer, during these ten Italian years, is a letter which shows that on one occasion the Koppernigk brothers (for Andreas had joined Nicolas as a student in Bologna) ran out of money and had to borrow a hundred ducats. They were lent to them by their Chapter's representative in Rome, one Bernard Sculteti, and were refunded to the latter by Uncle Lucas. It is the only episode with a flicker of human interest in Canon Koppernigk's uneventful youth, and his starved biographers understandably tried to squeeze the last drop out of it. But Sculteti's letter to Bishop Lucas, which is the source of the story, merely reports the bare facts of the financial transaction – and adds that Andreas has threatened 'to offer his services to Rome'[13] unless he could repay at once the debts which the brothers had contracted *scholarium more*, after the habit of students. By reporting Andreas' blackmailing threat and passing over Nicolas in silence, the diplomatic Sculteti (who later became Leo X's private chaplain and chamberlain) obviously meant to lay the blame for the affair on the elder brother; so that, whatever interest the episode contains, it primarily concerns Andreas, the rake.

## 4. *Brother Andreas*

Since he evidently exerted a strong and lasting influence on Nicolas, it will be of interest to know a little more about Andreas. Every single fact that has emerged about him confirms the contrast in character between the brothers. Andreas is the older one, but he immatriculates at Cracow University some time, and at Bologna two full years, later than Nicolas; and pays only part of the fee at Cracow, whereas Nicolas pays the full fee. Nicolas is made a Canon by Uncle Lucas in 1497; the older brother again two years later, in 1499. In 1501, both apply for a three-year extension of their leave of absence. Nicolas is readily granted his request: having promised to study medicine, it is hoped 'that he will later on be useful to the revered head of the diocese and the Canons of the Chapter'; whereas at the same session, Andreas' request is granted with the dry motivation 'because he is considered capable of continuing his studies'.

Everything seems to indicate that Andreas was the type of young man of whom, in the respectable world of small-town wholesale merchants, it is prophesied that he will come to an evil end. He did. At the termination of their Italian studies, Andreas returned to Frauenburg infected with an incurable disease, which the records of the Chapter describe as *lepra*. This expression was, at the time, used on the Continent as loosely as 'the pox' was in England, and may either have really meant leprosy, or, more likely, syphilis – which was ravaging Italy, whereas leprosy was on the decline.

It made, in fact, very little difference whether Canon Andreas had leprosy or the syphilis, for both spread horror and disrepute. A couple of years after his return, Andreas' condition began to deteriorate rapidly, and he asked for leave to go back to Italy and seek treatment there. This was granted in 1508. Yet four years later, Andreas was back in Frauenburg, by now so repulsive in appearance that the terrified Chapter decided to get rid of him by every means. In September 1512, a meeting was held of the assembled Chapter, including brother Nicolas, which resolved to break off all personal relations with Canon Andreas; to ask him to account for the sum of twelve hundred Hungarian gold florins which had been confided to him for ecclesiastical purposes; to seize his prebend and all other revenues; and to grant him a small annuity on condition that he took himself off from their midst.

Andreas refused to submit to this decision; he fought back simply by remaining in Frauenburg, and displaying his leprous countenance as a *memento mori* among his smug and pleasure-loving brethren in Christ. In the end they had to give in: the seizure was lifted, and a higher annuity granted to Andreas pending the final decision by the Apostolic See – always provided that 'the mortally infected, contagious leper' left the town. Andreas accepted the settlement, yet he lingered on in Frauenburg for another two or three months, and put in at least two more stage-appearances at sessions of the Chapter to spite his colleagues, including beloved brother Nicolas. Then he went back to the congenial Rome he had first known under the rule of the Borgias.

Yet even in his 'mortally infected' state, he took an active part in the intrigues at the Papal Court concerning the succession at the Bishopric of Ermland; and it is a tribute to his remarkable character that, at one stage, when Sigismund of Poland felt moved to protest against the machinations of the

Chapter, he addressed his letter not to its official delegates in Rome, but to the exiled and ostracized leper, Andreas. He died a few years later, under unknown circumstances, at an unknown date.

Canon Nicolas never mentioned Andreas' illness, nor his scandalous life and death. All that Rheticus has to say on the subject is that the astronomer had 'a brother called Andreas who had been acquainted with the famous mathematician Georg Hartman in Rome'.[14] The later biographers were equally discreet on the subject of brother Andreas. Not until A.D. 1800 did one Johan Albrecht Kries mention the illness of Andreas in an obscure journal.[15] But he quickly repented; and three years later, when Kries edited an earlier Copernicus biography by Lichtenberg, he too kept quiet on the subject.

Had the Koppernigks been born in Italy instead of a Prussian backwater, Andreas would have been a reckless condottière, and Uncle Lucas the autocratic ruler of a city state. Hemmed in between these two powerful and headstrong characters, bullied by the first, despised and disgraced by the second, Nicolas took refuge in secretiveness, caution, obliquity. The earliest engravings, and the later portraits of doubtful authenticity, all show a strong face with a weak expression: high cheekbones, wide-set dark eyes, square chin, sensuous lips; but the glance is uncertain and suspicious, the lips curve into a sour pout, the face is closed, on the defensive.

It was toward the end of his Italian studies that the heliocentric system began to take shape in Nicolas' mind. The idea was, of course, not new, and it was much discussed in Italy at that time; I shall return to this point later on. Nicolas had taken an active interest in astronomy at an early stage of his Italian studies; it became the main solace of his frustrated life. When he became acquainted with the Aristarchian idea of the sun-centred universe, he grasped at it and never let go again. For thirty-six years, on his own testimony, he hugged his theory to his anxious heart, and only agreed, reluctantly, to divulge its secret on the doorstep of death.

## 5. *The Secretary*

In 1506, at the age of thirty-three, Canon Koppernigk, Doctor of Canon Law, terminated his studies in Italy and returned home to Prussia. The next six years he spent with Uncle Lucas at Heilsberg Castle, the residence of the Bishops of Ermland.

Thirteen years had gone by since he had been elected a Canon of Frauenburg Cathedral, and he had as yet neither exercised his functions, nor paid more than two fleeting visits to his Chapter. The new, indefinite leave of absence was granted on the official grounds that he should act as private physician to Uncle Lucas. In fact, the Bishop wanted his *fidus Achates* in constant attendance, and to the end of his life kept Nicolas at his court.

However, the appointment of Nicolas as a house physician was not solely an official pretext. Though he never took his medical degree, he had studied medicine, as befitting in those days for a gentleman of the clergy, at the renowned University of Padua. One of his teachers had been the famed Marcus Antonius de la Torre, for whom Leonardo had drawn his anatomical studies of horses and men. Whether Nicolas had occasion to minister as a physician to Uncle Lucas, is not recorded; but later on he did treat Lucas' successors, Bishops Ferber and Dantiscus, for various ailments – partly in person, partly by mail; and he was summoned by Duke Albert of Prussia to attend one of his counsellors. In fact, Copernicus was far better known in Ermland as a physician than as an astronomer.

The nature of his approach to medicine one can gather from the prescriptions he copied out from various text-books. It was as conservative as his approach to science in general; he believed as unquestioningly in the doctrines of Avicenna, as he believed in the physics of Aristotle, and in the epicycles of Ptolemy. One of the prescriptions which he copied out twice (once on the back cover of Euclid's *Elements of Geometry*, and a second time on the margin of a surgical volume) contains the following ingredients: Armenian sponge, cinnamon, cedar wood, bloodroot, dittany, red sandalwood, ivory shavings, crocus (or saffron), spodumene, camomile in vinegar, lemon rind, pearls, emerald, red jacinth and sapphires; a deer's heart-bone or pulped heart, a beetle, the horn of a unicorn, red coral, gold, silver, and sugar.[16] It was a typical prescription of the age, together with lizards boiled in olive oil and earthworms washed in wine, calf's gall and donkey's urine. But it was also the age which saw the rise of Paracelsus, Servetus, and Vesalius, the overthrow of Avicenna and the medieval Arab school. There is a type of genius: Bacon and Leonardo, Kepler and Newton, who, as if they were charged with electricity, draw an original spark from any subject they touch, however remote from their proper field; Copernicus was not one of them.

His main duties, however, during his six years at Heilsberg Castle were not of a medical, but of a diplomatic nature. Little Ermland, a border territory, was an object of constant friction, intrigues and wars – as the neighbouring Danzig was to be four hundred years later. The principal cities of Ermland were Frauenburg, the Cathedral town; Heilsberg, where the Bishop resided; and, further inland, Allenstein – each of them centred on a medieval castle on a hill, and fortified by wall and moat. It was the largest of the four Prussian dioceses, and the only one which, thanks to Bishop Lucas' astuteness, successfully maintained its independence both against the Teutonic Order and the Polish King. While, politically, he sided with the latter, Bishop Lucas never surrendered his autonomous rights, and ruled his remote territory in the grand style of a Renaissance Prince.

A fifteenth century 'Ordinance of the Castle of Heilsberg'[17] describes in minute detail the personnel of the Bishops' Court, their order of precedence, and the table etiquette. At the sound of the dinner bell, all the residents and guests have to wait at the doors of their apartments till the Bishop enters the paved court, announced by the baying of his hounds which are released at that moment. When the Bishop, with mitre, staff, and purple gloves, appears in the court, a procession is formed which follows him into the Hall of Knights. The servants hand round wash-basins and towels, and after grace has been said, the Bishop ascends the raised dais to the principal table, reserved for the highest ranking dignitaries and guests. There are altogether nine tables: the second is reserved for the higher, the third for the lower officials, the fourth for the principal servants, the fifth for feeding the poor, the sixth, seventh, and eighth for the lower servants and the servants' servants, the ninth for the jugglers, jesters, and mountebanks who entertain the company.

It is not recorded to which of the tables Canon Nicolas was assigned; presumably to the second. He was now getting on to forty. His duties included accompanying Uncle Lucas on his journeys and diplomatic missions to Cracow and Torun, to the Prussian and Polish diets, to King Sigismund's Coronation and wedding; also the drafting of letters and political documents. He presumably assisted the Bishop with two of the latter's pet projects: to get rid of the Teutonic Knights by sending them on a crusade against the Turks, and to found a Prussian university at Elbing; both of which came to naught.

Yet the pulse of time in Ermland was of a leisurely rhythm, and his duties left Canon Koppernigk sufficient freedom to pursue his personal interests. Observing the sky was not one of them – during his six years at Heilsberg he did not record a single observation. But he was preparing two manuscripts: one a Latin translation, the other an outline of the Copernican system of the universe. The first he had printed, the second not.

The unpublished astronomical manuscript is known as the *Commentariolus* [18] or *Brief Outline*; it will be discussed later. The other manuscript was printed in Cracow A.D. 1509, when Copernicus was thirty-six, and is, apart from the *Revolutions*, the only book he published in his life. It also represents his only excursion into the field of *belles lettres*, and as such sheds a light on his personality and tastes.

The booklet is Canon Koppernigk's translation into Latin of the Greek epistles of one Theophylactus Simocatta. Theophylactus was a Byzantine historian of the seventh century, whose best-known work is a History of the reign of Emperor Mauritius. Of his literary merits, Gibbon says that he was voluble on trifles, short on essentials; [19] and Bernhardy remarks that 'Theophylactus' style, shallow yet inflated by meaningless flourishes ... reveals, earlier and more completely than one would have imagined, the emptiness and effete nature of his time'.[20] He also published a volume of eighty-five *Epistles* in the form of fictitious letters exchanged between various Greek characters; it was this work which Copernicus chose to translate into Latin, as his contribution to the literature of the Renaissance.

Simocatta's *Epistles* are classified under three headings: 'moral', 'pastoral', and 'amorous'. The following samples (unabridged) of each of the three *genres* are re-translations from Copernicus' Latin version.[21] They are the last three of the collection:

*The 83rd Epistle – Anthinus to Ampelinas (pastoral)*

The grape harvest is close and the grapes are full of sweet juice. Guard, then, closely the road, and take as a companion an able dog from Crete. For the vagrant's hands are only too willing to grab, and to deprive the farmer of the fruits of his sweat.

*The 84th Epistle – Chrysippa to Sosipater (amorous)*

Thou art caught in the nets of love, Sosipater, thou lovest Anthusia. Well deserving of praise are the eyes that turn in love to a beautiful maiden. Do not complain that thou hast been conquered by love; for greater is the delight that will reward thy labour of love. Though tears pertain to grief, those of love are sweet, for they are mixed with joy and pleasure. The gods of love bring delight at the same time as sadness; with manifold passions is Venus girded.

*The 85th Epistle – Plato to Dionysius (moral)*

If thou wouldst wish to obtain mastery over thy grief, wander among graves. There thou wilt find the cure for thine ailment. At the same time thou wilt realize that even the greatest happiness of man does not survive the grave.

What on earth, or in the skies, moved Canon Koppernigk to spend his labours just on this collection of pompous platitudes? He was not a schoolboy but a mature man; not an uncouth provincial, but a humanist and a courtier who had spent ten years in Italy. This is what he has to say in explanation of this curious choice – in his dedicatory preface to Uncle Lucas:

TO THE MOST REVEREND BISHOP LUCAS
OF ERMLAND
DEDICATED BY NICOLAUS COPERNICUS

MOST REVEREND LORD AND FATHER OF THE
FATHERLAND

With great excellence, so it seems to me, has Theophylactus the scholar compiled these moral, pastoral, and amorous epistles. He was certainly guided in his work by the consideration that variety is pleasing and should therefore be preferred. Very varied are the inclinations of men, and very divers matters delight them. One likes weighty thoughts, the other responds to levity; one likes earnestness, the other is attracted by the play of fancy. Because the public takes pleasure in such different things. Theophylactus alternates light subjects with weighty ones, frivolity with earnestness, so that the reader, as if in a garden, may choose the flower which pleases him best. But everything he offers yields so much profit that his prose poems appear to be not so much epistles as rather rules and precepts for the useful ordering of human life. Proof of this is their substantiality and brevity. Theophylactus took his material from various writers and compiled it in a short and very edifying manner. The value of the moral and pastoral epistles will hardly be denied by anyone. A different judgment is perhaps invited by the epistles on love, which, because of their subject, may seen lighthearted and frivolous. But as the physician softens the bitter medicine by the admixture of sweet ingredients to make it more agreeable to the patient: even so are the lighthearted epistles added; they are, incidentally, kept so pure that they could just as well be called moral epistles. Under these circumstances I considered it unfair that Theophylactus' epistles could only be read in the Greek language. To make them more generally accessible, I have tried to translate them, according to my powers, into the Latin.

To thee, most reverend Lord, I dedicate this small offering which, to be sure, bears no relation to the benefactions I received from thee. Whatever I achieve through my mind's capacity, I regard as thy property by right; for true beyond doubt is what Ovid once wrote to Caesar Germanicus:

'According to the direction of thy glance, falls and rises my spirit.' [22]

One must remember that this was an age of spiritual fermentation and intellectual revolution. It is depressing to compare

Canon Koppernigk's taste and style with that of his illustrious contemporaries – Erasmus and Luther, Melanchton and Reuchlin, or Bishop Dantiscus in Copernicus' own Ermland. However, the translating enterprise was not a random whim; and if we look closer into the matter, the choice of the obscure Theophylactus was a shrewd one. For this was a time when translating the rediscovered Greek texts of antiquity was regarded as one of the foremost and noblest tasks of the humanists. It was the time when Erasmus' translation of the Greek New Testament, by revealing the corruptions of the Roman vulgate, 'contributed more to the liberation of the human mind from the thraldom of the Clergy than all the uproar and rage of Luther's many pamphlets'; [23] and when a different kind of intellectual liberation was effected through the rediscovery of the Hyppocratics and Pythagoreans.

Yet in Northern Europe, the more bigoted minority of the clergy was still fighting a rearguard action against the revival of antique learning. In Copernicus' youth, Greek was not taught at any German or Polish university; the first teacher of Greek at Cracow, Georg Libanius, complained that religious zealots were trying to prohibit his lectures and to excommunicate all who learned Hebrew and Greek. Some German Dominicans were particularly vociferous in denouncing as heretical all research into the unexpurgated Greek and Hebrew texts. One of them, the monk Simon Grunau, grumbled in his chronicle: 'Some have not seen a Jew or Greek in all their days, and yet could read Jewish and Greek from books – they are obsessed.' [24]

This obscure Grunau and the aforementioned Libanius are often quoted in the literature on Copernicus, in order to prove that it needed great courage on the Canon's part to publish a translation from the Greek; and that by this symbolic gesture he took demonstratively the side of the humanists against the obscurantists. The gesture was certainly a calculated one, but insofar as it implied a taking of sides, Copernicus was siding with the victors: at the time when he published his booklet, Erasmus and the humanists seemed to be carrying the day. It was the time of the great European revival before the Western world split into two hostile camps, before the horrors of the

Reformation and Counter-Reformation, before Rome countered the advance of the printing press with its *index librorum prohibitorum*. Erasmus was still the undisputed intellectual leader, who could write without boasting that his disciples included

> the Emperor, the Kings of England, France, and Denmark, Prince Ferdinand of Germany, the Cardinal of England, the Archbishop of Canterbury, and more princes, more bishops, more learned and honourable men than I can name, not only in England, Flanders, France, and Germany, but even in Poland and Hungary.[25]

These considerations may help to explain the peculiar choice of the text. It was a *Greek* text, and its translation therefore meritorious in the eyes of the humanists; yet it was not an *antique* Greek text, but written by a Byzantine Christian of the seventh century, with such unimpeachable dullness and piety that not even a fanatical monk could object to it. In short, the Epistles of Theophylactus were both fish and fowl, Greek and Christian, and generally speaking, safe as houses. They attracted no attention whatsoever, either among humanists or obscurantists, and were soon forgotten.

## 6. *The Canon*

In 1512, Bishop Lucas suddenly died. He had journeyed to Cracow to attend the marriage of the Polish King, and had attended the ceremonies in full vigour. On the return journey he suddenly developed food poisoning and died in his native Torun. His faithful secretary and house physician, elusive as always, was not near him at his death; the reasons for his absence are unknown.

Soon after the Uncle's death Copernicus, now a man of forty, left Heilsberg Castle and, after a delay of fifteen years, took up his duties as a Canon of Frauenburg Cathedral – which he carried out faithfully to the end of his life.

The duties were not exacting. The sixteen Canons led the leisurely, worldly, and opulent life of provincial noblemen. They carried arms (except at the meetings of the Chapter), and were required to uphold its prestige by keeping at least two servants and three horses a head. Most of them came from the patrician families of Torun and Danzig, and were related to each other by inter-marriage. They each had a house or *curia* allotted to them inside the fortified walls – one of these was Copernicus' tower – and also two additional *allodia*, or small private estates in the countryside. Apart from all this, each Canon enjoyed the benefices of one or several additional prebends, and their income was considerable.

Only one of the sixteen Canons had taken the higher vows and was entitled to officiate mass; the remainder were merely bound, when not absent on some official mission, to attend, and occasionally to assist in, the morning and evening services. The rest of their duties were of a worldly nature: the administration of the Chapter's vast estates, over which they exercised nearly absolute power. They levied taxes, collected rent and tithe, appointed the mayors and officials in the villages, sat in court, made and administered the Law. These activities must have appealed to Canon Koppernigk's thrifty and methodical nature, for during four years he held the appointment of Administrator of the Chapter's outlying domains, at Allenstein and Mehlsack, and for another stretch he was General Administrator of all the Chapter's possessions in Ermland. He kept a ledger and a business journal, in which all transactions with tenants, serfs, and labourers are meticulously reported.

In between – in 1519 – the feud between Poles and Teutonic Knights flared up again. There were not major battles, but the countryside of Ermland was devastated by the plundering soldiery of both sides. They killed the peasants, raped their women, and set fire to their farms, but did not attack fortified towns. Fourteen out of the sixteen Canons spent that turbulent year in Torun or Danzig; Koppernigk preferred to remain, in the company of an aged confrater, in his tower behind the safe walls of Frauenburg, where he looked after the affairs of the Chapter. Subsequently he administered, for another year, Allenstein, and also seemed to have taken part in an abortive attempt at mediation between the hostile parties. When peace returned at last, in 1521, he was nearly fifty. His remaining twenty years, outwardly uneventful, were spent mainly in his tower.

He had plenty of leisure. In 1530 or thereabouts,[26] he completed the manuscript of the *Book of Revolutions* and locked it away, making only occasional corrections in it. He did nothing else of much consequence. He wrote, by request of a friend, a critique of the theories of a fellow astronomer,[27] which, like the *Commentariolus,* was circulated in manuscript; he drew up a memorandum on the damages caused by the Teutonic Knights during the war; and he wrote a treatise on monetary reform for the Prussian diet.[28] No great philosopher or scientist has ever published less.

In all these years he had acquired only one intimate friend, a fellow Canon at Frauenburg, later Bishop of Kulm and of Ermland, Tiedemann Giese. Canon Giese was a gentle and learned man who, though seven years younger than Copernicus, took a protective and affectionate interest in him. It was Giese who, after years of effort, and assisted by young Rheticus, finally talked his reluctant confrater into allowing the *Book of the Revolutions* to be published, and who, when Koppernigk became involved in a sordid conflict with his new Bishop, smoothed things out through his influence. Nicolas always needed a stronger personality to lean on; but while Uncle Lucas and brother Andreas had bullied and intimidated him, Giese guided him through the remaining years of his life with patience and gentle persuasions. He was, before Rheticus' last-minute arrival on the scene, the only one who had recognized the morose and unloved, ageing man's genius; who accepted the weaknesses of his friend's character and understood his tortuous ways, without letting them interfere with his intellectual admiration. It was a remarkable feat of charity and imagination, for in that age a man's intellect and his character were still perceived as an indivisible entity. A person was accepted or rejected as a whole; and most people who came into contact with Canon Koppernigk chose the second alternative. Tiedemann Giese, the firm yet tender protector, guide, and spur, is one of the silent heroes of history, who smooth its path but leave no personal mark on it.

There is a typical episode in the relationship between the two friends, which bears on their attitude to the central issue of their time: the Reformation of the Church they served.

Copernicus was forty-four when, in 1517, Martin Luther nailed his ninety-five theses on the door of the castle church at Wittenberg. No more than five years had to pass, and 'behold, the whole world is dragged into the fight, storming to wild struggle and slaughter, and all Churches are defiled by abuse as if Christ, on returning to Heaven, had bequeathed us not peace but war' – as the gentle Giese wrote in despair.[29] From its very beginnings, the Lutheran movement spread rapidly through Prussia and even into Poland. The former Grand Master of the Teutonic Knights who, when the Order was at last dissolved in 1525, assumed the title of Duke of Prussia, embraced the new faith; the King of Poland, on the other side, remained faithful to Rome and forcibly quelled a Lutheran rising in Danzig. Thus little Ermland became once again a no-man's-land between two hostile camps. Bishop Fabian von Lossainen, the successor of Uncle Lucas, could still observe an attitude of benevolent neutrality toward Luther, whom he called 'a learned monk who has his own *opiniones* regarding the Scriptures; he must be a daring man who will stand up in disputation against him'. But his successor, Bishop Mauritius Ferber, no sooner installed, started a determined fight against Lutheranism; his first edict, issued in 1524, threatened that all who listened to the schismatics 'will be cursed for eternity and smitten with the sword of anathema'. In the same week in which this edict was issued in Ermland, the Bishop of the neighbouring diocese of Samland also published an edict in which he admonished his clergy to read Luther's writings diligently and, following Lutheran practice, to preach and baptize in the language of the common people.

Two years later Canon Giese published a little book.[30] Its ostensible purpose was the refutation of a tract by that Lutheran fellow-traveller next door, the Bishop of Samland; in fact, it was a plea for tolerance and reconciliation, written entirely in the Erasmian vein. In the preface, Canon Giese said bluntly, 'I reject the battle'; and he ended the book with the plea:

> Oh, if only the Christian spirit informed the Lutheran attitude to the Romans, and the Romans' toward the Lutherans – verily, then our Churches would be spared these tragedies of which no end can be seen. ... Verily, the wild beasts deal more kindly with each other than Christian deals with Christian.

Now, at the beginning of his book Giese, in a rather deliberate manner, brings in Copernicus' name. The curious passage is contained in a prefatory letter by Giese to another Canon, one Felix Reich. Giese begs Reich not to let personal affection interfere with his critical judgement 'as, I believe, was the case with Nicolao Copphernico [*sic*] who advised me to have this my writing printed, though otherwise he is of discerning taste'. No doubt Canon Giese had obtained his friend's consent to this mention of his name, as a way of indicating that Copernicus endorsed his views. No doubt Giese and Copernicus – and the rest of the Chapter – had endlessly discussed the great schism and their attitude to it; it is probable, in view of the intimate friendship between the two men, and the passage in the preface, that Copernicus had directly or indirectly collaborated on Giese's book. Its contents were so irreproachable that Giese eventually became a bishop. However, there were a few passages in it – such as the opening 'I reject the battle', and certain admissions regarding corruption in the clergy – which, in the view of an over-cautious mind, might incur the disfavour of one's superiors. The tortuous reference in the preface was probably a compromise formula arrived at after long discussions between the gently persuasive Giese and his anxiety-ridden friend.[31]

But, though Canon Giese prevailed in extracting from Canon Koppernigk an indirect public statement of his religious views, he did not succeed for another fifteen years in persuading him to publish his views on astronomy. And when the first version of the Copernican system appeared in print it was, as a climax of Copernican obliqueness, not written or signed by him, but by a disciple, Joachim Rheticus.

## 7. *The* Commentariolus

The first intimation of the Copernican system was contained in the short treatise which Canon Nicolas wrote at Heilsberg Castle, or at the beginning of his stay in Frauenburg.[32] It was, as I mentioned before, circulated in manuscript only, and bore the title

A brief outline of Nicolai Copernicus'
hypotheses on the heavenly motions.[33]

The treatise begins with a historical production, in which Copernicus explains that the Ptolemaic system of the universe was unsatisfactory, because it did not fulfil the basic demand of the ancients that each planet should move with uniform speed in a perfect circle. Ptolemy's planets move in circles, but not with uniform velocity.[34] 'Having become aware of these defects, I often considered whether there could perhaps be found a more reasonable arrangement of circles ... in which everything would move uniformly about its proper centre, as the rule of absolute motion requires.' Copernicus then claims that he has constructed a system which solves 'this very difficult and almost insoluble problem' in a manner much simpler than Ptolemy's. provided that certain basic assumptions or axioms, seven in number, are granted to him. He then sets down, without further ado, his seven revolutionary axioms, which, translated into modern language, are:

1. That the heavenly bodies do not all move round the same centre;
2. That the earth is not the centre of the universe, only of the moon's orbit and of terrestrial gravity;
3. That the sun is the centre of the planetary system and therefore of the universe;
4. That, compared to the distance of the fixed stars, the earth's distance from the sun is negligibly small;
5. That the apparent daily revolution of the firmament is due to the earth's rotation on its own axis;
6. That the apparent annual motion of the sun is due to the fact that the earth, like the other planets, revolves around the sun, and

7. That the apparent 'stations and retrogressions' of the planets are due to the same cause.

Then, in seven short chapters, the new circles and epicycles of sun, moon, and the planets are described in crude outline, but without proof or mathematical demonstrations 'reserving these for my larger work'. The last paragraph of the treatise proudly announces:

> Then Mercury runs on seven circles in all; Venus on five; the earth on three; and round it the moon on four; finally, Mars, Jupiter, and Saturn on five each. Altogether, therefore, thirty-four circles suffice to explain the entire structure of the universe and the entire ballet of the planets.

I shall discuss the scientific relevance of the *Commentariolus* in the next chapter; at present we are only concerned with its repercussions. The names of the scholars to whom Canon Koppernigk sent his manuscript are unknown, and so is their number; but its reception was disappointing, and its echo, at the beginning, practically nil. Nevertheless, the first pebble had fallen into the pond and gradually, in the course of the following years, the ripples spread by rumour and hearsay in the Republic of Letters. This led to the paradoxical result that Canon Koppernigk enjoyed a certain fame, or notoriety, among scholars for some thirty years without publishing anything in print, without teaching at a university or recruiting disciples. It is a unique case in the history of science. The Copernican system spread by evaporation or osmosis, as it were.

Thus in 1514, Canon Koppernigk was invited, among a number of other astronomers and mathematicians, to participate in a Lateran Council on the reform of the calendar. The invitation was sent by Canon Sculteti, the benefactor who had arranged the famous loan for the Koppernigk brothers, and who had meanwhile become domestic chaplain to Leo X. Copernicus refused to attend on the grounds that the calendar could not be satisfactorily reformed until the motions of the sun and moon were more precisely known; but he mentioned the fact of the invitation nearly thirty years later, in the dedication of the *Book of Revolutions*.

The next ripple on record is a request, in 1522, by the learned Canon Bernhard Wapowsky in Cracow, for Copernicus' expert opinion on Johann Werner's astronomical treatise, *On the Motion of the Eighth Sphere*. Copernicus complied.

Ten years later, the personal secretary of Pope Leo X gave a lecture on the Copernican system, to a select company in the Vatican gardens, which was favourably received.

Another three years later, Cardinal Schoenberg, who enjoyed the Pope's special confidence, urgently entreated Copernicus 'to communicate your discoveries to the learned world' by word of print.

Yet, in spite of these encouragements, Canon Koppernigk hesitated for another six years, before he had his book printed. Why?

## 8. *Rumour and Report*

News travelled fast and far in the sixteenth century. The pulse of all humanity was quickening as if our planet, after traversing, on its journey through space, some somnolent and bemused zone of the Universe, were now emerging into a region bathed in vivifying rays, or filled with cosmic benzedrine in the interstellar dust. It seemed to act simultaneously on all levels of the nervous system of mankind, on the higher as well as on the lower centres, as a stimulant and aphrodisiac, manifesting itself as a thirst of the spirit, an itch of the brain, a hunger of the senses, a toxic release of passions. The human glands seemed to produce a new hormone which caused the sudden surge of a novel greed: curiosity – the innocent, lecherous, creative, destructive, cannibalistic curiosity of the child.

The new machines – type foundry and printing press – ministered to this devouring curiosity by a flood of broadsheets, news letters, almanacs, *libellea*, pasquils, pamphlets, and books. They spread the news at a hitherto unknown speed, increased the range of human communication, broke down isolation. The broadsheets and brochures were not necessarily read by all the people on whom they exercised their influence;

rather, each printed word of information acted like a pebble dropped into a pond, spreading its ripples of rumour and hearsay. The printing press was only the ultimate source of the dissemination of knowledge and culture; the process itself was complex and indirect, a process of dilution and diffusion and distortion, which affected ever increasing numbers, including the backward and illiterate. Even three and four centuries later, the teachings of Marx and Darwin, the discoveries of Einstein and Freud, did not reach the vast majority of people in their original, printed text, but through second- and third-hand sources, through hearsay and echo. The revolutions of thought which shape the basic outlook of an age are not disseminated through text-books – they spread like epidemics, through contamination by invisible agents and innocent germ-carriers, by the most varied form of contact, or simply by breathing the common air.

There are slow-spreading epidemics, like polio, and others that strike swiftly, like the plague. The Darwinian revolution struck like lightning, the Marxian took three quarters of a century to hatch. The Copernican revolution, which so decisively affected the fate of man, spread in a slower and more devious manner than all. Not because the printing press was new, or the subject obscure: Luther's theses created an immediate all-European turmoil, though they were less easy to compress into a single slogan than: 'The Sun does not go round the Earth, but the Earth goes round the Sun.' The reason why it took Rome three-quarters of a century to ban Canon Koppernigk's book, and why the book itself had almost no impact upon his contemporaries, is of a different order.

What we call the Copernican revolution was not made by Canon Koppernigk. His book was not intended to cause a revolution. He knew that much of it was unsound, contrary to evidence, and its basic assumption unprovable. He only half believed in it, in the split-minded manner of the Middle Ages. Besides, he was denied the essential qualities of the prophet: awareness of a mission, originality of vision, the courage of conviction.

The relationship between Canon Koppernigk as a person,

and the event known as the Copernican revolution, is summed up in the dedication of his book to Pope Paul III. The relevant passage reads:

I may well presume, most Holy Father, that certain people, on learning that in this my book *On the Revolutions of the Heavenly Spheres* I ascribe certain movements to the Earth, will cry out that, holding such views, I should at once be hissed off the stage. . . . Therefore I have doubted for a long time whether I should publish these reflections written to prove the earth's motion, or whether it would be better to follow the example of the Pythagoreans and others, who were wont to impart their philosophic mysteries only to intimates and friends, and then not in writing but by word of mouth, as the letter of Lysis to Hipparchus witnesses. . . . In considering this matter, fear of the scorn which my new and [apparently] absurd opinion would bring upon me, almost persuaded me to abandon my project.

He then goes on to explain that only the constant and reproachful admonitions of his friends persuaded him in the end to publish his book which he had kept to himself and withheld from the public 'not for nine years but for nearly four times nine years'.

Copernicus' infatuation with the Pythagorean cult of secrecy started early, and stems from the very roots of his personality. The letter by Lysis, which he mentions in his dedication, plays a curious part in it. It was a recent, apocryphal concoction; young Nicolas Koppernigk had found it in the same collection of Greek epistolography, published in 1499, which contained the work of Simocatta.[35] He had bought the book as a student in Padua, and later translated the Lysis letter into Latin. It is apparently, apart from Simocatta, the only lengthy translation from the Greek that Copernicus ever made – although a printed Latin version of the letter was already in existence, and in Copernicus' possession. This was contained in a work by Cardinal Bessarion, also published by Aldus in Padua;[36] the Lysis letter is specially marked in Copernicus' copy (another marked passage is in praise of celibacy). It is worth while to quote a few passages from this forgery which made such a deep impression on Copernicus.

Lysis greets Hipparchus.

After the death of Pythagoras, I could not believe that the ties between his pupils would be broken. Though against all expectation we were, as if by a shipwreck, cut adrift and dispersed hither and yon, it remains our sacred duty to remember the divine teaching of our master and not to divulge the treasures of philosophy to those who have not undergone preliminary purification of the mind. For it is not proper to divulge to all and sundry what we have acquired with such great effort, just as it is not permitted to let ordinary men into the sacred mysteries of the Elysian goddesses. . . . Let us remember how long it took us to purify our minds of their stains until, after five years had run their course, we became receptive to his teaching. . . . Some of his imitators achieve many and great things, but in the improper way and not in the manner in which youth should be taught; thus their audience is encouraged to ruthlessness and insolence, for they stain the pure tenets of philosophy with rash and impure demeanour. It is as if one were to pour clean, fresh water into a well filled with dirt – for the dirt will only get agitated, and the water will be wasted. This is what happens to those who teach and are taught in this manner. Thick and dark forests cover the minds and hearts of those who have not been initiated in the proper manner, and disturb the mild contemplation of ideas. . . . Many tell me that you teach philosophy in public, which was forbiddden by Pythagoras. . . . If you mend your ways I shall love you, if not, you are dead in my eyes. . . .[37]

Why did Copernicus, after ten years spent in the bubble-bath of Renaissance Italy, adopt this arrogantly obscurantist and anti-humanistic attitude? Why did he hug that apocryphal letter for forty years so close to his heart, like a talisman, make a new translation of it and quote it to the Pope? How could a Renaissance philosopher, a contemporary of Erasmus and Reuchlin, Hutten, and Luther, approve of the preposterous notion that one should not pour the clear water of truth into the muddy wells of the human mind? Why was Copernicus so afraid of the Copernican Revolution?

The answer is given in the text: *because the pure water would be wasted and the dirt would only get agitated.* Here is the core of that anxiety which paralysed his work and crippled his life. The hocus-pocus about the Pythagorean mysteries was a rationalization of his fear of getting sprayed with dirt if he published his theory. It was quite enough to be an orphan at ten, with a leper for a brother and a sombre bully for a ward. Was it necessary to expose oneself to the scorn and ridicule of one's contemporaries, to the risk of being 'hissed off the stage'?

It was not, as legend would have it, religious persecution that he had to fear. Legend pays little attention to dates; yet it is essential to remember that the *Book of Revolutions* was not put on the Index until seventy-three years after it was published, and that the notorious trial of Galileo took place ninety years after Copernicus' death. By then, owing to the Counter-Reformation and the Thirty Years War, the intellectual climate of Europe had radically changed – almost as radically as between the mid-Victorian and the Hitler–Stalin era. Canon Koppernigk's youth and middle years were spent in the golden age of intellectual tolerance: the age of Leo X, patron of learning and the arts; at a time when the highest dignitaries of the Church freely indulged in liberal, sceptical. revolutionary philosophizing. Savonarola was burnt and Luther was excommunicated, but only after they had openly defied the Pope, and after all attempts to appease them had been exhausted. Scholars and philosophers had no reason to fear persecution for their opinions so long as they refrained from directly and explicitly challenging the authority of the Church. If they exercised a minimum of discretion in their choice of words, they could not only say pretty well what they liked, but were even encouraged to do so by ecclesiastic patronage; and this is what actually happened to Copernicus himself. The astonishing proof of this is a document, included by Copernicus in the prefatory matter of the *Book of Revolutions*, and preceding his dedication to the Pope. It is a letter which I have already mentioned, written to Copernicus by Cardinal Schoenberg, who occupied a position of special trust under three succeeding Popes – Leo X, Clement VII, and Paul III.

Nicolaus Schoenberg, Cardinal of Capua, sends his greetings to Nicolaus Copernicus.

When several years ago I heard your diligence unanimously praised, I began to feel an increasing fondness for you and to deem our compatriots lucky on account of your fame. I have been informed that you not only have an exhaustive knowledge of the teachings of the ancient mathematicians, but that you have also created a new theory of the Universe according to which the Earth moves and the Sun occupies the basic and hence central position; that the eighth sphere [of the fixed stars] remains in an immobile and eternally fixed position and the Moon, together with the elements included in its sphere, placed between the spheres of Mars and Venus, revolves annually around the Sun; moreover, that you have written a treatise on this entirely new theory of astronomy, and also computed the movements of the planets and set them out in tables, to the greatest admiration of all. Therefore, learned man, without wishing to be inopportune, I beg you most emphatically to communicate your discovery to the learned world, and to send me as soon as possible your theories about the Universe, together with the tables and whatever else you have pertaining to the subject. I have instructed Dietrich von Rheden [another Frauenburg Canon] to make a fair copy of this at my expense and to send it to me. If you will do me these favours, you will find that you are dealing with a man who has your interests at heart, and wishes to do full justice to your excellence. Farewell.

Rome, 1 November 1536.[38]

It should be noted that this 'most emphatic' (*atque etiam oro vehementer*) request that Copernicus should publish his theory is expressed independently from the Cardinal's demand for a fair copy – there is no question of any preliminary vetting or censorship.

Moreover, it seems unlikely that the Cardinal would have gone as far as he did in urging publication of the book entirely on his own initiative; and there is further evidence of early benevolent interest in the Copernican theory shown by the Vatican. This has come to light through one of the bizarre hazards of history. There exists, at the Royal Library in Munich, a Greek manuscript – a treatise by one Alexander Aphrodisius *On the Senses and Sensibilities* which is of no interest to anyone whatsoever, except that the title page contains the following inscription:

Clement VII, High Pontiff, made me a present of this manuscript, A.D. 1533, in Rome, after I had, in the presence of Fra Urbino,

Cardinal Joh. Salviato, Joh. Petro, Bishop of Iturbo, and Mattias Curtio, Physician, explained to him, in the Vatican gardens, Copernicus' teaching about the movement of the Earth. Joh. Albertus Widmanstadius.

Cognominatus Lucretius.

Private and Personal Secretary to our serene Lord.[39]

In other words, Clement VII, who had followed Leo X's example in his liberal patronge of the Arts, gave the Greek manuscript to his learned Secretary as a reward for his lecture on the Copernican system. It seems fairly plausible to assume that his successor, Paul III, heard about Copernicus through Schoenberg or Widmanstad, and, his curiosity awakened, encouraged the Cardinal to write to the astronomer. At any rate, Copernicus himself perfectly understood the importance of the letter, otherwise he would not have printed it in the *Book of Revolutions.*

In spite of this semi-official encouragement which ought to have given him complete reassurance, Copernicus, as we saw, hesitated for another six years before he published his book. The whole evidence indicates that it was not martyrdom he feared but ridicule – because he was torn by doubt regarding his system, and knew that he could neither prove it to the ignorant, nor defend it against criticism by the experts. Hence the flight into Pythagorean secretiveness, and the reluctant, piecemeal yielding of his system to the public.

Yet, in spite of all his caution, the slowly spreading ripples did stir up some of the mud which Canon Koppernigk held in such dread. Not much, merely a few splashes – more exactly, three splashes, held carefully in evidence by his biographers. There is, firstly, Luther's coarse but harmless after-dinner joke about 'that new astrologer who wants to prove that the Earth goes round',[40] made about ten years before the publication of the *Revolutions*; secondly, a single remark in a similar vein contained in a private letter by Melanchton,[41] dated 1541; lastly, in 1531 or thereabouts, a carnival farce was enacted in the Prussian city of Elbing, in which the star-gazing Canon was included in a grotesque procession, ridiculing monks, prelates, and dignitaries, according to the custom of the time.

This is all the persecution which Canon Koppernigk had to endure in his lifetime – an after-dinner remark, a passage in a private letter, and a carnival joke. Yet even these harmless squirts from the dreaded bottom of the well were sufficient, notwithstanding all private and official encouragement, to keep his lips sealed. Until the one great dramatic turn in his life – the bursting on to the scene of George Joachim Rheticus.

### 9. *The Arrival of Rheticus*

Rheticus, like Giordano Bruno or Theophrastus Bombastus Paracelsus, was one of the knight errants of the Renaissance whose enthusiasm fanned borrowed sparks into flame; carrying their torches from one country to another, they acted as welcome incendiaries to the Republic of Letters. He was twenty-five when he arrived in Frauenburg 'at the extreme outskirts of the Earth', with a determined purpose to get the Copernican Revolution going which Copernicus tried to suppress; an *enfant terrible* and inspired fool, a *condottiere* of science, an adoring disciple and, fortunately, either homo- or bi-sexual, after the fashion of the time. I say 'fortunately' because the so afflicted have always proved to be the most devoted teachers and disciples, from Socrates to this day, and History owes them a debt. He was also a Protestant, a protégé of Melanchton, the *Preceptor Germaniae*, and held the most adventurous job a man could hold in the sixteenth century: that of a Professor of Mathematics and Astronomy.

Born in 1514 as Georg Joachim von Lauchen in the Austrian Tyrol, the ancient Rhaetia, he had latinized his name into Rheticus. As a child he had travelled with his wealthy parents in Italy; as a young man he had studied at the Universities of Zurich, Wittenberg, Nuremberg, and Goettingen. At the age of twenty-two, on Melanchton's recommendation, he was given one of the two professorships of Mathematics and Astronomy at the equally young University of Wittenberg, centre and glory of Protestant learning. The other Chair was held by a man only three years his senior, Erasmus Reinhold.

The two young Professors, Reinhold and Rheticus, were both converts to the sun-centred cosmology which they only knew by hearsay, and to which the great manitous of Wittenberg, Luther, and Melanchton, were opposed. Nevertheless, in the spring of 1539, Rheticus was granted leave of absence for the express purpose of visiting, in Catholic Ermland, Canon Koppernigk, whom Luther had called 'a fool who went against Holy Writ'.

Rheticus arrived in Frauenburg in the summer of 1539. He came laden with precious gifts: the first printed editions of Euclid and Ptolemy in the original Greek, and other mathematical books. He had planned to stay in Ermland for a few weeks – he stayed, with interruptions, for two years which left their mark on human history. His arrival in Ermland was beautifully timed: it almost coincided with an edict by the new Bishop, Dantiscus, in which all Lutherans were ordered to leave Ermland within a month, and threatened with forfeit of life and possessions if they returned. The edict was issued in March; three months later the Lutheran professor, arriving straight from the Capital of Heresy, paid his respects to the Chapter of Frauenberg, including Bishop Dantiscus, whom he described as 'famous for his wisdom and eloquence'. It all goes to show that Renaissance scholars were a species of sacred cow allowed to amble, ruminating and unmolested, through the turmoil of the bazaar.

A year later, Bishop Dantiscus issued a second, even more ferocious 'Edict Against Lutheranism' in which he ordered that 'All books, pamphlets ... and whatever else came from the poisoned places of heresy should be burned in the presence of officials'. At about the same time, the Professor who came from the most poisonous of all places of heresy wrote *In Praise of Prussia*:

So may the gods love me, ... it has not yet happened to me that I should enter the home of any distinguished man in this region – for the Prussians are a most hospitable people – without immediately seeing geometrical diagrams at the very threshhold or finding geometry present in their minds. Hence nearly all of them, being men of good will, bestow upon the students of these arts every possible benefit and service, since true knowledge and learning are never separated from goodness and kindness.[42]

It is a pity that Rheticus did not report, in his exuberant style, his first meeting with Canon Koppernigk. It was one of the great encounters of history, and ranks with the meetings of Aristotle and Alexander, Cortēs and Montezuma, Kepler and Tycho, Marx and Engels. On the part of the overstrung and expectant Rheticus, it was obviously love at first sight for the *Domine Praeceptor*, 'My Teacher', as he was always to call Copernicus, comparing him to Atlas who carries the earth on his back. On his side, the lonely and unloved old man was apparently swept off his feet by this onslaught, and prepared to tolerate the young fool. He was now sixty-six, and he felt his days drawing to a close. He had achieved a certain fame in the world of learning, but it was of the wrong kind – notoriety rather than repute, based on hearsay, not on evidence; for the manuscript of the *Revolutions* was still locked up in his tower and nobody knew what exactly its contents were. Only the *Commentariolus* was known to the handful who had seen it, and of whom few survived – for even that sketchy outline had been written and circulated a quarter-century before.

The old Canon felt that what he really needed was a young disciple in the Pythagorean tradition, who would hand the teaching down to the select few without stirring up the dirt at the bottom of the well. His only friend, the gentle Giese, no longer lived in Frauenburg; he had been made Bishop of the neighbouring Prussian diocese, Kulm. Besides, Giese too was by now nearly sixty, and merely an amateur astronomer who did not qualify as a disciple. The young, enthusiastic Professor from Goettingen did. It seemed that Providence herself had sent him – even if it was a Lutheran Providence. From the Catholic side there was not much to fear, as Schoenberg's letter proved: young Rheticus, on the other hand, was a protégé of Melanchton's; he would secure the Lutheran flank and carry the message straight into their headquarters, into Wittenberg and Goettingen.

Nevertheless, Copernicus hesitated. He could decide nothing without Giese. Besides, the presence of his Protestant guest in Frauenburg was an embarrassment, even if the guest was a sacred cow. A few weeks after Rheticus' arrival Canon Koppernigk packed him off, and they both went to stay with Bishop Giese at his residence in Loebau Castle.

For some time, master and disciple were the Bishop's guests. The cosmological triumvirate in the medieval castle must have argued endlessly through the milky nights of the Baltic summer about the launching of the Copernican system: Rheticus and Giese pressing for publication, the old Canon maintaining his stubborn opposition, yet forced to yield, step by step. Rheticus describes a few phases of the struggle with a kind of embarrassed restraint, oddly in contrast with his usual flamboyance. He quotes long passages of dialogue between his *domine praeceptor* and Bishop Giese, passing with modest silence over his own participation in the debate:

Since my Teacher was social by nature and saw that the scientific world also stood in need of improvement ... he readily yielded to the entreaties of his friend, the reverend Prelate. He promised that he would draw up astronomical tables with new rules, and that if this work had any value he would not keep it from the world. ... But he had long been aware that [the theory on which the tables were based] would overturn the ideas concerning the order of the motions and spheres ... that were commonly accepted and believed to be true; moreover, the required hypotheses would contradict our senses.

He therefore decided that he should ... compose tables with accurate rules but no proofs. In that way, he would provoke no dispute among philosophers ... and the Pythagorean principle would be observed that philosophy must be pursued in such a way that its inner secrets are reserved for learned men, trained in mathematics, etc.

Then His Reverence pointed out that such a work would be an incomplete gift to the world unless my Teacher set forth the reasons for his tables and also included, following the example of Ptolemy, the system or theory and the foundations and proofs upon which he relied. ... There was no place in science, he asserted, for the practice frequently adopted in kingdoms, conferences, and public affairs, where for a time plans are kept secret until the subjects see the fruitful results. ... As for the uneducated, whom the Greeks call 'those who do not know theory, music, philosophy, and geometry', their shouting should be ignored. ...[43]

In other words, the wily Canon, hard pressed by Rheticus and Giese, proposed to publish his planetary tables but to withhold the theory on which they were based; the motion of the earth was not to be mentioned.

This manoeuvre of evasion having failed, the struggle in the triumvirate was resumed. The next stage in an astonishing compromise, a triumph of Copernican obliqueness. Judging by the results, the terms of the agreement must have been as follows:

Copernicus' *Book of Revolutions* was not to be printed. But Rheticus was to write an account of the contents of the unpublished manuscript, and publish this account – on condition that he nowhere mentioned Copernicus by name. Rheticus was to call the author of the unpublished manuscript simply *domine praeceptor*; and on the title page, where mentioning some name could not be avoided, he was to refer to Copernicus as 'the learned Dr Nicolas of Torun'.[44]

In other words, Rheticus was to stick out his neck; and the Canon was to retract his into his tortoise shell.

## 10. Narratio Prima

Thus came into being Rheticus' *Narratio prima – The First Account* of the Copernican theory in printed form. It was written in the guise of a letter from Rheticus to his former teacher in astronomy and mathematics, Johannes Schoener in Nuremberg. It has seventy-six pages in small quarto, and bears the following cumbersome title:

> To the most illustrious Dr Johannes Schoener, a *First Account* of the Book of Revolutions by the most learned and most excellent mathematician, the Reverend Father, Dr Nicolas of Torun, Canon of Ermland, from a young student of mathematics.

Rheticus' own name is only mentioned in the caption leading into the text of the letter: 'To the illustrious Johannes Schoener, as to his own revered Father, Georg Joachim Rheticus sends his greetings.'

After an apology for the delay in sending his report, Rheticus explains that so far he had only had ten weeks to study the manuscript of his Teacher; the manuscript embraces the whole realm of astronomy and is divided into six books, of which so far he has mastered three, understood the general idea of the fourth, but has only gained a sketchy notion of the last two. He then gives a skilful account of the Copernican system, showing his grasp of the subject and independence of mind by disregarding the sequence of chapters in Copernicus' manuscript and substituting for it a resumé of its essential contents. In between, Rheticus inserted an astrological digression in which the rise and fall of the Roman and Moslem Empires, and the second coming of Christ, are made directly dependent on changes in the eccentricity of the Earth's orbit. He also gave his estimate of the total duration of the world as six thousand years, in conformity with a prophecy by Eliah.

Copernicus himself did not seem to have believed in astrology, but Rheticus did, Melanchton and Schoener did, and so did most of the scholars of the age; and since the digression about Eliah and the second coming was calculated to please them, Copernicus apparently raised no objection.

Interspersed in Rheticus' account are the usual quotations from Aristotle and Plato, eulogies on the divine wisdom of the ancients, and protestations that his Teacher never meant to go against their authority:

> If I have said anything with youthful enthusiasm (we young men are always endowed, as he says, with high rank rather than useful spirits), or if I have inadvertently let fall any remark which may seem directed against venerable and sacred antiquity, more boldly perhaps than the importance and dignity of the subject demanded, you surely, I have no doubt, will put a kind construction on the matter and will bear in my mind my feeling towards you rather than my fault. As for my learned Teacher, I should like you to know and be fully convinced that for him there is nothing better or more important than to walk in the footsteps of Ptolemy and to follow, as Ptolemy did, the ancients and those who were much earlier than himself. However, when the phenomena, which control the astronomer ... compelled him to make certain assumptions even against his wishes, it was enough, he thought, if he aimed his arrows

by the same method to the same target as Ptolemy, even though he employed a bow and arrows of far different type of material from Ptolemy's.[45]

But then Rheticus continues with a delightful *non sequitur*: 'At this point we shall recall the saying: "Free in mind must he be who desires to have understanding." '

The treatise is full of pious protestations that his Teacher 'is far from thinking that he should rashly depart, in a lust for novelty, from the sound opinions of the ancients', followed by '... except for good reasons and when the facts themselves force him to do so'.[46] These apologies were probably intended to reassure Copernicus rather than Melanchton and Luther, who, too shrewd to be fooled, persisted in their opposition to the Copernican theory, yet kept its young prophet in their favour.

For, within a few weeks, the disciple had indeed grown into a prophet; the most moving passages in the *narratio prima* which crop up unexpectedly in the scientific text, sound like sermons to an as yet non-existent congregation:

> Thus the astronomy of my Teacher may rightly be called eternal as the observations of past ages testify and the observations of posterity will doubtless confirm. ...[47] A boundless Kingdom in astronomy has God granted to my learned Teacher. May he rule, guard, and increase it, to the restoration of astronomic truth. Amen.[48]

Rheticus had arrived in Frauenburg in the summer of 1539; by the end of September the *narratio prima* was completed and dispatched; a few months later it appeared in print. Rarely have ten weeks been better spent. In that span of time he had worked through the bulky manuscript of the *Revolutions* bristling with astronomical tables, rows of figures, involved diagrams, and a host of computing errors. He had distilled its essence, put it into writing, and in the evenings, supported by Giese, had carried on the interminable negotiations with the obstinate old man who always thought of new evasions. The combined effect of strain and frustration seems to have been too much even for the irascible young prophet, for it is reported that at a certain point – while he was struggling with the particularly intricate theory of the orbit of Mars – his mind became temporarily unhinged. Two generations later, when the events at Loebau Castle were already becoming a kind of Homeric saga among scholars, Johannes Kepler wrote in the Dedication of his *New Astronomy* to the Emperor Rudolph:

Concerning Georg Joachim Rheticus, the well-famed disciple of Copernicus in our forefathers' days ... the following story is told: when on one occasion he became perplexed and got stuck in the theory of Mars and could no longer see his way out, he appealed as a last resort to his guardian angel as an Oracle. The ungracious spirit thereupon seized Rheticus by the hair and alternately banged his head against the ceiling, then let his body down and crashed it against the floor; to which treatment he added the following oracular pronouncement: 'These are the motions of Mars.' Rumour has an evil tongue ... yet one can very well believe that Rheticus, his mind deranged by deadlocked speculation, rose in a rage and himself crashed his head against the wall.[49]

The episode must have been well known in Kepler's and Galileo's day, as the following passage in one of Kepler's letters to a colleague further shows:[50]

You tease me with the example of Rheticus. I laugh with you. I have seen how miserably the Moon has tortured you and sometimes me too, I remember. If now things are going badly with my Mars, it would be fitting for you, who suffered similar vexations, to show pity for me.

Rheticus himself described in the *narratio prima* his mental torment – the torment of a scientist at the junction of the Middle Ages and the Renaissance, who intuitively feels that there must be a beautiful and luminous solution to the cosmic mystery, yet cannot escape the nightmare of the whirling epicycles:

The astronomer who studies the motion of the stars is surely like a blind man who, with only a staff [mathematics] to guide him, must make a great, endless, hazardous journey that winds through innumerable desolate places. What will be the result? Proceeding anxiously for a while and groping his way with his staff, he will at some time, leaning upon it, cry out in despair to Heaven, Earth and all the Gods to aid him in his misery.[51]

As an annex to the *narratio*, Rheticus wrote, according to a usage of the time, a eulogy on the country and people who had received him so hospitably: *Encomium Borussiae*. 'In Praise of Prussia' is a gushing effusion in the worst purple style of the humanists, teeming with Greek gods and far-fetched allegories. It starts with a flourish:

Pindar celebrates in an ode – which was reportedly written in golden letters on a tablet and exhibited in the Temple of Minerva – the prowess of Dyagoras of Rhodes who won the boxing competition at the Olympic Games. The ode calls the island of Rhodes a daughter of Venus, and the beloved wife of the Sun. Jupiter, it says, let much gold rain on Rhodes because its people worshipped his daughter Minerva. For the same reason Minerva herself made the Rhodeans famous for their wisdom and education to which they were devoted. I am not aware of any country in our days more suited to inherit the ancient fame of the Rhodeans than Prussia"

– and so on.[52] The concoction is of interest only because of its description of Giese's struggles with Copernicus, and because of its revealing omissions. It includes a eulogy on Giese in which the Apostle Paul is invoked, and another eulogy on the Mayor of Danzig, who is compared to Achilles; also a description of Giese's astronomical instruments: an armillary sphere made of bronze, and 'a truly princely *gnomon* [sundial] which he had brought from England and which I contemplated with the greatest delight'.[53] But there is no mention of Copernicus' instruments. Nor of his observatory; nor where or how he lives; nor what he is like.

To appreciate the paradox of this silence, it must be borne in mind that the book represents Rheticus' account of his pilgrimage to Copernicus, in a letter addressed to his former teacher in Nuremberg. One can hear the addressee's indignant exclamations: 'But where does he live, this new master of yours? How old is he? What is he like? What instruments does he use? You say that this Bishop has a *gnomon* and an armillary sphere – but what has *he* got?' The reason for these glaring omissions was probably the same which compelled Rheticus to omit mentioning 'my learned Master' by name: Copernicus' mania for secrecy. It cannot be explained by sensible caution, for if anybody wished to persecute the anonymous astronomer of Ermland, he would have had no difficulty in identifying Canon Nicolai of Torun.

## 11. *Preparations for the Printing*

Rheticus wrote the *narratio prima* under the watchful eyes of Copernicus. From Loebau Castle, master and disciple had returned to Frauenburg, from where the *narratio* is dated: 23 September A.D. 1539. When the manuscript was finished, Rheticus went off to Danzig, where the nearest printing press could be found, to have it published.

The first copies of the first printed account of the Copernican system were dispatched from Danzig in February 1540. Melanchton received one of them; another copy was sent, by Giese, to the Protestant Duke Albert of Prussia, who subsequently did much to help promoting the Copernican system. Rheticus also sent a copy to a scholarly friend of his, by name of Achilles Perminius Gassarus, who at once caught fire and arranged for an independent edition of the book to be printed in Basle, only a few weeks after the Danzig edition had come out. Thus the *narratio prima* made its inroads simultaneously from the North and the South, and it caused a certain stir in the learned world. The gentle Giese was no longer alone in pleading with his obstinate friend; Canon Koppernigk was urged from every side to publish his book.

He held out for another six months. He probably thought of more subterfuges and evasions. Yet, once he had permitted that a resumé of his manuscript be published by another hand, his continued refusal to have the manuscript itself printed would have exposed him to an even greater risk of ridicule than publication could entail.

As soon as the printing of the *narratio* was completed,

Rheticus had hurried back from Danzig to Wittenberg, to resume his lectures at the University. When the summer term ended, he rushed again to Frauenburg, at the opposite end of Germany, ostensibly for the purpose of adding a 'Second Account' to the 'First'. In fact, he was preparing the final onslaught on Copernicus, which would wrest the *Revolutions* from his trembling hands. This time he succeeded. Some time after Rheticus' second arrival at Frauenburg, Canon Koppernigk's resistance at long last collapsed.

Rheticus stayed with him from the summer of 1540 to September 1541. He spent this time copying out, in his own hand, the entire manuscript of the *Revolutions*, checking and correcting dubious figures, and making various minor alterations.[54] He also did other chores for his master. More than ten years earlier, the previous Bishop of Ermland had asked Canons Koppernigk and Sculteti to draw up a map of Prussia.[55] Copernicus had started on the task, but never finished it. Rheticus did it for him; and since he was an incorrigible enthusiast, he not only drew up a map, but added to it a gazetteer and a treatise on the art of map-making. He sent these to Duke Albert of Prussia, accompanied by a letter of dedication, in which he took pains to bring in a reference to the forthcoming publication of his Teacher's *magnum opus.*

Rheticus also made for the Duke 'a little instrument' – *ein Instrumentlein* – 'indicating the length of the day throughout the year'. The Duke thanked him warmly, sent him a Portugal ducat as a gift, but later complained that he could not make head or tail of the *Instrumentlein*, and added that 'in my opinion the master goldsmith who made it did not show much subtlety'. He asked Rheticus to give his, the Duke's love to Luther, Melanchton, and all other German Protestants in Wittenberg. Throughout these amiable transactions, Rheticus stubbornly pursued one aim: to enlist the Duke's support for the publication of the *Revolutions.* A few days after sending off the map and the *Instrumentlein*, he let the cat out of the bag: he asked the Duke for letters to the Protestant Elector of Saxony and to the University of Wittenberg, recommending that Rheticus should be permitted to put the book of Canon

Koppernigk into print. The reason for this request was that Rheticus wanted the *Revolutions* to be printed in the famous printing shop of Petreius, who specialized in works of astronomy, in Lutheran Nuremberg. Since Luther and Melanchton were opposed to the Copernican theory, and since the Duke of Prussia carried much weight in the Protestant world, it might be just as well to have his support in writing. The Duke willingly complied; but owing to some muddle in the ducal Chancellery, the two identical letters to Johan Friedrich of Saxony, and to the University of Wittenberg, recommended that Rheticus should be given permission and help to print *his own* 'admirable book on astronomy'. Perhaps the scribe in the Chancellery thought he had misunderstood his instructions, for no astronomer could be crazy enough to want to publish another astronomer's book. However, the mistake was explained, and the letters had their effect.

In August 1541, some fifteen months after Rheticus' return to Frauenburg, the copying of the 424 pages in small handwriting was completed; with the priceless text in his bag, the faithful disciple once more rode post-haste across Germany back to Wittenberg, to arrive in time for the beginning of the winter term. He would have preferred to go straight to Nuremberg and to start with the printing, which could not be done without his personal supervision. But he had been absent long enough from his duties; moreover, no sooner was he back, he was elected Dean of his faculty – another proof of the large-mindedness of an age which, alas, was now approaching its end.

To fill in the time of waiting, he had two chapters of the *Revolutions* separately printed in Wittenberg.[56] They were chapters dealing with trigonometry in general, and with no direct bearing on the Copernican theory: but Rheticus probably thought that the publication of this small treatise might help to draw attention to his Teacher and pave the way for the *magnum opus*. In the dedication, he congratulated the sixteenth century on the privilege of having Copernicus among the living.

In the spring he was, at last, free. On 2 May 1542, Rheticus set out for Nuremberg, equipped with several letters of recommendation by Melanchton to the leading patricians and Protestant clerics of that town.

A few days later, Petreius the printer started the setting up of the Book *On the Revolutions of the Heavenly Orbs.*

## 12. *The Scandal of the Preface*

The printing made quick progress. On 29 June, less than two months after Rheticus' arrival in Nuremberg, a certain T. Forsther, citizen of Nuremberg, wrote to his friend J. Schrad in Reutlingen:

> Prussia has given us a new and marvellous astronomer, whose system is already being printed here, a work of approximately a hundred sheets' length, in which he asserts and proves that the Earth is moving and the stars are at rest. *A month ago I saw two sheets in print*; the printing being supervised by a certain *Magister* from Wittenberg [Rheticus].[57]

I have italicized the above words because they provide a clue to what became perhaps the greatest scandal in the history of science. If the printed sheets were circulated to interested persons like Herr Forsther, as soon as they came off the press, then we may reasonably assume that they were also sent on to the author; and that Copernicus was thus able to follow the progress of the printing. If this hypothesis (which is supported, as we shall see, by the testimony of Rheticus) were admitted, then it would follow that Copernicus knew the Preface by another hand which was added to his book, and which is the cause of the scandal.

It would never have arisen, had Rheticus been able to finish the job he had started with such enthusiasm and devotion. But, unfortunately, he had to leave Nuremberg before the printing was completed. In the spring, he had applied for a new post: the important Chair of Mathematics at Leipzig University. Melanchton again supported his application, and a private letter of Melanchton's to a friend darkly hints at the reason why Rheticus needed a change of universities: there were rumours (*fabulae*) current about him in Wittenberg 'which cannot be mentioned in writing'.[58] The rumours evidently concerned his homosexuality.

His application was successful, and in November Rheticus had to leave Nuremberg to take up his new post in Leipzig. He left the supervision of the printing of the *Revolutions* in the hands of a man whom he had every reason to consider reliable – the leading theologian and preacher of Nuremberg, Andreas Osiander, one of the co-founders of the Lutheran creed. In contrast to Luther and Melanchton, Osiander was not only favourably disposed towards Copernicus, but took an active interest in his work, and had been corresponding with him for the past two years.

In the belief that everything was thus arranged for the best, Rheticus left for Leipzig; whereupon Osiander, now in charge of the printing, promptly wrote an anonymous preface to the *Revolutions* and inserted it in the book. The preface was addressed TO THE READER, CONCERNING THE HYPOTHESES OF THIS WORK.[59] (Its full text is printed in Note 59.) It started by explaining that the ideas of the book need not be taken too seriously: 'For these hypotheses need not be true or even probable'; it is sufficient that they should save the appearances. The preface then went on to demonstrate the improbability 'of the hypotheses contained in this work' by pointing out that the orbit ascribed to Venus would make that planet appear sixteen times as large when closest to the earth as when farthest away – 'which is contradicted by the experiences of all ages'. The book furthermore, contained 'no less important absurdities, which there is no need to set forth at the moment'. On the other hand, these new hypotheses deserved to become known, 'together with the ancient hypotheses which are no more probable', because they are 'admirable and also simple, and bring with them a huge treasure of very skilful observations'. But by their very nature, 'so far as hypotheses are concerned, let no one expect anything certain from astronomy, which cannot furnish it, lest he accept as the truth ideas conceived for another purpose [i.e. as mere calculating aids], and depart from this study a greater fool than when he entered it. Farewell.'

No wonder that the emotional shock caused by reading this

preface (assuming that he *did* read it) is supposed to have hastened Copernicus' end. Yet there can be no doubt that Osiander was acting with the best of intentions. Two years earlier, when Copernicus was still hesitating whether to publish the book, he had written to Osiander to pour out his anxieties and to ask for advice.[60] Osiander had replied:

> For my part I have always felt about hypotheses that they are not articles of faith but bases of computation, so that even if they are false, it does not matter, provided that they exactly represent phenomena. . . . It would therefore be a good thing if you could say something on this subject in your preface, for you would thus placate the Aristotelians and the theologians whose contradictions you fear.[61]

On the same day, Osiander had written on the same lines to Rheticus, who was then in Frauenburg:

> The Aristotelians and theologians will easily be placated if they are told that several hypotheses can be used to explain the same apparent motions; and that the present hypotheses are not proposed because they are in reality true, but because they are the most convenient to calculate the apparent composite motions.

Prefatory remarks of this kind would induce in the opponents a more gentle and conciliatory mood; their antagonism will disappear 'and eventually they will go over to the opinion of the author'.[62]

Neither Copernicus' nor Rheticus' answer to Osiander's suggestion has been preserved. According to Kepler, who saw some of the correspondence before it was destroyed, Copernicus rejected Osiander's proposal: 'Strengthened by a stoical firmness of mind, Copernicus believed that he should publish his convictions openly.'[63] But Kepler did not quote the text of Copernicus' answer, and his remark, which occurs in a polemical text, should not be given undue weight.* Kepler fought fanatically for the heliocentric theory, worshipped Copernicus, and credited him with a 'stoical firmness' which he did not possess.

The wording of the preface was certainly most unfortunate.

* See below, p. 173.

For one thing, it did not make it sufficiently clear that it was not written by Copernicus himself. It is true that in one sentence it referred to the author of the book in the third person and in a laudatory manner; but the scholars of that age did not suffer from undue modesty, and it required close scrutiny of the text to discover that it was written by an alien hand. So much so, that although Osiander's authorship was discovered and revealed by Kepler in 1609, and mentioned in Gassendi's biography of 1647, the later editions of the *Revolutions* (Basle 1566, and Amsterdam 1617) took over Osiander's preface without comment, leaving the reader under the impression that it was by Copernicus. Only the Warsaw edition of 1854 mentioned Osiander's authorship.

The mystery of the preface, which survived three centuries, is of course, quite in keeping with Canon Koppernigk's oblique way, his cult of Pythagorean secrecy, and the esoteric motto of his book: *For mathematicians only*. Legend has it that Copernicus was the victim of a perfidious trick by Osiander; but the internal evidence, and also a statement by Rheticus, to which I shall come presently, speak against this. Since Osiander knew of Copernicus' hesitations to publish his manuscript for 'four times nine years'; [63a] of his insistence that in the *narratio prima* his authorship should remain anonymous; of his attempt to publish only his planetary tables without the theory behind them, he must have assumed that Copernicus would agree with his cautious and conciliatory approach, which was merely reiterating the classical doctrine that physics and sky-geometry were matters apart. We have no reason to doubt that Osiander acted in good faith, intending both to reassure the anxious Canon, and to smooth the path for his work.

The next question is, whether Copernicus actually read the preface, and what his reactions to it were. We have two contradictory statements on this point: one from Rheticus, one from Kepler. Kepler's text runs as follows:

It is a most absurd fiction, I admit, that the phenomena of nature can be explained by false causes. But this fiction is not in Copernicus. He thought that his hypotheses were true, no less than did those ancient astronomers of whom you speak. And he did not merely think so, but he proves that they are true. As evidence, I offer this work.

Do you wish to know the author of this fiction, which stirs you to such great wrath? Andreas Osiander is named in my copy, in the handwriting of Jerome Schreiber, of Nuremberg. Andreas, who supervised the printing of Copernicus' work, regarded the Preface, which you declare to be most absurd, as most prudent (as can be inferred from his letter to Copernicus) and placed it on the title page of the book when Copernicus was either already dead or certainly unaware [of what Osiander was doing].[64]

Rheticus' evidence is contained in a letter by the Professor of Mathematics, Johannes Praetorius, to a correspondent. Praetorius was an intimate friend of Rheticus and a reliable scholar. His letter says:

Concerning the Preface in Copernicus' book, there has been uncertainty about its author. However, it was Andreas Osiander ... who did the Preface. For it was under his charge that Copernicus' book was first printed in Nuremberg. And some of the first pages were sent to Copernicus, but a short while later Copernicus died, before he could see the whole work. Rheticus used to assert with seriousness that this Preface of Osiander's was clearly displeasing to Copernicus, and that he was more than a little irritated by it. This seems likely, for his own intention was different, and what he would have liked the Preface to say is clear from the contents of his Dedication [to Paul III]. ... The title also was changed from the original beyond the author's intentions, for it should have been: *De revolutionibus orbium mundi*, whereas Osiander made it: *Orbium coelestium*.[65]

Praetorius' letter was written in 1609. Kepler's *Astronomia Nova*, in which the quoted passage appears, was published in the same year. It was sixty-six years after the event. Which of the two opposite versions should we trust?

To solve the puzzle, we must compare (*a*) the contents, (*b*) the source, and (*c*) the motive behind each of the two statements. The content of Kepler's is vague: Copernicus was 'either already dead or certainly unaware of' Osiander's preface. It is based on hearsay: Kepler's source is his old teacher Michael

Maestlin, whose own knowledge of the events was third-hand.[66] Praetorius' statement is precise, the incidental detail about the changed title is convincing, and his information came straight from the horse's mouth, as it were: from Rheticus, whose guest he had been on two occasions, in 1569 and 1571.[67] As for motive, Kepler's statement on Copernicus' beliefs appears as a motto at the beginning of Kepler's *Astronomia Nova* (which is based on the Copernican hypothesis) and clearly serves a propagandistic purpose; [67a] whereas Praetorius' version is contained in a chatty letter, with no apparent motive at all.

The balance is thus clearly in favour of Praetorius, and the conclusion seems to be that, contrary to accepted opinion, Copernicus was acquainted with Osiander's Preface. Oddly enough, the Praetorius document, as far as I know, escaped the attention of all biographers, except the most recent and scholarly among them, the German astronomer Ernst Zinner. As I felt doubtful about my own conclusions, I wrote to Professor Zinner and received the following reply:

> I do not share your doubts. We can regard it as certain that Copernicus knew Osiander's preface for which he was prepared by Osiander's previous letters of 1540–41. Praetorius' statements are trustworthy for they were based on direct communications from Rheticus, who knew best. Praetorius . . . was a conscientious scholar and left us important information and works. At any rate, his testimony is more important than the vague testimony of Kepler, who derived his information from Maestlin, who in turn was too remote from the whole affair. . . . It is not self-evident that Rheticus, who had snatched the manuscript from Copernicus almost by force, as it were, did forward its galleys to its author? I imagine that all the galleys were sent to Copernicus in the course of time, so that at his death the whole book was assembled in print, as Giese states. . . .[68]

Canon Koppernigk had, of course, every reason to be furious about Osiander's unfortunate remarks that his Venus-orbit was 'contradicted by the experience of all ages', that the book contained other 'absurdities', and so forth. That indeed was carrying the diplomacy of appeasement too far. But on Osiander's more fundamental point that his system was merely a computing hypothesis, he had no grounds for complaint. Copernicus *did* believe that the earth really moved; but it was impossible for him to believe that either the earth or the planets moved *in the manner* described in his system of epicycles and deferents, which were geometrical fictions. And so long as the why and how of the heavenly motions rested on a purely fictional basis, with wheels-on-wheels which the astronomer manipulated with happy unconcern for physical reality, he could not object to Osiander's correct statement about the purely formal nature of his hypotheses.[69]

Whether Copernicus actually protested against the wording of the Preface we do not know; but it is hard to believe that Osiander would have refused to alter the wording in defiance of the author's wishes. Perhaps it was too late; the Preface was written around November 1542, and in that last winter of his life Canon Koppernigk was a very sick man. Perhaps considerations on the lines mentioned in the previous paragraph made him realize that he had really no cause for protest; more likely he procrastinated, as he had done all his life.[70]

There is a strangely consistent parallel between Copernicus' character, and the humble, devious manner in which the Copernican revolution entered through the back door of history, preceded by the apologetic remark: 'Please don't take seriously – it is all meant in fun, for mathematicians only, and highly improbable indeed.'

## 13. *The Betrayal of Rheticus*

There was a second, and more personal scandal aroused by the publication of the book; it concerned Rheticus.

The great moment in the disciple's life is the master's death. It is the moment when he reaches his full stature, and acquires a new dignity as the keeper of the tradition, the preserver of the legend. In this particular case the death of the master coincided, moreover, with the long-awaited publication of his book. One would have expected that Rheticus, the prime mover of this event, would now become more active than ever as a prophet and propagandist. What an opportunity to indulge in personal reminiscences and intimate detail, no longer restrained by the *domine praeceptor*'s mania for secrecy! During his last stay in Frauenburg, Rheticus had actually written a biography of the master, which was the more needed as virtually nothing was known in the learned world about Canon Koppernigk's person and career. Rheticus was the legitimate heir and executor of the Copernican doctrine – destined, so it seemed, to become to the departed what Plato had been to Socrates, Boswell to Dr Johnson, Max Brod to Kafka.

To the surprise of his contemporaries and the vexation of posterity, the moment Rheticus left Nuremberg and handed over the editorship to Osiander, he suddenly and completely lost interest in Copernicus and his teachings. His biography of Copernicus was never published, and its manuscript was lost. The same fate befell a pamphlet which he wrote to prove that the Copernican theory was not at variance with Holy Scripture. Professor Rheticus lived on for another thirty-odd years; but the Apostle Rheticus had died even before his Teacher. He had died, more precisely, at the age of twenty-eight, some time in the summer of 1542, while the *Book of Revolutions* was being printed.

What caused this sudden extinction of the flame? Again one can only guess, but there is a plausible guess at hand. Copernicus' own introduction to the book, in the form of a Dedication to Paul III, was written in June 1542,[71] and sent to Rheticus in Nuremberg, while he was still in charge of the printing. It was probably the text of this Dedication which killed the apostle in Rheticus. It explained how the book came to be written; how Copernicus hesitated to publish it, for fear of being ridiculed, and thought of abandoning the whole project. The Dedication then continued:

But my misgivings and protests were overcome by my friends. Foremost among them was Nicolaus Schoenberg, Cardinal of Capua, distinguished in every department of learning. Next was one who loved me well, Tiedemann Giese, Bishop of Kulm, a devoted student of sacred and all other good literature, who often urged and even importuned me to publish this work. ... The same request was made to me by many other eminent and learned men. ... Yielding then to their persuasion, I at last permitted my friends to publish that work which they have so long demanded....

Here the dedication trailed off to other matters. Rheticus' name was not mentioned in the Dedication – nor anywhere else in the book.

It must have been a nasty shock. The omission was so fantastic and preposterous that the gentle Giese wrote, after Copernicus' death, an embarrassed apology to Rheticus, referring to

the unpleasant oversight that thy teacher omitted to mention thee in the preface to his book. Truly this was not due to indifference towards thee but to his clumsiness and inattention; for his mind was already rather dulled, and paid, as thou knowest, scant attention to anything not pertaining to philosophy. I know very well how highly he esteemed thy constant helpfulness and self-sacrifice. ... Thou assistest like a Theseus his heavy labours. ... How much we all owe thee for thy relentless fervour is as clear as the day.[72]

But these well-meaning excuses carried no conviction, for Copernicus' Dedication to the Pope betrays neither 'clumsiness' nor 'dull-mindedness'. It is an extremely shrewd and calculated document. The deliberate omission of Rheticus' name can only be explained by the fear that the mention of a Protestant might create an unfortunate impression on Paul III. But if so, Copernicus could, of course, have mentioned Rheticus at some other place, either in the prefactory matter or anywhere in the text itself. To pass his name in complete silence was an act as abject as it was futile, since Copernicus' name was already publicly linked with Rheticus' by the *narratio prima*, and by the fact that the book was being printed in Protestant Nuremberg under Rheticus' editorship.

Copernicus' dedication must have reached Rheticus some time in June or July. On 15 August, a small booklet by Rheticus himself, containing two of his lectures on astronomy and physics,[73] was published by Petreius. In the preface to it Rheticus reminisced about his first acquaintance with the master:

When I heard about the great reputation of Dr Nicolai Copernicus in Northern Germany, I had just been appointed professor of these sciences at the University of Nuremberg, but I thought that I should not accept this post until I had acquired some additional knowledge from his teaching. No obstacle could prevent me from setting out on the journey, neither money, nor the itinerary, nor other molestations.* I placed a great value on seeing his work, for here was a man advanced in years who was driven by a youthful audacity to communicate his mature ideas on this science to the whole world. And all other scholars will judge as I did when the book which we now have in the press at Nuremberg is published.

How depressing that this last affirmation of the pupil's loyalty coincided in time with the master's betrayal of him.

## 14. *Bishop Dantiscus*

The previous sections were concerned with the prolonged labour pains and caesarian delivery of the *Revolutions*, which took place in Nuremberg. We must now return again to the Cathedral fortress of Frauenburg on the Baltic, to complete the story of Canon Koppernigk's last years.

They were even less happy than the earlier ones had been. In addition to the doubts and worries concerning the publication of the book, the Canon had become involved in an absurd conflict with his new Bishop. This Bishop, Johannes Dantiscus, weighed as heavily on the end of Canon Nicolas' life as Bishop Lucas had weighed on its beginning. In all other respects, the radiant Dantiscus was as complete a contrast to the sombre Lucas as could be invented.

He was one of the outstanding diplomats of the Renaissance, a poet laureate who composed erotic verse in his youth and religious hymns in his old age;† a traveller, humanist, conversational charmer, and altogether a character of great attractiveness and complexity. Bishop Lucas had been by twenty-six

* Probably an allusion to the fact that he risked the displeasure of Melanchton and Luther by visiting Copernicus, and that, on the other hand, he was travelling to a Catholic land whose Bishop had just published an edict against Lutheranism.

† The *Encyclopaedia Britannica* ranks his later work 'with the best Latin poetry of modern Europe'.[74]

years Nicolas' senior, Bishop Dantiscus was by twelve years his junior, yet Nicolas was as submissive to the latter as he had been to the former. This submission to authority – to Lucas and Dantiscus on the one hand, to Ptolemy and Aristotle on the other – is perhaps the main clue to Copernicus' personality. It undermined his independence of character and his independence of thought, kept him in self-imposed bondage, and singled him out as an austere relic of the Middle Ages among the humanists of the Renaissance.

Old age seems in some cases to repeat the pattern of youth, or rather to bring out again the pattern, which was blurred during the active years. If Dantiscus was a kind of revenant, stepping into the place of Uncle Lucas – was not Rheticus, the adventurer and firebrand, in some respects a reincarnation of brother Andreas? Andreas had been the black sheep of the family, Rheticus was a heretic; Andreas was a leper, Rheticus was a sodomite. Their recklessness and intrepidity both fascinated and frightened the timid Canon; and this ambivalent attitude may explain his betrayal of both.

Johannes Flachsbinder, destined to become the bane of Canon Koppernigk's old age, was the son of a brewer in Danzig, hence the name, Dantiscus. By the age of twenty, he had fought in campaigns against the Turks and the Tartars, had studied at the University in Cracow, travelled in Greece, Italy, Arabia, and the Holy Land. On his return, he became confidential secretary to the King of Poland, and at the age of twenty-three, the King's special envoy to various Prussian Diets. It was in that period that he first became acquainted with Canon Koppernigk, then secretary to Bishop Lucas on similar missions. But their orbits soon parted: Copernicus remained in Ermland for the rest of his life, whereas Dantiscus, during the next seventeen years, travelled all over Europe as Polish Ambassador to the Emperors Maximilian and Charles V. He was a favourite of both Emperors as well as of his own King; Maximilian appointed him poet laureate and made him a knight, Charles gave him a Spanish title, and both borrowed him occasionally for missions of their own – as Maximilian's special envoy to Venice, and Charles V's to

Francois I in Paris. Yet this son of a beer brewer from the outskirts of the civilized world, who succeeded in highly delicate diplomatic missions, was neither a snob nor even particularly ambitious. At the age of forty-five, at the height of his career, he retired, at his own request, to his provincial land of birth, and spent the rest of his life there – first as Bishop of Kulm, then of Ermland.

During his ambassadorial years, Dantiscus' main interests had been poetry, women, and the company of learned men, apparently in that order. His correspondence, of Erasmian dimensions, extended even to the new continent of America – he exchanged letters with Cortez in Mexico. His amorous relationships were equally cosmopolitan, ranging from his Tyrolian 'Grinea', in Innsbruck, to Ysope de Galda in Toledo, who bore him a beautiful daughter. His celebrated poem *ad Grineam* was a charming elegy on the splendours and decline of virility, but he was equally devoted to his Toledan paramour and their daughter, Dantisca; he sent them, after his return to Ermland, a regular allowance through the banking houses of the Fuggers and Welsers in Augsburg, and received a portrait of Dantisca through the good offices of the Emperor's Spanish Ambassador. He remained loyal to his former friends and mistresses even when he became a devout Catholic; and his warm friendship towards Melanchton, the Lutheran leader, remained equally unaffected by his conversion. In January 1533, when Dantiscus was already a Bishop in Kulm, Melanchton wrote to him, across the front-lines, as it were, that he would all his life remain in Dantiscus' debt; and he added that more than Dantiscus' brilliant gifts, he admired his profound humanity.[75] Another contemporary summed up the general opinion prevailing among Lutheran scholars on the Catholic Bishop of Kulm: *Dantiscum ipsam humanitatem esse* – he is humanity itself.[76] The subsequent conflict between Dantiscus and Copernicus must be judged against this background.

In 1532, then, Dantiscus was established in the bishopric of Kulm, about a day's journey on horseback from Frauenburg. Moreover, he had also been made a Canon of the Frauenburg Chapter, and thus a confrater to Canon Nicolas. One would have thought that the arrival of such an illustrious humanist, in the provincial backwoods hidden by the 'vapours of the Vistula', would become a joyous event in the lonely Copernicus' life. There was hardly a person in Ermland, let alone in Frauenburg, with whom he could talk science and astronomy, except for Giese, who in these matters was not a great light. Dantiscus, on the other hand, apart from his other attractions, was keenly interested in science, corresponded with several scholars (including the great mathematician Gemma Frisius), possessed several globes and astronomical instruments, a map of America, and even three timepieces, one of which he carried on a chain around his neck.

Immediately after settling down in Kulm, Dantiscus made overtures to Copernicus – which, for some unfathomable reason were primly rejected. Among the altogether sixteen private letters by Copernicus which are preserved, ten are addressed to Dantiscus.[77] They make depressing reading. The first is dated 11 April 1533, that is, a few months after Dantiscus had been installed in his Bishopric. The letter is a refusal, on the grounds of official business occupations, of Dantiscus' invitation to visit him at Loebau Castle.[78]

*Reverendissime in Christo Pater et Domine!*

I have received Your Most Reverend Lordship's letter and I understand well enough Your Lordship's grace and good will towards me; which he has condescended to extend not only to me, but to other men of great excellence. It is, I believe, certainly to be attributed not to my merits, but to the well-known goodness of Your Rev. Lordship. Would that some time I should be able to deserve these things. I certainly rejoice, more than can be said, to have found such a Lord and Patron. However, regarding Your Rev. Lordship's invitation to join him on the 20th of this month (and that I should most willingly do, having no little cause to visit so great a friend and patron), misfortune prevents me from doing so, as at that time certain business matters and necessary occasions compel both Master Felix and me to remain at this place. Therefore I request that Your Rev. Lordship take in good part my absence at that time. I am in other respects most ready, as it is fitting, to oblige Your Rev. Lordship, as I am in duty bound to do innumerable other things, in whatever way Your Rev. Lordship will indicate to me at another time what it is that he desires. To whom I now confess that I am bound not to grant his requests, but rather to obey his commands.

Since Dantiscus knew exactly the nature and amount of the 'official business' transacted at the Frauenburg Chapter, of which he himself was a member, the excuse was unconvincing. The second letter is dated three years later – 8 June 1536. It is again a refusal of an invitation from Dantiscus to attend the marriage celebrations of a female relative of the Bishop's. The Excuse is again 'official business': [79]

*Reverendissime in Christo Pater et Domine Domine Clementissme!*

I have Your Rev. Lordship's letter, full of humanity and favour, in which he reminds me of that familiarity and favour with Your Rev. Lordship which I contracted in my youth; [we remember that Copernicus is twelve years older than Dantiscus] which I know to have remained just as vigorous up to now. And since I am thus to be numbered among his intimates, he has deigned to invite me to the marriage of his kinswoman. Truly, Your Rev. Lordship, I ought to obey Your Lordship and present myself from time to time to so great a Lord and Patron. But now being in fact occupied with business, which the Most Reverend Lord of Ermland has imposed on me, I am unable to absent myself. Wherefore let him deign to take this my absence in good part, and to preserve that ancient opinion of me, though absent; since the union of souls is wont to count for more than that of bodies. Your Rev. Lordship, in all felicity, to whom I commend my humble duties, and to whom I wish perpetual good health.

The tone of these and all subsequent letters, compared to the correspondence between contemporary humanists, and particularly Dantiscus' own, is astonishing and pathetic. The man who removed the earth from the centre of the universe writes to the poet laureate and former Ambassador at large, in the style of an obsequious clerk, submissive yet sour, nagged by some obscure jealousy, or resentment, or mere inability to loosen up and enter into a human relationship.

The third letter, dated a year later, 9 August 1537, is in a different, but not in a brighter vein. It was written after the death of the Bishop of Ermland, Mauritius Ferber, when it was a foregone conclusion that Dantiscus would be elected his successor. It contains some indifferent political gossip which had reached Copernicus in letters from Breslau two full months earlier; among other items, a rumour about an armistice between the Emperor and Francois I, which happened to be unfounded. It is hard to see what reason Canon Koppernigk had for sending this out-dated, second-hand information on to Dantiscus who had correspondents at the four corners of the earth – except the reason that Dantiscus was on the point of becoming his immediate superior.

On 20 September 1537, the Canons of Frauenburg Chapter solemnly assembled in the Cathedral to elect their new Bishop. The privilege to nominate the candidates was held, according to the intricate ecclesiastical procedure in Ermland, by the Polish King, while the election itself was the privilege of the Chapter. In fact, however, the candidates on the royal list had been previously agreed upon between the Chapter and the Chancellery, with Dantiscus as go-between. The list comprised Dantiscus himself (on whose election all parties had previously agreed), and three other candidates. The others were Canons Zimmerman and von der Trank, who do not concern us, and Canon Heinrich Snellenburg.

Now this Canon Snellenburg had some twenty years earlier incurred a debt of a hundred Marks to Canon Koppernigk, and had only repaid ninety. Canon Kopernigk thereupon had written a dusty epistle (one of the sixteen precious extant letters) to the Bishop of that time, petitioning him to make Snellenburg pay up the ten Marks. The outcome of the affair we do not know; the years had gone by, and now the lazy debtor Snellenburg was nominated as a candidate for the Bishop's See. It was a purely formal nomination since Dantiscus was to be elected, yet it gave rise to a grotesque little episode. Teidemann Giese, the devoted, angelic Giese, wrote a letter to Dantiscus asking him to take Snellenburg off the list of candidates because he 'would expose the Chapter to ridicule', and to put Canon Koppernigk's name in his place. Dantiscus, who evidently could not care less, obliged. Copernicus had the satisfaction of being a candidate to the Bishopric, and Dantiscus was elected unanimously, including Copernicus' vote.

So now Bishop Dantiscus was installed at Heilsberg Castle, where Copernicus had spent six years of his life as secretary to Uncle Lucas. In the autumn of 1538, he made an official tour of the towns of his new Bishopric, accompanied by Canons Reich – and Koppernigk. This, says Prowe 'was the last friendly encounter between the former friends Dantiscus and Copernicus' [80] – though there is no evidence that they had ever been friends.

In the course of that official tour, or perhaps a little later, Dantiscus must have broached an embarrassing subject. It concerned a certain Anna Schillings, a distant relative of Canon Koppernigk's, and his *focaria*. According to Copernicus' biographers, '*focaria*' meant housekeeper. According to Baxter and Johnson's *Medieval Latin Word List*,[81] it meant 'housekeeper or concubine'. We know that one other Canon in Frauenburg, Alexander Sculteti,[82] also had a *focaria*, and several children by her. Now Dantiscus was anything but a prude, he kept sending money to his former mistress, and doted on portraits of his pretty daughter. But it was one thing to have amorous affairs in one's youth while travelling in distant countries, and another to live openly with a *focaria* in one's own diocese. Besides, not only the two men had aged, but their century too; the Counter Reformation was determined to restore clean living among the clergy, whose corruption had bred the Luthers and Savonarolas. Canon Koppernigk was sixty-three; it was time, both by the personal and historical clock, to say *vale* to his *focaria*.

However, it is not easy to change one's housekeeper and habits at sixty-three. Canon Koppernigk, understandably, hesitated and delayed, perhaps hoping that Dantiscus would forget about the matter. In November, Dantiscus reminded Copernicus of his promise. His letter is not preserved, but Copernicus' answer is:

*Reverendissime in Christo Pater et Domine Domine Clementissime mihique et omnibus observande!*

Your Most Reverend Lordship's warning is fatherly enough and more than fatherly, I admit; I have received it in my innermost heart. As for Your Rev. Lordship's earlier intimation on the same subject, it was far from me to forget it. I intended to act accordingly; though it was not easy to find a proper person of my own kin, I intended nevertheless to terminate the matter before Easter. However, as I do not wish Your Rev. Lordship to think that I am seizing on pretexts for procrastination, I have reduced the period to one month, that is to Christmas; it cannot be shorter, as Your Rev. Lordship will understand. I wish to do my best to avoid giving offence to good manners, much less to Your Rev. Lordship, who has deserved to be revered, honoured, and most of all to be loved by me; to which I devote myself with all my powers.

ex Gynopoli, 2 December 1538.

Your Rev. Lordship's most obedient Nicolas Copernicus.

Even the devoted Prowe remarks that the letter is 'repellent to read' and that 'making allowance for the devotional manners of the curial style . . . it remains humiliating enough'.[83]

Six weeks later, Copernicus wrote to Dantiscus a kind of *consummatum est*:

*Reverendissime in Christo Pater et Domine Domine Clementissime!*

I have done what I neither would nor could have left undone, whereby I hope to have given satisfaction to Your Rev. Lordship's warning. As for the information you required of me how long Your Rev. Lordship's predecessor, my uncle Lucas Waczelrodt of blessed memory had lived: he lived 64 years, 5 months; was Bishop for 23 years; died on the last but one day of March, *anno Christi* 1522. With him came to an end a family whose insignia can be found on the ancient monuments and many [public] works in Torun. I recommend my obedience to Your Rev. Lordship.

ex Frauenburg, 11 January of the year 1539.

Your Rev. Lordship's most devoted Nicolas Copernicus.

But the *focarias* were not so easy to get rid of. Sculteti's housekeeper, and mother of his children, 'threatened and heatedly promised to damage the Chapter's obedient servant and shamelessly used shameful words of abuse'.[84] As for

Copernicus' Anna, she seems to have flatly refused to leave Frauenburg, and was determined to make things as awkward as possible for everybody concerned. More than two months after Copernicus' last letter to Dantiscus, another Canon, Plotowski, wrote to the Bishop as follows:

> As regards the Frauenburg wenches, Alexander's hid for a few days in his house. She promised that she would go away together with her son. Alexander [Sculteti] returned from Loebau with a joyous mien; what news he brought I know not. He remains in his *curia* with Niederoff and with his *focaria,* who looks like a beer-waitress tainted with every evil. The woman of Dr Nicolas did send her things ahead to Danzig, but she herself stays on in Frauenburg. . . .[85]

A full six months later the matter was still not finished. Dantiscus apparently tired of sending paternal admonitions to Copernicus, and of getting dripping letters in return; so he privately asked Giese (now Bishop of Kulm) to use his influence with Copernicus to put an end to the old man's secret meetings with Anna, and to avoid further scandal.

On 12 September 1539, Giese answered as follows:

> . . . I have spoken earnestly to Dr Nicolas on the matter, according to Your Most Rev. Lordship's wish, and have set the facts of the matter before his eyes. He seemed not a little disturbed [to learn] that although he had unhesitatingly obeyed the will of Your Rev. Lordship, malicious people still bring trumped up charges of secret meetings, and so on. For he denies having seen that woman since he dismissed her, except that on a journey to the Market in Koenigsberg she spoke to him in passing. I have certainly ascertained that he is not as much affected as many think. Moreover, his advanced age and his never-ending studies readily convince me of this, as well as the worthiness and respectability of the man; nevertheless I urged him that he should shun even the appearance of evil and this I believe he will do. But again I think it would be as well that Your Rev. Lordship should not put too much faith in the informer, considering that envy attaches so easily to men of worth and is unafraid of troubling even Your Most Rev. Lordship. I commend myself, etc.[86]

Giese's last remark was in the nature of a friendly dig as from one Bishop to another. Though earlier on, they had been rivals for the see of Ermland, a compromise had been reached by giving Giese the Bishopric of Kulm, and they remained on excellent terms. This made it possible for Danticus to ask Giese on repeated occasions to intercede with Copernicus, in order to spare the old Canon further humiliation.

Concurrently with the unpleasantness concerning Anna, there was also political trouble in the Chapter. Its causes are extremely involved (for a brief outline, see note 87); but the central character was again the intrepid Canon Sculteti, who not only lived openly with his 'beer-waitress' and brood of children, but led the resistance against Dantiscus' efforts to make East Prussia safe for the Polish crown. It was a struggle for high political stakes, which, a year later, led to Sculteti's proscription and banishment. and several years later to the temporary excommunication of the majority of the Ermland Chapter. Since Canon Koppernigk had been on friendly terms with Sculteti, and was in the same boat with him in the *focaria* scandal, Dantiscus was anxious to keep the old man from getting involved in all this. On 4 July 1539, he wrote to Giese:

> I have been told that the Dr Nic. Copernicus whom as you know, I love as my own brother, is staying with you as your guest. He maintains close friendship with Sculteti. That is bad. Admonish him that such connexions and friendships are harmful to him, but do not tell him that the warning originates from me. I am sure you know that Sculteti has taken a wife and that he is suspected of atheism.[88]

It should be remembered that Dantiscus was Canon Koppernigk's immediate superior, and that Giese now ruled another diocese. The letter proves that Dantiscus went out of his way to save Copernicus embarrasment, to the extent of keeping his warning anonymous, since a direct admonishment from his ecclesiastic superior might be humiliating to the old Canon. Yet the Copernicus legend has it that Dantiscus 'ordered him abruptly to break off relations with his friend Sculteti'; and that he persecuted Copernicus to prevent him from finishing his book.[89]

The truth is that when, in 1541, Dantiscus learnt of Copernicus' decision to publish, at long last, his *Revolutions*, he immediately wrote a warm and very friendly letter to Copernicus, enclosing a poetic epigram to serve as a motto for the book. Canon Koppernigk wrote back:[90]

*Reverendissime in Christo Pater et Domine Domine Clementissime.*

I have received Your Most Rev. Lordship's most humane and entirely intimate letter, in which he has condescended to send an epigram addressed to the readers of my book, soberly elegant and suited, not to my deserts, but to the extraordinary benevolence with which Your Rev. Lordship is wont to honour scholars. I should place it, therefore, at the title page of my work, if only the work might be worthy of deserving to be so greatly adorned by Your Rev. Lordship, although nevertheless very learned men, with whom it is fitting to comply, declare that I am of some account. Indeed, I desire so far as I have powers to earn it, to gratify the extraordinary benevolence and paternal affection towards me whereby Your Rev. Lordship does not cease to honour me; and to serve him, as is my duty, in all things in which I am able.

Frauenberg, 27 June 1541.

Your Rev. Lordship's

Most obedient

NICOLAUS COPERNICUS.

This is the last extant letter from Copernicus to Dantiscus, and probably the last he actually wrote him. The poet laureate's contribution did not appear in the book, nor on Copernicus' manuscript, and is lost. After thanking Dantiscus for his 'extraordinary benevolence', Copernicus quietly dropped his epigram into the waste bin, as he had done with Dantiscus' earlier invitations. He really was an old sourpuss.

## 15. *Death of Copernicus*

The last months of his life must have been very lonely indeed. He had forsaken Rheticus, and Rheticus had forsaken him. Giese now lived away from Frauenburg; Sculteti was exiled.

One by one the Canons of his generation had died. He had not been much loved among his contemporaries; to the generation which now stepped into their place, he had even less appeal, They could not even regard the old man in his tower with the respectful boredom that decrepitude compels, for the scandal about Anna added to his reputation as a miser that of a lecher; and his past association with the Lutheran madman from Wittenberg did not help either. He was virtually ostracized.

The measure of his loneliness can be gleaned from a letter which, at the onset of Copernicus' last illness, Giese wrote from Loebau Castle to one of the Frauenburg Canons, George Donner:[91]

> ... Since he [Copernicus] loved solitude even in his healthy days, so, I think, he has few friends to help him with his troubles now that he is ill – although we are all in his debt for his personal integrity and excellent teachings. I know that he has always had you among the most faithful. I beg you, therefore, since his nature is so formed, would you be in the place of a guardian to him and undertake the protection of the man whom we have both always loved, that he may not lack brotherly help in this necessity, and that we may not appear ungrateful to him, deserving as he is. Farewell.
>
> Loebau, 8 December 1542.

Toward the end of 1542, Canon Koppernigk suffered a cerebral haemorrhage, followed by partial paralysis, and took permanently to his bed. At the beginning of 1543 Dantiscus wrote to the astronomer, Gemma Frisius in Louvain, that Copernicus was dying. But the end came only after several months, on 24 May. In a letter to Rheticus, written a few weeks later, Giese recorded the event in a single, tragic sentence:

> For many days he had been deprived of his memory and mental vigour; he only saw his completed book at the last moment, on the day he died.[92]

We know that mind has the power to hang on to life and, within limits, to postpone the body's death. Copernicus' mind had been wandering, yet there was perhaps just enough determination left to hold out until that moment when his hand could caress the cover of his book.

His state of mind in the last period is expressed in a reflection on a text by Thomas Aquinas, which he jotted down in a small, shaky writing on a bookmark: [93]

*Vita brevis, sensus ebes, negligentiae torpor et inutiles occupationes nos paucula scire permittent. Et aliquotients scita excutit ab animo per temporum lapsum fraudatrix scientiae et inimica memoriae praeceps oblivio.*

The shortness of life, the dullness of the senses, the numbness of indifference and unprofitable occupations allow us to know but very little. And again and again swift oblivion, the embezzler of knowledge and the enemy of memory, shakes out of the mind, in the course of time, even what we knew.

The earliest monument to Copernicus, in St John's Church in his native Torun, has a curious inscription which is assumed to have been copied from a note found in his possession.[94] It is a poem by Aeneas Silvius:

*Non parem Pauli gratiam requiro, Veniam Petri neque Posco, sed quam In crucis ligno dederas latroni, Sedulus oro.*

I crave not the Grace bestowed on Paul
Nor the remission granted to Peter
Only forgive me, I fervently pray
As thou forgavest the crucified thieves.

A more earthy epitaph appeared on a copper medallion coined in the seventeenth century by one Christian Wermuth in Gotha. Its face shows a bust with the inscription: '*Nicolaus Copernicus mathematicus natus* 1473, *D*. 1543.' On the reverse is a quatrain in German: [95]

*Der Himmel nicht die Erd umgeht*
*Wie die Gelehrten meynen*
*Ein jeder ist seines Wurms gewiss*
*Copernicus des seinen.*

The sky walks not around the earth
Though so the Doctors concluded;
Each man is sure to meet his worm
Copernicus included.

In the Frankonian local dialect, '*koeppernieksch*' still means a far-fetched, cockeyed proposition.

## 16. *Death of Rheticus*

Rheticus survived his teacher by more than thirty years. He led a restless, colourful, hectic life, but the purpose had gone out of it, the mainspring was broken, and his activities became increasingly more crankish and fantastic. He held his new post at the University of Leipzig for less than three years; in 1545, he left for Italy, and in spite of two demands from the University, refused to return on the grounds of ill-health. He seems to have studied medicine in Switzerland for a while, but nobody knew what had become of him; thus a Wittenberg scholar named Gauricus wrote under Rheticus' horoscope: 'Returned from Italy, became insane and died in April '47' – [95a] which reminds one of Kepler's description of Rheticus going off his head in Loebau Castle.

In '48, however, he returned to Leipzig and tried to turn over a new leaf. Within the next three years two works of his were published, an astronomical yearbook for 1550, and a work on trigonometry with extensive tables. He referred in them to Copernicus as his teacher, mentioned that he had supervized the publication of his work, and said that 'nothing should be altered in it'.[96] This was probably said in self-defence, because Rheticus was pressed from all sides to correct the errors of calculation in the *Revolutions* and to continue expounding his teacher's doctrine. He did nothing of the sort. Instead, his Preface to the work on trigonometry contains the astonishing suggestion that the Commentaries of Proclus on the Ptolemaic system should be taught at the German universities! About the teaching of the Copernican system there is not a word. Nor does the ambitious list of future publications, which he announced in the same Preface, contain any mention of his biography of Copernicus, which he had completed in manuscript.[97]

Two years after his return to Leipzig, Rheticus had to leave again, this time under more dramatic circumstances. An inscription in a book by one Jakob Kroeger provides the explanation: 'He [Rheticus] was a prominent mathematician, who for a while lived and taught in Leipzig, but flew from this town about 1550 because of sexual delicts (sodomy and the Italian perversion); I knew the man.'[98] It was a repetition of the incidents which, eight years earlier, had caused his migration from Wittenberg to Leipzig, and which put Osiander in control of the printing of the *Revolutions*.

For the next seven years, Rheticus' movements are obscure. He seems to have left Germany for fear of being arrested. In 1557, he turned up in Cracow. His conscience was nagging him, for he announced that, in compliance with the wishes of his late Teacher, who had insisted on more and better observations of the stars, he, Rheticus, had erected an obelisk forty-five feet high: 'for no device can be compared in excellence to the obelisk; armillaries, Jacob's staffs, astrolabes, and quadrants are human inventions, but the obelisk, erected on God's advice, surpasses them all.' He had chosen Cracow for his observations, 'because it lies on the same meridian as Frauenberg'.[99]

But the enterprise seems to have come to nothing. Six years later he was again pressed by various scholars to continue, and to expound the work of Copernicus. He toyed with the idea, asked a colleague for his assistance; then dropped the matter again.

In 1567, he wrote to a friend that he loved astronomy and chemistry, but made his living as a physician,[100] and that he inclined toward the teachings of Paracelsus. A year later, he wrote about his plans to Pierre Ramus, the great French mathematician, explaining that the wobbly theory of Ptolemy must be replaced by a true system based on observation, and more specially on the use the Egyptians had made of the obelisk. Thus he would creat 'a German astronomy for my Germans'.[101] He also mentioned numerous other projects: the completion of his monumental work on trigonometry on which he had spent twelve years; a work on astronomy in nine books, several books on astrology, and seven books on chemistry, which he had already drafted.

Of all these projects, only the trigonometrical tables were of scientific value; they were published posthumously by his pupil, Otho, and secured Rheticus an honourable place in the history of mathematics. They represented an enormous amount of dreary labour, and were evidently the occupational therapy which kept him within the borders of sanity.

He was now in his fifties, and still he could not settle down. He became house physician to a Polish Prince, then migrated to Cassovia in Hungary, where some Magyar noblemen provided for him. He died there in 1576, at the age of sixty-two.[102]

It was in that last year of his life that the young mathematician, Valentine Otho, travelled all the way from Wittenberg to Cassovia in the foothills of the Tatra mountains, to become his pupil – and to publish, twenty years later, the result of Rheticus' life work, *the Opus Palatinum de Triangulis*. Otho's preface to the book contains this epitaph on Georg Joachim Rheticus:

> ... When I returned to the University of Wittenberg, fortune willed that I should read a dialogue by Rheticus, who had been attached to the Canon. I was so excited and enflamed by this that I could not wait but had to journey at the first opportunity to the author himself and learn from him personally about these matters. I went, therefore, to Hungary where Rheticus was then working and was received by him in the kindest manner. We had hardly exchanged a few words on this and that when, on learning the cause of my visit, he burst forth with the words:
>
> 'You come to see me at the same age as I myself went to Copernicus. If I had not visited him, none of his works would have seen the light.'[103]

# 2 The System of Copernicus

## 1. *The Book That Nobody Read*

The *Book of the Revolutions of the Heavenly Spheres* was and is an all-time worst seller.

Its first edition, Nuremberg 1543, numbered a thousand copies, which were never sold out. It had altogether four reprints in four hundred years: Basle 1566, Amsterdam 1617, Warsaw 1854, and Torun 1873.[1]

It is a remarkable negative record, and quite unique among books which made history. To appreciate its significance, it must be compared with the circulation of other contemporary works on astronomy. The most popular among them was the textbook by a Yorkshireman, John Holywood (died 1256), known as *Sacrobosco*, which saw no less than fifty-nine editions.[2] The Jesuit father Christophe Clavius' *Treatise on the Sphere*, published in 1570, had nineteen reprints during the next fifty years. Melanchton's textbook, *Doctrines of Physics*, which was published six years after Copernicus' book and which attempted to refute Copernicus' theories, was reprinted nine times before the *Revolutions* was reprinted a single time (1566); and had a further eight editions later on. Kaspar Peucer's textbook on astronomy, published in 1551, was reprinted six times in the next forty years. The works just mentioned, plus Ptolemy's *Almagest* and Peurbach's *Planetary Theory* reached altogether about a hundred reprints in Germany till the end of the sixteenth century – the *Book of Revolutions*, one.[3]

The main reason for this neglect is the book's supreme unreadability. It is amusing to note that even the most conscientious modern scholars, when writing about Copernicus, unwittingly betray that they have not read him. The give-away is the number of epicycles in the Copernican system. At the end of his *Commentariolus*, Copernicus had announced (*see* p. 149 f.): 'altogether, therefore, thirty-four circles suffice to explain the entire structure of the universe and the entire ballet of the planets'. But the *Commentariolus* had merely been an optimistic preliminary announcement; when Copernicus got down to detail in the *Revolutions*, he was forced to add more and more wheels to his machinery, and their number grew to nearly fifty. But since he does not add them up anywhere, and there is no summary to his book, this face has escaped attention. Even the former Astronomer Royal, Sir Harold Spencer Jones, fell into the trap by stating in *Chamber's Encyclopaedia* that Copernicus reduced the number of epicycles 'from eighty to thirty-four'. The same mis-statement can be found in Professor Dingle's Copernicus Memorial Address to the Royal Astronomical Society in 1943,[4] and in a number of excellent works on the History of Science.* They obviously took the frequently quoted proud announcement in the last phrase of the *Commentariolus* at face value.

In fact, Copernicus uses altogether forty-eight epicycles – if I counted them correctly (see table, note 9).

Moreover, Copernicus had exaggerated the number of epicycles in the Ptolemaic system.[10] Brought up to date by Peurbach in the fifteenth century, the number of circles required in the Ptolemaic system was not eighty, as Copernicus said, but forty.[11]

In other words, contrary to popular, and even academic belief, Copernicus *did not reduce the number of circles, but increased them* (from forty to forty-eight).[12] How could this mistaken idea survive for so long, and be repeated by so many eminent authorities? The answer is that very few people, even among professional historians of science, have read Copernicus' book, because the Copernican system (as opposed to the heliocentric idea) is hardly worth bothering about. Not even Galileo seems to have read it, as we shall see.

The manuscript of the *Revolutions* consists of 212 sheets in small folio. It contains neither the author's name nor any of the prefatory matter.[13]

* Among them, Burtt's *The Metaphysical Foundations of Modern Science*,[5] Herbert Butterfield's *The Origins of Modern Science*,[6] in H. T. Pledge's *Science since* 1500,[7] and Ch. Singer's *A Short History of Science*.[8]

The first printed edition starts with Osiander's preface, followed by Cardinal Schoenberg's letter and by Copernicus' dedication to Paul III.

The work itself is divided into six books.

The first contains a broad outline of the theory, followed by two chapters on spherical trigonometry; the second is entirely devoted to the mathematical principles of astronomy. The third concerns the motions of the earth; the fourth, the motions of the moon; the fifth and sixth, the motions of the planets.

The basic principles and the programme of the work are all set out in the first eleven chapters of the first book. They may be summed up as follows. The universe occupies a finite space bounded by the sphere of the fixed stars. In the centre is the sun. Both the sphere of the stars and the sun are at rest. Around the sun revolve the planets Mercury, Venus, Earth, Mars, Jupiter, and Saturn, in that order. The moon revolves round the earth. The apparent daily revolution of the entire firmament is due to the rotation of the earth round its own axis. The apparent annual motion of the sun in the ecliptic is due to the annual revolution of the earth in its orbit. The stations and retrogressions of the planets are due to the same cause. The small irregularities of the seasons, and other minor irregularities, are due to the 'librations' (oscillations, wobbles) of the earth's axis.

This synopsis of the theory occupies less than twenty pages at the beginning of the book, or about five per cent of the whole. The remaining ninety-five per cent consists of the application of it. And when that is completed, there is hardly anything left of the original doctrine. It has, so to speak, destroyed itself in the process. This may be the reason why no summary, conclusions, or winding-up of any kind is found at the end of the book, although we are repeatedly promised it in the text.

At the beginning (Book I, chapter 10), Copernicus had stated: 'in the midst of all dwells the sun. ... Sitting on the royal throne, he rules the family of planets which turn around him. . . . We thus find in this arrangement an admirable harmony of the world.' But in Book III, when it comes to reconciling the doctrine with actual observation, the earth no longer turns round the sun, but round a point in space removed from the sun by a distance of about three times the sun's diameter. Nor do the planets revolve round the sun – as every schoolboy believes that Copernicus taught. The planets move on epicycles of epicycles, centred not on the sun, but on the centre of the *earth's orbit*. There are thus two 'royal thrones': the sun, and that imaginary point in space around which the earth moves. The year, that is, the duration of the earth's complete revolution round the sun, has a decisive influence on the motions of all other planets. In short, the earth appears equal in importance in governing the solar system to the sun itself, and in fact nearly as important as in the Aristotelian or Ptolemaic system.

The principal advantage of the Copernican system over the Ptolemaic is greater geometrical simplicity in one essential respect. By transferring the hub of the universe from the earth to somewhere in the vicinity of the sun, the retrograde motions of the planets, which had so much worried the ancients, disappeared. It will be remembered that during their annual procession along the Zodiacal lane, the planets occasionally come to a standstill, reversing their direction for a while, and then resume their progress again. So long as the earth was the hub of the universe, this phenomenon could be 'saved' by adding more epicycles to the clockwork, but there was no natural reason why the planets should behave as they did. But if the hub was near the sun, and the earth turned round it together with the other planets, it was obvious that each time the earth 'overtakes' one of the outer planets (which circle at a slower rate) that planet will appear to recede for a while; and each time the earth itself is overtaken by the faster moving inner planets, an apparent reversal of direction will again result.

This was an enormous gain in simplicity and elegance. On the other hand, the shifting of the centre of the universe to a place in the vicinity of the sun entailed an almost equal loss in plausibility. Previously, the universe had possessed a solid hub, the earth, a very solid and tangible hub indeed; now the whole world was hinged on a point in empty space. Moreover, that imaginary point was still defined by the orbit of the earth, and the motions of the whole system still depended on the motions of the earth. Not even the *planes* of the planetary orbits met in the sun; they oscillated in space, again according to the position of the earth. The Copernican system was not a truly heliocentric one; it was a vacuo-centric system, so to speak.

If it was to be considered merely as sky-geometry, without reference to physical reality – as Osiander's preface affirmed – this did not matter too much. But in his text Copernicus repeatedly affirmed that the earth *really* moved, and thereby exposed his whole system to judgement based on real, physical considerations. And from that point of view the system was untenable. Ptolemy's forty crystal-wheels on wheels had been bad enough, but at least the whole machinery was supported by the earth. Copernicus' machine had even more wheels, but it was supported neither by the earth, nor by the sun; it had no physical centre. Moreover, the centre of Saturn's orbit lay outside the sphere of Venus, and the centre of Jupiter's orbit near the sphere of Mercury. How could these spheres function without colliding and interfering with each other? Then again, Mercury, that most recalcitrant of all planets, had to be accorded an oscillatory motion along a straight line. But straight motion was considered by Aristotle, and Copernicus, as impossible for a heavenly body; hence it had to be resolved into the combined motion of two more spheres, one revolving inside the second; and the same artifice had to be employed to 'save' the wobbling motion of the earth's axis and all motions in latitude. By now the earth had no less than nine independent circular motions. But, the bewildered reader of Copernicus asked, if the earth's motion is *real*, then the nine wheels on which it turns must also be *real* – where are they?

Instead of the harmonious simplicity which the opening chapter of the *Revolutions* promised, the system had turned into a confused nightmare. To quote a modern historian, who trespassed into science with an unprejudiced eye:

> When you go down, so to speak, for the third time, long after you have forgotten everything else in this lecture, there will still float before your eyes that hazy vision, that fantasia of circles and spheres which is the trademark of Copernicus.[14]

## 2. *The Arguments for the Earth's Motion*

In fact, Copernicus carried orthodoxy regarding circles and spheres even further than Aristotle and Ptolemy. This becomes evident where he tries to prove the earth's motion by physical arguments. It may be objected, he says, that all heavy things gravitate towards the centre of the universe; but if the earth moves, it is no longer in the centre. This objection he answers as follows:[15]

> Now it seems to me gravity is but a natural inclination, bestowed on the parts of bodies by the Creator so as to combine the parts in the form of a sphere and thus contribute to their unity and wholeness. And we may believe this property present even in the Sun, Moon, and Planets, so that thereby they retain their spherical form notwithstanding their various paths.

Thus the parts of a whole stick together because of their desire to make a perfect shape; gravity, to Copernicus, is the nostalgia of things to become spheres.

The other classic objections were, mainly, that a falling body would be 'left behind' by the moving earth; that the atmosphere, too, would be left behind; and that the earth itself would fly apart owing to the disruptive force of its rotation. Copernicus counters these Aristotelian objections with an even more orthodox interpretation of Aristotle. Aristotle distinguished between 'natural' and 'violent' motion. Natural motion, says Copernicus, cannot lead to violent results. The natural motion of the earth is to turn; being spherical in form it simply cannot help turning. Its rotation is a natural consequence of its sphericity, just as gravity is the natural longing for sphericity.

> But if one holds that the earth *moves*, he will also say that this motion is natural, not violent. Things which happen according to nature produce the opposite effects to those due to force. Things subjected to violence or force will disintegrate and cannot subsist for long. But whatever happens by nature is done appropriately and preserves things in their best conditions. Idle, therefore, is Ptolemy's fear that the earth and everything on it would be disintegrated by rotation which is an act of nature, entirely different from an artificial act or anything contrived by human ingenuity. . . .[16]

In a word, the rotation of the earth engenders no centrifugal forces.

After this scholastic sleight-of-hand, Copernicus reverses the argument: if the universe were turning round the earth, with incomparably greater speed, would it not be in even greater danger of flying apart? But evidently on Copernicus' own argument that natural rotation is not disruptive, the universe in this case would be equally safe, and the question remains undecided.

He turns next to the objection that falling bodies and the air would be left behind by the earth's motion. His answer again is strictly Aristotelian: since the nearer atmosphere contains an admixture of earthy and watery matter, it follows the same natural law as the earth: 'bodies which fall because of their weight, must, because of their maximum of earthiness, doubtless participate in the nature of the whole to which they belong'. In other words, clouds and falling stones keep pace with the earth not because they share its physical momentum – a concept totally alien to Copernicus – but because they share in the metaphysical attribute of 'earthiness', and therefore circular motion is 'natural' to them. They follow the earth by affinity or sympathy.

Lastly,

> we conceive immobility to be nobler and more divine than mutability and instability, which lattter is therefore more appropriate to the earth than to the universe. I add to this that it would seem quite absurd to attribute motion to that which contains and locates, rather than to that which is contained and located – namely, the earth.

Apart from the greater geometrical simplicity of his system as a means of saving the phenomena, this is all that Copernicus has to say, by way of *physical* arguments, in support of the motion of the earth.

## 3. *The Last of the Aristotelians*

We have seen that Copernicus' ideas on physics were purely Aristotelian, and his methods of deduction followed strictly scholastic lines. At the time when the *Revolutions* was written, the authority of Aristotle was still very considerable in the conservative academic world, but rejected by more progressive scholars. At the Sorbonne, in 1536, Peter Ramus received an ovation when he took as his thesis 'Whatever is in Aristotle is false'. Erasmus called Aristotelian science sterile pedantry, 'looking in utter darkness for that which has no existence whatever'; Paracelsus compared academic education to 'a dog's being trained to leap through a ring' and Vives to 'orthodoxy defending the citadel of ignorance'.[17]

At the Italian universities where Copernicus had studied, he had come into contact with a new, post-Aristotelian breed of scholars: the new Platonists. For the decline of Aristotle overlapped with a new Platonic revival. I have called that perennial pair the twin stars; let me once more change the metaphor and compare them to that familiar couple in Victorian toy barometers – a top-coated gentleman with an open umbrella and a lady in gay summer dress, who, turning on a common pivot, alternately emerge from their cubby-holes to announce rain or shine. The last time it had been Aristotle's turn, now Plato pops out again – but a Plato entirely different from the pale, otherworldly figure of the early Christian centuries. After that first period of Plato's reign, when nature and science had been held in utter contempt, the reappearance of Aristotle, the chronicler of dolphins and whales, the acrobat of premiss and synthesis, the tireless logic-chopper, had been welcomed with relief. But in the long run there could be no healthy progress of thought on the dialectical tightrope; just at the time of Copernicus' youth, Plato emerged again from his cubby-hole and was greeted with even greater joy by the progressive humanists.

But this Platonism, which came from Italy in the second half of the fifteenth century, was almost in every respect the opposite of the Neoplatonism of the early centuries, and had little more in common with it than a hallowed name. The first had brought out the Parmenidian side of Plato, the second brought out the Pythagorean side. The first had divorced spirit from matter in its 'dualism of despair'; the second united the intellectual *ekstasy* of the Pythagoreans with Renaissance man's delight in nature, art, and craftsmanship. The bright-eyed young men of Leonardo's generation were Jacks-of-all-trades, with multiple interests and a devouring curiosity, with nimble fingers and nimble minds; impetuous, restless, sceptical about authority – the radical opposite to the stuffy, narrow-minded, orthodox, and pedantic schoolmen of the Aristotelian decline.

Copernicus was twenty years younger than Leonardo. During his ten years in Italy, he had lived among this new breed of men, yet he had not become one of them. He had returned to his medieval tower and to his medieval outlook on life. He took back with him one idea only which the Pythagorean revival had brought into fashion: the motion of the earth; and he spent the rest of his life trying to fit it into a medieval framework based on Aristotelian physics and Ptolemaic wheels. It was like trying to fit a turbo-prop engine on a ramshackle old stage-coach.

Copernicus was the last of the Aristotelians among the great men of science. In their attitude to nature, men like Roger Bacon, Nicolas of Cusa, William of Ockham, and Jean Buridan, who predated him by a century or two, were 'moderns' compared to Copernicus. The Ockhamist school in Paris, which flourished in the fourteenth century, and to which I have briefly referred before, had made considerable advances in the study of motion, momentum, acceleration, and the theory of falling bodies – all of which are basic problems of the Copernican universe. They had shown that Aristotelian physics with its 'unmoved movers', its 'natural' and 'violent' motion *et cetera*, was empty verbiage; and they had come very close to formulating Newton's Law of Inertia. In 1337 Nicolas of Oresme had written a Commentary on Aristotle's *De Coelo* – in fact, a refutation of it – in which he attributed the daily round of the heavens to the rotation of the earth, and based his theory on much sounder physical grounds than Copernicus, as an Aristotelian, could do. Copernicus was not acquainted with the discoveries in dynamics of the Paris school (which seem to have been ignored in Germany); but my point is that at Merton College and at the Sorbonne, a century and a half before him, a succession of men of lesser fame than Copernicus had shaken off the authority of Aristotelian physics to which he remained a life-long slave.

It was this almost hypnotic submission to authority which became Copernicus' undoing, both as a man and a scientist. As Kepler was to remark later on, 'Copernicus tried to interpret Ptolemy rather than nature'. His absolute reliance not only on the physical dogmata, but on the astronomic observations of the ancients was the main reason for the errors and absurdities of the Copernican system. When the Nuremberg mathematician Johannes Werner published a treatise *On the Motion of the Eighth Sphere*, in which he permitted himself to question the reliability of certain observations by Ptolemy and Timocharis, Copernicus attacked him with venom:

> ... It is fitting for us [he wrote] to follow the methods of the ancients strictly and to hold fast to their observations which have been handed down to us like a Testament. And to him who thinks that they are not to be entirely trusted in this respect, the gates of our Science are certainly closed. He will lie before that gate and spin the dreams of the deranged about the motion of the eighth sphere; and he will get what he deserved for believing that he can lend support to his own hallucinations by slandering the ancients.[18]

This was not the outburst of a youthful fanatic – Copernicus wrot this in 1524, when he was past fifty. Departing from his habitual caution and restraint, the unexpected vehemence of language stems from a desperate need to cling to his faith in the ancients which was already shaken. Ten years later he was to confide to Rheticus that the ancients had cheated him, that 'they had not shown disinterestedness, but had arranged many observations to fit their personal theories about the movements of the planets'.[19]

Apart from the twenty-seven observations of his own, the entire Copernican system was based on the observational data of Ptolemy, Hipparchus, and other Greek and Arab astronomers, whose statements he had uncritically accepted as Gospel truth, never pausing to consider the possibility of errors committed by careless scribes and translators in those notoriously corrupt texts, nor of mistakes and the doctoring of figures by the ancient observers themselves. When, at last, he realized the unreliability of the data on which he had built, he must have felt that the bottom had fallen out of his system. But by then it was too late to do anything about it.[20] Apart from his fear of ridicule, it must have been this realization of its basic unsoundness which made him so reluctant to publish the book. He did believe that the earth really moved. But he could no longer believe that either the earth, or the other planets, really moved in the manner, and along the orbits, which his book assigned to them.

The tragedy of blind faith in ancient authority, which makes Copernicus such a pathetic figure, is illustrated by a curious example. The point is highly technical, and I must simplify it. Trusting a handful of very precarious data on alleged observations by Hipparchus, Menelaus, Ptolemy, and Al Battani, dispersed over two thousand years, Copernicus was led to believe in a phenomenon that does not exist – a periodic change in the rate of the wobble of the earth's axis.[21] In reality the wobble goes on at the same, steady rate; the figures of the ancients were simply wrong. As a result, Copernicus felt obliged to construct an incredibly laborious theory, which attributed two independent oscillatory motions to the earth's axis. But oscillations along a straight line are 'violent' motions forbidden by Aristotelian physics; hence Copernicus devotes a whole chapter[22] to show how this motion in a straight line can be produced by a combination of two 'natural', that is circular, motions. The result of this phantom-chase was that he had to ascribe four more circular motions to the earth, in addition to the already existing five.

Towards the end of this painful chapter, where Copernicus' obsession with circles reaches its climax, as it were, the manuscript contains the lines: 'It should be noticed, by the way, that if the two circles have different diameters, other conditions remaining unchanged, then the resulting movement will not be a straight line but ... *an ellipse.*' * This is actually not true, for the resulting curve will be a cycloid merely resembling an ellipse – but the odd fact is that Copernicus had hit on the ellipse which is the form of all planetary orbits – had arrived at it for the wrong reasons and by faulty deduction – and

* My italics.

having done so, promptly dropped it: the passage is crossed out in the manuscript, and is not contained in the printed edition of the *Revolutions*. The history of human thought is full of lucky hits and triumphant *eurekas*; it is rare to have on record one of the anti-climaxes, the missed opportunity which normally leaves no trace.

## 4. *The Genesis of the Copernican System*

The figure of Copernicus, seen from the distance, is that of an intrepid, revolutionary hero of thought. As we come closer, it gradually changes into that of a stuffy pedant, without the flair, the sleepwalking intuition of the original genius; who, having got hold of a good idea, expanded it into a bad system, patiently plodding on, piling more epicycles and deferents into the dreariest and most unreadable among the books that made history.

To deny that Copernicus was an original thinker may sound paradoxical or blasphemous. Let us try to retrace the process of reasoning which led Nicolas Koppernigk to the Copernican system. It is a much-debated problem, and of a certain interest both to the psychology of discovery and the history of human thought.

Our starting point is his first astronomical treatise, the *Commentariolus*. It opens, characteristically:

> Our ancestors assumed a large number of celestial spheres for a special reason: to explain the apparent motion of the planets by the principle of regularity. For they thought it altogether absurd that a heavenly body should not always move with uniform velocity in a perfect circle.

Having stated his *credo*, Copernicus turns to Ptolemy, whose system, he says, is consistent with the observed facts, *but* ... and here follows a revealing passage which explains the reason that started Copernicus on his quest. It is his shocked realization of the fact that in Ptolemy's universe a planet moves on perfect circles, *but not really at uniform speed*. More precisely, the planet does not cover equal distances at equal times when seen from the centre of its circle – it only *appears* to do so when observed from a different point specially chosen for that purpose. This point is called the *punctum equans*, or 'equant' for short. Ptolemy invented this trick to save the principle of uniform motion – his *punctum equans* enabled him to say that there exists, after all, a point in space where an observer could enjoy the illusion that the planet's motion is a steady one. But, Copernicus remarks indignantly, 'a system of this sort seemed neither sufficiently absolute nor sufficiently pleasing to the mind.'[23]

It was the grievance of a perfectionist who could not tolerate this offence against his ideal of circular uniform motion. It was an imaginary grievance, for in reality the planets did not move in circles anyway, but on the epicycles of epicycles, producing oval curves; and whether uniformity was 'saved' relative to the centre of the imaginary epicycle, or to the equally imaginary equant, made hardly any difference except to an obsessional mind. Yet, as Copernicus himself explains, it was this grievance which started the whole chain-reaction:

> Having become aware of these defects, I often considered whether there could perhaps be found a more reasonable arrangement of circles ... in which everything would move uniformly about its proper centre, as the rule of absolute motion requires.[24]

Thus Copernicus' first impulse to reform the Ptolemaic system originated in his urge to remove a minor blemish from it, a feature which did not strictly conform to conservative Aristotelian principles. He was led to reversing the Ptolemaic system by his desire to preserve it – like the maniac who, pained by a mole on his beloved's cheek, cut off her head to restore her to perfection. Yet it happened not for the first time in history that a puritan reformer started by attacking a minor imperfection, and ended by realizing that it was a symptom of a deep-rooted and irremediable disease. Ptolemy's equants were nothing to get excited about, but they were symptomatic of the jarring artificiality of the system.

Once he had started to take the Ptolemaic clockwork to pieces, he was on the lookout for some useful hint how to rearrange the wheels in a different order. He did not have to look for long:

I therefore went to the trouble of reading anew the books of all philosophers on which I could lay hands to find out whether someone did not hold the opinion that there existed other motions of the heavenly bodies than assumed by those who taught the mathematical sciences in the schools. And thus I found first in Cicero that Hiketas had held the belief that the earth moves. Afterwards I found in Plutarch * that others have also held this opinion. I shall put down his own words so that everybody can read them:

'But others hold that the earth moves; thus Philolaus the Pythagorean held that it revolves round the Fire in an oblique circle like the sun and moon. Herakleides of Pontus and Ekphantus the Pythagorean also suppose the earth to move, though not in a progressive motion, but after the manner of a wheel, turning upon an axle about its own centre from west to east.'

And so, taking occasion from this, I too began to think about the mobility of the earth. And although it seemed an absurd opinion, yet, because I knew that others before me had been granted the liberty of supposing whatever orbits they chose in order to demonstrate the phenomena of the stars, I considered that I too might well be allowed to try whether sounder demonstrations of the revolutions of the heavenly orbs, might be discovered by supposing some motion of the earth.[25]

There are further references[26] to 'the Pythagoreans Herakleides and Ekphantus', and to 'Hiketas of Syracuse, who let the earth rotate in the centre of the world'. Then in Book I, chapter 10, called *About the order of the heavenly orbits*, Copernicus gives us his own version of the Genesis of his system:

Therefore it seemed to me that it would be wrong to ignore certain facts well-known to Martianus Capella, who wrote an encyclopaedia, and of some other Latins. He believed that Venus and Mercury do not go round the earth like other planets, but turn round the sun as their centre and therefore cannot go farther away from the sun than the sizes of their orbits permit. What else does this mean but that the sun is the centre of their orbits and that they

* He refers in fact to the pseudo-Plutarch's work *De Placiti Philosophorum*, III, 13.

turn round him? Thus the sphere of Mercury would be enveloped by that of Venus which is twice as large and would find sufficient space inside it. If we seize the opportunity to refer Saturn, Jupiter, and Mars to the same centre [i.e. the sun] ... then their motions will fall into a regular and explainable order. ... And as now all of them are arranged round the same centre, so it becomes necessary that the space which is left between the convex surface of the sphere of Venus and the concave sphere of Mars should be filled in by the earth and the moon which accompanies it and by all the matter to be found in the sublunary sphere. ... Therefore we do not hesitate to state that moon and earth describe annually a circular orbit placed between the outer and the inner planets round the sun, which rests immobile in the centre of the world; and that everything which appears as a motion of the sun is in truth a motion of the earth.

Now all this is familiar ground to us. Copernicus refers first to the so-called 'Egyptian' system of Herakleides,* that 'half-way house', in which the two inner planets circle round the sun, while the sun itself, and the outer planets, still go round the earth. He then takes the second step (of letting the outer planets too circle round the sun) which in antiquity was taken either by Herakleides or by Aristarchus; and finally, the third step to the complete heliocentric system, where *all* the planets, including the earth, revolve round the sun, as suggested by Aristarchus of Samos.

There can be no doubt that Copernicus was acquainted with Aristarchus' idea, and that he was following in his footsteps. The proof of this is to be found in Copernicus' own manuscript of the *Revolutions*, where he refers to Aristarchus – but, characteristically, this reference is crossed out in ink. Thus Aristarchus' forerunners are given credit in the book, but not Aristarchus himself – just as the names of Rheticus, of Brudzewski, and Novara, the teachers to whom Copernicus owed most, are omitted. He had to mention the fact that the heliocentric idea was known to the ancients, in order to prove its respectability, as it were; yet he confused the trail, as was his habit, by leaving out the most important among them.[27]

It is highly unlikely, though, that Copernicus should have

* See Part One, Chapter III, 2.

stumbled on his idea simply by browsing through the ancient philosophers. Talk of a moving earth, of the earth as a planet or star, was becoming increasingly frequent in the days of his youth. We have seen (p. 102 f.) that during the later Middle Ages the system of Herakleides had been favoured by most of the scholars who took an interest in astronomy. From the thirteenth century onward the influence of Ptolemy had reasserted itself, simply because there existed no other planetary theory as detailed and comprehensive as the *Almagest*; but soon afterwards a strong current of criticism and opposition arose. Earlier already, Averroes, the greatest Arab philosopher in Europe (1126–98) had commented: 'The Ptolemaic astronomy is nothing so far as existence is concerned; but it is convenient for computing the non-existent.'[28] He had no better alternative to offer; but his epigram could serve as a motto for the growing discontent with the prevailing double-think in cosmology.

This metaphysical *malaise* flared into open revolt in the first half of the century into which Copernicus was born. Nicolas of Cusa (1401–64), a German ecclesiastic, the son of a boatsman on the Moselle, who rose to the rank of Cardinal, was the first to kick against the lid of the medieval universe. In his *Learned Ignorance*,[29] written in 1440, printed in 1514, twenty years before the *Revolutions*, he asserted that the world had no boundaries, and consequently neither a periphery nor a centre. it was not infinite, merely 'interminate', that is unbounded, and everything in it was in flux:

> Since, then, the earth cannot be the centre, it cannot be entirely devoid of motion. . . It is clear to us that the earth is really in motion though this may not be apparent to us, since we do not perceive motion except by comparison with something fixed.[30]

Earth, moon and planets all move round in a centre, which is not defined; but Cusa expressly denies that they move either in perfect circles, or at uniform speed:

> Moreover, neither the sun, nor the moon, nor any sphere – though to us it seems otherwise – can in [its] motion describe a true circle, because they do not move around a fixed base. Nowhere is there a

true circle such that a truer one would not be possible, nor is [anything] ever at one time [exactly] as at another, neither does it move in a precisely equal [manner], nor does it describe an equally perfect circle, though we are not aware of it.[31]

By denying that the universe has either a centre or a periphery, Cusa also denied its hierarchic structure, denied the lowliness of the earth's position in the Chain of Being, denied that mutability is an evil confined to the sub-lunary sphere. 'The earth is a noble star', he proclaimed triumphantly, 'it is not possible for human knowledge to determine whether the region of the earth is in a degree of greater perfection or baseness in relation to the regions of the other stars. . . .'[32]

Lastly, Cusa was convinced that the stars were made of the same stuff as the earth, and that they were inhabited by beings neither better nor worse than man, but simply *different*:

. . . It cannot be said that this place of the world [is less perfect because it is] the dwelling place of men, and animals, and vegetables that are less perfect than the inhabitants of the region of the sun and of the other stars. . . . It does not seem that, according to the order of nature, there could be a more noble or more perfect nature than the intellectual nature which dwells here on this earth as in its region, even if there are in the other stars inhabitants belonging to another genus: man indeed does not desire another nature, but only the perfection of his own.[33]

Cusa was no practising astronomer, and he built no system; but his teaching shows that long before Copernicus not only the Franciscans at Oxford and the Ockhamists in Paris had broken loose from Aristotle and the walled-in universe, but that in Germany too there were men with a far more modern outlook than the Canon of Frauenburg. Cusanus died seven years before Copernicus was born; they had both been members of the German *natio* in Bologna, and Copernicus was acquainted with Cusa's teaching.

He was equally familiar with the work of his immediate predecessors: the German astronomer Peurbach and his pupil Regiomontanus, who, between them, had brought about the revival of astronomy as an exact science in Europe, after a millennium of stagnation. George Peurbach (1423—61) came from a small town on the Bavarian border, studied in Austria and Italy, where he knew Nicolas of Cusa, and became subsequently a professor at Vienna University, and Court Astronomer to the King of Bohemia. He wrote an excellent textbook on the Ptolemaic system, which had fifty-six later editions, and was translated into Italian, Spanish, French, and Hebrew.[34] During his professorship at Vienna, he presided over a public discussion for and against the motion of the earth;[35] and though Peurbach, in his textbook, took a conservative attitude, he underlined the fact that the motions of all the planets were governed by the sun. He also mentioned that the planet Mercury rides on an epicycle whose centre moves not on a circular, but an egg-shaped or oval orbit. A number of other astronomers from Cusanus down to Copernicus' first teacher, Brudzewski, had also spoken tentatively of oval orbits.[36]

Peurbach's work was continued by Johann Mueller from Koenigsberg, called Regiomontanus (1436–76), a Renaissance genius and infant prodigy, who, at the age of twelve, published the best astronomical yearbook for 1448, and at fifteen was asked by the Emperor Frederick III to cast a horoscope for the Imperial bride. He went to the University of Leipzig when he was eleven, and at sixteen became the pupil and associate of Peurbach in Vienna. Later he travelled with Cardinal Bessarion to Italy, to learn Greek and study Ptolemy in the original. After Peurbach's death, he edited the former's book on the planetary motions, then published his own treatise on spherical trigonometry, from which Copernicus is supposed to have heavily borrowed, without acknowledgement, in his own chapters on trigonometry.[36a]

Regiomontanus' later years reflect a growing discontent with traditional astronomy. A letter, written in 1464, contains this typical outburst:

> . . . I cannot get over my amazement at the mental inertia of our astronomers in general who, like credulous women, believe what they read in the books, tablets, and commentaries as if it were the divine and unalterable truth; they believe the authors and neglect the truth.[37]

In another context, he says:

> It is necessary to keep the stars doggedly before one's eyes, and to rid posterity from ancient tradition.[38]

It sounds like a polemic against the programme of Copernicus, who was not yet born, 'to follow the methods of the ancients strictly and to hold fast to their observations which have been handed down to us like a Testament'!

In his middle thirties, Regiomontanus held a profitable position in Hungary at the court of King Mathias Corvinus. But he convinced his royal patron that Ptolemy could no longer be relied on, and that it was necessary to put astronomy on new foundations by patient observations, making use of such recent inventions as the corrected sundial and the mechanical clock. Mathias agreed, and in 1471, Regiomontanus went to Nuremberg where, with the help of a rich patrician, Johann Walther, he installed the first European observatory, for which he partly invented the instruments.

The manuscripts and notes of Regiomontanus' last years are lost, and there remain only scant indications of the reform of astronomy that he planned. But we know that he had paid special attention to Aristarchus' heliocentric system, as a note on one of his manuscripts shows.[39] And much earlier he, too, had noted that the sun ruled the motions of the planets. Towards the end of his life, he wrote on a piece of paper enclosed in a letter the words: 'It is necessary to alter the motion of the stars a little because of the motion of the earth.' The wording, as Zinner has shown, seems to indicate that the 'motion of the earth' here refers not to the daily rotation but to its annual revolution round the sun;[40] in other words that Regiomontanus had arrived at the same conclusions as Aristarchus and Copernicus, but was prevented from going further by his untimely death. He died at forty, three years after Copernicus was born.

At the universities where Copernicus studied, the tradition of Cusa and Regiomontanus was very much alive. His principal teachers in astronomy: Brudzewski in Cracow, and Maria Novara in Bologna, both called themselves pupils of Regiomontanus. Finally, in Ferrara Copernicus met young Celio

Calcagnini, poet and philosopher, who later published a short book with a significant title: *Quomodo coelum stet, terra moveatur, vel de perenni motu terrae Commentario* – 'A Treatise concerning how the Heavens rest, the Earth moves, or on the perennial motions of the Earth'.[41] Calcagnini, who had written a pretty poem to greet the arrival of Lucretia Borgia in Ferrara, was not a profound intellect; his thesis that the heavens are at rest, the earth in eternal movement, was inspired by Cusa and simply echoed an idea that was, as we saw, much in the air. He probably owed his insight to his friend and contemporary in Ferrara, Jacob Ziegler, an astronomer of some merit, who wrote a commentary on Pliny which contains the lapidary statement: 'The motions of all planets depend on the sun.'

More examples of a similar kind could be quoted, but I have said enough to show that the ideas of a moving earth, and of the sun as the true ruler of the planetary system, belonged both to the antique tradition of cosmology, and were much discussed in Copernicus' own time. Yet Canon Koppernigk was undoubtedly the first to develop the idea into a comprehensive system. This is his lasting merit, regardless of the inconsistencies and shortcomings of his system. He was not an original thinker, but a crystallizer of thought; and the crystallizers often achieve more lasting fame and a greater influence on history than the initiators of new ideas.

There is a well-known process in chemistry which will illustrate what I mean by a crystallizer. If you put kitchen salt into a glass of water until the water is 'saturated' and will dissolve no more salt, and suspend a thread with a knot at its end in the solution, then after a while a crystal will form round the knot. The shape and texture of the knot are irrelevant; what matters is that the liquid has reached saturation point, and that a core has been provided round which it can start to crystallize. Cosmology at the end of the Middle Ages was saturated with vague notions of a spinning and moving earth, with echoes of the Pythagoreans, of Aristarchus and Herakleides, of Macrobius and Pliny, with the exciting suggestions thrown out by Cusa and Regiomontanus. Canon Koppernigk was the patient knot, suspended in the solution, who enabled it to crystallize.

I have tried to reconstruct the process from its starting point – Copernicus' discontent with Ptolemy's equants, which he regarded as an imperfection – to his re-shaping of the Ptolemaic system with the aid of an ancient idea which was being revived during his student days. But if it was really as simple as that, then the equally simple question arises why nobody before him had worked out a heliocentric system? It would be meaningless to ask why nobody before Shakespeare had written Hamlet; but if Copernicus was really as devoid of originality and imagination as I tried to make him out, then it is legitimate to ask why the task of 'crystallizing' fell to him – whereas, for instance, the intellectually more flexible, and 'modern' Regiomontanus left it at a few hints, but never developed a systematic sun-centred theory.

The key to the answer is perhaps Kepler's already quoted remark that Copernicus was interpreting Ptolemy (and Aristotle) rather than nature. To a fifteenth century 'modern' mind such an undertaking must have appeared partly impossible, and partly a waste of time. Only a conservative-minded person such as Copernicus could devote himself to the task of reconciling the irreconcilable doctrines of Aristotelian physics and Ptolemaic wheel-geometry on the one hand, with a sun-centred universe on the other. To arrive at a self-consistent, and physically plausible heliocentric system, it was necessary first to wrench the mind free of the hold of Aristotelian physics, to shake off the obsession with circles and spheres, to smash up the whole jarring machinery of fictitious wheels-on-wheels. The great discoveries of science often consist, as we saw, in the uncovering of a truth buried under the rubble of traditional prejudice, in getting out of the *cul-de-sacs* into which formal reasoning divorced from reality leads; in liberating the mind trapped between the iron teeth of dogma. The Copernician system is not a discovery in this sense, but a last attempt to patch up an out-dated machinery by reversing the arrangement of its wheels. As a modern historian put it, the fact that the earth moves is 'almost an incidental matter in the system of Copernicus which, viewed geometrically, is just the old Ptolemaic pattern of the skies, with one or two wheels interchanged and one or two of them taken out'.[42] There is a well-known saying that Marx 'turned Hegel upside down'. Copernicus did the same to Ptolemy; in both cases, the reversed authority remained the bane of the disciple.

From Roger Bacon in the thirteenth century to Peter Ramus in the sixteenth, there had been outstanding individuals and schools who realized, more or less consciously, more or less articulately, that Aristotelian physics and Ptolemaic astronomy had to be put out of the way before a new departure could be made. That may be the reason why Regiomontanus built himself an observatory instead of building himself a system. When he had completed the commentaries on Ptolemy which Peurbach had begun, he realized the need to put astronomy on a new basis by 'ridding posterity of ancient tradition'. In Copernicus' eyes, such an attitude amounted to blasphemy. If Aristotle had stated that God created only birds, Canon Koppernigk would have described *homo sapiens* as a bird without feathers and wings who hatches his eggs before laying them.

The Copernican system is precisely that kind of construction. Apart from the inconsistencies which I have mentioned before, it did not even succeed in remedying the specific faults of Ptolemy which it had set out to remedy. True, the 'equants' had been eliminated, but rectilinear motion, which Copernicus called 'worse than a disease', had to be imported in their stead. In his Dedication, he had mentioned, beside the equants, as the chief reason for his enterprise, the uncertainty of existent methods to determine the length of the year; but the *Revolutions* shows no progress in this specific respect. Ptolemy's orbit of Mars disagreed conspicuously with the observed data, but in the Copernican system it was equally faulty – so much so, that later on Galileo was to speak with admiration of Copernicus' courage in defending his system, although it was so evidently contradicted by the observed motions of Mars!

One last objection against the system, and perhaps the strongest of all, arose through no fault of its author. If the earth moves round the sun in a huge circle, with a diameter of about ten million miles,[43] then the pattern of the fixed stars ought to change continually according to the different positions which the earth occupies on its journey. Thus when we approach a certain group of stars, it ought to 'open up', for the distances between the members of that group should appear to grow with our approach, and to shrink as we recede from it on our journey. Such apparent displacements of objects due to a change in the position of the observer are called *parallax.*

But the stars belied this expectation. They showed no parallax – their pattern remained fixed and immutable.[44] It followed that either the theory of the earth's motion was wrong – or the distance of the fixed stars was so immense that, compared to it, the circle described by the earth shrank to nothingness, and produced no noticeable effect. This was, in fact, Copernicus' answer![45] but it was difficult to swallow and added to the inherent improbability of the system. As Burtt remarks: 'Contemporary empiricists, had they lived in the sixteenth century, would have been the first to scoff out of court the new philosophy of the universe.'[46]

## 5. *The First Repercussions*

No wonder, then, that the publication of the *Revolutions* attracted very little attention. It created less of a stir than Rheticus' *First Account* of it. Rheticus had promised that the book would be a revelation; it turned out to be a disappointment. For more than fifty years, until the beginning of the seventeenth century, it raised no particular controversy either in public or among professional astronomers. Whatever their philosophical convictions about the structure of the universe, they realized that Copernicus' book did not stand up to scientific scrutiny.

If his name nevertheless enjoyed a certain repute among the generation which immediately succeded him, this was due not to his theory of the universe, but to the astronomical tables which he had compiled. They were published in 1551 by Erasmus Reinhold, Rheticus' former associate at Wittenberg, and were welcomed by astronomers as a long overdue replacement for the Alfonsine Tables, which dated from the thirteenth century. Reinhold, after revising all figures and eliminating the frequent slips, paid in his preface generous tribute to Copernicus' labours as a practical astronomer, without mentioning at all the Copernican theory of the universe. The next generation of astronomers referred to the Tables as *Calculatio Coperniciano*, and this helped to keep the Canon's reputation alive, but it had little to do with the Copernican system. Leaving non-astronomers such as Thomas Digges, William Gilbert, and Giordano Bruno for the moment aside, the Copernican theory was practically ignored until the opening of the seventeenth century, when Kepler and Galileo enter the scene. Then and only then, did the heliocentric system burst upon the world – like a conflagration caused by a delayed-action bomb.

The reaction of the Churches during the half century following Copernicus' death was equally indifferent. On the Protestant side, Luther gave out a few uncouth growls, while Melanchton elegantly proved that the earth was at rest; but he did not withdraw his patronage from Rheticus. On the Catholic side, the initial reaction, as we have seen, was one of encouragement, and the *Revolutions* was put on the Index in 1616 only – seventy-three years after its publication. There were occasional discussions as to whether the motion of the earth was compatible with Holy Scripture or not, but until the decree of 1616 the question remained undecided.

The clerical attitude of ironical indifference towards the new system is reflected in John Donne's *Ignatius His Conclave*. Here Copernicus appears as one of the four pretenders to the principal place next to Lucifer's throne, the other contenders being Ignatius of Loyola, Macchiavelli, and Paracelsus. Copernicus stakes his claim by declaring that he has raised the Devil and his prison, the earth, into the heavens, while relegating the sun, the Devil's energy, into the lowest part of the universe: 'Shall these gates be shut against me who have turned the whole frame of the world, and am thereby almost a new Creator?'

The jealous Ignatius, who wants the place of honour in Hell for himself, debunks Copernicus:

But for you, what new thing have you invented, by which our *Lucifer* gets any thing? What cares hee whether the earth travell, or stand still? Hath your raising up of the earth into heaven, brought men to that confidence, that they build new towers or threaten God againe? Or do they out of this motion of the earth conclude, that there is no hell, or deny the punishment of sin? Do not men beleeve? do they not live just, as they did before? Besides, this detracts from the dignity of your learning, and derogates from your right and title of comming to this place, that those opinions of yours may very well be true. . . . But your inventions can scarce bee called yours, since long before you, *Heraclides*, *Ecphantus*, and *Aristarchus* thrust them into the world: who notwithstanding content themselves with lower roomes among the other Philosophers, and aspire not to this place, reserved onely for *Antichristian Heroes*. . . Let therefore this little *Mathematitian*, dread Emperour, withdraw himselfe to his owne company.

*Ignatius* was published in 1611. It reflects, broadly speaking, the attitude of the two generations between Copernicus and Donne. But those two generations who ignored Copernicus were mistaken; the 'little Mathematitian', that pale, sour, insignificant figure, ignored by his contemporaries and those who immediately succeeded them, was to throw a giant shadow on the history of mankind.

How is this last paradox in a paradoxical story to be explained? How was it possible that the faulty, self-contradictory Copernican theory, contained in an unreadable and unread book, rejected in its time, was to give rise, a century later, to a new philosophy which transformed the world? The answer is that the details did not matter, and that it was not necessary to read the book to grasp its essence. Ideas which have the power to alter the habits of human thought do not act on the conscious mind alone; they seep through to those deeper strata which are indifferent to logical contradictions. They influence not some specific concept, but the total outlook of the mind.

The heliocentric idea of the universe, crystallized into a system by Copernicus, and restated in modern form by Kepler, altered the climate of thought not by what it expressly stated, but by what it implied. Its implications were certainly not conscious in Copernicus' mind, and acted on his successors by equally insidious, subterranean channels. They were all negative, all destructive to the solid edifice of medieval philosophy, undermining the foundations on which it rested.

### 6. *The Delayed Effect*

The medieval Christian universe had hard, firm limits in space, time, and knowledge. Its extension in time was limited to the relatively short span between the creation of the world, which lay some five thousand years back, and the second coming of Christ which lay ahead, and which many expected to occur in the foreseeable future. Thus the history of the universe was thought to be limited to something of the order of two or three hundred generations from beginning to end. God had modelled his world on the art form of the short story.

In space the world was equally bounded by the ninth sphere, beyond which lay the heavenly Empyrean. It was not necessary for the sophisticated to believe strictly in all that was said about heaven and hell; but the existence of solid boundaries in time and space were a habit of thought as self-evident as the walls and ceiling of his room, as his own birth and death.

Thirdly, there were equally firm limits to the progress of knowledge, technology, science, social organization; all of which had been completed long ago. There was a final truth regarding every subject, as finite and bounded as the universe itself. The truth about religion was revealed in the Scriptures; the truth about geometry in Euclid, the truth about physics in Aristotle. The science of the ancients was taken as Gospel truth, not because of any particular respect for the pagan Greeks, but because it was obvious that since they had come so much earlier they had harvested all there was to harvest in these fields, and left nothing but a few stray stalks to pick in the way of tidying up. Since there was only one answer to every question, and the ancients had filled in all the answers, the edifice of knowledge was completed. If the answer did not happen to fit the facts, the error was blamed on the scribes who copied the ancient manuscript. The authority of the ancients did not rest on idolatry, but on the belief in the finite nature of knowledge.

From the thirteenth century onward, humanists, sceptics, and reformers had started making holes in the walls of this stable and static universe. They chipped off bits of it here and there, letting in draughts and loosening the structure. But it still held. Donne's 'little Mathematitian' did not ram his head against doors, he made no frontal attack, he was not even conscious of attacking at all. He was a conservative who felt quite at home in the medieval edifice, and yet he undermined its foundations more effectively than the thundering Luther. He let in the destructive notions of infinity and eternal change, which destroyed the familiar world like a dissolvent acid.

He did not state that the universe is infinite in space. He preferred, with his usual caution, 'to leave the question to the philosophers'.[47] But unwittingly he altered an unconscious habit of thought by making the earth rotate instead of the sky. So long as the rotation was attributed to the sky, the mind automatically assumed it to be a solid and finite sphere – how else could it go round as a unit every twenty-four hours? But once the apparent daily round of the firmament was explained by the earth's rotation, the stars could recede to any distance; putting them on a solid sphere became now an arbitrary, unconvincing act. The sky no longer had a limit, infinity opened its gaping jaws, and Pascal's 'libertin', seized by cosmic agoraphobia, was to cry out a century later: '*Le silence éternel de ce espaces infinis m'effraie!*'

Infinite space is not a part of the Copernican system. But it is implied in it; it irresistibly tended to push thought in that direction. This distinction between the explicit, and the unconsciously implied consequences becomes even more apparent in Copernicus' impact on the metaphysics of the universe. Aristotelian physics was, as we have seen, already discredited in parts, and Copernicus was one of the last orthodox defenders. But in one fundamental respect it still ruled the mind of man like a self-evident proposition or an act of faith: one may call this the grand topography of the universe. It was this fundamental pattern which Copernicus, the defender of Aristotle, unwittingly destroyed.

The Aristotelian universe was centralized. It had one centre of gravity, one hard core, to which all movement referred. Everything that had weight fell towards the centre, everything buoyant, like fire and air, tried to get away from it; while the stars, neither heavy nor buoyant and of an altogether different nature, moved in circles around it. The details of the scheme might be right or wrong, but it was a simple, plausible, reassuringly orderly scheme.

The Copernican universe is not only *expanded* towards the infinite, but at the same time *decentralized*, perplexing, anarchic. It has no natural centre of orientation to which everything else can be referred. The directions 'up and down' are no longer absolute, nor are weight and buoyancy. The 'weight' of a stone had meant, before, its tendency to fall towards the centre of the earth: that was the meaning of 'gravity'. Now the sun and the moon become centres of gravity of their own. There are no longer any absolute directions in space. The universe has lost its core. It no longer has a heart, but a thousand hearts.

The reassuring feeling of stability, of rest and order are gone; the earth itself spins and wobbles and revolves in eight or nine simultaneous different motions. Moreover, if the earth is a planet, the distinction between the sub-lunary region of change and the ethereal heavens disappears. If the earth is made up of four elements, the planets and stars may be of the same earthy, watery, fiery, and airy nature. They may even be inhabited by other kinds of men, as Cusa and Bruno asserted. Would in this case God have to become incarnate on every star? And could God have created this whole colossal multitude of worlds for the sake of the inhabitants of one single star among millions?

None of these questions is posed in the *Book of Revolutions.* All of them are implicit in it. All of them, inescapably, were asked sooner or later by the Copernicans.

From all pre-Copernican diagrams of the Universe there emerges, with minor variations, always the same reassuring, familiar picture: the earth at the centre, surrounded by the concentric shells of the hierarchy of spheres in space, and the hierarchy of values associated with it on the great Scale of Being. *Here be tygers* and here be seraphim: every item had its assigned place in the cosmic inventory. But in an unbounded universe without centre or circumference, no region or sphere ranked 'higher' or 'lower' than another either in space or on the scale of values. That scale was no longer. The Golden Chain was torn, its links scattered throughout the world; homogeneous space implied a cosmic democracy.

The notion of limitlessness or infinity, which the Copernican system implied, was bound to devour the space reserved for God on the medieval astronomer's charts. They had taken it for granted that the realms of astronomy and theology were contiguous, separated only by the thickness of the ninth crystal sphere. Henceforth, the space-and-spirit continuum would be replaced by a space-time continuum. This meant, among other things the end of intimacy between man and God. *Homo sapiens* had dwelt in a universe enveloped by divinity as by a womb; now he was being expelled from the womb. Hence Pascal's cry of horror.

But that cry was uttered a hundred years later. Canon Koppernigk in his tower in Frauenburg would never have understood why the Reverend John Donne made him a pretender to the seat next to Lucifer's throne. With his blessed lack of humour he foresaw none of these consequences when he published his book with the motto: 'For Mathematicians Only'. Nor did his contemporaries. During the remainder of the sixteenth century, the new system of the universe went, like an infectious disease, through a period of incubation. Only at the beginning of the seventeenth did it burst into the open and cause the greatest revolution in human thought since the heroic age of Greece.

A.D. 1600 is probably the most important turning point in human destiny after 600 B.C. Astride that milestone, born almost exactly a hundred years after Copernicus, with one foot in the sixteenth, the other in the seventeenth century, stands the founder of modern astronomy, a tortured genius in whom all the contradictions of his age seem to have become incarnate: Johannes Kepler.

## *Chronological Table to Part Three*

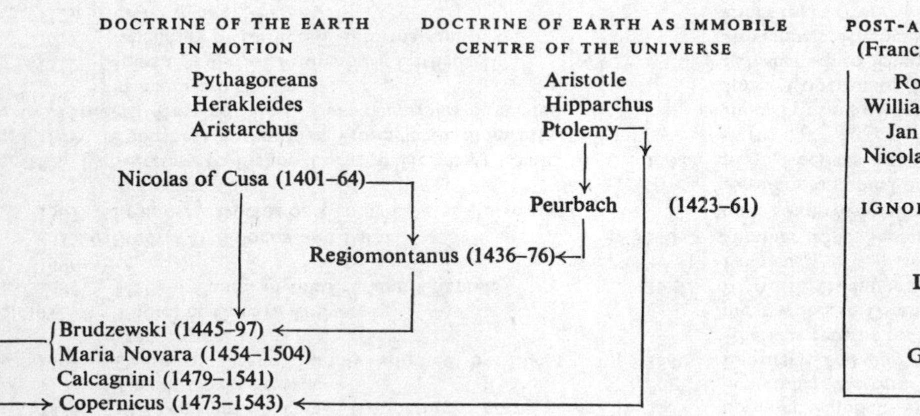

| | | |
|---|---|---|
| A.D. | 1473 | 19 February Nicolas Koppernigk born at Torun, Royal Prussia. |
| „ | 1483 | Death of father. Family adopted by Lucas Waczelrode. |
| „ | 1491–4 | Studies at Cracow University. |
| „ | 1496 | Made a Canon of the Ermland Chapter. |
| „ | 1496–c. 1506 | Studies at Bologna and Padua. |
| „ | 1503 | Promoted Doctor of Canon Law at University of Ferrara. |
| „ | 1506–12 | Secretary to Bishop Lucas at Heilsberg Castle. |
| „ | 1509 | Publishes translation of Teophylactus Simocatta. |
| „ | c. 1510 | (1514 at the latest) *Commentariolus* circulated in manuscript. |
| „ | 1512 | Joins Chapter of Frauenburg Cathedral. |
| „ | 1517 | Beginning of Lutheran Reformation. |
| „ | 1522 | *Letter against Werner.* |
| „ | 1533 | Widmanstad's Lecture in the Vatican Gardens. |
| „ | 1536 | Letter from Cardinal Schoenberg. |
| „ | 1537 | Dantiscus elected Bishop of Ermland. |
| A.D. | 1539 | Summer. Rheticus arrives in Frauenburg. September. *Narratio prima* completed. |
| „ | 1540 | February. *Narratio prima* published in Danzig. Rheticus returns to Wittenberg. 1 July. Copernicus writes to Osiander. |
| „ | 1541 | 20 April. Osiander's letters to Copernicus and Rheticus. |
| „ | 1540–1 | Summer 1540–September. 1541. Rheticus' second sojourn in Frauenburg; manuscript of *Revolutions* copied out. |
| „ | 1542 | May. Rheticus arrives in Nuremberg. Printing starts. June. Printing of first two sheets completed. June. Copernicus writes *Dedication to Paul III*, sends it to Rheticus. November. Rheticus leaves Nuremberg. Osiander takes over. |
| „ | 1543 | 24 May. Arrival of first printed copy of *Revolutions*. Death of Copernicus. |
| „ | 1576 | Death of Rheticus. |

# Part Four. The Watershed

# 1 The Young Kepler

## 1. *Decline of a Family*

Johannes Kepler, Keppler, Khepler, Kheppler, or Keplerus was conceived on 16 May A.D. 1571, at 4.37 a.m., and was born on 27 December at 2.30 p.m., after a pregnancy lasting 224 days, 9 hours and 53 minutes. The five different ways of spelling his name are all his own, and so are the figures relating to conception, pregnancy, and birth, recorded in a horoscope which he cast for himself.[1] The contrast between his carelessness about his name and his extreme precision about dates reflects, from the very outset, a mind to whom all ultimate reality, the essence of religion, of truth and beauty, was contained in the language of numbers.

He was born in the township of Weil in wine-happy Swabia, a blessed corner of south-west Germany between the Black Forest, the Neckar, and the Rhine. Weil-der-Stadt – a freak name, meaning Weil-the-Town, but with the masculine 'der' instead of the feminine 'die' – has beautifully succeeded in preserving its medieval character to our day.* It stretches along the top of a mound, long and narrow like the hull of a battleship, surrounded by massive, crenelated, ochre-coloured walls, and slender watch-towers topped by spire and weathercock. The gabled houses, with their irregular patterns of small, square windows, are covered with scarab green, topaz blue, and lemon yellow stucco on their cockeyed facades; where the stucco peels, the mud and lath peep through like weathered skin showing through a hole in a peasant's shirt. If, after fruitless knocking, you push open the door of a house, you are liable to be greeted by a calf or a goat, for the ground-floors of some old houses still serve as stables, with an inner staircase leading up to the family's living quarters. The warm smell of

* At least, to be precise, to the days of May 1955, when I visited Kepler's birthplace.

compost floats everywhere in the cobbled streets, but they are kept scrupulously, teutonically clean. The people speak a broad Swabian dialect and frequently address even the stranger with 'thou'; they are rustic and *gemuetlich*, but also alert and bright. There are places outside the walls still called 'God's Acre' and 'Gallows Hill'; and the old family names, down from the mayor, Herr Oberdorfer, to the watchmaker, Herr Speidel, are the same which appear on documents from Kepler's time, when Weil had only two hundred citizens. Though it produced some other distinguished men – among them the phrenologist Gall, who traced each faculty of the mind to a bump on the skull – Johannes Kepler is the town's hero, venerated like a patron saint.[2]

One of the entries, dated 1554, in the municipal ledger, refers to the lease of a cabbage patch to Johannes' grandfather, Sebaldus Kepler:

> Daniel Datter and Sebold Kepler, furrier, shall pay seventeen pennies at Martinmas out of their cabbage patch on the Klingelbrunner Lane between the fields of Joerg Rechten and those of Hans Rieger's children. Should they relinquish the cabbage patch, they shall cart six cartloads of compost into or onto it.

From this bucolic prelude one would expect a happy childhood for the infant Johannes. It was a ghastly one.

Grandfather Sebaldus, the furrier with the cabbage patch, was said to stem from a noble family,[3] and became mayor of Weil; but after him, the respectable Keplers went into decline. His offspring were mostly degenerates and psychopaths, who chose mates of the same ilk. Johannes Kepler's father was a mercenary adventurer who narrowly escaped the gallows. His mother, Katherine, an innkeeper's daughter, was brought up by an aunt who was burnt alive as a witch, and Katherine herself, accused in old age of consorting with the Devil, had as narrow an escape from the stake as the father had from the gallows.

Grandfather Sebaldus' house (burnt down in 1648 but rebuilt later in the same style) stood on a corner of the marketplace. Facing the house is a beautiful Renaissance fountain with four long, fluted copper spouts which issue from four human faces carved into the stone. Three of the faces are stylized masks; the fourth, turned towards the Town Hall and the Kepler house, looks like the realistic portrait of a bloated, coarse-featured man. There is a tradition in Weil according to which it is the likeness of old Sebaldus, the mayor. This may or may not be so, but it tallies with Kepler's own description of him:

My grandfather Sebald, mayor of the imperial city of Weil, born in the year 1521 about St James's day ... is now seventy-five years of age. ... He is remarkably arrogant and proudly dressed ... short-tempered and obstinate, and his face betrays his licentious past. It is a red and fleshy face, and his beard gives it much authority. He was eloquent, at least as far as an ignorant man can be. ... From the year 1578 onward his reputation began to decline, together with his substance. ...[4]

This thumbnail sketch, and the others which follow, are part of a kind of genealogical horoscope, embracing all members of his family (including himself) which Kepler drew up when he was twenty-six. It is not only a remarkable document, but also a precious contribution to the study of the hereditary background of genius, for it happens only rarely that the historian has such ample material at his disopsal.*

When Grandfather Sebald was twenty-nine, he married Katherine Mueller from the nearby village of Marbach. Kepler describes her as:

restless, clever, and lying, but devoted to religion; slim and of a fiery nature; vivacious, an inveterate troublemaker; jealous, extreme in her hatreds, violent, a bearer of grudges. ... And all her children have something of this. ...[5]

He also accuses his grandmother of pretending that she married at eighteen, when she was really twenty-two. However that may be, she bore Sebaldus twelve children in twenty-one years. The first three, named Sebaldus, Johan, and Sebaldus, all died in infancy. The fourth was Kepler's father, Heinrich,

* As the document is a horoscope, events and character traits are derived from planetary constellations, which I have mostly left out.

whom we leave aside for a moment. Of numbers 5 to 9 among his aunts and uncles, Kepler records:[6]

5. Kunigund, born 1549, 23 May. The moon could not have been worse placed. She is dead, the mother of many children, poisoned they think, in the year 1581, 17 July [Added later on: 'Otherwise she was pious and wise'].*

6. Katherine, born 1551, 30 July. She too is dead.

7. Sebaldus, born 1552, 13 November.† An astrologer and a Jesuit, he underwent the first and second ordinations for the priesthood; though a Catholic, he imitated the Lutherans and led a most impure life. Died in the end of dropsy after many earlier illnesses. Acquired a wife who was rich and nobly born, but one of many children. Contracted the French sickness. Was vicious and disliked by his fellow townsmen. In 1576, 16 August, he left Weil for Speyr where he arrived on the 18th; on the 22 December he left Speyr against the will of his superior and wandered in extreme poverty through France and Italy. [He was held to be kind and a good friend.]

8. Katherine, born 1554, 5 August. She was intelligent and skilful, but married most unfortunately, lived sumptuously, squandered her goods, now a beggar. [Died in 1619 or 1620.]

9. Maria, born 1556, 25 August. She too is dead.

Of Nos. 10 and 11, he has nothing to say; No. 12, the last born of his uncles and aunts, also died in infancy.‡

All this mis-shapen progeniture – except those who died in their cots – lived with old choleric Sebaldus and his shrewish wife, crowded into the narrow Kepler house, which, in fact, was rather a cottage. Kepler's father, Heinrich, though the fourth child, was the oldest among those who survived, and

* In later years, Kepler added a few remarks to his text, which soften, and sometimes contradict, the trenchant characterizations of his youth. I have put these addenda into brackets.

† The grandparents' third and last attempt to produce a Sebaldus who would survive.

‡ cf. Kretschmer: 'One is tempted to say, genius arises in the hereditary process particularly at that point where a highly gifted family begins to degenerate. . . . This degeneration often announces itself in the generation to which the genius belongs, or even in the preceding one, and generally in the form of psychopathic and psychotic conditions.'[6a]

thus inherited the house, producing seven children in his turn. Kepler describes him thus:

4. Heinrich, my father, born 1547, 19 January. ... A man vicious, inflexible, quarrelsome, and doomed to a bad end. Venus and Mars increased his malice. Jupiter combust[7] in descension made him a pauper but gave him a rich wife. Saturn in VII made him study gunnery; many enemies, a quarrelsome marriage ... a vain love of honours, and vain hopes about them; a wanderer. ... 1577: he ran the risk of hanging. He sold his house and started a tavern. 1578: a hard jar of gunpowder burst and lacerated my father's face. ... 1589: treated my mother extremely ill, went finally into exile and died.

There is not even the usual mitigating addendum at the end. The story behind the entries is briefly this:

Heinrich Kepler married at the age of twenty-four. He seems to have studied no trade or craft, except 'gunnery', which refers to his later military adventures. Seven months and two weeks after his marriage to Katherine Guldenmann, Johannes Kepler was born. Three years later, after the birth of his second son, Heinrich took the Emperor's shilling and went off to fight the Protestant insurgents in the Netherlands – an act the more ignominious as the Keplers were among the oldest Protestant families in Weil. The next year, Katherine joined her husband, leaving her children in the care of the grandparents. The year after they both returned, but not to Weil, where they were disgraced; instead, Heinrich bought a house in nearby Leonberg; but in a short time left again for Holland, to join the mercenary hordes of the Duke of Alba. It was apparently on this journey that he 'ran the risk of hanging' for some unrecorded crime. He returned once more, sold the house in Leonberg, ran a tavern in Ellmendingen, again went back to Leonberg, and in 1588 vanished for ever from the sight of his family. Rumour has it that he enlisted in the Neapolitan fleet.

His wife Katherine, the innkeeper's daughter, was an equally unstable character. In the family horoscope, Kepler describes her as: 'small, thin, swarthy, gossiping, and quarrelsome, of a bad disposition'. There was not much to choose between the two Katherines, the mother and the grandmother; and yet the mother was the more frightening of the two, with an aura of magic and witchcraft about her. She collected herbs and concocted potions in whose powers she believed; I have already mentioned that the aunt who brought her up, had ended her days at the stake, and that Katherine nearly shared the same fate, as we shall hear.

To complete the survey of this idyllic family, I must mention our Johannes' brothers and sisters. There were six of them; of whom three again died in childhood, and two became normal, law-abiding citizens (Gretchen, who married a vicar, and Christopher, who became a pewterer). But Heinrich, the next in age to Johannes, was an epileptic and a victim of the psychopathic streak running through the family. An exasperating problem child, his youth seems to have been a long succession of beatings, misadventures, and illnesses. He was bitten by animals, nearly drowned, and nearly burnt alive, He was apprenticed to a draper, then a baker, and finally ran away from home when his loving father threatened to sell him. In subsequent years, he was a camp follower with the Hungarian army in the Turkish wars, a street singer, baker, nobleman's valet, beggar, regimental drummer, and halberdier. Throughout this chequered career, he remained the hapless victim of one misadventure after another – always ill, sacked from every job, robbed by thieves, beaten up by highwaymen – until he finally gave up, begged his way home to his mother, and hung to her apron strings until he died at forty-two. In his childhood and youth, Johannes conspicuously shared some of his younger brother's attributes, particularly his grotesque accident-proneness, and constant ill-health combined with hypochondria.

## 2. *Job*

Johannes was a sickly child, with thin limbs and a large, pasty face surrounded by dark curly hair. He was born with defective eyesight – myopia plus anocular polyopy (multiple vision). His stomach and gall bladder gave constant trouble; he suffered from boils, rashes, and probably from piles, for he tells us that he could never sit still for any length of time and had to walk up and down.

The gabled house on the market-place in Weil, with its crooked beams and dolls-house windows, must have been bedlam. The bullying of red-faced old Sebaldus; the high-pitched quarrels of mother Katherine and grandmother Katherine; the brutality of the weak-headed, swashbuckling father; the epileptic fits of brother Heinrich; the dozen or more of seedy uncles and aunts, parents and grandparents, all crowded together in that unhappy little house.

Johannes was four years old when his mother followed her husband to the wars; five, when the parents returned and the family began its restless wanderings to Leonberg, Ellmendingen, and back to Leonberg. He could attend school only irregularly, and from his ninth to his eleventh year did not go to school at all but was 'put to hard work in the country'. As a result, and in spite of his precocious brilliance, it took him twice as long as it took normal children to complete the three classes of the elementary Latin school. At thirteen, he was at last able to enter the lower theological seminary at Adelberg.

The notes on his own childhood and youth, in the family horoscope, read like the diary of Job:

On the birth of Johann Kepler. I have investigated the matter of my conception, which took place in the year 1571, 16 May, at 4.37 a.m. ... My weakness at birth removes the suspicion that my mother was already pregnant at the marriage, which was the 15th of May. ... Thus I was born premature, at thirty-two weeks, after 224 days, ten hours. ... 1575 [aged four] I almost died of smallpox, was in very ill health, and my hands were badly crippled. ... 1577 [aged six]. On my birthday I lost a tooth, breaking it off with a string which I pulled with my hands. ... 1585–6 [fourteen-fifteen]. During these two years, I suffered continually from skin ailments, often severe sores, often from the scabs of chronic putrid wounds in my feet which healed badly and kept breaking out again. On the middle finger of my right hand I had a worm, on the left a huge sore. ... 1587 [sixteen]. On 4 April I was attacked by a fever. ... 1589 [nineteen]. I began to suffer terribly from headaches and a disturbance of my limbs. The mange took hold of me. ... Then there was a dry disease. ... 1591 [twenty]. The cold brought on prolonged mange. ... A disturbance of body and mind had set in because of the excitement of the Carnival play in which I was playing Marianne. ... 1592 [twenty-one]. I went down to Weil and lost a quarter florin at gambling. ... At Cupinga's I was offered union with a virgin; on New Year's Eve I achieved this with the greatest possible difficulty, experiencing the most acute pains of the bladder. ...

Only two brief memories mitigate the gloom and squalor of this childhood. At the age of six:

I heard much of the comet of that year, 1577, and was taken by my mother to a high place to look at it.

And at the age of nine:

I was called outdoors by my parents especially, to look at the eclipse of the moon. It appeared quite red.

So much for the sunny side of life.

No doubt, some of his miseries and ailments existed only in his imagination; while others – all these cold sores, worms on the finger, scabs, and manges – seem like the stigma of his self-detestation, physical projections of the image he had formed of himself: the portrait of a child as a mangy dog. He meant this literally, as we shall see.

## 3. *Orphic Purge*

There are always compensations. In Kepler's case, the compensations offered by destiny were the exceptional educational facilities in his native land.

The Dukes of Wuerttemberg, after embracing the Lutheran creed, had created a modern educational system. They needed erudite clergymen who could hold their own in the religious controversy that was raging across the country, and they needed an efficient administrative service. The Protestant universities in Wittenberg and Tuebingen were the intellectual arsenals of the new creed; the confiscated monasteries and convents provided ideal accommodation for a network of elementary and secondary schools, which fed the universities and chancelleries with bright young men. A system of scholarships and grants for 'the children of the poor and faithful who are of a diligent, Christian, and god-fearing disposition' vouchsafed an efficient selection of candidates. In this respect, Wuerttemberg before the Thirty Years War was a modern welfare state in miniature. Kepler's parents would certainly not have bothered about his education; the precocious brilliance of the child automatically guaranteed his progress from school to seminary and from there to university, as on a moving belt.

The curriculum at the seminary was in Latin, and the pupils were rigorously held to use only Latin even among themselves. In the elementary school already, they were made to read the comedies of Plautus and Terence, to add colloquial fluency to scholarly precision. The German vernacular, though it had acquired a new dignity through Luther's Bible translation, was not yet considered a worthy medium of expression for scholars. As a happy result of this, Kepler's style, in those pamphlets and letters which he wrote in German, has an enchantingly naïve and earthy quality which, in contrast to the dehydrated medieval Latin, sounds like the joyous din of a country fair after the austerities of the lecture room. Canon Koppernigk's German was modelled on the stilted and devious 'Chancellery Style' of the bureaucracy; Kepler's German seems modelled on Luther's pronouncement: 'One should not imitate those asses who ask the Latin language how German should be spoken; but should ask the mother in her home, the children in the gutters, the common man at the fair, and watch their big mouths as they speak, and do accordingly.'

When he had passed the Elementary Latin School, Johannes' good brains, bad health, and interest in religion made the career of a clergyman the obvious choice. The theological seminary which he attended from his thirteenth to his seventeenth year, was divided into a lower (Adelberg) and a higher course (Maulbronn). The curriculum was broad and rounded, adding Greek to Latin, and embracing, besides theology, the study of the pagan classics, rhetorics and dialectics, mathematics, and music. Discipline was strict: classes started in summer at four, in winter at five o'clock in the morning; the seminarists had to wear a sleeveless, shapeless cloak reaching below their knees, and were hardly ever allowed out on leave. Young Kepler recorded two of his most daring and paradoxical utterances from his seminarist days: that the study of philosophy was a symptom of Germany's decline; and that the French language was worthier of study than the Greek. No wonder his fellows regarded him as an intolerable egghead and beat him up at every opportunity.

He was, indeed, as unpopular among his schoolmates as he was beloved by his friends in later years. In his horoscope record, the entries relating his physical afflictions alternate with others which reveal his moral misery and loneliness:

> February, 1586. I suffered dreadfully and nearly died of my troubles. The cause was my dishonour and the hatred of my school fellows whom I was driven by fear to denounce. ... 1587. On 4 April I was attacked by a fever from which I recovered in time, but I was still suffering from the anger of my schoolmates with one of whom I had come to blows a month before. Koellin became my friend; I was beaten in a drunken quarrel by Rebstock; various quarrels with Koellin ... 1590. I was promoted to the rank of Bachelor. I had a most iniquitous witness, Mueller, and many enemies among my comrades. ...

The narrative of the horoscope was continued in the same year (his twenty-sixth) in another remarkable document, a self-analysis more unsparing than Rousseau's.[8] Written in the year when his first book was published, when he had undergone a kind of orphic purge and found his final vocation, it is perhaps the most introspective piece of writing of the Renaissance. Several pages of it describe his relations with colleagues and teachers at the seminary, and later at the University of Tuebingen. Referring to himself in the third person, as he mostly does in this document, the passage begins: 'From the time of his arrival [at the seminary] some men were his adversaries.' He lists five of them, then continues: 'I record the most lasting enemies.' He lists another seventeen, 'and many other such'. He explains their hostility mainly on the grounds that 'they were always rivals in worth, honours, and success'. There follows a monotonous and depressing record of these enemities and quarrels. Here are samples:

> Kolinus did not hate me, rather I hated him. He started a friendship with me, but continually opposed me. . . . My love of pleasure and other habits turned Braunbaum from being a friend into an equally great enemy. . . . I willingly incurred the hatred of Seiffer because the rest hated him too, and I provoked him although he had not harmed me. Ortholphus hated me as I hated Kolinus, although I on the contrary liked Ortholphus, but the rivalry between us was manysided. . . . I have often incensed everyone against me through my own fault: at Adelberg it was my treachery [in denouncing his schoolmates]; at Maulbronn, my defence of Graeter; at Tuebingen, my violent request for silence. Lendlinus I alienated by foolish writings, Spangenburg, by my temerity in correcting him when he was my teacher; Kleberus hated me as a rival. . . . The reputation of my talent annoyed Rebstock and also my frivolousness. . . . Husalius opposed my progress. . . . With Dauber there was a secret rivalry and jealousy. . . . My friend Jaeger betrayed my trust: he lied to me and squandered much of my money. I turned to hatred and exercised it in angry letters during the course of two years.

And so on. The list of friends turned into enemies ends with the pathetic remark:

> Lastly, religion divided Crellius from me, but he also broke faith; henceforth I was enraged with him. God decreed that he should be the last. And so the cause was partly in me and partly in fate. On my part anger, intolerance of bores, an excessive love of annoying and of teasing, in short of checking presumptions. . . .

Even more pathetic is the one exception in the list:

> Lorhard never communicated with me. I admired him, but he never knew this, nor did anyone else.

Immediately following this dismal recital, Kepler put down, with acid amusement, this portrait of himself – where the past tense alternates revealingly with the present:[9]

> That man [i.e. Kepler] has in every way a dog-like nature. His appearance is that of a little lap-dog. His body is agile, wiry, and well-proportioned. Even his appetites were alike: he liked gnawing

bones and dry crusts of bread, and was so greedy that whatever his eyes chanced on he grabbed; yet, like a dog, he drinks little and is content with the simplest food. His habits were similar. He continually sought the goodwill of others, was dependent on others for everything, ministered to their wishes, never got angry when they reproved him and was anxious to get back into their favour. He was constantly on the move, ferreting among the sciences, politics, and private affairs, including the lowest kind; always following someone else, and imitating his thoughts and actions. He is bored with conversation, but greets visitors just like a little dog; yet when the last thing is snatched away from him, he flares up and growls. He tenaciously persecutes wrong-doers – that is, he barks at them. He is malicious and bites people with his sarcasms. He hates many people exceedingly and they avoid him, but his masters are fond of him. He has a dog-like horror of baths, tinctures, and lotions. His recklessness knows no limits, which is surely due to Mars in quadrature with Mercury, and in trine with the moon; yet he takes good care of his life. ... [He has] a vast appetite for the greatest things. His teachers praised him for his good dispositions, though morally he was the worst among his contemporaries. ... He was religious to the point of superstition. As a boy of ten years when he first read Holy Scripture ... he grieved that on account of the impurity of his life, the honour to be a prophet was denied him. When he committed a wrong, he performed an expiatory rite, hoping it would save him from punishment: this consisted in reciting his faults in public. ...

In this man there are two opposite tendencies: always to regret any wasted time, and always to waste it willingly. For Mercury makes one inclined to amusements, games, and other light pleasures. ... Since his caution with money kept him away from play, he often played by himself. [The word for 'play', *lusu*, may refer here either to gambling or to sex.] It must be noted that his miserliness did not aim at acquiring riches, but at removing his fear of poverty – although, perhaps avarice results from an excess of this fear. ...

Of love, there is no mention, with two scant exceptions: the painful episode with the virgin on New Year's Eve, and an isolated obscure entry, referring to his twentieth year:

1591. The cold brought on prolonged mange. When Venus went through the Seventh House, I was reconciled with Ortholphus: when she returned, I showed her to him; when she came back a third time, I still struggled on, wounded by love. The beginning of love: 26 April.

That is all. We are told no more about that nameless 'she'.

We remember that Kepler wrote this at the age of twenty-six. It would be a harsh self-portrait even for a modern young man, reared in the age of psychiatry, anxiety, masochism, and the rest; coming from a young German at the close of the sixteenth century, the product of a coarse, brutal, and callow civilization, it is an astonishing document. It shows the ruthless intellectual honesty of a man whose childhood was spent in hell and who had fought his way out of it.

With all its rambling inconsequences, its baroque mixture of sophistication and naïvety, it unfolds the timeless case-history of the neurotic child from a problem family, covered with scabs and boils, who feels that whatever he does is a pain to others and a disgrace to himself. How familiar it all is: the bragging, defiant, aggressive pose to hide one's terrible vulnerability; the lack of self-assurance, the dependence on others, the desperate need for approval, leading to an embarrassing mixture of servility and arrogance; the pathetic eagerness for play, for an escape from the loneliness which he carries with him like a portable cage; the vicious circle of accusations and self-accusations; the exaggerated standards applied to one's own moral conduct which turns life into a long series of Falls into the ninefold inferno of guilt.

Kepler belonged to the race of bleeders, the victims of emotional haemophilia, to whom every injury means multiplied danger, and who nevertheless must go on exposing himself to stabs and slashes. But one customary feature is conspicuously absent from his writings: the soothing drug of self-pity, which makes the sufferer spiritually impotent, and prevents his suffering from bearing fruit. He was a Job who shamed his Lord by making trees grow from his boils. In other words, he had that mysterious knack of finding original outlets for inner pressure; of transforming his torments into creative achievement, as a turbine extracts electric current out of the turbulent stream. His eye-deficiency seems the most perfidious trick that fate could inflict on a stargazer; but how is one to decide whether an inborn affliction will paralyse or galvanize? The myopic child, who sometimes saw the world doubled or quadrupled, became the founder of modern optics (the word 'dioptries' on the oculist's prescription is derived from the title of one of Kepler's books); the man who could only see clearly at a short distance, invented the modern astronomical telescope. We shall have occasion to watch the working of this magic dynamo, which transforms pain into achievement and curses into blessings.

### 4. *Appointment*

He graduated from the Faculty of Arts at the University of Tuebingen at the age of twenty. Then, continuing on the road of his chosen vocation, he matriculated at the Theological Faculty. He studied there for nearly four years, but before he could pass his final examinations, fate intervened. The candidate of divinity was unexpectedly offered the post of a teacher of mathematics and astronomy in Gratz, capital of the Austrian province of Styria.

Styria was a country ruled by a Catholic Hapsburg prince and its predominantly Protestant Estates. Gratz accordingly had both a Catholic university and a Protestant school. When, in 1593, the mathematicus of the latter died, the Governors asked, as they often did, the Protestant university of Tuebingen to recommend a candidate. The Tuebingen senate recommended Kepler. Perhaps they wanted to get rid of the querulous young man, who had professed Calvinist views and defended Copernicus in a public disputation. He would make a bad priest but a good teacher of mathematics.

Kepler was taken by surprise and at first inclined to refuse – 'not because I was afraid of the great distance of the place (a fear which I condemn in others) but because of the unexpected and lowly nature of the position, and my scant knowledge in this branch of philosophy'.[10] He had never thought of becoming an astronomer. His early interest in Copernicus had been one among many others; it had been aroused, not by an interest in astronomy proper, but by the mystical implications of the sun-centred universe.

Nevertheless, after some hesitations he accepted the offer – mainly, it seems, because it meant financial independence, and because of his inborn love of adventure. He made it a condition, however, that he should be allowed to resume his study of divinity at a later date – which he never did.

The new teacher of astronomy and 'Mathematicus of the Province' – a title that went with it – arrived in Gratz in April 1594, at the age of twenty-three. A year later he hit on the idea which would dominate the rest of his life, and out of which his revolutionary discoveries were born.

I have so far concentrated on the emotional life of his childhood and adolescence. I must now briefly speak of his intellectual development. Here again, we have his self-portrait to guide us:

> This man was born destined to spend much time on difficult tasks from which others shrunk. As a boy he precociously attempted the science of versifying. He tried to write comedies and chose the longest poems to learn by heart. . . . His efforts were at first devoted to acrostics and anagrams. Later on he set about various most difficult forms of lyric poetry, wrote a pindaric lay, dithyrambic poems and compositions on unusual subjects, such as the resting-place of the sun, the sources of rivers, the sight of Atlantis through the clouds. He was fond of riddles and subtle witticisms and made much play with allegories which he worked out to the most minute detail, dragging in far-fetched comparisons. He liked to compose paradoxes and . . . loved mathematics above all other studies.
>
> In philosophy he read the texts of Aristotle in the original. . . . In theology he started at once on predestination and fell in with the Lutheran view of the absence of free will. . . . But later on he opposed it. . . . Inspired by his view of divine mercy, he did not believe that any nation was destined to damnation. . . . He explored various fields of mathematics as if he were the first man to do so [and made a number of discoveries], which later on he found to have already been discovered. He argued with men of every profession for the profit of his mind. He jealously preserved all his writings and kept any book he could lay hands on with the idea

that they might be useful at some time in the future. He was the equal of Crusius* in his attention to detail, far inferior to Crusius in industry, but his superior in judgement. Crusius collected facts, he analysed them; Crusius was a hoe, he a wedge. .

In his Horoscope he further reports that during his first year at the University he wrote essays on 'the heavens, the spirits, the Genii, the elements, the nature of fire, the tides, the shape of the continents, and other things of the same kind'.

The last remark about his student days reads:

At Tuebingen I often defended the opinions of Copernicus in the disputation of the candidates, and I composed a careful disputation on the first motion, which consists in the rotation of the earth; then I was adding to this the motion of the earth around the sun for physical, or if you prefer, metaphysical reasons.

If there are living creatures on the moon (a matter about which I took pleasure in speculating after the manner of Pythagoras and Plutarch in a disputation written in Tuebingen in 1593), it is to be assumed that they should be adapted to the character of their particular country.

None of this points as yet in any definite direction. Indeed, his main complaint against himself, which he repeats over and over again, is his 'inconsistency, thoughtlessness, lack of discipline, and rashness'; his 'lack of persistence in his undertakings, caused by the quickness of his spirit'; his 'beginning many new tasks before the previous one is finished'; his 'sudden enthusiasms which do not last, for, however industrious he may be, nevertheless he is a bitter hater of work'; his 'failure to finish things he has begun'.

Again we see that magic dynamo of the psyche at work. The streak of irresponsibility and restlessness in the blood, which turned his father, brother, and uncles into vagabonds who could never settle down in any place or profession, drove Kepler into his unorthodox, often crankish intellectual enterprises, made him into the most reckless and erratic spiritual adventurer of the scientific revolution.

The lectures of this new teacher must have been quite an

* One of Kepler's teachers.

experience. He thought himself a poor pedagogue because, as he explains in his self-analysis, whenever he got excited – which was most of the time – he 'burst into speech without having time to weigh whether he was saying the right thing'. His 'enthusiasm and eagerness is harmful, and an obstacle to him', because it continually leads him into digressions, because he always thinks of 'new words and new subjects, new ways of expressing or proving his point, or even of altering the plan of his lecture or holding back what he intended to say'. The fault, he explains, lies in his peculiar kind of memory which makes him promptly forget everything he is not interested in, but which is quite wonderful in relating one idea to another. 'This is the cause of the many parentheses in his lectures when everything occurs to him at once and, because of the turmoil of all these images of thought in his memory, he must pour them out in his speech. On these grounds his lectures are tiring, or at any rate perplexing and not very intelligible.'

No wonder that in his first year he had only a handful of students in his class, and in his second, none at all. Barely twelve months after his arrival in Gratz he wrote to his old teacher of astronomy in Tuebingen, Michael Maestlin, that he could not hope to last for another year, imploring Maestlin to get him a job back at home He felt unhappy, an exile from his sophisticated *alma mater* among the provincial Styrians. On his arrival, he had been promptly attacked by 'Hungarian fever'. Besides, religious tension was growing in the town, and made prospects even gloomier.

However, the directors of the school took a more optimistic view. In their report on the new teacher[11] they explained that the absence of students should not be blamed on him, 'because the study of mathematics is not every man's affair'. They made him give some additional lectures on Virgil and rhetorics 'so that he should not be paid for nothing – until the public is prepared to profit from his mathematics too'. The remarkable thing about their reports is their unmitigated approval not only of Kepler's intellect, but also of his character. He had 'at first *perorando*, then *docendo*, and finally also *disputando*, given such account of himself that we cannot judge otherwise but that he is, in spite of his youth, a learned and *in moribus* a modest, and to this school of a respected Province a fitting magister and professor'. This praise contradicts Kepler's own statement that the head of the school was his 'dangerous enemy', because 'I did not respect him sufficiently as my superior and disregarded his orders'.[12] But young Kepler was as hypochondriacal about his relations to others as he was about his health.

## 5. *Astrology*

Another onerous duty, which he secretly enjoyed, during his four years in Gratz, was the publication of an annual calendar of astrological forecasts. This was a traditional obligation imposed on the official mathematicus in Styria and brought an additional remuneration of twenty florins per calendar – which Kepler direly needed at his miserable salary of a hundred and fifty florins *per annum*.

With his first calendar, Kepler was decidedly lucky. He had prophesied, among other things, a cold spell and an invasion by the Turks. Six months later he reported smugly to Michael Maestlin:

> By the way, so far the calendar's predictions are proving correct. There is an unheard-of cold in our land. In the Alpine farms people die of the cold. It is reliably reported that when they arrive home and blow their noses, the noses fall off. ... As for the Turks, on 1 January they devastated the whole country from Vienna to Neustadt, setting everything on fire and carrying off men and plunder.[13]

The successful prophecies of the first calendar contributed more to the popularity of the new mathematicus than his enthusiastic and garbled lectures before an empty class-room. As always in times of crisis, belief in astrology was again on the increase in the sixteenth century, not only among the ignorant, but among eminent scholars. It played an important, and at times a dominant part in Kepler's life. His attitude to it was typical of the contradictions in his character, and of an age of transition.

He started his career with the publication of astrological calendars, and ended it as Court Astrologer to the Duke of Wallenstein. He did it for a living, with his tongue in his cheek, called astrology 'the step-daughter of astronomy', popular prophecies 'a dreadful superstition' and 'a sortilegous monkey-play'.[14] In a typical outburst he wrote: 'A mind accustomed to mathematical deduction, when confronted with the faulty foundations [of astrology] resists a long, long time, like an obstinate mule, until compelled by beating and curses to put its foot into that dirty puddle.'[15]

But while he despised these crude practices, and despised himself for having to resort to them, he at the same time believed in the possibility of a new and true astrology as an exact empirical science. He wrote a number of serious treatises on astrology as he would understand it, and the subject constantly intrudes even in his classic scientific works. One of these treatises carries, as a motto, 'a warning to certain Theologians, Physicians, and Philosophers . . . that, while justly rejecting the stargazers' superstitions, they should not throw out the child with the bathwater'.[16] For 'nothing exists nor happens in the visible sky that is not sensed in some hidden manner by the faculties of Earth and Nature: [so that] these faculties of the spirit here on earth are as much affected as the sky itself'.[17] And again: 'That the sky does something to man is obvious enough; but what it does specifically remains hidden.'[18] In other words, Kepler regarded the current astrological practices as quackery, but only to the extent to which a modern physician distrusts an unproven slimming diet, without doubting for a moment the influence of diet on health and figure. 'The belief in the effect of the constellations derives in the first place from experience, which is so convincing that it can be denied only by people who have not examined it.'[19]

We have seen that in his self-analysis, in spite of its astoundingly modern introspective passages and acute characterizations of his family, all main events and character-attributes were derived from the planetary constellations. But on reflection, what other explanation was there available at the time? To a questing mind without an inkling of the processes by which heredity and environment shape a man's character, astrology, in one form or another, was the obvious means of relating the individual to the universal whole, by making him reflect the all-embracing constellation of the world, by establishing an intimate sympathy and correspondence between microcosmos and macrocosmos: 'The natural soul of man is not larger in size than a single point, and on this point the form and character of the entire sky is potentially engraved, even if it were a hundred times larger.'[20] Unless predestination alone were to account for everything, making further inquiry into the Book of Nature pointless, it was only logical to assume that man's condition and fate were determined by the same celestial motions which determine the weather and the seasons, the quality of the harvest, the fertility of animal and plant. In a word, astrological determinism, to a scientific mind like Kepler's, was the forerunner of biological and psychological determinism.

Already as a child he was fascinated by the problem of why he had become what he had become. We remember the passage in his self-analysis: 'In theology I started at once on predestination and fell into the Lutheran view of the absence of free will.' But he quickly repudiated it. When he was thirteen, 'I wrote to Tuebingen asking that a certain theological treatise be sent to me, and one of my comrades upbraided me thus: "Bachelor, does't thou too suffer from doubts about predestination?" '[21] The mystery of 'why am I what I am?' must have been experienced with particular intensity by a precocious and unhappy adolescent in that century of awakening, when individual consciousness was emerging from the collective consciousness of the medieval beehive-hierarchy, where queens and warriors, workers and drones, had all inhabited their ordained cubby-holes in existence. But if there was no predestination, how was one to explain the differences in character and personality, talent and worth, between members of the same race, all descended from Adam; or between young Johannes himself, the infant prodigy, and his epileptic brother? Modern man has an explanation of sorts in terms of chromosomes and genes, adaptive responses and traumatic experiences;

sixteenth-century man could only search for an explanation in the state of the universe as a whole at the moment of his conception or birth, as expressed by the constellation of earth, planets, and stars.

The difficulty was to find out how exactly this influence worked. That 'the sky does something to man' was self-evident; but specifically what? 'Truly in all my knowledge of astrology I know not enough with certainty that I should dare to predict with confidence any specific thing.'[21a] Yet he never gave up hope:

> No man should hold it to be incredible / that out of the astrologers' foolishness and blasphemies / some useful and sacred knowledge may come / that out of the unclean slime / may come a little snail / or mussel / or oyster or eel, all useful nourishments; / that out of a big heap of lowly worms / may come a silk worm / and lastly / that in the evil-smelling dung / a busy hen may find a decent corn / nay, a pearl or a golden corn / if she but searches and scratches long enough.[22]

There is hardly a page in Kepler's writings – some twenty solid volumes in folio – that is not alive and kicking.

And gradually, a vision did indeed emerge out of the confusion. At twenty-four, he wrote to a correspondent:

> In what manner does the countenance of the sky at the moment of a man's birth determine his character? It acts on the person during his life in the manner of the loops which a peasant ties at random around the pumpkins in his field: they do not cause the pumpkin to grow, but they determine its shape. The same applies to the sky: it does not endow man with his habits, history, happiness, children, riches, or a wife, but it moulds his condition. . . .[23]

Thus only the pattern is cosmically determined, not any particular event; within that pattern, man is free. In his later years, this *Gestalt* concept of cosmic destiny became more abstract and purified from dross. The individual soul, which bears the potential imprint of the entire sky, reacts to the light coming from the planets according to the angles they form with each other, and the geometrical harmonies or disharmonies that result – just as the ear reacts to the mathematical harmonies of music, and the eye to the harmonies of colour. This capacity of the soul to act as a cosmic resonator has a mystic and a causal aspect: on the one hand it affirms the soul's affinity with the *anima mundi*, on the other, it makes it subject to strictly mathematical laws. At this point, Kepler's particular brand of astrology merges into his all-embracing and unifying Pythagorean vision of the Harmony of the Spheres.

# 2 The 'Cosmic Mystery'

## 1. *The Perfect Solids*

From the frustrations of his first year in Gratz, Kepler escaped into the cosmological speculations which he had playfully pursued in his Tuebingen days. But now these speculations were becoming both more intense, and more mathematical in character. A year after his arrival – more precisely on 9 July 1595, for he has carefully recorded the date – he was drawing a figure on the blackboard for his class, when an idea suddenly struck him with such force that he felt he was holding the key to the secret of creation in his hand. 'The delight that I took in my discovery,' he wrote later, 'I shall never be able to describe in words.'[1] It determined the course of his life, and remained his main inspiration throughout it.

The idea was, that the universe is built around certain symmetrical figures – triangle, square, pentagon, etc. – which form its invisible skeleton, as it were. Before going into detail, it will be better to explain at once that the idea itself was completely false; yet it nevertheless led eventually to Kepler's Laws, the demolition of the antique universe on wheels, and the birth of modern cosmology. The pseudo-discovery which started it all is expounded in Kepler's first book, the *Mysterium Cosmographicum*,* which he published at the age of twenty-five.

In the Preface to the work, Kepler explained how he came to make his 'discovery'. While still a student in Tuebingen, he had heard from his teacher in astronomy, Maestlin, about Copernicus, and agreed that the sun must be in the centre of

* The full title reads: *A Forerunner (Prodromus) to Cosmographical Treatises, containing the Cosmic Mystery of the admirable proportions between the Heavenly Orbits and the true and proper reasons for their Numbers, Magnitudes, and Periodic Motions*, by Johannes Kepler, Mathematicus of the Illustrious Estates of Styria, Tuebingen, *anno* 1596.

the universe 'for physical, or if you prefer, for metaphysical reasons'. He then began to wonder why there existed just six planets 'instead of twenty or a hundred', and why the distances and velocities of the planets were what they were. Thus started his quest for the laws of planetary motion.

At first he tried whether one orbit might perchance be twice, three or four times as large as another. 'I lost much time on this task, on this play with numbers; but I could find no order either in the numerical proportions or in the deviations from such proportions.' He warns the reader that the tale of his various futile efforts 'will anxiously rock thee hither and thither like the waves of the sea'. Since he got nowhere, he tried 'a startlingly bold solution': he inserted an auxiliary planet between Mercury and Venus, and another between Jupiter and Mars, both supposedly too small to be seen, hoping that now he would get some sensible sequence of ratios. But this did not work either; nor did various other devices which he tried.

I lost almost the whole of the summer with this heavy work. Finally I came close to the true facts on a quite unimportant occasion. I believe Divine Providence arranged matters in such a way that what I could not obtain with all my efforts was given to me through chance; I believe all the more that this is so as I have always prayed to God that he should make my plan succeed, if what Copernicus had said was the truth.[2]

The occasion of this decisive event was the aforementioned lecture to his class, in which he had drawn, for quite different

purposes, a geometrical figure on the blackboard. The figure showed (I must describe it in a simplified manner) a triangle fitted between two circles; in other words, the outer circle was circumscribed around the triangle, the inner circle inscribed into it.

As he looked at the two circles, it suddenly struck him that their ratios were the same as those of the orbits of Saturn and Jupiter. The rest of the inspiration came in a flash. Saturn and Jupiter are the 'first' (i.e. the two outermost) planets, and 'the triangle is the first figure in geometry. Immediately I tried to inscribe into the next interval between Jupiter and Mars a square, between Mars and Earth a pentagon, between Earth and Venus a hexagon. . . .'

It did not work – not yet, but he felt that he was quite close to the secret. 'And now I pressed forward again. Why look for two-dimensional forms to fit orbits in space? One has to look for three-dimensional forms – and, behold dear reader, now you have my discovery in your hands! . . .'

The point is this. One can construct any number of regular polygons in a two-dimensional plane; but one can only construct a limited number of regular solids in three-dimensional space. These 'perfect solids', of which all faces are identical, are: (1) the tetrahedron (pyramid) bounded by four equilateral triangles; (2) the cube; (3) the octahedron (eight equilateral triangles); (4) the dodecahedron (twelve pentagons) and (5) the icosahedron (twenty equilateral triangles).

They were also called the 'Pythagorean' or 'Platonic' solids. Being perfectly symmetrical, each can be *inscribed* into a sphere, so that all of its vertices (corners) lie on the surface of the sphere. Similarly, each can be *circumscribed* around a sphere, so that the sphere touches every face in its centre. It is a curious fact, inherent in the nature of three-dimensional space, that (as Euclid proved) the number of regular solids is limited to these five forms. Whatever shape you choose as a face, no other perfectly symmetrical solid can be constructed except these five. Other combinations just cannot be fitted together.

So there existed only five perfect solids – and five intervals between the planets! It was impossible to believe that this should be by chance, and not by divine arrangement. It provided the complete answer to the question why there were just six planets 'and not twenty or a hundred'. And it also answered the question why the distances between the orbits were as they were. They had to be spaced in such a manner that the five solids could be exactly fitted into the intervals, as an invisible skeleton or frame. And lo, they fitted! Or at least, they seemed

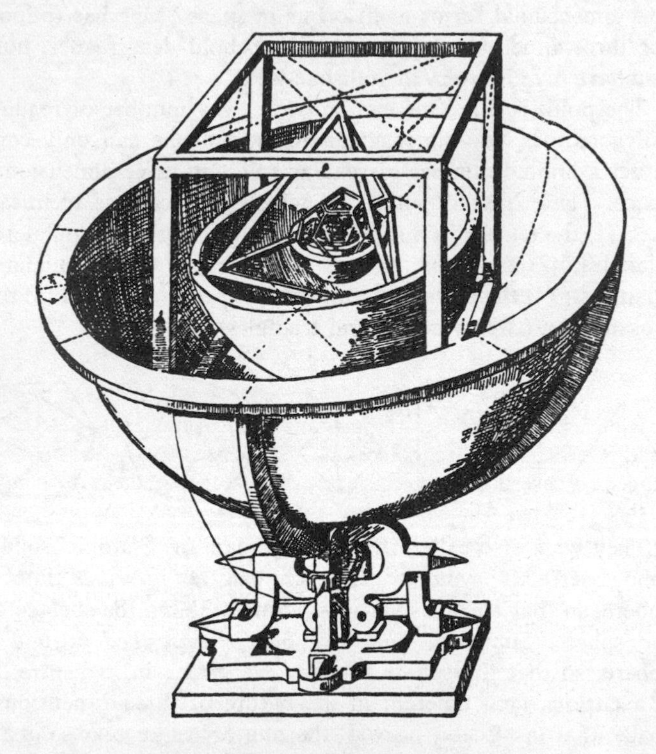

Model of the universe; the outermost sphere is Saturn's.
Illustration in Kepler's *Mysterium Cosmographicum.*

to fit, more or less. Into the orbit, or sphere, of Saturn he inscribed a cube; and into the cube another sphere, which was that of Jupiter. Inscribed in that was the tetrahedron, and inscribed in it the sphere of Mars. Between the spheres of Mars and Earth came the dodecahedron; between Earth and Venus the icosahedron; between Venus and Mercury the octahedron. Eureka! The mystery of the universe was solved by young Kepler, teacher at the Protestant school in Gratz.

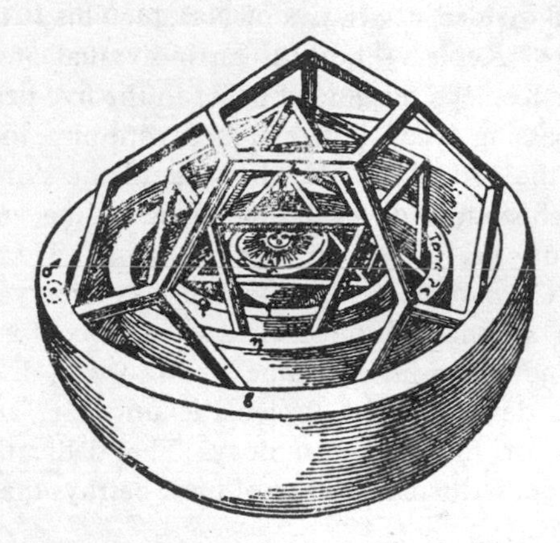

Detail, showing the spheres of Mars, Earth, Venus and Mercury with the Sun in the centre.

It is amazing! [Kepler informs his readers] although I had as yet no clear idea of the order in which the perfect solids had to be arranged, I nevertheless succeeded ... in arranging them so happily that later on, when I checked the matter over, I had nothing to alter. Now I no longer regretted the lost time; I no longer tired of my work; I shied from no computation, however difficult. Day and night I spent with calculations to see whether the proposition that I had formulated tallied with the Copernician orbits or whether my joy would be carried away by the winds. ... Within a few days everything fell into its place. I saw one symmetrical solid after the other fit in so precisely between the appropriate orbits, that if a peasant were to ask you on what kind of hook the heavens are fastened so that they don't fall down, it will be easy for thee to answer him. Farewell![3]

We had the privilege of witnessing one of the rare recorded instances of a false inspiration, a supreme hoax of the Socratic *daimon*, the inner voice that speaks with such infallible, intuitive certainty to the deluded mind. That unforgettable moment before the figure on the blackboard carried the same inner conviction as Archimedes' *Eureka* or Newton's flash of insight about the falling apple. But there are few instances where a delusion led to momentous and true scientific discoveries and yielded new Laws of Nature. This is the ultimate fascination of Kepler – both as an individual and as a case history. For Kepler's misguided belief in the five perfect bodies was not a passing fancy, but remained with him, in a modified version, to the end of his life, showing all the symptoms of a paranoid delusion; and yet it functioned as the *vigor motrix*, the spur of his immortal achievements. He wrote the *Mysterium Cosmographicum* when he was twenty-five, but he published a second edition of it a quarter-century later, towards the end, when he had done his life work, discovered his three Laws, destroyed the Ptolematic universe, and laid the foundations of modern cosmology. The dedication to this second edition, written at the age of fifty, betrays the persistence of the *idée fixe*:

> Nearly twenty-five years have passed since I published the present little book. . . . Although I was then still quite young and this publication my first work on astronomy, nevertheless its success in the following years proclaims with a loud voice that never before has anybody published a more significant, happier, and in view of its subject, worthier first-book. It would be mistaken to regard it as a pure invention of my mind (far be any presumption from my intent, and any exaggerated admiration from the reader's, when we touch the seven-stringed harp of the Creator's wisdom). For as if a heavenly oracle had dictated it to me, the published booklet was in all its parts immediately recognized as excellent and true throughout (as it is the rule with obvious acts of God).

Now, Kepler's style is often exuberant and sometimes bombastic, but rarely to this extent. The apparent presumption is in fact the radiance of the *idée fixe*, an emanation of the immense emotive charge which such ideas carry. When the patient in a mental home declares that he is the mouthpiece of the Holy Ghost, he means it not as a boast but as a flat statement of fact.

Here we have, then, a young man of twenty-four, an aspirant of theology, with only a sketchy knowledge of astronomy, who hits upon a crank idea, convinced that he has solved the 'cosmic mystery'. 'There is no great ingenuity,' to quote Seneca, 'without an admixture of dementedness,' but as a rule the dementedness devours the ingenuity. Kepler's history will show how exceptions to this rule may occur.

## 2. *Contents of the* Mysterium

Leaving its crankish *leitmotif* aside, Kepler's first book contains the seeds of his principal future discoveries. I must therefore briefly describe its content.

The *Mysterium* has an overture, a first and a second movement. The overture consists of the *Introduction to the Reader*, which I have already discussed, and the first chapter, which is an enthusiastic and lucid profession of faith in Copernicus.[4] It was the first unequivocal, public commitment by a professional astronomer which appeared in print fifty years after Canon Koppernigk's death, and the beginning of his posthumous triumph.[5] Galileo, by six years Kepler's senior, and astronomers like Maestlin, were still either silent on Copernicus, or agreed with him only in cautious privacy. Kepler had intended to add to his chapter a proof that there was no contradiction between the teaching of Copernicus and Holy Scripture; but the head of the theological faculty in Tuebingen, whose official consent to the publication of the book had to be obtained, directed him to leave out any theological reflections and – in the tradition of the famous Osiander preface – to treat the Copernican hypothesis as a purely formal, mathematical one.* Kepler accordingly postponed his theological apologia to a later work, but

* It was, as we know, Kepler himself who, a few years later, discovered that the preface to the Revolutions was written by Osiander and not by Copernicus.

otherwise did the exact opposite of what he was advised to do, by proclaiming the Copernican system to be literally, physically, and incontrovertibly true, 'an inexhaustible treasure of truly divine insight into the wonderful order of the world and all bodies therein'. It sounded like a fanfare in praise of the brave new heliocentric world. The arguments in its favour which Kepler adduced could mostly be found in Rheticus' *Narratio Prima*, which Kepler reprinted as an appendix to the *Mysterium*, to save his readers the labour of toiling through Copernicus' unreadable book.

After this overture, Kepler gets down to his 'principal proof' that the planetary spheres are separated from each other, or fenced in, as it were, by the five perfect solids. (He does not mean, of course, that the solids are really present in space, nor does he believe in the existence of the spheres themselves, as we shall see.) The 'proof' consists, roughly, in the deduction that God could only create a perfect world, and since only five symmetrical solids exist, they are obviously meant to be placed between the six planetary orbits 'where they fit in perfectly'. In fact, however, they do not fit at all, as he was soon to discover to his woe. Also, there are not six planets but nine (not to mention the small fry of asteroids between Jupiter and Mars), but at least Kepler was spared in his lifetime the discovery of the three others, Uranus, Neptune, and Pluto.

In the next six chapters (III to VIII), it is explained to us why there are three planets outside and two inside the earth's orbit; why that orbit is placed just where it is; why the cube lies between the two outermost planets and the octahedron between the two innermost; what affinities and sympathies exist between the various planets and the various solids, and so on – all this by *a priori* deductions derived straight from the Creator's secret thoughts, and supported by reasons so fantastic that one can hardly believe one is listening to one of the founders of modern science. Thus, for instance,

> the regular solids of the first order [i.e. those which lie outside the earth's orbit] have it in their nature to stand upright, those of the second order to float. For, if the latter are made to stand on one of

their sides, the former on one of their corners, then in both cases the eye shies from the ugliness of such a sight.

By this kind of argument young Kepler succeeds in proving everything that he believes and in believing everything that he proves. The ninth chapter deals with astrology, the tenth with numerology, the eleventh with the geometrical symbolism of the Zodiac; in the twelfth, he alludes to the Pythagorean harmony of the spheres, searching for correlations between his perfect solids and the harmonic intervals in music – but it is merely one more arabesque to the dream. On this note ends the first half of the book.

The second is different. I have talked of a work in two movements, because they are written in different moods and keys, and are held together only by their common *leitmotif*. The first is medieval, aprioristic, and mystical; the second modern and empirical. The *Mysterium* is the perfect symbol of the great watershed.

The opening paragraph of the second half must have come as a shock to his readers:

> What we have so far said served merely to support our thesis by arguments of probability. Now we shall proceed to the astronomical determination of the orbits and to geometrical considerations. If these do not confirm the thesis, then all our previous efforts have doubtless been in vain.[6]

So all the divine inspiration and *a priori* certitude were merely 'probabilities'; and their truth or falsehood was to be decided by the observed facts. Without transition, in a single startling jump, we have traversed the frontier between metaphysical speculation and empirical science.

Now Kepler got down to brass tacks: the checking of the proportions of his model of the universe against the observed data. Since the planets do not revolve around the sun in circles but in oval-shaped orbits (which Kepler's First Law, years later, identified as ellipses), each planet's distance from the sun varies within certain limits. This variation (or eccentricity) he accounted for by allotting to each planet a spherical shell of sufficient thickness to accommodate the oval orbit between its walls (*see* the model on p. 252). The inner wall represents the planet's minimum distance from the sun, the outer wall its maximum distance. The spheres, as already mentioned, are not considered as physically real, but merely as the limits of space allotted to each orbit. The thickness of each shell and the intervals between them, were laid down in Copernicus' figures. Were they spaced in such a way that the five solids could be exactly fitted between them? In the Preface, Kepler had confidently announced that they could. Now he found that they could not. There was fairly good agreement for the orbits of Mars, Earth, and Venus, but not for Jupiter and Mercury. The trouble with Jupiter Kepler dismissed with the disarming remark that 'nobody will wonder at it, considering the great distance'. As for Mercury, he frankly resorted to cheating.[7] It was a kind of Wonderland croquet through mobile celestial hoops.

In the following chapters Kepler tried various methods to explain away the remaining discrepancies. The fault must lie either in his model or in the Copernican data; and Kepler naturally preferred to blame the latter. First, he discovered that Copernicus had placed into the centre of the world not really the sun, but the centre of the earth's orbit, 'in order to save himself trouble and so as not to confuse his diligent readers by dissenting too strongly from Ptolemy'.[8] Kepler undertook to remedy this, hoping thereby to obtain more favourable *Lebensraum* for his five solids. His mathematical knowledge was as yet insufficient for this task, so he turned for help to his old teacher, Maestlin, who willingly complied. The new figures did not help Kepler at all; yet he had at one stroke, and almost inadvertently, shifted the centre of the solar system where it belonged. It was the first momentous by-product of the phantom chase.

His next attempt to remedy the disagreement between his dream and the observed facts concerned the moon. Should her orbit be included into the thickness of the earth's sphere, or not? He explained frankly to his dear readers that he would choose the hypothesis which best fits his plan; he will tuck the moon into the earth's shell, or banish her into the outer darkness, or let her orbit stick half-way out, for there are no *a priori* reasons in favour of either solution. (Kepler's *a priori* proofs were mostly found *a posteriori*.) But fiddling with the moon did not help either, so young Kepler proceeded to a frontal attack against the Copernican data. He declared them with admirable impertinence to be so unreliable that Kepler's own figures would be strongly suspect if they agreed with Copernicus'. Not only were the tables unreliable; not only was Copernicus inexact in his observations, as reported by Rheticus (from whom Kepler quotes long, damning passages); the old Canon also cheated:

> How human Copernicus himself was in adopting figures which within certain limits accorded with his wishes and served his purpose; this the diligent reader of Copernicus may test by himself. . . . He selects observations from Ptolemy, Walter, and others with a view to making his computations easier, and he does not scruple to neglect or to alter occasional hours in observed time and quarter degrees of angle.[9]

Twenty-five years later, Kepler himself amusedly commented on his first challenge of Copernicus:

> After all, one approves of a toddler of three who decides that he will fight a giant.[10]

So far, in the first twenty chapters of his book, Kepler had been concerned with finding reasons for the number and spatial distribution of the planets. Having satisfied himself (if not his readers) that the five solids provided all the answers, and that existing discrepancies were due to Copernicus' faulty figures, he now turned to a different, and more promising problem, which no astronomer before him had raised. He began to look for a mathematical relation between a planet's distance from the sun, and the length of its 'year' – that is, the time it needed for a complete revolution.

These periods were, of course, known since antiquity with considerable precision. In round figures, Mercury needs three months to complete a revolution, Venus seven and a half months, the earth a year, Mars two years, Jupiter twelve years, and Saturn thirty years. Thus the greater the planet's distance from the sun, the longer it takes to complete a revolution, but this is only roughly true: an exact mathematical ratio was lacking. Saturn, for instance, is twice as far out in space as Jupiter, and should therefore take twice as long to complete a circuit, that is twenty-four years; but Saturn in fact takes thirty. The same is true of the other planets. As we travel from the sun outward into space, the motion of the planets along their orbits gets slower and slower. (To get the point quite clear: they not only have a longer way to travel to complete a circuit, but they also travel at a slower rate along it. If they travelled at the same rate, Saturn, with a circuit twice as long as Jupiter's would take twice as long to complete it; but it takes two and a half times as long.)

Nobody before Kepler had asked the question *why* this should be so, as nobody before him had asked why there are just six planets. As it happens, the latter question proved scientifically sterile,* the former immensely fertile. Kepler's answer was, that there must be *a force emanating from the sun* which drives the planets round their orbits. The outer planets move slower because this driving force diminishes in ratio to distance 'as does the force of light'.

It would be difficult to over-estimate the revolutionary significance of this proposal. For the first time since antiquity, an attempt was made not only to *describe* heavenly motions in geometrical terms, but to assign them a *physical cause*. We have arrived at the point where astronomy and physics meet again, after a divorce which lasted for two thousand years. This reunion of the two halves of the split mind produced explosive results. It led to Kepler's three Laws, the pillars on which Newton built the modern universe.

Again we are in the fortunate position of being able to watch, as in a slow-motion film, how Kepler was led to taking that decisive step. In the key passage from the *Mysterium Cosmographicum* which follows, the index-numbers are Kepler's own, and refer to his Notes in the second edition:

* At least, our mathematical tools are as yet inadequate for tackling the genesis and morphology of the solar system. Much depends on asking the right question at the right time.

> If we want to get closer to the truth and establish some correspondence in the proportions [between the distances and velocities of the planets] then we must choose between these two assumptions: either the souls[ii] which move the planets are the less active the farther the planet is removed from the sun, or there exists only one moving soul[iii] in the centre of all the orbits, that is the sun, which drives the planet the more vigorously the closer the planet is, but whose force is quasi-exhausted when acting on the outer planets because of the long distance and the weakening of the force which it entails.[11]

To this passage Kepler made, in the second edition, the following notes:

> (ii). That such souls do not exist I have proved in my *Astronomia Nova*.
> (iii) If we substitute for the word 'soul' the word 'force' then we get just the principle which underlies my physics of the skies in the *Astronomia Nova*. ... For once I firmly believed that the motive force of a planet was a soul. ... Yet as I reflected that this cause of motion diminishes in proportion to distance, just as the light of the sun diminishes in proportion to distance from the sun, I came to the conclusion that this force must be something substantial – 'substantial' not in the literal sense but ... in the same manner as we say that light is something substantial, meaning by this an unsubstantial entity emanating from a substantial body.[12]

We are witnessing the hesitant emergence of the modern concepts of 'forces' and 'radiating energies' which are both material and non-material, and, generally speaking, as ambiguous and bewildering as the mystical concepts which they have come to replace. As we watch the working of the mind of Kepler (or Paracelsus, Gilbert, Descartes) we are made to realize the fallacy of the belief that at some point between the Renaissance and the Enlightenment, man shook off the 'superstitions of medieval religion' like a puppy getting out of the water, and started on the bright new road of Science. Inside these minds, we find no abrupt break with the past, but a gradual transformation of the symbols of their cosmic experience – from *anima motrix* into *vis motrix*, moving spirit into moving force, mythological imagery into mathematical hieroglyphics – a transformation which never was, and, one hopes, never will be entirely completed.

The details of Kepler's theory were again all wrong. The driving force which he attributed to the sun has no resemblance to gravity; it is rather like a whip which lashes the sluggish planets along their paths. As a result, Kepler's first attempt to formulate the law relating planetary distances with periods was so obviously wrong, that he had to admit it.[13] He added wistfully:

Though I could have foreseen this from the beginning, I nevertheless did not want to withhold from the reader this spur to further efforts. Oh, that we could live to see the day when both sets of figures agree with each other! ... My only purpose was that others may feel stimulated to search for that solution toward which I have opened the path.[14]

But it was Kepler himself who found the correct solution, toward the end of his life: it is his Third Law. In the second edition of the *Mysterium*, he added a Note to the phrase, 'Oh, that we could live to see the day....' It reads:

We have lived to see this day after twenty-two years and rejoiced in it, at least I did; I trust that Maestlin and many other men ... will share in my joy.[15]

The closing chapter of the *Mysterium* is a return to the medieval shore of the Keplerian torrent of thought. It is described as 'the dessert after this substantial meal,' and concerns the constellations of the sky on the first and last days of the world. We are given a fairly promising horoscope for the Creation – which started on Sunday, 27 April 4977 B.C.; but about the last day Kepler says modestly: 'I did not find it possible to deduce an end of the motions from inherent reasons.'

On this childish note ends Kepler's first book, the dream of five perfect solids determining the scheme of the universe. The history of thought knows many barren truths and fertile errors. Kepler's error turned out to be of immense fertility. 'The direction of my whole life, of my studies and works, has been determined by this one little book,' he wrote a quarter-century later.[16] 'For nearly all the books on astronomy which I have published since then were related to one or the other of the main chapters in this little book and are more thorough expositions or completions of it.'[17] Yet he also had an inkling of the paradoxical nature of all this, for he added:

> The roads by which men arrive at their insights into celestial matters seem to me almost as worthy of wonder as those matters in themselves.[18]

## 3. *Back to Pythagoras*

One crucial question was left unexplained in the previous chapters. What exactly was it that so forcefully attracted Kepler, when he was still a student of theology, to the Copernican universe? In his self-analysis he expressly stated that it was not interest in astronomy proper, that he was converted 'by physical, or if you prefer, metaphysical reasons'; and he repeats this statement almost verbatim in the preface to the *Mysterium*. These 'physical or metaphysical reasons' he explains differently in different passages; but the gist of them is that the sun must be in the centre of the world because he is the symbol of God the Father, the source of light and heat, the generator of the force which drives the planets in their orbits, and because a sun-centred universe is geometrically simpler and more satisfactory. These seem to be four different reasons, but they form a single, indivisible complex in Kepler's mind, a new Pythagorean synthesis of mysticism and science.

We remember that to the Pythagoreans and Plato the animating force of the deity radiated from the centre of the world outward, until Aristotle banished the First Mover to the periphery of the universe. In the Copernican system, the sun again occupied the place of the Pythagorean Central Fire, but God remained outside, and the sun had neither divine attributes, nor any physical influence on the motions of the planets. In Kepler's universe, all mystic attributes and physical powers are centralized in the sun, and the First Mover is returned to the focal position where he belongs. The visible universe is the symbol and 'signature' of the Holy Trinity: the sun represents the Father, the sphere of the fixed stars the Son, the invisible forces which, emanating from the Father, act through interstellar space, represent the Holy Ghost:

> The sun in the middle of the moving stars, himself at rest and yet the source of motion, carries the image of God the Father and Creator. ... He distributes his motive force through a medium which contains the moving bodies even as the Father creates through the Holy Ghost.[19]

The fact that space has three dimensions is itself a reflection, a 'signature' of the mystic Trinity:

> And thus are bodily things, thus are *materia corporea* represented in *tertia quantatis specie trium dimensionum*.[20]

The unifying truth between the mind of God and the mind of man is represented for Kepler, as it was for the Pythagorean Brotherhood, by the eternal and ultimate truths of 'divine Geometry'.

> Why waste words? Geometry existed before the Creation, is co-eternal with the mind of God, *is God himself* (what exists in God that is not God himself?); geometry provided God with a model for the Creation and was implanted into man, together with God's own likeness – and not merely conveyed to his mind through the eyes.[21]

But if God created the world after a geometrical model, and endowed man with an understanding of geometry, then it must be perfectly feasible, young Kepler thought, to deduce the whole blueprint of the universe by pure *a priori* reasoning, by reading the mind of the Creator, as it were. The astronomers are 'the priests of God, called to interpret the Book of Nature', and surely priests have a right to know the answers.

If Kepler's evolution had stopped here, he would have remained a crank. But I have already pointed out the contrast between the *a priori* deductions in the first part of the book and the modern scientific approach of the second. This coexistence of the mystical and the empirical, of wild flights of thought and dogged, painstaking research, remained, as we shall see, the main characteristic of Kepler from his early youth to his old age. Other men living on the watershed displayed the same dualism, but in Kepler it was more pointed and paradoxical, carried to extremes verging on insanity. It accounts for the incredible mixture in his works of recklessness and pedantic caution, his irritability and patience, his naïvety and philosophical depth. It emboldened him to ask questions which nobody had dared to ask without trembling at their audacity, or blushing at their apparent foolishness. Some of them appear to the modern mind as meaningless. The others led to the reconciliation of earth-physics with sky-geometry, and were the beginning of modern cosmology. That some of his own answers were wrong does not matter. As in the case of the Ionian philosophers of the heroic age, the philosophers of the Renaissance were perhaps more remarkable for the revolutionary nature of the questions they asked than for the answers they proposed. Paracelsus and Bruno, Gilbert and Tycho, Kepler and Galileo formulated some answers which are still valid; but first and foremost they were giant question-masters. *Post factum*, however, it is always difficult to appreciate the originality and imagination it required to ask a question which had not been asked before. In this respect, too, Kepler holds the record.

Some of his questions were inspired by a medieval brand of mysticism, and yet proved to be amazingly fertile. The shifting of the First Mover from the periphery of the universe into the physical body of the sun, symbol of the Godhead, prepared the way to the concept of a gravitational force, symbol of the Holy Ghost, which controls the planets. Thus a purely mystical inspiration was the root out of which the first rational theory of the dynamics of the universe developed, based on the secular trinity of Kepler's laws.

Equally astonishing was the fertility of Kepler's errors – starting with a universe built around the five solids, and ending with a universe governed by musical harmonies. This process, of error begetting truth, is illuminated by Kepler's own comments on the *Mysterium Cosmographicum*. They are contained in his Notes to the second edition, to which I have repeatedly referred, written twenty-five years later. In complete contrast to his claim that the book was written as if under the dictation of a 'heavenly oracle', and represented 'an obvious act of God', Kepler's Notes castigate its errors with acid sarcasm. The book starts, as we remember, with an 'Outline of my Principal Proof' and Kepler's comment starts with 'Woe to me, here I blundered'. The ninth chapter deals with the 'sympathies' between the five solids and the individual planets; in the Notes it is dismissed as a mere 'astrological fancy'. Chapter 10, 'On the Origin of Privileged Numbers', is described in the Notes as 'empty chatter'; Chapter 11, 'Concerning the Positions of the Regular Solids and the Origin of the Zodiac', is qualified in the Notes as 'irrelevant, false, and based on illegitimate assumptions'. On Chapter 17, concerning the orbit of Mercury, Kepler's comments are: 'this is not at all true', 'the reasoning of the whole chapter is wrong'. The important twentieth chapter, 'On the Relation between the Motions and Orbits', in which the Third Law is foreshadowed, is dismissed as faulty 'because I used uncertain ambiguous words instead of arithmetical method'. The twenty-first chapter, which discusses discrepancies between theory and observation, is attacked in the Notes in an almost unfairly petulant manner; e.g.: 'This question is superfluous. ... Since there is no discrepancy, why did I have to invent one?'

Yet the Notes to this chapter contain two remarks in a different key:

> If my false figures came near to the facts, this happened merely by chance. ... These comments are not worth printing. Yet it gives me pleasure to remember how many detours I had to make, along how many walls I had to grope in the darkness of my ignorance until I found the door which lets in the light of truth. ... In such manner did I dream of the truth.[22]

By the time he had finished with his Notes to the second edition (which amount to approximately the same length as the original work) the old Kepler had demolished practically every point in the book of the young Kepler – except its subjective value to him as the starting-point of his long journey, a vision which, though faulty in every detail, was 'a dream of truth': 'inspired by a friendly God'. The book indeed contained the dreams, or germs, of most of his later discoveries – as by-products of its erroneous central idea. But in later years, as the Notes show, this *idée fixe* was intellectually neutralized by so many qualifications and reserves, that it could do no harm to the working of his mind; while his irrational belief in its basic truth remained, emotionally, the motive power behind his achievements. The harnessing to a rational pursuit of the immense psychic energies derived from an irrational obsession seems to be another secret of genius, at least of genius of a certain type. It may also explain the distorted view of their own achievements so frequently found among them. Thus in Kepler's Notes to the *Mysterium* he proudly refers to some minor discoveries in his later works, but there is not one single mention of the first and second of his immortal Laws, which every schoolboy associates with his name. The Notes are chiefly concerned with the planetary orbits, yet the fact that these are ellipses (Kepler's First Law) is nowhere mentioned; it was as if Einstein, in his old age, had been discussing his work without mentioning relativity. Kepler set out to prove that the solar system was built like a perfect crystal around the five divine solids, and discovered, to his chagrin, that it was dominated by lopsided and undistinguished curves; hence his unconscious taboo on the word 'ellipse', his blind spot for his greatest achievement, and his clinging to the shadow of the *idée fixe*.[23] He was too sane to ignore reality, but too mad to value it.

A modern scholar remarked about the scientific revolution: 'One of the most curious and exasperating features of this whole magnificent movement is that none of its great representatives appears to have known with satisfying clarity just what he was doing or how he was doing it.'[24] Kepler, too, discovered his America, believing that it was India.

But the urge that drove him on was not aimed at any practical benefit. In the labyrinth of Kepler's mind, the thread of Ariadne is his Pythagorean mysticism, his religious-scientific quest for a harmonious universe governed by perfect crystal shapes or perfect chords. It was this thread that led him, through abrupt turns and dizzy gyrations, in and out of *culs-de-sac*, to the first exact laws of nature, to the healing of the millennial rift between astronomy and physics, to the mathematization of science. Kepler said his prayers in the language of mathematics, and distilled his mystic faith into a mathematician's Song of Songs:

> Thus God himself / was too kind to remain idle / and began to play the game of signatures / signing his likeness unto the world: therefore I chance to think / that all nature and the graceful sky are / symbolized in the art of Geometria. ... / Now, as God the maker play'd / he taught the game to Nature / whom he created in his image: / taught her the selfsame game / which he played to her. ...[25]

Here at last was the jubilant refutation of Plato's cave. The living world no longer is a dim shadow of reality but Nature's dance to which God sets the tune. Man's glory lies in his understanding of the harmony and rhythm of the dance, an understanding made possible through his divine gift of thinking in numbers:

> ... these figures pleased me because they are quantities, that is, something which existed before the skies. For quantities were created at the beginning, together with substance; but the sky was only created on the second day. ...[26] The ideas of quantities have been and are in God from eternity, they are God himself; they are therefore also present as archetypes in all minds created in God's likeness. On this point both the pagan philosophers and the teachers of the Church agree.[26a]

By the time Kepler wrote down this credo, the first stage in the young pilgrim's progress was completed. His religious doubts and anxieties had been transformed into the mystic's mature innocence – the Holy Trinity into a universal symbol, his craving for the gift of prophecy into the search for ultimate causes. The sufferings of a mange-eaten, chaotic childhood had left a sober thirst for universal law and harmony; memories of a brutal father may have influenced his vision of an abstract God, without human features, bound by mathematical rules which admitted of no arbitrary acts.

His physical appearance had undergone an equally radical change: the adolescent with the bloated face and spindly limbs had grown into a sparse, dark wiry figure, charged with nervous energy, with chiselled features and a somewhat Mephistophelian profile, belied by the melancholia of the soft, short-sighted eyes. The restless student who had never been able to finish what he began, had changed into a scholar with a prodigious capacity for work, for physical and mental endurance, and a fanatical patience unequalled in the annals of science.

In the Freudian universe, Kepler's youth is the story of a successful cure of neurosis by sublimation, in Adler's, of a successfully compensated inferiority complex, in Marx's, History's response to the need of improved navigational tables, in the geneticist's, of a freak combination of genes. But if that were the whole story, every stammerer would grow into a Demosthenes, and sadistic parents ought to be at a premium. Perhaps Mercury in conjunction with Mars, taken with a few cosmic grains of salt, is as good an explanation as any other.

# 3 Growing Pains

## 1. *The Cosmic Cup*

The inspiration about the five perfect solids had come to Kepler when he was twenty-four, in July 1595. During the next six months he had worked feverishly on the *Mysterium*. He reported on every stage of his progress to Maestlin in Tuebingen, pouring out his ideas in long letters and asking for his former teacher's help, which Maestlin gave in a grumbling but generous manner.

Michael Maestlin was a kind of inverted Rheticus to Kepler. He was twenty years Kepler's senior, yet was to outlive him. A contemporary engraving shows him as a bearded worthy with a jovial and somewhat vacant face. He had held the chair of mathematics and astronomy at Heidelberg, then at his native Tuebingen, and was a competent teacher with a solid academic reputation. He had published a textbook of astronomy of the conventional type, based on the Ptolemaic system, although in his lectures he spoke with admiration of Copernicus, and thus ignited the spark in young Kepler's inflammable mind. After the manner of good-natured mediocrities who know and accept their own limitations, he had a naïve admiration for the genius of his former pupil and went to considerable trouble to help him, though with an occasional growl at Kepler's unceasing demands. When the book was finished and the Senate of Tuebingen asked for Maestlin's expert opinion, he enthusiastically recommended that it should be published; and when permission was granted, he supervised the printing himself. This, in those days, was practically a full-time job; as a result Maestlin was reprimanded by the University Senate for neglecting his own work. He complained about this to Kepler in understandably peeved tones; Kepler replied, among his usual effusions of gratitude, that Maestlin shouldn't worry about the reprimand since, by seeing the *Mysterium* through print, Maestlin had acquired immortal fame.

By February 1596, the rough draft of the book was completed and Kepler asked his superiors in Gratz for leave of absence to visit his native Wuerttemberg and make arrangements for its publication. He asked for two months, but stayed away for seven, as he had become involved in a typically Keplerian chimera. He had persuaded Frederick, Duke of Wuerttemberg, to have a model of the universe, incorporating the five perfect solids, made in the shape of a drinking cup. 'A childish or fatal craving for the favour of princes,' as he later confessed, had driven him to Stuttgart, to Frederick's court, to whom he explained his idea in a letter:

> Since the Almighty granted me last summer a major *inventum* in astronomy, after lengthy, unsparing toil and diligence; which same *inventum* I have explained in a special booklet which I am willing to publish any time; which whole work and demonstration thereof can be fittingly and gracefully represented by a drinking cup of an ell in diameter which then would be a true and genuine likeness of the world and model of the creation insofar as human reason may fathom, and the like of which has never before been seen or heard of by any man; therefore I have postponed the preparation of such a model or its showing to any man to the present time of my arrival from Styria, intending to put this true and correct model of the world before the eyes of your Grace, as my natural sovereign, for him to see it as the first man on earth.[1]

Kepler went on to suggest that the various parts of the cup should be made by different silversmiths, and then fitted together, to make sure that the cosmic secret would not leak out. The signs of the planets could be cut in precious stones – Saturn in diamond, Jupiter in jacinth, the moon a pearl, and so on. The cup would serve seven different kinds of beverage, conducted by concealed pipes from each planetary sphere to seven taps on its rim. The sun will provide a delicious *aqua vita*, Mercury brandy, Venus mead, the moon water, Mars a strong Vermouth, Jupiter 'a delicious new white wine', and Saturn 'a bad old wine or beer', 'whereby those ignorant in astronomical matters could be exposed to shame and ridicule'. Assuring

Frederick that in ordering the cup he would do a favour to the arts and a service to God Almighty, Kepler remained Frederick's obedient servant, hoping for the best.

The Duke wrote on the margin of Kepler's letter: 'Let him first make a model of copper and when we see it and decide that it is worth being made in silver, the means shall not want.' Kepler's letter was dated 17 February and the Duke's answer was transmitted to him on the next day; Frederick's imagination had obviously caught on. But Kepler had no money to make a copper model, as he resentfully conveyed to the Duke in his next letter; instead, he settled down to the Herculean task of making a paper model of all the planetary orbits and the five perfect solids in between. He laboured day and night for a week; years later he nostalgically remarked that it had been quite a pretty model, made out in paper of different colours, with all the orbits in blue.

When the paper monster was finished, he sent it to the Duke, apologizing for its clumsiness and huge dimensions. Again promptly on the next day, the Duke ordered his chancellery to ask for the expert opinion of Professor Maestlin. The good Maestlin wrote to Frederick that Kepler's cup would represent a 'glorious work of erudition', and the Duke wrote on the margin: 'Since this is so, we are content that the work should be executed.'

But apparently it had been easier for God to build the world around the five polyhedra than for the silversmiths to execute a copy of it. Besides, Frederick did not want the cosmic mystery in the form of a drinking cup, but to have it encased in a celestial globe. Kepler made another paper model, left it with the silversmith, and in September returned to Gratz, having wasted nearly six months at Frederick's court. But the Duke would not drop the project, and it dragged on for several years. In January '98 Kepler wrote to poor Maestlin (who now served as the go-between): 'If the Duke agrees, it would be best to break up the whole junk and refund the silver to him. . . . The thing is hardly worth while. . . . I started it too ambitiously.'[2] But six months later, he submitted via Maestlin a new project. The cup, which had turned into a globe, was now to turn into a mobile planetarium, driven by a clockwork. The description of it occupied ten printed pages in folio. Kepler informed the Duke that a Frankfurt mathematician, Jacob Cuno, had offered to construct a planetarium which would reproduce the heavenly motions 'within an error of one degree for the next six or ten thousand years'; but, Kepler explained, such a machine would be too large and costly, and proposed a more modest one, guaranteed for a century only. 'For it is not to be hoped (apart from the Last Judgement) that such a work would remain undisturbed in one place over a hundred years. Too many wars, fires, and other changes are wont to occur.'[3]

The correspondence went on for another two years; then the subject was at last mercifully forgotten. But this quixotic escapade inevitably reminds one of the ill-fated vagabondages of his father, uncle, and brother. He worked off his innate restlessness in bold imaginations and plain drudgery; but from time to time some residual poison in his blood would make him break out in a rash and momentarily turn the sage into a clown. This fact is painfully evident in the tragicomedy of Kepler's first marriage.

## 2. *Marriage*

Before his journey to Wuerttemberg, Kepler's friends in Gratz had found a prospective bride for the young mathematicus in the daughter of a rich mill-owner, twice widowed at the age of twenty-three. Barbara Muehleck had been married at sixteen, against her wish, to a middle-aged cabinet-maker who had died after two years; then to an elderly, widowed pay-clerk who brought into the marriage a bunch of mis-shapen children, chronic illness, and, after his timely demise, was found to have defrauded money in his trust. Barbara, described by Kepler as 'simple of mind and fat of body', now lived with her parents, who could not have very high expectation of her future. Yet when Kepler presented his suit through two respectable middlemen (a school inspector and a deacon) the proud miller refused on the grounds that he could not entrust Barbara and her dowry to a man of such lowly standing and miserable pay. This was the beginning of long and sordid negotiations conducted by Kepler's friends with the family.

When he left for Stuttgart nothing was settled. but in the spring his friends wrote to him that his suit had been accepted, advised him to hurry home, and to bring with him from Ulm 'some truly good silk cloth, or at least of the best double taffeta, sufficient for complete robes for thyself and the bride'. But Kepler was too busy with his cosmic silver cup, delayed his return, and by the time he got back to Gratz, Frau Barbara's father had changed his mind again. Kepler seems not to have been unduly perturbed, but the indefatigable friends continued their efforts; the Dean of the school and even the Church authorities joined in – 'and so they vied with one another to assault the minds now of the widow, now of her father, took them by storm and arranged for me a new date for the nuptials. Thus, with one blow, all my plans for beginning another life collapsed.'[4]

The marriage took place on 27 April 1597, 'under a calamitous sky', as the horoscope indicated. He was somewhat comforted by the arrival of the first printed copies of the *Mysterium Cosmographicum*, but not even that event was all joy; he had to buy two hundred copies of the book for cash to compensate the printer for the risk; and the author's name in the catalogue of the Frankfurt Book Fair was transformed, by misprint, from Keplerus into Repleus.

Kepler's attitude to marriage in general, and to his own wife in particular, is expressed in several letters with shocking frankness. The first is addressed to Maestlin and dated a week before the wedding. It occupies nearly six pages in folio, of which only the last speaks of the impending great event:

I ask you only one favour, that you should be close to me in your prayers on my wedding day. My financial situation is such that should I die within the next year hardly anybody could leave a worse situation after him. I am obliged to spend a big sum of my own, for it is the custom here to celebrate marriages splendidly. If, however, God prolongs my life, I shall be bound and constricted to this place. ... For my bride possesses here estates, friends, and a prosperous father; it seems that after a few years I would not need my salary any longer. . . . Thus I shall be unable to leave this province except if a public or a private misfortune intervened. A public misfortune it would be if the country ceased to be safe for Lutherans, or if the Turks, who have already massed six hundred thousand men, invaded it. A private misfortune it would be if my wife died.[5]

Not a word is said about the person of his betrothed or his feelings for her. But in another letter written two years later, he blames her horoscope for her 'rather sad and unlucky fate. . . . In all dealings she is confused and inhibited. Also she gives birth with difficulty. Everything else is in the same vein.' [6]

After her death, he described her in even more depressing terms. She had known how to make a favourable impression on strangers, but at home she had been different. She resented her husband's lowly position as a stargazer and understood nothing of his work. She read nothing, not even stories, only her prayer book which she devoured day and night. She had 'a stupid, sulking, lonely, melancholy complexion'. She was always ailing and weighed down with melancholia. When his salary was withheld, she refused to let him touch her dowry, even to pawn a cup or to put her hand into her private purse.

And because, due to her constant illness, she was deprived of her memory, I made her angry with my reminders and admonitions for she would have no master and yet often was unable to cope herself. Often I was even more helpless than she, but in my ignorance persisted in the quarrel. In short she was of an angry nature, and uttered all her wishes in an angry voice; this incited me to provoke her, I regret it, for my studies sometimes made me thoughtless; but I learnt my lesson, I learnt to have patience with her. When I saw that she took my words to heart, I would rather have bitten my own finger than to give her further offence. . . .[7]

Her avarice made her neglect her appearance; but she lavished everything on the children because she was a woman 'entirely imprisoned by maternal love'; as for her husband 'not much love came my way'. She nagged not only him but also the servant wenches and 'could never keep a wench'. When he was working she would interrupt him to discuss her household problems. 'I may have been impatient when she failed to understand and went on asking me questions, but I never called her a fool, though it may have been her understanding that I considered her a fool, for she was very touchy.'[8] There is not much left to be added to this portrait of the perennial Xanthippe.

Nine months after the wedding their first child was born, a little boy, with his genitals so deformed that 'their composition looked like a boiled turtle in its shell'[9] – which, Kepler explains, was due to turtles being his wife's favourite dish. After two months the child died of cerebral meningitis, and the next, a little girl, died after a month of the same disease. Frau Barbara bore three more children, of which one boy and one girl survived.

Altogether, their marriage lasted fourteen years; Barbara died at the age of thirty-seven, with a distraught mind. The marriage horoscope had shown a *coelo calamitoso*, and in predicting disaster Kepler's horoscopes were nearly always right.

## 3. *Limbering Up*

When, in the spring of 1597, the *Mysterium* at last appeared in print, the proud young author sent copies to all leading scholars he could think of, including Galileo and Tycho de Brahe. There existed as yet no scientific journals nor, happy days, book reviewers; on the other hand, there was an intensive exchange of letters among scholars and a luxuriant international academic grapevine. By these means the unknown young man's book created a certain stir; though not the earthquake which its author expected, yet remarkable enough if we consider that the average number of scientific (and pseudo-scientific) books published in Germany in a single year was well over a thousand.[10]

But the response was not surprising. Astronomy, from Ptolemy to Kepler, had been a purely descriptive geography of the sky. Its task was to provide maps of the fixed stars, timetables of the motions of sun, moon, and planets, and of such special events as eclipses, oppositions, conjunctions, solstices, equinoxes, and the rest. The physical causes of the motions, the forces of nature behind them, were not the astronomer's concern. Whenever necessary, a few epicycles were added to the existing machinery of wheels – which did not matter much since they were fictional anyway, and nobody believed in their physical reality. The hierarchy of cherubim and seraphim who were supposed to keep the wheels turning was, since the end of the Middle Ages, regarded as another polite, poetic fiction. Thus the physics of the sky had become a complete blank. There were events but no causes, motions but no moving forces. The astronomer's task was to observe, describe, and predict, not to search for causes – 'theirs not to reason why'. Aristotelian physics, which made any rational and causal approach to the heavenly phenomena unthinkable, was on the wane, but it had left only a vacuum behind it. Ears were still ringing with the vanished song of the star-spinning angels, but all was silence. In that fertile silence the unformed, stammering voice of the young theologian-turned-astronomer obtained an immediate hearing.

Opinions were divided, according to the philosophy of the scholars. The modern and empirically minded, such as Galileo in Padua and Praetorius in Altdorf, rejected Kepler's mystical *a priori* speculations and with them the whole book, without realizing the explosive new ideas hidden among the chaff. Galileo, especially, seems to have been prejudiced from the beginning against Kepler, of which we shall hear more later on.

Those, however, who lived on the other side of the watershed, who believed in the ageless dream of an *a priori* deduction of the cosmic order, were enthusiastic and delighted. Most of all, of course, the endearing Maestlin, who wrote to the Tuebingen Senate:

> The subject is new and has never before occurred to anybody. It is most ingenious and deserves in the highest degree to be made known to the world of learning. Who has ever dared before to think, and much less to try to expose and explain *a priori* and, so to speak, out of the hidden knowledge of the Creator, the number, order, magnitude, and motion of the spheres? But Kepler has undertaken and successfully done just this. ... Henceforth [astronomers] shall be freed from the necessity of exploring the dimensions of the spheres *a posteriori*, that is by the method of observations

(many of which are inexact and not to say doubtful) after the manner of Ptolemy and Copernicus, because now the dimensions have been established *a priori*. ... Whereby the computation of the movements will become much more successful. ...[11]

In a similar vein enthused Limneus in Jena, who congratulated Kepler, all students of astronomy and the whole learned world that 'at last the old and venerable [Platonic] method of philosophy had been resurrected'.[12]

In a word, the book which contained the seeds of the new cosmology was welcomed by the 'reactionaries' who did not see its implications, and rejected by the 'moderns', who did not see them either. Only one man took a middle course and, while rejecting Kepler's wild speculations, immediately realized his genius: the most outstanding astronomer of the day, Tycho de Brahe.

But Kepler had to wait for three years until he met Tycho, became his assistant, and started on his true lifework. During these three years (1597–9) he at last got down to a serious study of mathematics, of which he had still been shockingly ignorant when he wrote the *Mysterium*, and undertook a motley variety of scientific and pseudo-scientific researches. It was a kind of limbering up before the great contest.

The first task he set himself was to find direct confirmation of the earth's motion round the sun by proving the existence of stellar parallax, that is, a shift in the apparent position of the fixed stars according to the earth's position on its annual journey. He pestered, in vain, all his correspondents to help him with observations, and at last decided to take a peep for himself; but his 'observatory' consisted of a self-made staff suspended on a rope from the ceiling: 'it comes from a workshop like the huts of our forebears – hold your laughter, friends, who are admitted to this spectacle'.[13] Even so, it would have been sufficiently precise to show the variation of half a degree, which Kepler expected, in the positions of the polar star as seen from extreme points of the earth's path. But there was no variation; the starry sky remained immutable, poker-faced. This meant either that the earth stood still, or that the size of the universe (that is, the radius of the sphere of fixed stars) was much larger than previously assumed. To be precise, its radius must be at least five hundred times the distance of the earth from the sun. This works out at 2,400 million miles, a trifle by our, but not a lot even by Kepler's, standards; only about five times more than he had expected.[14] However, assuming that even much better instruments failed to show a parallax, meaning that the stars are quite inconceivably distant, in the eyes of God the universe would still have a reasonable size, only man's physical stature would shrink. But this would not diminish his moral stature, 'otherwise the crocodile or the elephant would be nearer to His heart than man, because they are larger. With the help of this and similar intellectual pills, we shall perhaps be able to digest this monstrous bite.'[15] In fact, no pill has been discovered since to digest the lump of infinity.

Other problems which occupied him were his first researches into optics, out of which eventually a new science was to emerge; investigations of the moon's orbit, of magnetism, of meteorology – he started a weather diary which he kept up for twenty or thirty years; of Old Testament chronology, and the like. But dominating all these interests was his search for a mathematical law of the harmony of the spheres – a further development of his *idée fixe*.

In the *Mysterium*, Kepler had tried to build his universe around the five Pythagorean solids. Since the theory did not quite fit the facts, he now tried to build it round the musical harmonies of the Pythagorean scale. The combination of these two ideas led, twenty years later, to his great work *Harmonice Mundi*, which contains the third of Kepler's laws; but the ground work to it was laid during his last years in Gratz.

The moment this new idea had occurred to him, his letters resounded with jubilant *eurekas*: 'Fill the skies with air, and they will produce true and real music.' But as he began to compute the details of his cosmic musical box, he ran into increasing difficulties. He was never short of an excuse for ascribing to any pair of planets the musical interval which approximately happened to fit it; when things became sticky, he asked the shadow of Pythagoras for help – 'unless the soul of Pythagoras has migrated into mine.' He managed to construct a system of sorts, but its inadequacies were obvious to himself. The principal trouble was that a planet does not move at uniform speed, but faster when it is close to the sun, slower when away from it. Accordingly it does not 'hum' on a steady pitch, but alternates between a lower and a higher note. The interval between the two notes depends on the lopsidedness or 'eccentricity of the planet's orbit. But the eccentricities were only inaccurately known. It was the same difficulty he had come up against when he had tried to define the thickness of the spherical shells between his perfect solids, which also depended on the eccentricities. How could you build a series of crystals, or a musical instrument, without knowing the measurements? There was only one man alive in the world who possessed the exact data which Kepler needed: Tycho de Brahe.

All his hopes became now focused on Tycho, and his observatory at Uraniburg, the new wonder of the world:

> Let all keep silence and hark to Tycho who has devoted thirty-five years to his observations. ... For Tycho alone do I wait; he shall explain to me the order and arrangement of the orbits. ... Then I hope I shall one day, if God keeps me alive, erect a wonderful edifice.[16]

Thus he knew that the building of that edifice still lay in the distant future, though in his euphoric moments he claimed to have it already completed. During his manic periods, the discrepancies between theory and fact appeared to him as contemptible details, which could be smoothed over by a little cheating; yet the other half of his divided self humbly acknowledged the duty of pedantic accuracy and patient observation. With one eye he was reading the thoughts of God; the other squinted enviously at Tycho's shining armillary spheres.

But Tycho refused to publish his observations until he had completed his own theory. He jealously guarded his treasure, volumes of figures, the result of a lifetime of work.

> Any single instrument of his [young Kepler wrote bitterly] cost more than my and my whole family's fortune put together. ... My opinion of Tycho is this: he is superlatively rich, but he knows not

how to make proper use of it as is the case with most rich people. Therefore, one must try to wrest his riches from him.[17]

In this outcry, Kepler had revealed his intentions towards Tycho de Brahe a year before they met for the first time.

## 4. *Waiting for Tycho*

Had Kepler not succeeded in getting hold of Tycho's treasure, he could never have discovered his planetary laws. Now Newton was born only twelve years after Kepler's death, and without the planetary laws he could not have arrived at his synthesis. No doubt somebody else would have done so, but it is at least possible that the scientific revolution would have carried different metaphysical undertones if it had been fathered not by an Engish empiricist, but, say, a Frenchman with Thomist inclinations, or a German mystic.

The point of such idle speculation is merely to insert a question mark here and there against the supposed logical inevitability and cast-iron determinism of the evolution of scientific thought. The shape of Cleopatra's nose influences not only wars, but ideologies. The mathematics of the Newtonian universe would have been the same whoever worked them out, but its metaphysical climate might have been quite different.[18]

Yet it was touch and go whether Kepler's laws would be ready for Newton.[19] They could only be discovered with Tycho's help; and by the time Kepler met him, Tycho had only eighteen months left to live. If it was divine providence which timed their meeting, it chose a rather perverse method: Kepler was hounded out of Gratz and into the arms of Tycho, by religious persecution. Though he always endeavoured to read the thoughts of God, he never offered his thanks for this Machiavellian stratagem.

That last year in Gratz – the last of the century – was indeed not easy to endure. The young Archduke Ferdinand of Hapsburg (later Emperor Ferdinand II) was determined to cleanse the Austrian provinces of the Lutheran heresy. In the summer of 1598, Kepler's school was closed down, and in September all Lutheran preachers and schoolmasters were ordered to leave the Province within eight days or forfeit their lives. Only one among them received permission to return, and that was Kepler. His exile, the first, lasted less than a month.

The reasons why an exception was made with him are rather interesting. He himself says [20] that the Archduke was 'pleased with my discoveries' and that this was the reason for his favour at his court; besides, as a mathematicus he occupied a 'neutral position' which set him apart from the other teachers. But it was not as simple as that. Kepler had a powerful ally behind the scenes: the Jesuit order.

Two years previously the Catholic Chancellor of Bavaria, Herwart von Hohenburg, amateur philosopher and patron of the arts, had asked Kepler among other astronomers, for his opinion on certain chronological problems. It was the beginning of a life-long correspondence and friendship between the two men. Herwart tactfully advertised his protective interest in the Protestant mathematicus by sending his letters to Kepler via the Bavarian envoy at the Emperor's court in Prague, who forwarded them to a Capuchin father at Ferdinand's court in Gratz; and he instructed Kepler to use the same channels. In his first letter to Herwart,[21] Kepler wrote delightedly: 'Your letter so impressed some men in our Government that nothing more favourable to my reputation could have happened.'

It was all done with great subtlety; yet on later occasions Catholic and especially Jesuit influences were more openly active on behalf of Kepler's welfare. There seem to have been three reasons for this benevolent cabal. Firstly, a scholar was still to some extent regarded as a sacred cow amidst the turmoil of religious controversy – one remembers how Rheticus was fêted in Catholic Ermland at the time of Bishop Dantiscus' edicts against the Lutheran heresy. Secondly, the Jesuits, following in the steps of the Dominicans and Franciscans, were beginning to play a leading part in science and specially in astronomy – quite apart from the fact that it enabled their missionaries in distant countries to make a great impression by predicting eclipses and other celestial events. And lastly, Kepler himself disagreed with certain points of Lutheran doctrine, which made his Catholic friends hope – though in vain – that he might become a convert. He was repelled by the clerics of both warring churches who, from their pulpits, screamed at each other like fishwives – or like his parents and kin in old Sebaldus' house. His attitude was the same as the gentle Bishop Giese's: 'I refuse the battle'; and he also did a certain amount of fence-straddling. Yet he refused to change sides, even when he was excommunicated by his own Church, as we shall hear; and when he suspected that Herwart was counting on his conversion, Kepler wrote to him:

I am a Christian, the Lutheran creed was taught me by my parents, I took it unto myself with repeated searchings of its foundations, with daily questionings, and I hold fast to it. Hypocrisy I have never learnt. I am in earnest about Faith and I do not play with it.[22]

It was the outburst of a man of basic integrity, forced to swim in the troubled waters of his time. He was as sincere in matters of religion as circumstances permitted him to be; at any rate, his deviations from the straight path were perhaps not greater than those of his orbits from God's five perfect solids.

Kepler was, then, made an exception and permitted to return from exile in October 1599. Since his school had been closed down he could devote most of his time to his speculations on the harmony of the spheres; yet he knew that the reprieve was only a temporary one; and that his days in Gratz were numbered. He sank into a profound depression, deepened by the death of his second child; in a despairing letter he asked Maestlin, in August 1599, for his help in finding a job at home in Protestant Wuerttemberg.

The hour could not have been more propitious; but God has offered this fruit, too, only to take it away again. The child died of a cerebral meningitis (exactly as its brother a year ago) after thirty-five days. ... If its father should follow soon, his fate would not be unexpected. For everywhere in Hungary bloody crosses have appeared on the bodies of men and similar bloody signs on the gates of houses, on benches, and walls, which history shows to be a sign of a general pestilence. I am, as far as I know, the first person in our town to see a small cross on my left foot, the colour of which passes from bloody red to yellow. The spot is on the foot, where the back of the foot curves into the instep, half-way between the toes and the end of the shin-bone. I believe it is just the spot where the nail was hammered into the foot of Christ. Some carry, I am told, marks in the shape of drops of blood in the hollow of the hand. But so far this form has not appeared on me. . . .

The ravages of dysentery kill people of all ages here, but particularly children. The trees stand with dry leaves on their crowns as if a scorching wind had passed over them. Yet it was not the heat that so disfigured them, but worms. . . .[23]

He had the worst fears. There was talk of torture for heretics, even of burnings. He was fined ten dalers for burying his child according to Lutheran rites: 'half of it was remitted at my request, but the other half I had to pay before I was allowed to carry my little daughter to her grave'. If Maestlin cannot get him a job at once, would he at least let him know about the present cost of living in Wuerttemberg: 'How much wine costs and how much wheat, and how things stand regarding the supply of delicatessen (for my wife is not in the habit of living on beans).'

But Maestlin knew that his university would never give a job to the unruly Kepler, and he was getting thoroughly fed up with Kepler's unceasing demands and badgerings; the more so, as Kepler had followed up his S.O.S. with the foolish remark:

Of course, nobody would expel me; the most intelligent among the members of the Diet are the most fond of me, and my conversation at meals is much sought after.[24]

No wonder that Maestlin underestimated the urgency of the situation, and delayed for five months before he answered with an evasive and grumpy epistle: 'If only you had sought the advice of men wiser and more experienced in politics than I, who am, I confess, as unexperienced in such matters as a child.' [25]

Only one hope remained: Tycho. The previous year, Tycho, in a letter, had expressed the hope that Kepler would 'some day' visit him. Though Kepler was panting and pining for 'Tycho's treasure', the invitation was couched in too general terms, and the journey too long and costly. Now, however, it was no longer a matter of scientific curiosity for Kepler, but of the urgent necessity of finding a new home and a livelihood.

Tycho, in the meantime, had been appointed Imperial Mathematicus by Rudolph II, and had taken up residence near Prague. Kepler's long-awaited opportunity came when a certain Baron Hoffmann, Councillor to the Emperor, had to return from Gratz to Prague, and agreed to take him along in his suite. The date of Kepler's departure for the meeting with Tycho is, by the courtesy of History, easy to remember: it was 1 January, *anno domini* 1600.

# 4 Tycho de Brahe

## 1. *The Quest for Precision*

Johannes Kepler was a pauper who came from a family of misfits; Tycho de Brahe was a *grand seigneur* from the Hamlet country, the scion of truculent and quixotic noblemen of pure Danish stock. His father had been Governor of Helsingborg Castle, which faces Elsinor across the Sund; his Uncle Joergen, a country squire and vice-admiral.

This Uncle Joergen, being childless, had extracted a promise from his brother, the Governor, that if he had a son, Joergen could adopt him and bring him up as his own. Nature seemed to sanction this arrangement, for, in 1546, the Governor's wife bore him twin sons; but unfortunately one of them was still-born, and the father went back on his promise. Joergen, a true, headstrong Brahe, waited until another son was born to his brother, then kidnapped the firstborn, Tyge – Tycho. The Governor, also in true Brahe fashion, threatened murder, but quickly cooled off and generously consented to the *fait accompli*, knowing that the child would be well looked after and inherit some of Joergen's fortune. This came indeed to pass and sooner than expected, for while Tyge was still a student, his foster-father met an untimely and glorious end. He had just returned from a naval battle against the Swedes, and was riding in the suite of his King over the bridge joining Copenhagen to the royal castle, when the good King, Frederick II, fell into the water. Joergen, the vice-admiral, jumped after him, saved his King and died of pneumonia.

Whether Tyge suffered a traumatic shock by being kidnapped from his cradle we cannot know; but the blood of the Brahes and his education by the irascible vice-admiral must have sufficed to turn him into an eccentric in the grand style. This was visible at first glance, even in his physical appearance: for if Tycho was born with a silver spoon in his mouth, he later acquired a nose of silver and gold. As a student, he fought a duel with another noble Danish youth in the course of which part of Tyge's nose was sliced off. According to a contemporary account,[1] the quarrel originated in a dispute as to which of the two noble Danes was the better mathematician. The lost piece, which seems to have been the bridge of the nose, was replaced by a gold and silver alloy, and Tycho is said to have always carried a kind of snuff-box 'containing some ointment or glutinous composition which he frequently rubbed on his nose'.[2] In his portraits, the nose appears as a too rectilinear, cubistic feature among the curves of a large, bald, egg-shaped head, set between the cold, haughty eyes and the aggressively twirled handlebar moustache.

True to the family tradition, young Tyge was intended to take up the career of a statesman, and was accordingly sent at thirteen to study rhetorics and philosophy at the University of Copenhagen. But at the end of his first year, he witnessed an event which made an overwhelming impression on him and decided the whole future course of his life. It was a partial eclipse of the sun which, of course, had been announced beforehand, and it struck the boy as 'something divine that men could know the motions of the stars so accurately that they were able a long time beforehand to predict their places and relative positions'.[3] He immediately began to buy books on astronomy, including the collected works of Ptolemy for the considerable sum of two Joachims-Thaler. From now onward his course was set and he never swerved from it.

Why did that partial eclipse, which was not at all spectacular as a sight, have such a decisive impact on the boy? The great revelation for him, Gassendi tells us, was the *predictability* of astronomical events – in total contrast, one might speculate, to the unpredictability of a child's life among the temperamental Brahes. It is not much of a psychological explanation, but it is worth noting that Brahe's interest in the stars took from the beginning a quite different, in fact almost opposite direction from both Copernicus' and Kepler's. It was not a speculative interest, but a passion for exact observation. Starting on Ptolemy at fourteen, and making his first observation at seventeen, Tycho took to astronomy at a much earlier age than those two. The timid Canon had found a refuge from a life of frustrations in the secret elaboration of his system; Kepler resolved the unbearable miseries of his youth in his mystic harmony of the spheres. Tycho was neither frustrated nor unhappy, only bored and irritated by the futility of a Danish nobleman's existence among, in his own words, 'horses, dogs, and luxury'; and he was filled with naïve wonder at the contrasting soundness and reliability of the stargazers' predictions. He took to astronomy not as an escape or metaphysical lifebelt, but rather as a full-time hobby of an aristocrat in revolt against his *milieu*. His later life seems to confirm this interpretation, for he entertained kings on his wonder island, but the mistress of the house, with whom he begot a large family of children, was a woman of low caste to whom he was not even married in church.

After three years at Copenhagen, the Vice-Admiral thought that it was time for Tyge to go to a foreign university, and sent him, accompanied by a tutor, to Leipzig. The tutor was Anders Soerensen Vedel, who later became famous as the first great Danish historian, translator of *Saxo Grammaticus* and collector of Nordic sagas. Vedel was then twenty, only four years older than his charge; he had received instructions to cure young Tyge of his unseemly preoccupation with astronomy and lead him back to studies more fitting for a nobleman. Tyge had bought a small celestial globe to learn the names of the constellations, but he had to hide it under his blanket; and when he added to this a cross-staff, he could use it only when his tutor was asleep. After a year of this, however, Vedel realized that Tyge was star-struck beyond remedy, gave in, and the two remained lifelong friends.

After Leipzig, Tycho continued his studies at the Universities of Wittenberg, Rostock, Basle, and Augsburg until his twenty-sixth year, all the time collecting, and later designing, bigger and better instruments for observing the planets. Among these was a huge quadrant of brass and oak, thirty-eight feet in diameter and turned by four handles – the first of a series of fabulous instruments which were to become the wonder of the world. Tycho never made any epoch-making discovery except one, which made him the father of modern observational astronomy; but that one discovery has become such a truism to the modern mind that it is difficult to see its importance. The discovery was that astronomy needed *precise* and *continuous* observational data.

It will be remembered that Canon Koppernigk recorded only twenty-seven observations of his own in the whole *Book of Revolutions*; for the rest he relied on the data of Hipparchus, Ptolemy, and others. This had been the general practice up to Tycho. It was taken for granted that planetary tables must be exact, as far as possible, for calendrical and navigational purposes; but apart from the limited data required for these practical reasons, the necessity for precision was not at all realized. This attitude, which is all but incomprehensible to the modern mind, was partly due to the Aristotelian tradition with its emphasis on qualities instead of quantitative measurement; within that mental framework only a crank could be interested in precision for precision's sake. Besides, and more specifically, a geometry of the skies consisting of cycles and epicycles did not require a great many, or even very precise, observational data, for the simple reason that a circle is defined when its centre and a single point of its circumference are known, or, if the centre is unknown, by three points of its circumference alone. Hence it was, by and large, sufficient to determine the positions of a planet at a few characteristic points of its orbit, and then to arrange one's epicycles and deferents in the way most favourable to 'save the phenomena'. If one projects one's mind back to the other side of the watershed, Tycho's devotion to measurements, to fractions of minutes of arc, appears as highly original. No wonder that Kepler called him the Phoenix of Astronomy.

On the other hand, if Tycho was ahead of his time, he was only a step ahead of Kepler. We have seen how Kepler was pining for Tycho's observations, for precise data on mean distances and eccentricities. A century earlier, Kepler would probably have rested on the laurels of his solution of the cosmic mystery without bothering about those small disagreements with observed facts; but this metaphysical cavalier-attitude towards facts was on the wane among the advanced minds of the time. Ocean navigation, the increasing precision of magnetic compasses and clocks, and the general progress in technology created a new climate of respect for hard fact and exact measurement. Thus, for instance, the debate between the Copernican and Ptolemaic systems was no longer pursued by theoretical arguments alone; both Kepler and Tycho independently decided to let experiment be the arbiter, and tried to determine by measurement whether a stellar parallax existed or not.

One of the reasons for Tycho's quest for precision was, in fact, his desire to check the validity of the Copernican system. But this was perhaps rather the rationalization of a deeper urge. Meticulous patience, precision for precision's sake was for him a form of worship. His first great experience had been the awestricken realization that astronomic events could be exactly predicted; his second was of the opposite kind. On 17 August 1563, at the age of seventeen, while Vedel was asleep, he noticed that Saturn and Jupiter were so close together as to be almost indistinguishable. He looked up his planetary tables and discovered that the Alphonsine tables were a whole month in error regarding this event, and the Copernican tables by several days. This was an intolerable and shocking state of affairs. If the stargazers, of whose low company his family so disapproved, could not do better, let a Danish nobleman show them how a proper job is done.

And show them he did, by methods and gadgets the like of which the world had never seen.

## 2. *The New Star*

At the age of twenty-six, Tycho considered his education complete, and returned to Denmark. For the next five years, till 1575, he lived first on the family estate at Knudstrup, then with an uncle, Steen Bille, the only one in the family who approved of Tycho's perverse hobby. Steen had founded the first paper mill and glassworks in Denmark, and dabbled a lot in alchemy in which Tycho assisted him.

Like Kepler, Tycho stood with one foot in the past and was devoted both to alchemy and astrology. Like Kepler, he became a court astrologer and had to waste much of his time with the casting of horoscopes for patrons and friends; like Kepler, he did it with his tongue in his cheek, despised all other astrologers as quacks, and yet was profoundly convinced that the stars influenced man's character and destiny though nobody quite knew how. Unlike Kepler's, however, his belief in astrology derived not from mysticism – which was completely alien to his domineering nature – but from stark superstition.

The great event of these years, an event that was discussed all over the world and which established, at a single stroke, Tycho's fame as the leading astronomer of his time, was the new star of 1572. In Tycho's life, all the decisive landmarks were sky-marks: the eclipse of the sun when he was fourteen which brought him to astronomy, the conjunction of Jupiter and Saturn when he was seventeen, which made him realize its insufficiencies; the new star when he was twenty-six, and the comet of 1577, five years later. Of all these, the new star was the most important.

On the evening of 11 November 1572, Tycho was walking from Steen's alchemist laboratory back to supper when, glancing at the sky, he saw a star brighter than Venus at her brightest, in a place where no star had been before. The place was a little to the north-west of the familiar 'W' – the constellation of Cassiopeia, which then stood near the Zenith. The sight was so incredible that he literally did not believe his eyes; he called at first some servants, and then several peasants to confirm the fact that there really was a star where no star had any business to be. It was there all right, and so bright that later on people with sharp eyes could see it even in the middle of the day. And it remained in the same spot for eighteen months.

Other astronomers besides Tycho had seen the new star in the first days of November. It was then in full blaze; in December it began very slowly to fade, but ceased to be visible only by the end of March, the year after next. The world had never seen or heard the like since the year 125 B.C when Hipparchus, according to the second book of Pliny's *Natural History*, had seen a new star appear in the sky.

The sensational importance of the event lay in the fact that it contradicted the basic doctrine – Aristotelian, Platonic, and Christian – that all change, all generation and decay were confined to the immediate vicinity of the earth, the sub-lunary sphere; whereas the distant eighth sphere in which the fixed stars were located, was immutable from the day of Creation to eternity. The only known exception in history was the appearance of the above-mentioned new star of Hipparchus; but that had been very long ago, and one could explain it away by assuming that Hipparchus had merely seen a comet (which was then considered an atmospheric phenomenon in the sub-lunary region).

Now, what distinguishes a fixed star from a planet, or a comet, or a meteor, is the fact that it is 'fixed': apart from its participation in the daily rotation of the firmament as a whole, it does not move. No sooner did that bright new cuckoo egg appear on the tip of the celestial 'W', far out-shining the legitimate stars in its nest, stargazers all over Europe feverishly tried to determine whether it moved or not. If it did, it was not a real star and academic science was saved; if it did not, the world had to be thought afresh.

Maestlin in Tuebingen who, though one of the leading astronomers of the time, seems to have possessed no instruments whatsoever, held a thread at arm's length from his eyes in such a way that it passed through the new star and two other fixed stars. When, after a few hours, the three were still in the same straight line, he concluded that the new star did not move.[4] Thomas Digges in England used a similar method, and came to the same result; others found a displacement, but only a small one, due, of course, to the errors of their coarse instruments. This was Tycho's great opportunity, and he fully rose to it. He had just finished a new instrument – a sextant with arms five and a half feet long, joined by a bronze hinge, with a metallic arc scale graduated to single minutes and, as a novelty, a table of figures designed to correct the errors of the instrument. It was like a heavy gun compared to the slings and catapults of his colleagues. The result of Tycho's observations was unequivocal: the new star stood still in the sky.

All Europe was agog, both with the cosmological and astrological significance of the event. The new star had appeared only about three months after the massacre of French protestants on St Bartholomew's night; no wonder that in the flood of pamphlets and treatises on the star, it was mostly regarded as a sinister omen. The German painter, George Busch, for instance, explained that it was really a comet condensed from the rising vapours of human sins, which had been set afire by the wrath of God. It created a kind of poisonous dust (rather like the fall-out from a Hydrogen bomb) which was drifting down on people's heads and caused all sorts of evil, such as 'bad weather, pestilence, and Frenchmen'. The more serious astronomers, with few exceptions, tried to explain the star away from the eighth sky by calling it a tailless comet, ascribing to it a slow motion, and using other subterfuges which made Tycho contemptuously talk of *O caecos coeli spectatores* – oh blind watchers of the sky.

The next year, he published his first book: *De Nova Stella.* He hesitated some time before publishing it, because he had not yet quite overcome the idea that the writing of books was an undignified occupation for a nobleman. The book is a hodgepodge of tedious prefatory letters, calendrical and meteorological diaries, astrological predictions, and versified outpourings, including an eight-page 'Elegy to Urania'; but it contained in twenty-seven pages an exact description of Tycho's observation of the New Star, and of the instrument with which the observations were made – twenty-seven pages of 'hard, obstinate facts', which alone would suffice to establish his lasting fame.

Five years later, he gave Aristotelian cosmology the *coup de grâce*, by proving that the great comet of 1577 was also not a sub-lunary phenomenon as comets had previously been regarded, but must be 'at least six times' as far in space as the moon.

About the physical nature of the new star, and how it was

created, Tycho wisely professed ignorance. Contemporary astronomy calls 'new stars' *novae*, and explains their sudden increase in brightness by an explosive process. There had doubtless been other *novae* between 125 B.C. and A.D. 1572; but man's new consciousness of the sky, and the new attitude to precise observation, gave the star of 1572 a special significance: the explosion which caused its sudden flaring up shattered the stable, walled-in universe of the ancients.

### 3. *Sorcerer's Island*

King Frederick II of Denmark, whose life had been saved by Tycho's foster-father, the late Vice-Admiral, was a patron of philosophy and the arts. When Tycho was still a student of twenty-four, the King's attention had been called to the brilliant young man, and he had promised him, as a sinecure, the prebend from the first canonry to become vacant. In 1575, when his reputation was already established, Tycho, who loved travelling and did it like everything else in the grand style, made a tour of Europe, visiting friends, mostly astronomers, in Frankfurt and Basle, Augsburg, Wittemberg, and Venice, among them the *Landgraf* Wilhelm IV in Cassel. The *Landgraf* was more than an aristocratic dilettante; he had built himself an observatory on a tower in Cassel, and was so devoted to astronomy that, when told that his house was on fire while he was observing the new star, he calmly finished his observation before giving his attention to the flames.

He and Tycho got on so well that, after the visit, the *Landgraf* urged King Frederick to provide Tycho with the means for building his own observatory. When Tycho returned to Denmark, Frederick offered him various castles to choose from; but Tycho declined because he had set his heart on taking up residence in Basle, the charming and civilized old town which had captured the love of Erasmus, Paracelsus, and other illustrious humanists. Now Frederick became really eager to preserve Tycho for Denmark, and in February '76, sent a messenger – a youth of noble birth with orders to travel day and night – bearing a royal order for Tycho to come and see the King at once. Tycho obeyed, and the King made him an offer that sounded like a fairy tale: an island in the Sund between Copenhagen and Elsinor Castle, three miles in length, extending over two thousand acres of flat tableland rising on sheer white cliffs out of the sea. Here Tycho should build his house and observatory at Denmark's expense, and in addition receive an annual grant, plus various sinecures, which would make his income one of the highest in Denmark. After a further week's hesitation, Tycho graciously accepted the island of Hveen, and the fortune that went with it.

Accordingly, a royal instrument, signed on 23 May 1576, decreed that:

> We, Frederick the Second, &c., make known to all men, that we of our special favour and grace have conferred and granted in fee, and now by this our open letter confer and grant in fee, to our beloved Tyge Brahe, Otto's son, of Knudstrup, our man and servant, our land of Hveen, with all our and the crown's tenants and servants who thereon live, with all rent and duty which comes from that, and is given to us and to the crown, to have, use and hold, quit and free, without any rent, all the days of his life, and as long as he lives and likes to continue and follow his *studia mathematices*. . . .[5]

Thus came into existence the fabulous Uraniburg on the island of Hveen, where Tycho lived for twenty years and taught the world the methods of exact observation.

Tycho's new domain, which he called 'the island of Venus, vulgarly named Hveen', had an old tradition of its own. It was often referred to as the 'Scarlet Island' – for reasons which a sixteenth-century English traveller explains in his account:

> The Danes think this Island of Wheen to be of such importance, as they have an idle fable, that a King of England should offer for the possession of it, as much scarlet cloth as would cover the same, with a Rose-noble at the corner of each cloth.[6]

It also had some thirteenth century ruins, to which Danish folklore attached a Niebelung saga all of its own. Its inhabitants, distributed over some forty farms grouped around a small village, became Tycho's subjects, who lorded over them like an oriental despot.

Tycho's observatory, the Uraniburg, built by a German architect under Tycho's supervision, was a symbol of his character, in which meticulous precision combined with fantastic extravagance. It was a fortress-like monster which is said to have been 'epoch-making in the history of Scandinavian architecture', but on the surviving woodcuts looks rather like a cross between the Palazzo Vecchio and the Kremlin, its Renaissance façade surmounted by an onion-shaped dome, flanked by cylindrical towers, each with a removable top housing Tycho's instruments, and surrounded by galleries with clocks, sundials, globes, and allegorical figures. In the basement were Tycho's private printing press, fed by his own paper mill, his alchemist's furnace, and private prison for unruly tenants. He also had his own pharmacy, his game preserves, and artificial fishponds; the only thing he was missing was his tame elk. It had been dispatched to him from his estate but never reached the island. While spending a night in transit at Landskroner Castle, the elk wandered up the stairs to an empty apartment where it drank so much strong beer that on its way downstairs it stumbled, broke its leg, and died.

In the library stood his largest celestial globe, five feet in diameter, made of brass, on which, in the course of twenty-five years, the fixed stars were engraved one by one, after their correct positions had been newly determined by Tycho and his assistants in the process of re-mapping the sky; it had cost five thousand dalers, the equivalent of eighty years of Kepler's salary. In the south-west study, the brass arc of Tycho's largest quadrant – fourteen feet in diameter – was fastened to the wall; the space inside the arc was filled with a mural depicting Tycho himself surrounded by his instruments. Later on, Tycho added to the Uraniburg a second observatory, the 'Starburg', which was built entirely underground to protect the instruments from vibration and wind, only the dome-shaped roofs rising above ground level; so that 'even from the bowels of the earth he could show the way to the stars and the glory of God'.[7] Both buildings were full of gadgets and automata, including statues turning on hidden mechanisms, and a communication system that enabled him to ring a bell in the room of any of his assistants – which made his guests believe that he was convoking them by magic. The guests came in an unceasing procession, savants, courtiers, princes, and royalty, including King James VI of Scotland.

Life at Uraniburg was not exactly what one would expect to be the routine of a scholar's family, but rather that of a Renaissance court. There was a steady succession of banquets for distinguished visitors, presided over by the indefatigable, hard-drinking, gargantuan host, holding forth on the variations in the eccentricity of Mars, rubbing ointment on his silver nose, and throwing casual titbits to his fool Jepp, who sat at the master's feet under the table, chattering incessantly amidst the general noise. This Jepp was a dwarf, reputed to have second sight, of which he seemed to give spectacular proof on several occasions.

Tycho is really a refreshing exception among the sombre, tortured, neurotic geniuses of science. He was, it is true, not a creative genius, only a giant of methodical observation. Still, he displayed all the vanity of genius in his interminable poetic outpourings. His poetry is even more dreadful than Canon Koppernigk's, and more abundant in quantity – Tycho was never in want of a publisher, since he had his own paper mill and printing press. Even so, his verses and epigrams overflowed on to the murals and ornaments of Uraniburg and Stjoerneburg, which abounded in mottoes, inscriptions, and allegorical figures. The most impressive of these, adorning the wall of his chief study, represented the eight greatest astronomers in history, from Timocharis to Tycho himself, followed by 'Tychonides', a yet unborn descendant – with a caption expressing the hope that he would be worthy of his great ancestor.

## 4. *Exile*

Tycho stuck it out on his Scarlet Island for twenty years; then, at fifty-one, he took up his wanderings again. But by that time the bulk of his life's work was done.

In looking back at it, he divided his observations into 'childish and doubtful ones' (during his student days at Leipzig), into 'juvenile and habitually mediocre ones' (up to his arrival at Hveen), and into 'virile, precise, and absolutely certain ones' (made at the Uraniburg).[8] The Tychonic revolution in astronomical method consists in the previously unequalled precision and continuity of his observations. The second point is perhaps even more important than the first: one could almost say that Tycho's work compares with that of earlier astronomers as a cinematographic record with a collection of still photographs.

In addition to his remarkable survey of the solar system, his re-mapping of the firmament comprised a thousand fixed stars (of which the positions of 777 were determined accurately, and the remaining 223 places were hastily thrown in just before he left Uraniburg, to make up a round thousand). His proof that the *nova* of '72 was a true star, and that the comet of '77 moved in an orbit far outside the moon's, disposed of the already shaken belief in the immutability of the skies and the solidity of the celestial spheres. Lastly, his system of the world, which he offered as an alternative to the Copernican, though without much scientific value, played, as we shall see, a historically important part.[8a]

The reasons which made Tycho abandon his island realm were of a rather sordid character. Tyge, the Scandinavian squire, was as high-handed in his dealing with men as he was humble towards scientific fact; as arrogant towards his like as he was delicate and tender in handling his instruments. He treated his tenants appallingly, extracting from them labour and goods to which he was not entitled, and imprisoning them when they demurred. He was rude to all who evoked his displeasure, including the young King, Christian IV. The good King Frederick had died in 1588 (of too much drink, as Vedel dutifully pointed out in his funeral oration) and his successor, though well disposed to Tycho, on whose sorcerer's island he had spent a delightful day as a boy, was unwilling to close his eyes to Tycho's scandalous rule of Hveen. By this time, Tycho's arrogance seemed to be verging on mania of grandeur. He left several letters of the young King unanswered, flaunted the decisions of the provincial courts, and even of the High Court of Justice, by holding a tenant and all his family in chains. As a result, the great man who had been Denmark's glory, became a personage thoroughly disliked throughout the country. No direct steps were taken against him, but his fantastic sinecures were reduced to more reasonable proportions, and this gave Tyge, who was becoming increasing bored and restless on his Scarlet Island, the needed pretext to resume his wanderings again.

He had been preparing his emigration for several years, and when he left Hveen around Easter 1597, he did it in his customary grandiose manner, travelling with a suite of twenty – family, assistants, servants, and the dwarf Jepp – his baggage comprising the printing press, library, furniture, and all the instruments (except the four largest, which followed later). Ever since, as a student, he had ordered his first quadrant at Augsburg, he had been careful to have all his instruments made in a way that they could be dismantled and transported. 'An astronomer,' he declared, 'must be cosmopolitan because ignorant statesmen can not be expected to value their services.'[9]

The first station of the Tychonic caravan was Copenhagen, the next Rostock from where, having left Danish territory, Tycho wrote a rather impertinent letter to King Christian, complaining about the treatment he had received from his ungrateful country, and declaring his intention 'to look for help and assistance from other princes and potentates', yet graciously expressing his willingness to return 'if it could be done on fair conditions and without injury to myself'. Christian wrote back a remarkable letter which soberly refuted Tycho's complaints point by point, and made it clear that the condition of his return to Denmark was 'to be respected by you in a different manner if you are to find in us a gracious lord and King'.[10]

For once Tyge had found his match. There were only two men in his life who got the better of him, King Christian of Denmark, and Johannes Kepler from Weil-der-Stadt.

His bridges burnt, Tycho and his private circus continued their wanderings for another two years – to Wandsbeck Castle near Hamburg, to Dresden, to Wittenberg. Lastly, in June 1599, they arrived in – or rather made their entry into – Prague, residence of the Emperor Rudolph II, to whom, by the grace of God, Tycho de Brahe had been appointed Imperial Mathematicus. He was again to have a castle of his choice, and a salary of three thousand florins a year (Kepler in Gratz had two hundred), in addition to some 'uncertain income which might amount to some thousands'.[11]

Had Tycho remained in Denmark, it is highly unlikely that Kepler could have afforded the expense to visit him during the short remaining span of Tycho's life. The circumstances which made them both exiles, and guided them towards their meeting, can be attributed to coincidence or providence, according to taste, unless one assumes the existence of some hidden law of gravity in History. After all, gravity in the physical sense is also merely a word for an unknown force acting at a distance.

## 5. *Prelude to the Meeting*

Before they met in the flesh at Benatek Castle, near Prague, Kepler and Tycho had been corresponding for two years.

From the very beginning, the relationship had started on the wrong foot, owing to an innocent blunder which young Kepler committed. The episode involved Tycho's lifelong bitter enemy, Ursus the Bear, and makes the fathers of astronomy appear like actors in an *opera buffo*.

Reymers Bear,* who came from Ditmar, had started as a swineherd, and ended up as Imperial Mathematicus – at which post Tycho was to succeed him, and Kepler was to succeed Tycho. To achieve, in the sixteenth century, such a career, certainly required considerable gifts – which, in Ursus, were combined with a dogged and ferocious character, always ready to crush his victims' bones in a bear-like hug. In his youth he had published a Latin grammar and a book on land surveying, then entered the service of a Danish nobleman called Erik Lange. In 1584, Lange visited Tycho at Uraniburg, and took Ursus with him. It must have been a rather hectic encounter, as will presently be seen.

Four years after that visit, Ursus published his *Fundaments*

* German for bear, hence his latinized name, Ursus.

*of Astronomy*[12] in which he explained his system of the universe. It was, except for some details, the same system which Tycho had worked out in secret, but had not published yet, since he wanted more data to elaborate it. In both systems the earth was reinstated as the centre of the world, but the five planets were now circling round the sun and, with the sun, round the earth.[13] This was obviously a revival of the intermediary system between those of Herakleides and Aristarchus of Samos (*see* Fig. C, p. 48).

Tycho's system was, therefore, by no means very original; but it had the advantage of a compromise between the Copernican universe and the traditional one. It automatically recommended itself to all those who were reluctant to antagonize academic science, and yet desirous to 'save the phenomena', and was to play an important part in the Galileo controversy. Actually, the Tychonic system was 'discovered' quite independently by yet a third scholar, Helisaeus Roeslin, as it so often happens with inventions that 'lie in the air'. But Tycho, who was as proud of his system as Kepler of his five perfect solids, was convinced that Ursus had stolen it, by snooping through his manuscripts during that visit in 1584. He collected evidence to prove that Ursus had been prying among his papers; that he had taken the precaution of letting his pupil Andreas share a room with Ursus; that while Ursus was asleep, the faithful pupil 'had taken a handful of papers out of one of his breeches pockets, but was afraid to search the other pocket for fear of waking him'; and that Ursus, on discovering what had happened, 'behaved like a maniac', whereupon all papers which did not concern Tycho were restored to him.

According to Ursus, on the other hand, Tycho had been haughty and arrogant to him; had tried to shut him up by remarking that 'all these German fellows are half-cracked'; and had been so suspicious about his observations 'which he was able to take through his nose, without needing other sights' that he got somebody to search his, Ursus', papers the night before his departure.

The long and short of it is that the Bear had probably been snooping among Tycho's observations, but there is no proof that he had stolen Tycho's 'system', nor that there was any need for him to steal it.

It was into this hornets' nest that young Kepler blundered when he had just hit upon the idea of the *Mysterium* and felt the urgent need to share his joy with the whole world of learning. Ursus was then the Imperial Mathematicus at Prague; so Kepler dashed off a fan letter to him, starting in typical fan-mail style: 'There exist curious men who, unknown, write letters to strangers in distant lands'; and continuing with Keplerian effusion that he was familiar 'with the bright glory of thy fame which makes thee rank first among the *mathematici* of our time like the sun among the minor stars'.[14]

This was written in November 1595. The Bear never answered the unknown young enthusiast's letter; but two years later, when Kepler was already well known, Ursus printed the letter, without asking for Kepler's permission, in a book[15] in which he claimed the priority of the 'Tychonic' system, and abused Tycho in most ferocious language. The book bore the motto 'I will meet them [meaning Tycho and Co.] as a bear bereaved of her whelps – Hosiah 13'. Thus Tycho, of course, got the impression that Kepler was siding with the Bear – which was precisely what the Bear had intended. The situation was all the more embarrassing for poor Kepler as he had in the meantime also written a fan letter to Tycho, in which he called him 'the prince of mathematicians not only of our time but of all times'.[16] Moreover, unaware of the Homeric battle between the two, he had asked Ursus, of all people, to forward a copy of the *Mysterium* to Tycho!

Tycho reacted with unusual diplomacy and restraint. He acknowledged Kepler's letter and book with great courtesy, praised him for the ingenuity of the *Mysterium* while expressing certain reserves, and expressed the hope that Kepler would now make an effort to apply his theory of the five solids to Tycho's own system of the universe. (Kepler wrote on the margin: 'Everybody loves himself, but one can see his high opinion of my method.')[17] Only in a postscript did Tycho complain about Kepler's praise of Ursus. A little later Tycho wrote another letter to Maestlin,[18] in which he criticized Kepler's book much more severely, and repeated his complaint. The intention behind this was obvious: Tycho had immediately realized young Kepler's exceptional gifts, wanted to win him over to his side, and hoped that Maestlin would exert his authority with his former pupil in this sense. Maestlin duly transmitted Tycho's complaint to Kepler, who realized only now into what a frightful tangle he had got himself – and, of all people, with Tycho, who was his only hope. So he sat down and penned a long and agonized epistle to Tycho in true Keplerian style, bubbling with sincerity, cheating a little about the facts, pathetic and brilliant and slightly embarrassing, all at the same time:

> How come? Why does he [Ursus] set such value on my flatteries? ... If he were a man he would despise them, if he were wise he would not display them on the market place. The nonentity which I then was, searched for a famous man who would praise my new discovery. I begged him for a gift and behold, it was he who extorted a gift from the beggar. ... My spirit was soaring and melting away with joy over the discovery I had just made. If, in the selfish desire to flatter him I blurted out words which exceeded my opinion of him, this is to be explained by the impulsiveness of youth.[19]

And so on. But there is one staggering admission in the letter: when Kepler read Ursus' *Fundaments of Astronomy* he had believed that the trigonometrical rules in it were Ursus' original discoveries and did not realize that most of them could be found in Euclid![20] One feels the ring of truth in this admission of young Kepler's abysmal ignorance of mathematics at a time when, guided by intuition alone, he had mapped out the course of his later achievements in the *Mysterium*.

Tycho replied briefly, and with a gracious condescension which must have been rather galling to Kepler, that he had not required such an elaborate apology. Thus the incident was patched up, though it kept rankling in Tycho who, later on, when Kepler became his assistant, would force him to write a pamphlet *In Defence of Tycho against Ursus* – a chore which Kepler detested.

But for the time being, Tycho was willing to forget the

unfortunate episode, and anxious to get Kepler as his collaborator. He found it difficult to get the new observatory at Benatek Castle going, and his former assistants were in no hurry to rejoin the former despot of Hveen. So he wrote to Kepler in December 1599:

You have no doubt already been told that I have been most graciously called here by his Imperial Majesty and that I have been received in the most friendly and benevolent manner. I wish that you would come here, not forced by the adversity of fate, but rather on your own will and desire for common study. But whatever your reason, you will find in me your friend who will not deny you his advice and help in adversity, and will be ready with his help. But if you come soon we shall perhaps find ways and means so that you and your family shall be better looked after in future. *Vale.*

Given at Benatek, or the Venice of Bohemia, on 9 December 1599, by your very sympathetic Tycho Brahe's own hand.[21]

But by the time this letter arrived in Gratz, Kepler was already on his way to Tycho.

# 5 Tycho and Kepler

## 1. *The Gravity of Fate*

The town and castle of Benatek were twenty-two miles, a six-hours' journey, to the north-east of Prague. They overlooked the river Iser, which often flooded the surrounding orchards, hence the name 'Bohemian Venice'. Tycho had selected Benatek among the three castles which the Emperor had offered for his choice – perhaps because the watery surroundings reminded him of Hveen. He had taken possession of the castle in August 1599 – six months before Kepler's arrival – and had started at once tearing down walls and erecting new ones, intending to build another Uraniburg, and announcing his intention in high-flung poems which were inscribed over the entrance of the future observatory. There was also to be a separate gate for the Emperor, who had reserved an adjoining building for his visits.

But everything seemed to go wrong from the beginning. The salary of three thousand florins which the Emperor had granted Tycho beat all previous records; 'there was nobody at Court, not even among counts and barons of long service, who enjoyed so large an income'.[1] Both the mind of Rudolph, and his finances, were in a highly disordered state; and his court officials effectively sabotaged the carrying out of his extravagant royal promises. Tycho had to fight for his salary and to be content if he could squeeze half of it out of the Exchequer; when Kepler succeeded him, he would get only a dribble of what was due to him.

By the time Kepler arrived at Benatek, Tycho had already quarrelled with the Director of the Crown Estates who held the purse-strings, complained to the Emperor, threatened to leave Bohemia, and to tell the world the reasons. Also, several of Tycho's assistants, who had promised to join him at the new Uraniburg, had failed to turn up; and the largest instruments were still delayed on the long trek from Hveen. Towards the end of the year the plague had broken out, obliging Tycho to sit it out with Rudolph at the imperial residence in Girsitz, and supply him with a secret elixir against epidemics. To add to Tycho's worries, Ursus, who had disappeared from Prague on Tycho's arrival, now returned again, trying to create trouble; and Tycho's second daughter, Elisabeth, was having an illicit love affair with one of his assistants, the Junker Tengnagel. Young Kepler, in the provincial backwoods of Gratz, had dreamed of Benatek as a serene temple of Urania; he arrived at a mad-house. The castle was teeming with workmen, surveyors, visitors, and the formidable de Brahe clan including the sinister dwarf Jepp – huddled under the table during the interminable, tumultuous meals, and finding an easy butt for his sarcasms in that timid scarecrow of a provincial mathematicus.

Kepler had arrived in Prague in the middle of January. He had at once written to Benatek, and a few days later received an answer from Tycho, regretting that he could not welcome Kepler in person because of a forthcoming opposition of Mars and Jupiter, to be followed by a lunar eclipse; and inviting him to Benatek 'not so much as a guest, than as a very welcome friend and colleague in the contemplation of the skies'. The bearers of the letter were Tycho's eldest son and the Junker Tengnagel, both of whom were jealous of Kepler from the start, and remained hostile to the end. It was in their company that Kepler completed the last lap of his journey to Tycho – but only after a further delay of nine days. Tengnagel and Tycho Junior were probably having a good time in Prague, and were in no hurry to get back.

At last, then, on 4 February 1600, Tycho de Brahe and Johannes Keplerus, co-founders of a new universe, met face to face, silver nose to scabby cheek. Tycho was fifty-three, Kepler twenty-nine. Tycho was an aristocrat, Kepler a plebeian; Tycho a Croesus, Kepler a church-mouse; Tycho a Great Dane, Kepler a mangy mongrel. They were opposites in every respect but one: the irritable, choleric disposition which they shared. The result was constant friction, flaring into heated quarrels, followed by half-hearted reconciliations.

But all this was on the surface. In appearance, it was a meeting of two crafty scholars, each determined to make use of the other for his own purposes. But under the surface, they both knew, with the certainty of sleepwalkers, that they were born to complete each other; that it was the gravity of fate which had drawn them together. Their relationship was to alternate all the time between these two levels: *qua* sleepwalkers, they strolled arm in arm through uncharted spaces; in their waking contacts they brought out the worst in the other's character, as if by mutual induction.

Kepler's arrival led to a reorganization of work at Benatek. Previously, Tycho's younger son, Joergen, had been in charge of the laboratory, the senior assistant, Longomontanus, was assigned the study of the orbit of Mars, and Tycho had intended to put Kepler in charge of the next planet to be taken up for systematic observation. But his eagerness, and the fact that Longomontanus got stuck with Mars, led to a re-distribution of the planetary realm among the Tychonites: Kepler was given Mars, the notoriously most difficult planet, while Longomontanus was switched to the moon. This decision proved of momentous importance. Kepler, proud to be entrusted with Mars, boasted that he would solve the problem of its orbit in eight days, and even made a bet with this deadline. The eight days grew into nearly eight years; but out of the struggle of these years with the recalcitrant planet emerged Kepler's *New Astronomy or Physics of the Skies.*

He knew, of course, nothing of what lay ahead of him. He had come to Tycho to wrench from him the exact figures of the eccentricities and mean distances, in order to improve his model of the universe built around the five solids and the musical harmonies. But though he never discarded his *idée fixe,* it was now relegated into the background. The new problems which arose out of Tycho's data 'took such a hold of me that I nearly went out of my mind'.[2] Himself no more than an amateur observer with the coarsest of instruments, an armchair astronomer with the intuition of genius but still lacking in intellectual discipline, he was overwhelmed by the wealth and precision of Tycho's observations, and only now began to realize what astronomy really meant. The hard facts embodied in Tycho's data, the scrupulousness of Tycho's method, acted like a grindstone on Kepler's fantasy-prone intellect. But although Tycho did the grinding, and the process seemed to be more painful for Kepler than for him, in the end it was the grindstone which was worn down, while the blade emerged sharp and shining from it.

Soon after his arrival in Benatek, Kepler wrote:

> Tycho possesses the best observations, and thus so-to-speak the material for the building of the new edifice; he also has collaborators and everything he could wish for. He only lacks the architect who would put all this to use according to his own design. For although he has a happy disposition and real architectural skill, he is nevertheless obstructed in his progress by the multitude of the phenomena and by the fact that the truth is deeply hidden in them. Now old age is creeping upon him, enfeebling his spirit and his forces.[3]

There could be no doubt regarding the identity of the architect in Kepler's mind. Nor was it difficult for Tycho to guess Kepler's true opinion of him. He had amassed a treasure of data as nobody before him; but he was old, and lacking the boldness of imagination to build, out of his wealth of raw material, the new model of the universe. Its laws were there, in his columns of figures; but 'too deeply hidden' in them for him to decipher. He must also have felt that only Kepler was capable of succeeding in this task – and that nothing could prevent him from succeeding; that it would be this grotesque upstart, and not Tycho himself, nor the hoped-for Tychonides of the Uraniburg mural, who would reap the fruit of his lifelong labours. Half resigned to, half appalled by his own fate, he wanted at least to make it as difficult for Kepler as possible. He had always been most reluctant to disclose his treasured observations; if Kepler had thought he could simply grab them, he was woefully mistaken – as the indignant complaints in his letters show:

> Tycho gave me no opportunity to share in his experiences. He would only, in the course of a meal, and in between conversing

about other matters mention, as if in passing, today the figure for the apogee of one planet, tomorrow the nodes of another.[4]

One might add: as if he were handing bones to Jepp under the table. Nor would he allow Kepler to copy out his figures. In exasperation, Kepler even asked Tycho's Italian rival, Magini, to offer his own data in exchange for some of Tycho's. Only gradually, step by step, did Tycho yield; and when he put Kepler in charge of Mars, he was forced to disgorge his Mars data.

Kepler had spent barely a month at Benatek when Tycho, in a letter, first hinted at difficulties that had arisen between them; another month later, on 5 April, the tension blew up in an explosion which might have shattered the future of cosmology.

The immediate cause of the row was a document which Kepler had drafted, and in which the conditions of his future collaboration with Tycho were laid down in unpleasant detail. If he and his family were to live permanently at Benatek, Tycho must provide them with a detached apartment because the noise and disorder of the household were having a terrible effect on Kepler's gall, and provoked him to violent outbursts of temper. Next, Tycho must obtain a salary for Kepler from the Emperor, and in the meantime pay him fifty florins a quarter. He must also provide the Keplers with specified quantities of firewood, meat, fish, beer, bread, and wine. As for their collaboration, Tycho must leave Kepler his freedom to choose the time and subject of his work, and only ask him to undertake such researches which were directly connected with it; and since Kepler was 'not in need of a spur but rather of a brake to prevent the threat of galloping consumption due to overwork',[5] he must be allowed to rest in the day-time if he had worked deep into the night. And so on, for several pages.

This document was not meant for Tycho's perusal. Kepler handed it to a guest, a certain Jessenius, professor of medicine in Wittenburg, who was to serve as an intermediary in the negotiations between Tycho and himself. But whether by chance or intrigue, Tycho got hold of the document, which he could hardly regard as flattering to himself. Nevertheless he took it with that good-humoured magnanimity which lived side by side in the Danish *grand seigneur* with jealousy and bullying. He remained a benevolent despot so long as nobody challenged his rule; and socially, Kepler was so much his inferior that his carping and bickering demands did not affect Tycho as a challenge. One of the reasons for Kepler's bitterness was, incidentally, that he had been assigned an inferior position at the dinner table.

But above all, Tycho needed Kepler, who alone could put his life-work into proper shape. Hence he sat down to negotiate with Kepler in the presence of Jessenius, patiently rubbing ointment on his nose, a paragon of paternal moderation. This attitude grated even more on Kepler's inferiority complex, and he attacked Tycho, in the latter's words 'with the vehemence of a mad dog, to which creature he, Kepler himself, so much likes to compare himself in irritability.'[6]

Immediately after the stormy session Tycho, who always had an eye on posterity, wrote down the minutes of it, and requested Jessenius to endorse them. However, when his temper had cooled down, he entreated Kepler to stay on, at least for another few days, until an answer arrived from the Emperor, whom Tycho had approached concerning Kepler's employment. But Kepler refused to listen, and on the next day departed in the company of Jessenius to Prague, where he took quarters with Baron Hoffman. Just before his departure, Kepler had another choleric outburst; at the moment of farewell, he was overcome with remorse and apologized; while Brahe whispered into Jessenius' ear that he should try to bring the *enfant terrible* back to reason. But as soon as they arrived in Prague, Kepler wrote another abusive letter to Tycho.

He must have been in a dreadful state of hysteria. He was suffering from one of his recurrent obscure fevers; his family was in faraway Gratz; the persecution of the Protestants in Styria, and the debacle at Benatek had made a shambles of his future; and the data on Mars remained inaccessibly in Tycho's hands. Within a week, the pendulum swung to the other extreme: Kepler wrote a letter of apologies to Tycho which sounds like the ravings of a masochist against his own guilty ego:

The criminal hand which, the other day, was quicker than the wind in inflicting injury, hardly knows how to set about it to make amends. What shall I mention first? My lack of self-control, which I can only remember with the greatest pain, or your benefactions, noblest Tycho, which can neither be enumerated nor valued according to merit? For two months you have most generously provided for my needs ... you have extended to me every friendliness, you have allowed me to share in your most cherished possession. ... Taken all in all, neither to your children, nor to your wife, nor to yourself did you devote yourself more than to me. ... Therefore I think with the deepest dismay that God and the Holy Ghost delivered me to such an extent to my impetuous attacks and to my sick mind that instead of displaying moderation, I indulged during three weeks with closed eyes in sullen stubbornness against you and your family; that instead of thanking you, I displayed blind rage; that instead of showing you respect I displayed the greatest insolence against your person which by noble descent, prominent learning, and great fame deserves all respect; that instead of sending you a friendly greeting I let myself be carried away by suspicion and insinuation when I was itching with bitterness. ... I never considered how cruelly I must have hurt you by this despicable behaviour. ... I come to you as a postulant to ask, in the name of Divine pity, for your forgiveness of my terrible offences. What I have said or written against your person, your fame, your honour, and your scientific rank. ... I retract in all parts, and declare it voluntarily and freely as invalid, false, and unsound. ... I also promise sincerely that henceforth at whatever place I shall be I shall not only refrain from such foolish acts, words, deeds, and writings, but I shall also never and in no way unjustly and deliberately offend you. ... But since the ways of men are slippery, I ask you that whenever you notice in me any tendency towards such unwise manner of behaviour, to remind me of myself; you will find me willing. I also promise ... to oblige you by all kinds of services and ... thus to prove by my acts that my attitude towards your person is different, and always was different, from what one may conclude from the reckless condition of my heart and body during these last three weeks. I pray that God may help to fulfil this promise.[7]

I have quoted this letter at some length, because it reveals the tragic core of Kepler's personality. These turns of phrase do not seem to come from a scholar of repute, but from a tortured adolescent, begging to be forgiven by a father whom he hates and loves. Tycho had replaced Maestlin. At the base of his iridescent, complex character, Kepler always remained a waif and stray.

But Tycho was no less dependent on Kepler than Kepler on Tycho. In their worldly contacts, Tycho was the old man of the tribe, Kepler the nagging, ill-mannered adolescent. But on the other level, the rules were reversed: Kepler was the magician from whom, Tycho hoped, would come the solution of his problems, the answer to his frustrations, the salvation from ultimate defeat; and however foolishly they both behaved, *qua* sleepwalkers, they both knew all this.

Therefore, three weeks after the row, Tycho turned up in Prague and drove Kepler back to Benatek in his coach – one can almost see Tycho's great fat arm in the leg-of-mutton sleeve, crushing in an affectionate embrace Kepler's skimpy bones.

## 2. *The Inheritor*

Altogether the association between Kepler and Tycho lasted for eighteen months, until Tycho's death. Fortunately for both, and for posterity, they were only part of this time in personal contact, for Kepler twice returned to Gratz and spent a total of eight months there to settle his affairs and get his wife's property out.

He left for Gratz the first time shortly after his reconciliation with Tycho, in June 1600. Though peace had been re-established, nothing definite had been settled regarding their future collaboration,[7a] and Kepler was in two minds whether he would return to Tycho or not. He still hoped either to save his position and salary in Gratz by being granted a long leave of absence, or to obtain a chair in his native Wuerttemberg – his lifelong ambition. He wrote to Maestlin and Herwart, his adoptive fathers Nos. One and Two, hinting that No. Three was rather a disappointment; but nothing came of it. He sent to the Archduke Ferdinand a treatise on a solar eclipse, also to no avail; but in that treatise he hit on something for which he had not looked: that there was 'a force in the earth' which influenced the moon's motion, a force which diminished in proportion to distance. As he had already attributed a physical force to the sun as an explanation of the motions of the planets, the dependence of the moon on a similar force in the earth was the next important step towards the concept of universal gravity.

But such trifles could not deter the Archduke from his plan to stamp out heresy in his lands. On 31 July and the following days, all Lutheran citizens of Gratz, a little over a thousand in number, had to appear, one by one, before an ecclesiastical commission, and either to declare their willingness to return to the Roman faith, or to suffer expulsion. This time no exemption was made, not even for Kepler – though he was let off paying half of the exit levy and granted other financial privileges. The day after he appeared before the commission, a rumour was rife in Gratz that he had changed his mind and declared his readiness to become a Catholic. Whether he had really wavered or not, is impossible to know; but in any case, he overcame the temptation and accepted exile with all its consequences.

He sent a last S.O.S. to Maestlin.[8] It starts with a dissertation on the eclipse of the sun on 10 July, which he had observed through a *camera obscura* of his own construction, erected in the middle of the market place in Gratz – with the twofold result that a thief stole his purse containing thirty florins, while Kepler himself discovered an important new optical law. The letter continues with the threat that Kepler plus his family would travel down the Danube into Maestlin's arms and a professorship (even if only a small one) which Maestlin would no doubt provide; and ends with the request that Maestlin should pray for him. Maestlin answered that he would gladly pray, but could do nothing else for Kepler, 'the steadfast and valiant martyr of God';[9] and after that, answered none of Kepler's letters for four years. He probably thought that he had done his share, and that it was now Tycho's turn to look after the infant prodigy.

Tycho himself was delighted with the sad news. He had

doubted whether Kepler would return to him, and welcomed the prospect all the more as his senior assistant, Longomontanus, had in the meantime left. When Kepler informed him of his impending expulsion, he wrote back that Kepler should come at once; 'do not hesitate, make haste, and have confidence'.[10] He added that during a recent audience with the Emperor he had requested that Kepler should be officially attached to his observatory, and that the Emperor had nodded his consent. But in a postscript to the long and affectionate letter, Tycho could not refrain from alluding to a subject which had been one of the main reasons for Kepler's unhappiness at Benatek. On his arrival there, Tycho had imposed on him the irksome chore of writing a pamphlet refuting the claims of Ursus; and though Ursus had in the meantime died, Tycho still insisted on persecuting him beyond the grave. Moreover, Kepler was also to write a refutation of a pamphlet by John Craig, physician to James of Scotland, in which Craig had dared to doubt Tycho's theories about comets. It was not a joyous prospect for Kepler to waste his time on these futile labours to serve Tycho's vanity; but now he had no other choice.

In October, he arrived back in Prague with his wife – but without his furniture and chattels, which he had to leave behind in Linz as he had no money to pay for the transport. He was again ill with intermittent fever, and again thought that he was suffering from consumption. The imperial nod of consent to his employment was not followed by concrete action, so Kepler and his wife had to live entirely on Tycho's bounty. At the Emperor's request, who wanted his mathematicus close at hand, Tycho had given up the splendours of Benatek and moved to a house in Prague, where the Keplers, having no money for rent, were forced to take up quarters. During the next six months, Kepler had little time for astronomy, as he was fully occupied with writing the accursed polemics against Ursus and Craig, and nursing his real and imaginary ailments. Frau Barbara, who even in better days had not been a cheerful soul, hated the alien ways and narrow, winding streets of Prague, whose stench was strong enough 'to drive back the Turks', as a contemporary English traveller wrote.[11] The Keplers were drinking the bitter cup of refugee existence to the dregs.

In the spring of 1601, Frau Barbara's rich father died back in Styria – he had paid the price of conversion to die in his country. This gave Kepler a welcome pretext to leave his family in Tycho's charge, and to go back to Gratz to save the inheritance. In this he did not succeed; but he stayed in Gratz for another four months, and seems to have had a wonderful time, dining in the houses of the Styrian nobles as a kind of distinguished exile on home leave, climbing mountains to measure the curvature of the earth, writing infuriating letters to Tycho whom he reproached for not giving enough money to Frau Barbara, and thoughtfully asking her whether Elisabeth Brahe, who was at last allowed to marry the Junker Tengnagel, was 'showing signs of the baby' – which arrived three months after the ceremony. He returned to Prague in August, his mission unaccomplished, but his health fully restored, and in radiant spirits. He now only had to mark time for another two months till the decisive turn in his life.

On 13 October 1601, Tycho was a guest at supper at Baron Rosenberg's table in Prague. Among the other guests was an Imperial Councillor, so it must have been an illustrious company; but since Tycho had been in the habit of entertaining royalty, and was accustomed to vast amounts of drink, it is difficult to understand why he was unable to cope with the predicament in which he found himself. Kepler has carefully recorded what happened in the Diary of Observations – a kind of logbook where all important events of the Brahe household were entered:

On 13 October, Tycho Brahe, in the company of Master Minkowitz, had dinner at the illustrious Rosenberg's table, and held back his water beyond the demands of courtesy. When he drank more, he felt the tension in his bladder increase, but he put politeness before his health. When he got home, he was scarcely able to urinate.

At the beginning of his illness, the moon was in opposition to Saturn ... [follows the horoscope of the day].

After five sleepless nights, he could still only pass his water with the greatest pain, and even so the passage was impeded. The insomnia continued, with internal fever gradually leading to delirium; and the food he ate, and from which he could not be kept, exacerbating the evil. On 24 October, his delirium ceased for several hours; nature conquered and he expired peacefully among the consolations, prayers, and tears of his people.

So from this date the series of celestial observations was interrupted, and his own observations of thirty-eight years have come to an end.

On his last night in his gentle delirium, he repeated over and again these words, like someone composing a poem:

*Let me not seem to have lived in vain.*

No doubt he wished that these words should be added to the title-page of his works, thus dedicating them to the memory and uses of posterity.[12]

During his last days, whenever the pain subsided, the great Dane had refused to keep to a diet, ordered and ate ravenously whatever dish came to his mind. When delirium set in again, he kept repeating softly that he hoped his life had not been wasted (*ne frusta vixisse videar*). The meaning of these words becomes clear through his last wish addressed to Kepler.[13] It was the same wish which he had expressed in his first letter to him: that Kepler should build the new universe not on the Copernican, but on Tycho's system. Yet he must have known, as his delirious complaint revealed, that Kepler would do just the opposite, and put the Tychonic legacy to his own use.

Tycho was buried with great pomp in Prague, his coffin carried by twelve imperial Gentlemen-at-Arms, preceded by his coat of arms, his golden spurs, and favourite horse.

Two days later, on 6 November 1601, the Emperor's privy councillor, Barwitz, called on Kepler at his lodgings, to appoint him, as Tycho's successor, to the post of Imperial Mathematicus.

# 6 The Giving of the Laws

## 1. Astronomia Nova

Kepler stayed in Prague as Imperial Mathematicus from 1601 to 1612, to the death of Rudolph II.

It was the most fruitful period of his life, and brought him the unique distinction of founding two new sciences: instrumental optics, which does not concern us, and physical astronomy. His *magnum opus*, published in 1609, bears the significant title:

A NEW ASTRONOMY *Based on Causation*
*or* A PHYSICS OF THE SKY
*derived from Investigations of the*
MOTIONS OF THE STAR MARS
*Founded on Observations of* THE NOBLE TYCHO BRAHE[1]

Kepler worked on it, with interruptions, from his arrival at Benatek in 1600, to 1606. It contains the first two of Kepler's three planetary laws: (1) that the planets travel round the sun not in circles but in elliptical orbits, one focus of the ellipse being occupied by the sun; (2) that a planet moves in its orbit not at uniform speed but in such a manner that a line drawn from the planet to the sun always sweeps over equal areas in equal times. The third law, published later, does not concern us at this point.

On the surface, Kepler's laws look as innocent as Einstein's $E=Mc^2$, which does not reveal, either, its atom-exploding potentialities. But the modern vision of the universe is shaped, more than by any other single discovery, by Newton's law of universal gravitation, which in turn was derived from Kepler's three laws. Although (owing to the peculiarities of our educational system), a person may never have heard of Kepler's laws, his thinking has been moulded by them without his knowledge; they are the invisible foundation of a whole edifice of thought.

Thus the promulgation of Kepler's laws is a landmark in history. They were the first 'natural laws' in the modern sense: precise, verifiable statements about universal relations governing particular phenomena, expressed in mathematical terms. They divorced astronomy from theology, and married astronomy to physics. Lastly, they put an end to the nightmare that had haunted cosmology for the last two millennia: the obsession with spheres turning on spheres, and substituted a vision of material bodies not unlike the earth, freely floating in space, moved by physical forces acting on them.

The manner in which Kepler arrived at his new cosmology is fascinating; I shall attempt to re-trace the zigzag course of his reasoning. Fortunately, he did not cover up his tracks, as Copernicus, Galileo, and Newton did, who confront us with the result of their labours, and keep us guessing how they arrived at it. Kepler was incapable of exposing his ideas methodically, text-book fashion; he had to describe them in the order they came to him, including all the errors, detours, and the traps into which he had fallen. The *New Astronomy* is written in an unacademic, bubbling baroque style, personal, intimate, and often exasperating. But it is a unique revelation of the ways in which the creative mind works.

> What matters to me [Kepler explained in his Preface] is not merely to impart to the reader what I have to say, but above all to convey to him the reasons, subterfuges, and lucky hazards which led me to my discoveries. When Christopher Colombus, Magelhaen, and the Portuguese relate how they went astray on their journeys, we not only forgive them, but would regret to miss their narration because without it the whole, grand entertainment would be lost. Hence I shall not be blamed if, prompted by the same affection for the reader, I follow the same method.[1a]

Before embarking on the story, it will be prudent to add my own apology to Kepler's. Prompted by the same 'affection for the reader' I have tried to simplify as far as possible a difficult subject: even so, the present chapter must of necessity be slightly more technical than the rest of this book. If some passages tax his patience, even if occasionally he fails to grasp a point or loses the thread, he will, I hope, nevertheless get a general idea of Kepler's odyssey of thought, which opened up the modern universe.

## 2. *Opening Gambits*

It will be remembered that at the partitioning of the cosmos which followed young Kepler's arrival at Benatek Castle, he was allotted the study of the motions of Mars which had defeated Tycho's senior assistant, Longomontanus, and Tycho himself.

> I believe it was an act of Divine Providence [he commented later on] that I arrived just at the time when Longomontanus was occupied with Mars. For Mars alone enables us to penetrate the secrets of astronomy which otherwise would remain forever hidden from us.[2]

The reason for this key position of Mars is that, among the outer planets, his orbit deviates more than the others' from the circle; it is the most pronouncedly elliptical. It was precisely for that reason that Mars had defied Tycho and his assistant: since they expected the planets to move in circles, it was impossible to reconcile theory with observation:

> He [Mars] is the mighty victor over human inquisitiveness, who made a mockery of all the strategems of astronomers, wrecked their tools, defeated their hosts; thus did he keep the secret of his rule safe throughout all past centuries and pursued his course in unrestrained freedom; wherefore that most famous of Latins, the priest of nature Pliny, specially indicted him: MARS IS A STAR WHO DEFIES OBSERVATION.[3]

Thus Kepler, in his dedication of the *New Astronomy* to the Emperor Rudolph II. The dedication is written in the form of an allegory of Kepler's war against Mars, begun under 'Tycho's supreme command', patiently pursued in spite of the warning example of Rheticus who went off his head over Mars, in spite of other dangers and terrible handicaps, such as a lack of supplies owing to Rudolph's failure to pay Kepler's salary – and so on to the triumphant end when the Imperial Mathematicus, riding a chariot, leads the captive enemy to the Emperor's throne.

Thus Mars held the secret of all planetary motion, and young Kepler was assigned the task of solving it. He first attacked the problem on traditional lines; when he failed, he began to throw out ballast and continued doing so until, by and by, he got rid of the whole load of ancient beliefs on the nature of the universe, and replaced it by a new science.

As a preliminary, he made three revolutionary innovations to gain elbow room, as it were, for tackling his problem. It will be remembered that the centre of Copernicus' system was not the sun, but the centre of the earth's orbit; and that already in the *Mysterium Cosmographicum* Kepler had objected to this assumption as physically absurd. Since the force which moved the planets emanated from the sun, the whole system should be centred on the body of the sun itself.[4]

But in fact it was not. The sun occupies not the exact centre

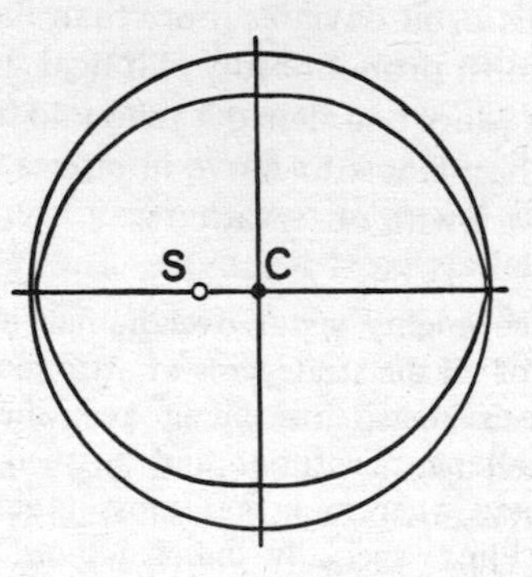


of the orbit at *C*; it occupies one of the two foci of the ellipse at S.

Kepler did not know as yet that the orbit was an ellipse; he still regarded it as a circle. But even so, to get approximately correct results, the centre of the circle had to be placed at C, and not in the sun. Accordingly, the question arose in his mind: if the force which moves the planets comes from S, why do they insist on turning round C? Kepler answered the question by the assumption that each planet was subject to two conflicting influences: *the force of the sun, and a second force located in the planet itself.* This tug-of-war caused it now to approach the sun, now to recede from him.

The two forces are, as we know, gravity and inertia. Kepler, as we shall see, never arrived at formulating these concepts. But he prepared the way for Newton by postulating two dynamic forces to explain the eccentricity of the orbits. Before him, the need for a physical explanation was not felt; the phenomenon of eccentricity was merely 'saved' by the introduction of an epicycle or eccenter, which made C turn round S. Kepler replaced the fictitious wheels by real forces.

For the same reason, he insisted on treating the sun as the centre of his system not only in the physical but in the geometrical sense, by making the distances and positions of the planets relative to the sun (and not relative to the earth or the centre C) the basis of his computations. This shift of emphasis, which was more instinctive than logical, became a major factor in his success.

His second innovation is simpler to explain. The orbits of all planets lie very nearly, but not entirely, in the same plane; they form very small angles with each other – rather like adjacent pages of a book which is nearly, but not entirely closed. The planes of all planets pass, of course, through the sun – a fact which is self-evident to us, but not to pre-Keplerian astronomy. Copernicus, once again misled by his slavish devotion to Ptolemy, had postulated that the plane of the Martian orbit *oscillates in space*; and this oscillation he made to depend on the position of the earth – which, as Kepler remarks, 'is no business of Mars'. He called this Copernican idea 'monstrous' (though it was merely due to Copernicus' complete indifference to physical reality) and set about to prove that the plane in which Mars moves passes through the sun, and does not oscillate, but forms a fixed angle with the plane of the earth's orbit. Here he met, for once, with immediate success. He proved, by several independent methods, all based on the Tychonic observations, that the angle between the planes of Mars and Earth remained always the same, and that it amounted to 1° 50′. He was delighted, and remarked smugly that 'the observation took the side of my preconceived ideas, as they often did before'.[5]

The third innovation was the most radical. To gain more elbow room, he had to get out of the strait-jacket of 'uniform motion in perfect circles' – the basic axiom of cosmology from Plato up to Copernicus and Tycho. For the time being, he still let circular motion stand, but he threw out uniform speed. Again he was guided mainly by physical considerations: if the sun ruled the motions, then his force must act more powerfully on the planet when it is close to the source, less powerfully when away from it; hence the planet will move faster or slower, in a manner somehow related to its distance from the sun.

This idea was not only a challenge to antique tradition; it also reversed the original purpose of Copernicus. It will be remembered that Copernicus' original motive for embarking on a reform of the Ptolemaic system was his discontent with the fact that, according to Ptolemy, a planet did not move at uniform speed around the centre of its orbit, but only around a point at some distance from the centre. This point was called the *punctum equans* – the point in space, from which the planet gave the illusion of 'equal motion'. Canon Koppernigk regarded this arrangement as an evasion of the command of uniform motion, abolished Ptolemy's equants, and added, instead, more epicycles to his system. This did not make the planet's *real* motion either circular, or uniform, but each wheel in the imaginary clockwork which was supposed to account for it, did turn uniformly – if only in the astronomer's mind.

When Kepler renounced the dogma of uniform motion, he was able to throw out the epicycles which Copernicus had introduced to save it. Instead, he reverted to the equant as an important calculating device (see diagram on p. 323).

Let the circle be the track of a toy train chugging round a room. When near the window it runs a little faster, near the door a little slower. Provided that these periodic changes of speed follow some simple, definite rule, then it is possible to find a *punctum equans*, 'E', from which the train *seems* to move at uniform speed. The closer we are to a moving train, the faster it seems to move; hence the *punctum equans* will be somewhere between the centre C of the track and the door, so that the speed-surplus of the train when passing the window will be eliminated by distance, its speed-deficiency at the door compensated by closeness. The advantage gained by the introduction of the imaginary *punctum equans* is that, *seen from E*, the train seems to move uniformly, that is, it will cover equal

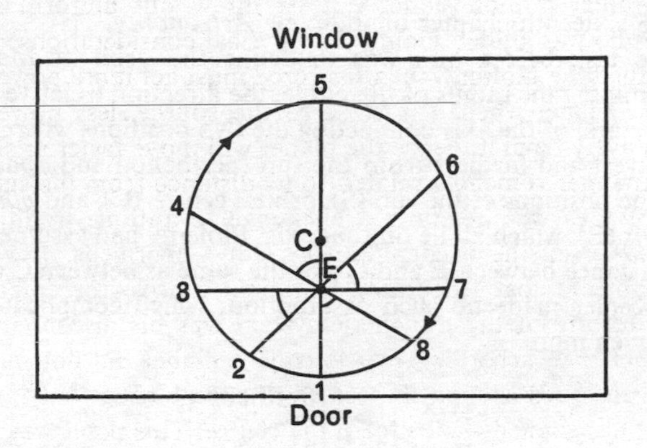


angles at equal times – which makes it possible to compute its various positions 1, 2, 3, etc., at any given moment.

By these three preliminary moves: (a) the shifting of the system's centre into the sun; (b) the proof that the orbital planes do not 'oscillate' in space, and (c) the abolition of uniform motion, Kepler had cleared away a considerable amount of the rubbish that had obstructed progress since Ptolemy, and made the Copernican system so clumsy and unconvincing. In that system Mars ran on five circles; after the clean-up, a single eccentric circle must be sufficient – if the orbit was really a circle. He felt confident that victory was just around the corner, and before the final attack wrote a kind of obituary notice for classical cosmology:

Oh, for a supply of tears that I may weep over the pathetic diligence of Apianus [author of a very popular textbook] who, relying on Ptolemy, wasted his valuable time and ingenuity on the construction of spirals, loops, helixes, vortices, and a whole labyrinth of convolutions, in order to represent that which exists only in the mind, and which Nature entirely refuses to accept as her likeness. And yet that man has shown us that, with his penetrating intelligence, he would have been capable of mastering Nature.[6]

## 3. *The First Assault*

Kepler's first attack on the problem is described in great detail in the sixteenth chapter of the *New Astronomy*.

The task before him was to define the orbit of Mars by determining the radius of the circle, the direction (relative to the fixed stars) of the axis connecting the two positions where Mars is nearest and farthest from the sun (perihelion and aphelion), and the positions of the sun (S), orbital centre (C), and *punctum equans* (E), which all lie on that axis. Ptolemy had assumed that the distance between E and C was the same as between C and S, but Kepler made no such assumption, which complicated his task even more.[7]

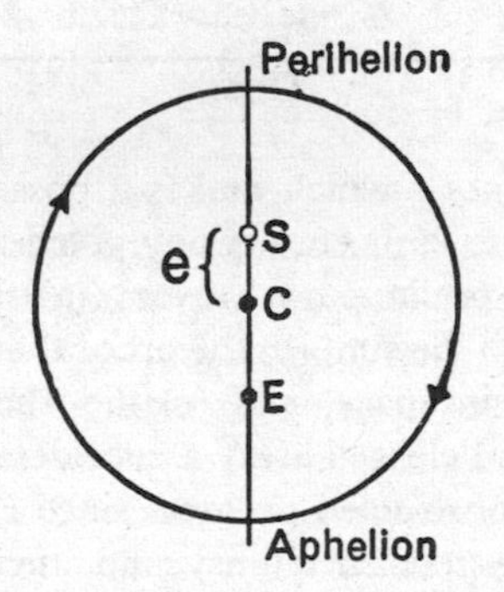


He chose out of Tycho's treasure four observed positions of Mars at the convenient dates when the planet was in opposition to the sun.[8] The geometrical problem which he had to solve was, as we saw, to determine, out of these four positions, the radius of the orbit, the direction of the axis, and the position of the three central points on it. It was a problem which could not be solved by rigorous mathematics, only by approximation, that is, by a kind of trial-and-error procedure which has to be continued until all the pieces in the jig-saw puzzle fit together tolerably well. The incredible labour that this involved may be gathered from the fact that Kepler's draft calculations (preserved in manuscript) cover nine hundred folio pages in small handwriting.

At times he was despairing; he felt, like Rheticus, that a demon was knocking his head against the ceiling, with the shout: 'These are the motions of Mars.' At other times, he appealed for help to Maestlin (who turned a deaf ear), to the Italian astronomer, Magini (who did the same), and thought of sending an S.O.S. to Francois Vieta, the father of modern algebra: 'Come, oh Gallic Apollonius, bring your cylinders and spheres and what other geometer's houseware you have. ...'[8a] But in the end he had to slog it out alone, and to invent his mathematical tools as he went along.

Half-way through that dramatic sixteenth chapter, he bursts out:

> If thou [dear reader] art bored with this wearisome method of calculation, take pity on me who had to go through with at least seventy repetitions of it, at a very great loss of time; nor wilst thou be surprised that by now the fifth year is nearly past since I took on Mars. ...

Now, at the very beginning of the hair-raising computations in chapter sixteen, Kepler absentmindedly put three erroneous figures for three vital longitudes of Mars, and happily went on from there, never noticing his error. The French historian of astronomy, Delambre, later repeated the whole computation, but, surprisingly, his correct results differ very little from Kepler's faulty ones. The reason is, that towards the end of the chapter Kepler committed several mistakes in simple arithmetic – errors in division which would bring bad marks to any schoolboy – and these errors happen very nearly to cancel out his earlier mistakes. We shall see, in a moment, that, at the most crucial point of the process of discovering his Second Law, Kepler again committed mathematical sins which mutually cancelled out, and 'as if by miracle' (in his own words), led to the correct result.

At the end of that breathtaking chapter, Kepler seems to have triumphantly achieved his aim. As a result of his seventy-odd trials, he arrived at values for the radius of the orbit and for the three central points which gave, with a permissible error of less than 2′, the correct positions of Mars for all the ten oppositions recorded by Tycho. The unconquerable Mars seemed at last to have been conquered. He proclaimed his victory with the sober words:

> Thou seest now, diligent reader, that the hypothesis based on this method not only satisfies the four positions on which it was based, but also correctly represents, within two minutes, all the other observations. . . .[9]

There follow three pages of tables to prove the correctness of his claim; and then, without further transition, the next chapter starts with the following words:

> Who would have thought it possible? This hypothesis, which so closely agrees with the observed oppositions, is nevertheless false. . . .

## 4. *The Eight Minutes Arc*

In the two following chapters Kepler explains, with great thoroughness and an almost masochistic delight, how he discovered that the hypothesis is false, and why it must be rejected. In order to prove it by a further test, he had selected two specially rare pieces from Tycho's treasury of observations, and lo! they did not fit; and when he tried to adjust his model to them, this made things even worse, for now the observed positions of Mars differed from those which his theory demanded by magnitudes up to eight minutes arc.

This was a catastrophe. Ptolemy, and even Copernicus could afford to neglect a difference of eight minutes, because their observations were only accurate within a margin of ten minutes, anyway.

> But, [the nineteenth chapter concludes] but for us, who, by divine kindness were given an accurate observer such as Tycho Brahe, for us it is fitting that we should acknowledge this divine gift and put it to use. . . . Henceforth I shall lead the way towards that goal according to my own ideas. For, if I had believed that we could ignore these eight minutes, I would have patched up my hypothesis accordingly. But since it was not permissible to ignore them, those

eight minutes point the road to a complete reformation of astronomy: they have become the building material for a large part of this work. . . .[10]

It was the final capitulation of an adventurous mind before the 'irreducible, obstinate facts'. Earlier on, if a minor detail did not fit into a major hypothesis, it was cheated away or shrugged away. Now this time-hallowed indulgence had ceased to be permissible. A new era had begun in the history of thought: an era of austerity and rigour. As Whitehead has put it:

All the world over and at all times there have been practical men, absorbed in 'irreducible and stubborn facts': all the world over and at all times there have been men of philosophic temperament who have been absorbed in the weaving of general principles. It is this union of passionate interest in the detailed facts with equal devotion to abstract generalization which forms the novelty in our present society.[11]

This new departure determined the climate of European thought in the last three centuries, it set modern Europe apart from all other civilizations in the past and present, and enabled it to transform its natural and social environment as completely as if a new species had arisen on this planet.

The turning point is dramatically expressed in Kepler's work. In the *Mysterium Cosmographicum* the facts are coerced to fit the theory. In the *Astronomia Nova*, a theory, built on years of labour and torment, was instantly thrown away because of a discord of eight miserable minutes arc. Instead of cursing those eight minutes as a stumbling block, he transformed them into the cornerstone of a new science.

What caused this change of heart in him? I have already mentioned some of the general causes which contributed to the emergence of the new attitude: the need of navigators, and engineers, for greater precision in tools and theories; the stimulating effects on science of expanding commerce and industry. But what turned Kepler into the first law-maker of nature was something different and more specific. It was *his introduction of physical causality into the formal geometry of the skies* which made it impossible for him to ignore the eight minutes arc. So long as cosmology was guided by purely geometrical rules of the game, regardless of physical causes, discrepancies between theory and fact could be overcome by inserting another wheel into the system. In a universe moved by real, physical forces, this was no longer possible. The revolution which freed thought from the stranglehold of ancient dogma, immediately created its own, rigorous discipline.

The Second Book of the *New Astronomy* closes with the words:

> And thus the edifice which we erected on the foundation of Tycho's observations, we have now again destroyed. ... This was our punishment for having followed some plausible, but in reality false, axioms of the great men of the past.

## 5. *The Wrong Law*

The next act of the drama opens with Book Three. As the curtain rises, we see Kepler preparing himself to throw out more ballast. The axiom of *uniform* motion has already gone overboard; Kepler feels, and hints[12] that the even more sacred one of *circular* motion must follow. The impossibility of constructing a circular orbit which would satisfy all existing observations, suggests to him that the circle must be replaced by some other geometrical curve.

But before he can do that, he must make an immense detour. For if the orbit of Mars is not a circle, its true shape can only be discovered by defining a sufficient number of points on the unknown curve. A circle is defined by three points on its circumference; every other curve needs more. The task before Kepler was to construct Mars's orbit without any preconceived ideas regarding its shape; to start from scratch, as it were.

To do that, it was first of all necessary to re-examine the motion of the earth itself. For, after all, the earth is our observatory; and if there is some misconception regarding its motion, all conclusions about the motions of other bodies will be distorted. Copernicus had assumed that the earth moves at uniform speed – not, as the other planets, only 'quasi-uniformly' relative to some equant or epicycle, but *really* so. And since observation contradicted the dogma, the inequality of the earth's motion was explained away by the suggestion that the orbit periodically expanded and contracted, like a kind of pulsating jellyfish.[13] It was typical of those improvisations which astronomers could afford so long as they felt free to manipulate the universe as they pleased on their drawing boards. It was equally typical that Kepler rejected it as 'fantastic',[14] again on the grounds that no physical cause existed for such a pulsation.

Hence his next task was to determine, more precisely than Copernicus had done, the earth's motion round the sun. For that purpose he designed a highly original method of his own. It was relatively simple, but it so happened that nobody had thought of it before. It consisted, essentially, in the trick of transferring the observer's position from earth to Mars, and to compute the motions of the earth exactly as an astronomer on Mars would do it.[15]

The result was just as he had expected: the earth, like the other planets, did not revolve with uniform speed, but faster or slower according to its distance from the sun. Moreover, at the two extreme points of the orbit, the aphelion and perihelion (see Figure on p. 324) the earth's velocity proved to be, simply and beautifully, inversely proportional to distance.

At this decisive point,[16] Kepler flies off the tangent and becomes air-borne, as it were. Up to here he was preparing, with painstaking patience, his second assault on the orbit of Mars. Now he turns to a quite different subject. 'Ye physicists, prick your ears,' he warns, 'for now we are going to invade your territory.'[17] The next six chapters are a report on that invasion into celestial physics, which had been out of bounds for astronomy since Plato.

A phrase seems to have been humming in his ear like a tune one cannot get rid of; it crops up in his writings over and again: there is a force in the sun which moves the planet, there is a force in the sun, there is a force in the sun. And since there is a force in the sun, there must exist some beautifully simple relation between the planet's distance from the sun and its speed. A light shines the brighter the nearer we are to its source, and the same must apply to the force of the sun: the closer the planet to it, the quicker it will move. This is his instinctive conviction, already expressed in the *Mysterium Cosmographicum*; but now, at last, he has succeeded in proving it.

In fact he has not. He has proved the inverse ratio of speed to distance only for the *two extreme points* of the orbit; and his extension of this 'Law' to the *entire* orbit was a patently incorrect generalization. Moreover, Kepler knew this, and admitted it at the end of the thirty-second chapter,[18] before he became airborne; but immediately afterwards, he conveniently forgot it. This is the first of the critical mistakes which 'as if by a miracle' cancelled out, and led Kepler to the discovery of his Second Law. It looks as if his conscious, critical faculties were anaesthetized by the creative impulse, by his impatience to get to grips with the physical forces in the solar system.

Since he had no notion of the *momentum* which makes the planet persist in its motion, and only a vague intuition of *gravity* which bends that motion into a closed orbit, he had to find, or invent, a force which, like a broom, sweeps the planet around its path. And since the sun causes all motion, he let the sun handle the broom. This required that the sun rotate round its own axis – a guess which was only confirmed much later; the force which it emitted rotated with it, like the spokes of a wheel, and swept the planets along. But if that were the only force acting on them, the planets would all have the same angular velocity, they would all complete their revolutions in the same period – which they do not. The reason, Kepler thought, was the laziness or 'inertia' of the planets, who desire to remain in the same place, and resist the sweeping force. The 'spokes' of that force are not rigid; they allow the planet to lag behind; it works rather like a vortex or whirlpool.[19] The power of the whirlpool diminishes with distance, so that the farther away the planet, the less power the sun has to overcome its laziness, and the slower its motion will be.

It still remained to be explained, however, why the planets moved in eccentric orbits instead of always keeping the same distance from the centre of the vortex. Kepler first assumed that apart from being lazy, they performed an epicycle motion in the opposite direction under their own steam, as it were, apparently out of sheer cussedness. But he was dissatisfied with this, and at a later stage assumed that the planets were 'huge round magnets' whose magnetic axis pointed always in the same direction, like the axis of a top; hence the planet will periodically be drawn closer to, and be repelled by the sun, according to which of its magnetic poles faces the sun.

Thus, in Kepler's physics of the universe, the roles played by gravity and inertia are reversed. Moreover, he assumed that the sun's power diminishes in direct ratio to distance. He sensed that there was something wrong here, since he knew that the intensity of light diminishes with the *square* of distance; but he had to stick to it, to satisfy his theorem of the ratio of speed to distance, which was equally false.

## 6. *The Second Law*

Refreshed by this excursion into the *Himmelsphysik*, our hero returned to the more immediate task in hand. Since the earth no longer moved at uniform speed, how could one predict its position at a given time? (The method based on the *punctum equans* had proved, after all, a disappointment.) Since he believed to have proved that its speed depended directly on its distance from the sun, the time it needed to cover a small fraction of the orbit was always proportionate to that distance. Hence he divided the orbit (which, forgetting his previous resolve, he still regarded as a circle) into 360 parts, and computed the distance of each bit of arc from the sun. The sum of all distances between, say 0° and 85°, was a measure of the time the planet needed to get there.

But this procedure was, as he remarked with unusual modesty, 'mechanical and tiresome'. So he searched for something simpler:

> Since I was aware that there exists an infinite number of points on the orbit and accordingly an infinite number of distances [from the sun] the idea occurred to me that the sum of these distances is

contained in the *area* of the orbit. For I remembered that in the same manner Archimedes too divided the area of a circle into an infinite number of triangles.[20]

Accordingly, he concluded, the area swept over by the line connecting planet and sun AS–BS is a measure of the time required by the planet to get from A to B; *hence the line will sweep out equal areas in equal times.* This is Kepler's

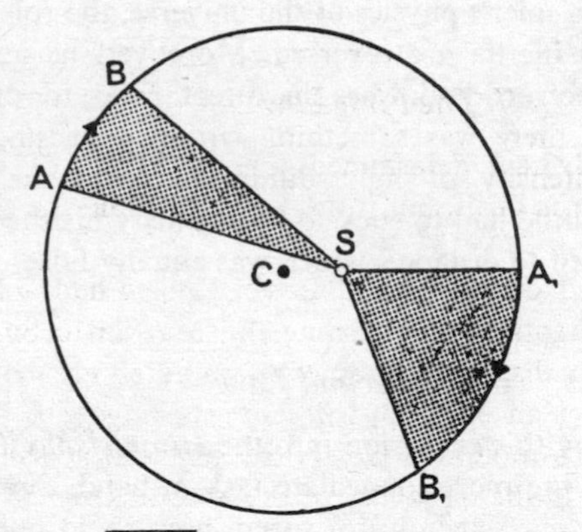


Area $\overline{ABS}$ = Area $\overline{A_1B_1S}$

immortal Second Law (which he discovered before the First) – a law of amazing simplicity at the end of a dreadfully confusing labyrinth.

Yet the last step which had got him out of the labyrinth had once again been a faulty step. For it is not permissible to equate an area with the sum of an infinite number of neighbouring lines, as Kepler did. Moreover, he knew this well, and explained at length why it was not permissible.[21] He added that he had also committed a second error, by assuming the orbit to be circular. And he concluded: 'But these two errors – it is like a miracle – cancel out in the most precise manner, as I shall prove further down.'[22]

The correct result is even more miraculous than Kepler realized, for his explanation of the reasons *why* his errors cancel out was once again mistaken, and he got, in fact, so hopelessly confused that the argument is practically impossible to follow – as he himself admitted. And yet, by three incorrect steps and their even more incorrect defence, Kepler stumbled on the correct law.[23] It is perhaps the most amazing sleep-walking performance in the history of science – except for the manner in which he found his First Law, to which we now turn.

## 7. *The First Law*

The Second Law determined the variations of the planet's speed along its orbit, but it did not determine the shape of the orbit itself.

At the end of the Second Book, Kepler had acknowledged defeat in his attempts to define the Martian orbit – a defeat caused by a discrepancy of eight minutes arc. He had then embarked on an enormous detour, starting with the revision of the earth's motion, followed by physical speculations, and terminating in the discovery of the Second Law. In the Fourth Book he resumed his investigation of the Martian orbit where he had left off. By this time, four years after his first, frustrated attempts, he had become even more sceptical of orthodox dogma, and gained an unparalleled skill in geometry by the invention of methods all his own.

The final assault took nearly two years; it occupies chapters 41 to 60 of the *New Astronomy*. In the first four (41–4), Kepler tried for the last time, with savage thoroughness, to attribute a circular orbit to Mars and failed: this section ends with the words:

> The conclusion is quite simply that the planet's path is not a circle – it curves inward on both sides and outward again at opposite ends. Such a curve is called an oval. The orbit is not a circle, but an oval figure.

But now a dreadful thing happened, and the next six chapters (45–50) are a nightmare journey through another labyrinth.

This oval orbit is a wild, frightening new departure for him. To be fed up with cycles and epicycles, to mock the slavish imitators of Aristotle is one thing; to assign an entirely new, lopsided, implausible path for the heavenly bodies is quite another.

Why indeed an oval? There is something in the perfect symmetry of spheres and circles which has a deep, reassuring appeal to the unconscious – otherwise it could not have survived two millennia. The oval lacks all such archetypal appeal. It has an abitrary form. It distorts that eternal dream of the harmony of the spheres, which lay at the origin of the whole quest. Who art thou, Johann Kepler, to destroy divine symmetry? All he has to say in his own defence is, that having cleared the stable of astronomy of cycles and spirals, he left behind him 'only a single cart-full of dung': his oval.[24]

At this point, the sleepwalker's intuition failed him, he seems to be overcome by dizziness, and clutches at the first prop that he can find. He must find a physical cause, a cosmic *raison d'être* for his oval in the sky – and he falls back on the old quack remedy which he has just abjured, the conjuring up of an epicycle! To be sure, it is epicycle with a difference: it has a physical cause. We have heard earlier on that while the sun's force sweeps the planet round in a circle, a second, antagonistic force, 'seated in the planet itself' makes it turn in a small epicycle in the opposite direction. It all seems to him 'wonderfully plausible',[25] for the result of the combined movement is indeed an oval. But a very special oval: it has the shape of an egg, with the pointed end at the perihelion, the broad end at the aphelion.

No philosopher had laid such a monstrous egg before. Or, in Kepler's own words of wistful hindsight:

> What happened to me confirms the old proverb: a bitch in a hurry produces blind pups. ... But I simply could not think of any other means of imposing an oval path on the planets. When these ideas fell upon me, I had already celebrated my new triumph over Mars without being disturbed by the question ... whether the figures tally or not. ... Thus I got myself into a new labyrinth. ... The reader must show tolerance to my gullibility.[26]

The battle with the egg goes on for six chapters, and took a full year of Kepler's life. It was a difficult year; he had no money, and was down with 'a fever from the gall'. A threatening new star, the *nova* of 1604 had appeared in the sky; Frau Barbara was also ill, and gave birth to a son – which provided Kepler with an opportunity for one of his dreadful jokes: 'Just when I was busy squaring my oval, an unwelcome guest entered my house through a secret doorway to disturb me.'[27]

To find the area of his egg, he again computed series of one hundred and eighty Sun-Mars distances and added them together; and this whole operation he repeated no less than forty times. To make the worthless hypothesis work, he temporarily repudiated his own, immortal Second Law – to no avail. Finally, a kind of snowblindness seemed to descend on him: he held the solution in his hand without seeing it. On 4 July 1603, he wrote to a friend that he was unable to solve the geometrical problems of his egg; but 'if only the shape were a perfect ellipse all the answers could be found in Archimedes' and Appollonius' work'.[28] A full eighteen months later, he again wrote to the same correspondent that the truth must lie somewhere half-way between the egg-shape and circular-shape 'just as if the Martian orbit were a perfect ellipse. But regarding that I have so far explored nothing.'[29] What is even more astonishing, he constantly used ellipses in his calculations – but merely as an *auxiliary* device to determine, by approximation, the area of his egg-curve – which by now had become a veritable fixation. Was there some unconscious biological bias behind it? Apart from the association between the squaring of the egg and the birth of his child, there is nothing to substantiate that attractive hypothesis.*

And yet, these years of wandering through the wilderness were not entirely wasted. The otherwise sterile chapters in the *New Astronomy* devoted to the egg hypothesis, represent an important further step towards the invention of the infinitesimal

* It will be remembered that Copernicus, too, stumbled on the ellipse and kicked it aside; but Copernicus, who firmly believed in circles, had much less reason to pay attention to it than Kepler, who had progressed to the oval.

calculus. Besides, Kepler's mind had by now become so saturated with the numerical data of the Martian orbit, that when the crucial hazard presented itself, it responded at once like a charged cloud to a spark.

This hazard is perhaps the most unlikely incident in this unlikely story. It presented itself in the shape of a number which had stuck in Kepler's brain. The number was 0·00429.

When he at last realized that his egg had 'gone up in smoke'[30] and that Mars, whom he had believed a conquered prisoner 'securely chained by my equations, immured in my tables', had again broken loose, Kepler decided to start once again from scratch.

He computed very thoroughly a score of Mars-Sun distances at various points of the orbit. They showed again that the orbit was some kind of oval, looking like a circle flattened inward at two opposite sides, so that there were two narrow sickles or 'lunulae' left between the circle and the Martian orbit. The width of the sickle, where it is thickest, amounted to 0·00429 of the radius:

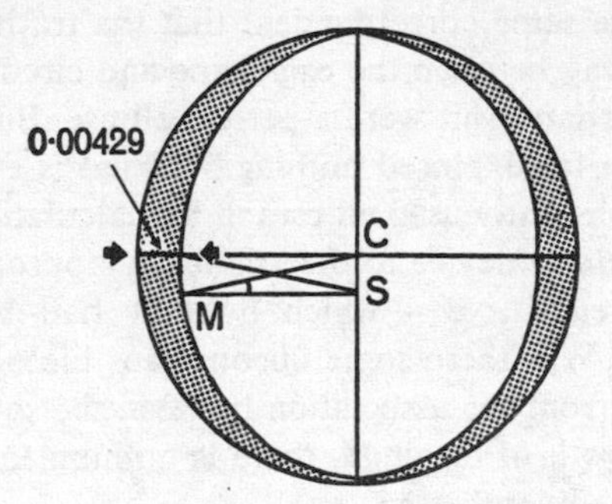


At this point Kepler, for no particular reason, became interested in the angle at M – the angle formed between the sun and the centre of the orbit, as seen from Mars. This angle was called the 'optical equation'. It varies, of course, as Mars moves along its orbit; its maximum value is 5° 18′. This is what happened next, in Kepler's own words:[31]

... I was wondering why and how a sickle of just this thickness (0.00429) came into being. While this thought was driving me around, while I was considering again and again that ... my apparent triumph over Mars had been in vain, I stumbled entirely by chance on the secant* of the angle 5° 18′, which is the measure of the greatest optical equation. When I realized that this secant equals 1.00429, I felt as if I had been awakened from a sleep. ...

It had been a true sleepwalker's performance. At the first moment, the reappearance of the number 0·00429 in this unexpected context must have appeared as a miracle to Kepler. But he realized in a flash that the apparent miracle must be due to a fixed relation between the angle at M and the distance to S, a relation which must hold true for any point of the orbit; only the manner in which he had stumbled on that relation was due to chance. 'The roads that lead man to knowledge are as wondrous as that knowledge itself.'

Now at last, at long last, after six years of incredible labour, he held the secret of the Martian orbit. He was able to express the manner in which the planet's distance from the sun varied with its position, in a simple formula, a mathematical Law of Nature. *But he still did not realize that this formula specifically defined the orbit as an ellipse.*† Nowadays, a student with a little knowledge of analytical geometry would realize this at a glance; but analytical geometry came after Kepler. He had discovered his magic equation empirically, but he could no more identify it as the shorthand sign for an ellipse than the average reader of this book can; it was nearly as meaningless to him. He had reached his goal, but he did not realize that he had reached it.

The result was that he went off on one more, last, wild goose chase. He tried to construct the orbit which would correspond to his newly discovered equation; but he did not know how, made a mistake in geometry, and arrived at a curve which was too bulgy; the orbit was a *via buccosa*, chubby-faced, as he noted with disgust.

What next? We have reached the climax of the comedy. In

* The 'secant' of the angle at M is the ratio MC:MS.

† In modern denotation, the formula is: $R=1+e\cos\beta$ where R is the distance from the sun, $\beta$ the longitude referred to the centre of the orbit, and e the eccentricity.

his despair, Kepler threw out his formula (which denoted an elliptic orbit) because he wanted to try out an entirely new hypothesis: to wit, an elliptic orbit. It was as if the tourist had told the waiter, after studying the menu: 'I don't want *côtelette d'agneau*, whatever that is; I want a lamb chop.'

By now he was convinced that the orbit must be an ellipse, because countless observed positions of Mars, which he knew almost by heart, irresistibly pointed to that curve; but he still did not realize that his equation, which he had found by chance-plus-intuition, *was* an ellipse. So he discarded that equation, and constructed an ellipse by a different geometrical method. And then, at long last, he realized that the two methods produced the same result.

With his usual disarming frankness, he confessed what had happened:

> Why should I mince my words? The truth of Nature, which I had rejected and chased away, returned by stealth through the backdoor, disguising itself to be accepted. That is to say, I laid [the original equation] aside, and fell back on ellipses, believing that this was a quite different hypothesis, whereas the two, as I shall prove in the next chapter, are one and the same ... I thought and searched, until I went nearly mad, for a reason, why the planet preferred an elliptical orbit [to mine]. ... Ah, what a foolish bird I have been![32]

But in the List of Contents, in which he gives a brief outline of the whole work, Kepler sums up the matter in a single sentence:

> I show [in this chapter] how I unconsciously repair my error.

The remainder of the book is in the nature of a mopping-up operation after the final victory.

## 8. *Some Conclusions*

It was indeed a tremendous victory. The great Ferris wheel of human delusion, with its celestial catwalks for the wandering planets, this plantasmagoria which had blocked man's approach to nature for two thousand years, was destroyed, 'banished to the lumber-room'. Some of the greatest discoveries, as we saw, consist mainly in the clearing away of psychological road-blocks which obstruct the approach to reality; which is why, *post factum*, they appear so obvious. In a letter to Longomontanus[33] Kepler qualified his own achievement as the 'cleansing of the Augean stables'.

But Kepler not only destroyed the antique edifice; he erected a new one in its place. His Laws are not of the type which appear self-evident, even in retrospect (as, say, the Law of Inertia appears to us); the elliptic orbits and the equations governing planetary velocities strike us as 'constructions' rather than 'discoveries'. In fact, they make sense only in the light of Newtonian Mechanics. From Kepler's point of view, they did not make much sense; he saw no logical reason why the orbit should be an ellipse instead of an egg. Accordingly, he was more proud of his five perfect solids than of his Laws; and his contemporaries, including Galileo, were equally incapable of recognizing their significance. The Keplerian discoveries were not of the kind which are 'in the air' of a period, and which are usually made by several people independently; they were quite exceptional one-man achievements. That is why the way he arrived at them is particularly interesting.

I have tried to re-trace the tortuous progress of his thought. Perhaps the most astonishing thing about it is the mixture of cleanness and uncleanness in his method. On the one hand, he throws away a cherished theory, the result of years of labour, because of those wretched eight minutes of arc. On the other hand he makes impermissible generalizations, knows that they are impermissible, yet does not care. And he has a philosophical justification for both attitudes. We heard him sermonizing about the duty to stick rigorously to observed fact. But on the other hand he says that Copernicus 'sets an example for others by his contempt for the small blemishes in expounding his wonderful discoveries. If this had not been always the usage, then Ptolemy would never have been able to publish his *Almagest*, Copernicus his *Revolutions*, and Reinhold his *Prutenian Tables*. . . . It is not surprising that, when he dissects the universe with a lancet, various matters emerge only in a rough manner.'[34]

Both precepts have, of course, their uses. The problem is to know when to follow one, when the other. Copernicus had a one-track mind; he never flew off at a tangent; even his cheatings were heavy-handed. Tycho was a giant as an observer, but nothing else. His leanings toward alchemy and astrology never fused, as in Kepler, with his science. The measure of Kepler's genius is the intensity of his contradictions, and the use he made of them. We saw him plod, with infinite patience, along dreary stretches of trial-and-error procedure, then suddenly become airborne when a lucky guess or hazard presented him with an opportunity. What enabled him to recognize instantly his chance when the number 0·00429 turned up in an unexpected context was the fact that not only his waking mind, but his sleepwalking unconscious self was saturated with every conceivable aspect of his problem, not only with the numerical data and ratios, but also with an intuitive 'feel' of the physical forces, and of the Gestalt-configurations which it involved. A locksmith who opens a complicated lock with a crude piece of bent wire is not guided by logic, but by the unconscious residue of countless past experiences with locks, which lend his touch a wisdom that his reason does not possess. It is perhaps this intermittent flicker of an overall vision which accounts for the mutually compensatory nature of Kepler's mistakes, as if some balancing reflex or 'backfeed' mechanism had been at work in his unconscious mind.

Thus, for instance, he *knew* that his inverse ratio 'law' (between a planet's speed and solar distance) was incorrect. His thirty-second chapter ends with a short, almost off-hand admission of this. But, he argues, the deviation is so small, that it can be neglected. Now this is true for Earth with its small eccentricity, yet not at all true for Mars, with its large eccentricity. Yet even toward the end of the book (in chapter 60), long after he had found the correct law, Kepler speaks of the inverse-ratio postulate as if it were true not only for earth, but also for Mars. He could not deny, even to himself, that the hypothesis was incorrect; he could only forget it. Which he promptly did. Why? Because, though he knew that the postulate was bad geometry, it made good physics to him, and therefore ought to be true. The problem of the planetary orbits had been hopelessly bogged down in its purely geometrical frame of reference, and when Kepler realized that he could not get it unstuck, he tore it out of that frame and removed it into the field of physics. This operation of removing a problem from its traditional context and placing it into a new one, looking at it through glasses of a different colour as it were, has always seemed to me of the very essence of the creative process.[35] It leads not only to a revaluation of the problem itself, but often to a synthesis of much wider consequences, brought about by a fusion of the two previously unrelated frames of reference. In our case, the orbit of Mars became the unifying link between the two formerly separate realms of physics and cosmology.

It may be objected that Kepler's ideas of physics were so primitive that they ought to be regarded merely as a subjective stimulus to his work (like the five perfect solids), without objective value. In fact, however, his was the first serious attempt at explaining the mechanism of the solar system in terms of physical forces, and once the example was set, physics and cosmology could never again be divorced. And secondly, whereas the five solids were indeed merely a psychological spur, his sky-physics played, as we saw, a direct part in the discovery of his laws.

For, although the functions of gravity and inertia are reversed in the Keplerian cosmos, his intuition that there are *two antagonistic forces* acting on the planets, guided him in the right direction. A single force, as previously assumed (that of the Prime Mover or kindred spirits) could never produce oval orbits and periodic changes of speed. These could only be the result of some dynamic tug-of-war going on in the sky – as indeed they are; though his ideas about the nature of the sun's force' and the planet's 'laziness' or 'magnetism' were pre-Newtonian.

## 9. *The Pitfalls of Gravity*

I have tried to show that without his invasion into the territory of physics Kepler could not have succeeded. I must now discuss briefly Kepler's particular brand of physics. It was, as to be expected, physics-on-the-watershed, half-way between Aristotle and Newton. The essential concept of impetus or momentum, which makes a moving body persist in its motion without the help of an external force, is absent from it; the planets must still be dragged through the ether like a Greek oxcart through the mud. In this respect Kepler had not advanced further than Copernicus, and both were unaware of the progress made by the Ockhamists in Paris.

On the other hand, he came very near to discovering universal gravity, and the reasons for his failure to do so are not only of historical, but also of topical interest. Over and over again he seems to balance on the brink of the idea and yet, as if pulled back by some unconscious resistance, to shrink from the final step. One of the most striking passages is to be found in the introduction to the *Astronomia Nova.* There Kepler starts by demolishing the Aristotelian doctrine that bodies which are by nature 'heavy' strive toward the centre of the world, whereas those which are 'light' strive towards its periphery. His conclusions are as follows:

It is therefore clear that the traditional doctrine about gravity is erroneous. ... Gravity is the mutual bodily tendency between cognate [i.e. material] bodies towards unity or contact (of which kind the magnetic force also is), so that the earth draws a stone much more than the stone draws the earth. ...

Supposing that the earth were in the centre of the world, heavy bodies would be attracted to it, not because it is in the centre, but because it is a cognate [material] body. It follows that regardless where we place the earth ... heavy bodies will always seek it. ...

*If two stones were placed anywhere in space near to each other, and outside the reach of force of a third cognate body, then they would come together, after the manner of magnetic bodies, at an intermediate point, each approaching the other in proportion to the other's mass* [my italics].

If the earth and the moon were not kept in their respective orbits by a spiritual or some other equivalent force, the earth would ascend towards the moon one fifty-fourth part of the distance, and the moon would descend the remaining fifty-three parts of the interval, and thus they would unite. But the calculation presupposes that both bodies are of the same density.

If the earth ceased to attract the waters of the sea, the seas would rise and flow into the moon. . . .

If the attractive force of the moon reaches down to the earth, it follows that the attractive force of the earth, all the more, extends to the moon and even farther. . . .

Nothing made of earthy substance is absolutely light; but matter which is less dense, either by nature or through heat is relatively lighter. . . .

Out of the definition of lightness follows its motion; for one should not believe that when lifted up, it escapes to the periphery of the world, or that it is not attracted by the earth. It is merely less attracted than heavier matter, and is therefore displaced by heavier matter, so that it comes to rest and is kept in its place by the earth. . . .[35a]

In the same passage, Kepler gives the first correct explanation of the tides as a motion of the waters 'towards the regions where the moon stands in the zenith'. In a later work (the *Somnium*) he explained the tides, not by the attraction of the moon alone, but of moon and sun combined; he thus realized that the attraction of the sun reached as far as the earth!

Yet in spite of this, the sun in his cosmology is not an attracting force, but acts like a sweeping broom. In the text of the *New Astronomy* he seems to have forgotten all that he said in the Preface about the mutual attraction between two bodies in empty space, and his strikingly correct definition of gravity being proportional to the attracting mass. The definitions of gravity in the Preface are indeed so striking that Delambre exclaims:[36]

*Violà qui était neuf, vraiment beau, et qui n' avait besoin que de quelques developpements et que de quelques explications. Violà les fondaments de la Physique moderne, céleste et terrestre.**

But when he tried to work out the mechanics of the solar system, all these beautiful new insights were lost again in con-

* 'Here was something new and truly beautiful, which only needed a little development and explanation. Here were the foundations of modern physics, both of the earth and the skies.'

fusion. Could some similar paradox be responsible for the crisis in modern physics – some unconscious blockage which prevents us from seeing the 'obvious', and compels us to persist in our own version of wavemechanical double-think?

At any rate, most twentieth-century physicists will feel a sneaking sympathy for the man who nibbled at the concept of gravity, and yet was unable to swallow it. For Newton's concept of a 'gravitational force' has always lain as an undigested lump in the stomach of science; and Einstein's surgical operation, though easing the symptoms, has brought no real remedy. The first to sympathize with Kepler would have been Newton himself, who, in a famous letter to Bentley, wrote:

It is inconceivable, that inanimate brute matter should, without the mediation of something else, which is not material, operate upon, and affect other matter without mutual contact; as it must do, if gravitation, in the sense of Epicurus, be essential and inherent in it. And this is one reason, why I desired you would not ascribe innate gravity to me. That gravity should be innate, inherent, and essential to matter, so that one body may act upon another, at a distance through a vacuum, without the mediation of anything else, by and through which their action and force may be conveyed from one to another, is to me so great an absurdity, that I believe no man who has in philosophical matters a competent faculty of thinking, can ever fall into it.[37]

Newton, in fact, could only get over the 'absurdity' of his own concept by invoking either an ubiquitous ether (whose attributes were equally paradoxical) and/or God in person. The whole notion of a 'force' which acts instantly at a distance without an intermediary agent, which traverses the vastest distances in zero seconds, and pulls at immense stellar objects with ubiquitous ghost-fingers – the whole idea is so mystical and 'unscientific', that 'modern' minds like Kepler, Galileo, and Descartes, who were fighting to break loose from Aristotelian animism, would instinctively tend to reject it as a relapse into the past.[38] In their eyes, the idea of 'universal gravity' would amount to much the same kind of thing as the *anima mundi* of the ancients. What made Newton's postulate nevertheless a modern Law of Nature, was his mathematical formulation of the mysterious entity to which it referred. And that formulation Newton deduced from the discoveries of Kepler – who had intuitively glimpsed gravity, and shied away from it. In such crooked ways does the tree of science grow.

## 10. *Matter and Mind*

In a letter to Herwart, which he wrote when the book was nearing completion,[39] Kepler defined his programme:

> My aim is to show that the heavenly machine is not a kind of divine, live being, but a kind of clockwork (and he who believes that a clock has a soul, attributes the maker's glory to the work), insofar as nearly all the manifold motions are caused by a most simple, magnetic, and material force, just as all motions of the clock are caused by a simple weight. And I also show how these physical causes are to be given numerical and geometrical expression.

He had defined the essence of the scientific revolution. But he himself never completed the transition from a universe animated by purposeful intelligence to one moved by inanimate, 'blind' forces. The very concept of a physical 'force' devoid of purpose, which we take so much for granted, was only just emerging from the womb of animism, and the word for it – *virtus* or *vis* – betrays its origin. It was (and is) indeed much easier to talk about a 'simple, magnetic, material force' than to form a concrete idea of its working. The following passage will illustrate the enormous difficulty which the notion of the 'moving force' emanating from the sun presents to Kepler's mind:

> Though the light of the sun cannot itself be the moving force ... it may perhaps represent a kind of vehicle, or tool, which the moving force uses. But the following considerations seem to contradict this. Firstly, the light is arrested in regions that lie in shade. If then, the moving force were to use light as a vehicle, then darkness would bring the planets to a standstill. ...
>
> Since there is as much of this force present in the wider distant orbits as in nearer and narrower ones, it follows that nothing of this force is lost on the journey from its source, nothing is dispersed between the source and the star. This emanation is therefore un-

substantial as light is, and not accompanied by a loss of substance as are the emanations of odours, or of the heat which goes out from a glowing stove, and the like, where the intervening space is filled [by the emanation]. We must, therefore, conclude that, just as the light which lights up everything on earth is a non-substantial variety of the fire in the solar body, likewise this force which grips and carries the planet-bodies is a non-substantial variety of the force which has its seat in the sun itself; and that it has immeasurable strength, and thus gives the first impulse to all motion in the world....

This kind of force, just as the kind of force which is light ... cannot be regarded as something which expands into the space between its source and the movable body, but as something which the movable body receives out of the space which it occupies. ...* It is propagated through the universe ... but it is nowhere received except where there is a movable body, such as a planet. The answer to this is: although the moving force has no substance, it is aimed at substance, i.e., at the planet-body to be moved....

Who, I ask, will pretend that light has substance? Yet nevertheless it acts and is acted upon in space, it is refracted and reflected, and it has quantity, so that it may be dense or sparse, and can be regarded as a plane where it is received by something capable of being lit up. For, as I said in my *Optics*, the same thing applies to light as to our moving force: it has no present existence in the space between the source and the object which it lights up, although it has passed through that space in the past; it 'is' not, it 'was', so to speak.[40]

The contemporary physicist grappling with the paradoxa of relativity and quantum mechanics will find here an echo of his perplexities. In the end, Kepler managed to get to terms with his 'moving force' by visualizing it as a vortex, 'a raging current which tears all the planets, and perhaps all the celestial ether, from West to East'.[41] But he was nevertheless compelled to ascribe to each planet a kind of mind which enables it to recognize its position in space, and to adjust its reactions accordingly. To careless readers of the *Astronomia Nova* this looked as if the animal spirits had gained re-admission into the

* Note that this description is closer to the modern notion of the gravitational or electro-magnetic *field* than to the classic Newtonian concept of *force*.

model which he intended to be a purely mechanical clockwork – like ghosts who cannot resign themselves to their final banishment from the world of the living. But Kepler's planetary minds bear in fact no resemblance to those medieval planet-moving angels and spirits. They have no 'souls', only 'minds'; no sense organs, and no will of their own; they are rather like the computing machines in guided missiles:

O Kepler, does't thou wish then to equip each planet with two eyes? Not at all. For it is not necessary, either, to attribute them feet or wings to enable them to move. ... Our speculations have not yet exhausted all Nature's treasures, to enable us to know, how many senses exist. ...

The subtle reflections of some people concerning the blessed angels' and spirits' nature, motions, places, and activities, do not concern us here. We are discussing natural matters of much lower rank: forces which do not exercise free will when they change their activities, intelligences which are by no means separate from, but attached to, the stellar bodies to be moved, and are one with them.[42]

Thus the function of the planet's mind is confined to responding in a lawful, orderly, and therefore 'intelligent' manner to the various forces tugging at him. It is really a superior kind of electronic brain with an Aristotelian bias. Kepler's ambiguity is, in the last analysis, merely a reflection of the mind-matter dilemma, which becomes particularly acute in periods of transition – including our own. As his most outstanding German biographer has put it:

The physical expositions of Kepler have a special message to those who feel the need to inquire into the first beginnings of the mechanistic explanation of nature. He touches, indeed, on the profoundest questions of the philosophy of nature when he confronts, in his subtle manner, the concepts of *mens* and *natura*, compares their pragmatic values and delimits their fields of application. Have we outgrown this antithesis in our day? Only those will believe that who are unaware of the metaphysical nature of our concept of physical force. ... At any rate, Kepler's explanations may serve as a stimulus to a wholesome contemplation of the axioms and limits of mechanistic philosophy in our time of widespread and disastrous scientific dogmatism.[43]

Though Kepler was unable to solve the dilemma, he clarified it and polished its horns, as it were. The angels, spirits, and unmoved movers, were banished from cosmology; he sublimated and distilled the problem to a point where only the ultimate mystery remains. Though he was always attracted, with a mixture of disgust and fascination, by theological disputes, he uncompromisingly and indeed vehemently rejected the incursion of the theologians into science. On this point he made his position very clear in a statement – or rather a battle-cry – in the introduction to the *New Astronomy*:

So much for the authority of Holy Scripture. Now as regards the opinions of the saints about these matters of nature, I answer in one word, that in theology the weight of Authority, but in philosophy the weight of Reason alone is valid. Therefore a saint was Lanctantius, who denied the earth's rotundity; a saint was Augustine, who admitted the rotundity, but denied that antipodes exist. Sacred is the Holy Office of our day, which admits the smallness of the earth but denies its motion: but to me more sacred than all these is Truth, when I, with all respect for the doctors of the Church, demonstrate from philosophy that the earth is round, circumhabited by antipodes, of a most insignificant smallness, and a swift wanderer among the stars.

# 7 Kepler Depressed

## 1. *Publishing Difficulties*

The writing of the *New Astronomy* had been an obstacle race over six years.

At the start there had been the quarrels with Tycho, the long sojourns in Gratz, illness, and the drudgery of the pamphlets against Ursus and Craig. When the Great Dane died, and Kepler was appointed his successor, he may have hoped to be able to work in peace; instead, his life became even more disorganized. His official and unofficial duties included the publication of annual calendars with astrological predictions; the casting of horoscopes for distinguished visitors at Court; the publication of comments on eclipses, comets, and a new star; answering at great length the queries on every subject under the sun, put to him by the various patrons with whom he corresponded; and above all petitioning, lobbying, and intriguing to obtain at least a fraction of the salaries and printing costs due to him. He had discovered his Second Law as early as 1602, one year after Tycho's death; but the next year he was almost entirely occupied with other labours, among them the great work on optics, published in 1604; the year after that, he became stuck on the egg-shaped orbit, fell ill and again thought that he was dying; and only round Easter in 1605 was the *New Astronomy* completed in outline.

But it took another four years to get it published. The reason for this delay was lack of money to pay for the printing, and a harassing feud with Tycho's heirs, led by the swashbuckling Junker Tengnagel. This character, it will be remembered, had married Tycho's daughter Elisabeth after putting her in the family way – the only achievement on which he could base his claim to the Tychonic heritage. He was determined to cash in on it, and sold Tycho's observations and instruments to the Emperor for the sum of twenty thousand Thalers. But the imperial treasury never paid the Junker; he had to content himself with an annual five per cent interest on the debt – which was still twice the amount of Kepler's salary. As a result, Tycho's instruments, the wonder of the world, were kept by Tengnagel behind lock and key; within a few years they decayed to scrap metal. A similar fate would no doubt have befallen the treasure of Tycho's observations, if Kepler had not hurriedly pinched them, for the benefit of posterity. In a letter to one of his English admirers[1] he calmly reported:

> I confess that when Tycho died, I quickly took advantage of the absence, or lack of circumspection, of the heirs, by taking the observations under my care, or perhaps usurping them. . . .

It had always been his avowed intention to get possession of Tycho's treasure, and he had succeeded.

The Tychonides were understandably furious; Kepler, the introspective grave-robber, quite saw their point:

> The cause of this quarrel lies in the suspicious nature and bad manners of the Brahe family, but on the other hand also in my own passionate and mocking character. It must be admitted that Tengnagel had important reasons for suspecting me. I was in possession of the observations and refused to hand them over to the heirs. . . .[2]

The negotiations dragged on for several years. The Junker, ambitious, stupid, and vain, proposed a dirty deal: he would keep his peace if all Kepler's future works were published under their joint names. Surprisingly, Kepler agreed: he was always strangely indifferent to the fate of his published works. But he asked that the Junker should, in exchange, hand over a quarter of the annual thousand Thalers which he drew from the treasury. This Tengnagel refused, considering two hundred and fifty a year to be too high a price for immortality. He thus deprived future scholars of a delightful subject of controversy on the question by which of the two partners the Tengnagel-Kepler Laws were discovered.

In the meantime, the Junker had embraced the Catholic faith and been made an Appellate Counsellor at court. This enabled him to impose his conditions on Kepler, which made it impossible for him to publish his book without Tengnagel's consent. Thus Kepler found himself 'tied hands and feet', while the Junker 'sits like a dog in the manger, unable to put the treasure to use, and preventing others from doing so'.[2a] A compromise was reached at last: Tengnagel gave his gracious consent to the printing of the *New Astronomy* on condition that it should carry a preface from his own pen. Its text is printed in Note 3. If Osiander's preface to the *Book of the Revolutions* displayed the wisdom of a gentle snake, in Tengnagel's preface to the *New Astronomy*, we hear the braying of a pompous ass echoing down the centuries.

At last, in 1608, the printing of the book could begin; it was finished in the summer of 1609, in Heidelberg, under Kepler's supervision. It was a beautifully printed volume in folio, of which only a few copies survive. The Emperor claimed the whole edition as his property and forbade Kepler to sell or give away any copy of it 'without our foreknowledge and consent'. But since his salary was in arrears, Kepler felt at liberty to do as he liked, and sold the whole edition to the printers. Thus the story of the *New Astronomy* begins and ends with acts of larceny, committed *ad majorem Dei gloriam*.

## 2. *Reception of* Astronomia Nova

How far ahead of his time Kepler was – not merely by his discoveries, but in his whole manner of thought – one can gather from the negative reactions of his friends and correspondents. He received no help, no encouragement; he had patrons and well-wishers, but no congenial spirit.

Old Maestlin had been silent for the last five years, in spite of a persistent stream of letters from Kepler, who kept his old teacher informed of every important event in his life and researches. Just before the completion of the *New Astronomy*, Maestlin broke his silence with a very moving letter which, however, was a complete let-down insofar as Kepler's hopes for guidance, or at least of shared interests were concerned:

Tuebingen, 28 January 1605.

Although I have for several years neglected writing to you, your steadfast attachment, gratitude, and sincere affection have not weakened but become rather stronger, albeit you have reached such a high step and distinguished position that you could, if you wished, look down on me. ... I do not wish to apologize further, and say only this: I have nothing of the same value to offer in writing to such an outstanding mathematician. ... I must further confess that your questions were sometimes too subtle for my knowledge and gifts, which are not of the same stature. Hence I could only keep silent. ...You will wait in vain for my criticism of your book on optics, which you request so urgently; it contains matters too lofty for me to permit myself to judge it. ... I congratulate you. The frequent and most flattering mention of my name [in that book] is a special proof of your attachment. But I fear that you credit me with too much. If only I were such as your praise makes me appear. But I understand only my modest craft.[4]

And that was the end of it, though Kepler persisted in the one-sided correspondence, and also in his miscellaneous requests – Maestlin should make inquiries about the suitor of Kepler's sister; Maestlin should find him an assistant, and so forth – which the old man steadfastly ignored.

The most detailed letters about the progress of the *New Astronomy* Kepler wrote to David Fabricius, a clergyman and amateur astronomer in Friesland. Some of these letters cover over twenty, and up to forty foolscap pages. Yet he could never persuade Fabricius to accept the Copernican view; and when Kepler informed him of his discovery of the First Law, Fabricius' reaction was:

With your ellipse you abolish the circularity and uniformity of the motions, which appears to me the more absurd the more profoundly I think about it. ... If you could only preserve the perfect circular orbit, and justify your elliptic orbit by another little epicycle, it would be much better.[5]

As for the patrons and wellwishers, they tried to encourage him, but were unable to comprehend what he was up to. The most enlightened among them, the physician Johannes Brengger, whose opinion Kepler particularly valued, wrote:

When you say that you aim at teaching both a new physics of the sky and a new kind of mathematics, based not on circles but on magnetic and intelligent forces, I rejoice with you, although I must frankly confess that I am unable to imagine, and even less to comprehend, such a mathematical procedure.[6]

This was the general reaction of Kepler's contemporaries in Germany. It was summed up by one of them:

In trying to prove the Copernican hypothesis from physical causes, Kepler introduces strange speculations which belong not in the domain of astronomy, but of physics.[7]

Yet a few years later the same man confessed:

I no longer reject the elliptical form of the planetary orbits and allowed myself to be persuaded by the proofs in Kepler's work on Mars.[8]

The first to realize the significance and implications of Kepler's discoveries, were neither his German compatriots, nor Galileo in Italy but the British: the traveller Edmund Bruce, the mathematician Thomas Harriot, tutor of Sir Walter Raleigh; the Reverend John Donne, the astronomical genius Jeremiah Horrocks, who died at twenty-one; and lastly, Newton.

### 3. *Anticlimax*

Delivered from his monumental labours, the usual anti-climax set in for Kepler.

He turned back to his persistent dream, the harmony of the spheres, convinced that the whole *New Astronomy* was merely a stepping-stone towards the ultimate aim in his 'sweating and panting pursuit of the Creator's tracks'.[9] He published two polemical works on astrology, a pamphlet on comets, another about the shape of snow crystals, conducted a voluminous correspondence on the true date of the birth of Christ. He continued with his calendars and weather predictions: on one occasion, when a violent thunder-storm darkened the sky at noon, as he had predicted a fortnight earlier, the people in the streets of Prague yelled, pointing at the clouds: 'It's that Kepler coming.'

He was by now an internationally famous scholar, a member

of the Italian *Accadèmia dei Lincei* (a forerunner of the Royal Society), but even more pleased about the distinguished society in which he moved in Prague:

The Imperial Counsellor and First Secretary, Johann Polz, is very fond of me. [His wife and] the whole family are conspicuous here in Prague for their Austrian elegance and their distinguished and noble manners; so that it would be due to their influence if on some future day I made some progress in this respect, though, of course, I am still far away from it. . . Notwithstanding the shabbiness of my household and my low rank (for they are considered to belong to the nobility), I am free to come and go in their house as I please.[10]

His rise in social status was reflected in the personalities of the godparents to the two children who were born to him in Prague: the wives of halberdiers to the first, Counts of the Palatinate and Ambassadors to the second. There was an endearing Chaplinesque quality about Kepler's efforts to display social graces: 'What a job, what an upheaval to invite fifteen to sixteen women to visit my wife in child-bed, to play host to them, to compliment them to the door!'[10a] Though he wore fine cloth and Spanish ruffles, his salary was always in arrears: 'My hungry stomach looks up like a little dog to its master who used to feed it.'[11]

Visitors to Prague were invariably impressed by his dynamic personality and quicksilvery mind; yet he was still suffering from lack of self-assurance – a chronic ill, on which success acted as a temporary sedative, but never as a complete cure. The turbulent times increased his feeling of insecurity; he lived in constant fear of penury and starvation, complicated by his obsessive hypochrondria:

You inquire after my illness? It was an insidious fever which originated in the gall and returned four times because I repeatedly sinned in my diet. On 29 May my wife forced me, by her pesterings, to wash, for once, my whole body. She immersed me in a tub (for she has a horror of public baths) with well heated water; its heat afflicted me and constricted my bowels. On 31 May I took a light laxative, according to habit. On 1 June, I bled myself, also according to habit: no urgent disease, not even the suspicion of one, compelled me to do it, nor any astrological consideration. ... After losing blood, I felt for a few hours well; but in the evening an evil sleep threw me on my mattress and constricted my guts. Sure enough the gall at once gained access to my head, bypassing the bowels. ... I think I am one of those people whose gall-bladder has a direct opening into the stomach; such people are shortlived as a rule.[12]

Even without hypochrondria, there were sufficient reasons for anxiety. His imperial patron sat on a quaking throne – though, in truth, Rudolph rarely sat on it, preferring to hide from his odious fellow-creatures among his mechanical clocks and toys, gems and coins, retorts and alembics. There were wars and rebellions in Moravia and Hungary, and the treasury was empty. As Rudolph progressed from eccentricity to apathy and melancholia, his brother was depriving him piecemeal of his domains; in a word, Rudolph's final abdication was only a question of time. Poor Kepler, already expelled from his livelihood in Gratz, saw a second exile looming before him, and had to start once more pulling wires, stretching out feelers, and clutching at straws. But the Lutheran worthies in his beloved Wuerttemberg would have nothing to do with their *enfant terrible*, and Maximilian of Bavaria turned a politely deaf ear, as did other Princes whom he approached. The year after the publication of the *New Astronomy* saw him at his lowest ebb, unable to do any serious work, 'my mind prostrate in a pitiful frost'.

Then came an event which not only thawed it, but set it to bubble and boil.

## 4. *The Great News*

One day in March 1610, a certain Herr Johannes Matthaeus Wackher von Wackenfels, Privy Counsellor to his Imperial Majesty, Knight of the Golden Chain and of the order of St Peter, amateur philosopher and poet, drove up in his coach to Kepler's house, and called for him in great agitation. When Kepler came down, Herr Wackher told him the news had just arrived at Court that a mathematician named Galileus in Padua had turned a Dutch spy-glass at the sky, and discovered through its lenses four new planets in addition to the five which had always been known.

I experienced a wonderful emotion while I listened to this curious tale. I felt moved in my deepest being.... [Wackher] was full of joy and feverish excitement; at one moment we both laughed at our confusion, the next he continued his narrative and I listened intently – there was no end to it....[13]

Wackher von Wackenfels was twenty years older than Kepler, and devoted to him. Kepler sponged on the Privy Counsellor's excellent wine, and had dedicated to him his treatise on the snow crystals as a New Year's gift. Wackher, though a Catholic convert, believed in the plurality of worlds; accordingly, he thought that Galileo's discoveries were planets to other stars, outside our solar system. Kepler rejected this idea; but he equally refused to admit that the new heavenly bodies could be revolving round the sun, on the grounds that since there were only five perfect solids, there could only be six planets – as he had proved to his own satisfaction in the *Cosmic Mystery*. He accordingly deduced *a priori*, that what Galileo had seen in the sky could only be secondary satellites, which circled round Venus, Mars, Jupiter, and Saturn, as the moon circled round the earth. Once again he had guessed nearly right for the wrong reasons: Galileo's discoveries were indeed moons, but all the four of them were moons of Jupiter.

A few days later, authentic news arrived in the shape of Galileo's short but momentous booklet, *Sidereus Nuncius – The Message from the Stars*.[14] It heralded the assault on the universe with a new weapon, an optic battering ram, the telescope.

# 8 Kepler and Galileo

## 1. *A Digression on Mythography*

It was indeed a new departure. The range and power of the main sense organ of *homo sapiens* had suddenly started to grow in leaps to thirty times, a hundred times, a thousand times its natural capacity. Parallel leaps and bounds in the range of other organs were soon to transform the species into a race of giants in power – without enlarging his moral stature by an inch. It was a monstrously one-sided mutation – as if moles were growing to the size of whales, but retaining the instincts of moles. The makers of the scientific revolution were individuals who in this transformation of the race played the part of the mutating genes. Such genes are *ipso facto* unbalanced and unstable. The personalities of these 'mutants' already foreshadowed the discrepancy in the next development of man: the intellectual giants of the scientific revolution were moral dwarfs.

They were, of course, neither better nor worse than the average of their contemporaries. They were moral dwarfs only in proportion to their intellectual greatness. It may be thought unfair to judge a man's character by the standard of his intellectual achievements, but the great civilizations of the past did precisely this; the divorce of moral from intellectual values is itself a characteristic development of the last few centuries. It is foreshadowed in the philosophy of Galileo, and became fully explicit in the ethical neutrality of modern determinism. The indulgence with which historians of science treat the Founding Fathers is based on precisely that tradition which the Fathers introduced – the tradition of keeping intellect and character as strictly apart as Galileo taught us to separate the 'primary' and 'secondary' qualities of objects. Thus moral assessments are thought to be essential in the case of Cromwell or Danton, but irrelevant in the case of Galileo, Descartes, or Newton. However, the scientific revolution produced not only discoveries, but a new attitude to life, a change in the philosophical climate. And on that new climate, the personalities and beliefs of those who initiated it had a lasting influence. The most pronounced of these influences, in their different fields, were Galileo's and Descartes'.

The personality of Galileo, as it emerges from works of popular science, has even less relation to historic fact than Canon Koppernigk's. In his particular case, however, this is not caused by a benevolent indifference towards the individual as distinct from his achievement, but by more partisan motives. In works with a theological bias, he appears as the nigger in the woodpile; in rationalist mythography, as the Maid of Orleans of Science, the St George who slew the dragon of the Inquisition. It is, therefore, hardly surprising that the fame of this outstanding genius rests mostly on discoveries he never made, and on feats he never performed. Contrary to statements in even recent outlines of science, Galileo did not invent the telescope; nor the microscope; nor the thermometer; nor the pendulum clock. He did not discover the law of inertia; nor the parallelogram of forces or motions; nor the sun spots. He made no contribution to theoretical astronomy; he did not throw down weights from the leaning tower of Pisa, and did not prove the truth of the Copernican system. He was not tortured by the Inquisition, did not languish in its dungeons, did not say '*eppur si muove*'; and he was not a martyr of science.

What he *did* was to found the modern science of dynamics, which makes him rank among the men who shaped human destiny. It provided the indispensable complement to Kepler's laws for Newton's universe: 'If I have been able to see farther,' Newton said, 'it was because I stood on the shoulders of giants.' The giants were, chiefly, Kepler, Galileo, and Descartes.

## 2. *Youth of Galileo*

Galileo Galilei was born in 1564 and died in 1642, the year Newton was born. His father, Vincento Galilei, was an impoverished scion of the lower nobility, a man of remarkable culture, with considerable achievements as a composer and writer on music, a contempt for authority, and radical leanings. He wrote, for instance (in a study on counter-point): 'It appears to me that those who try to prove an assertion by relying simply on the weight of authority act very absurdly.'[1]

One feels at once the contrast in climate between the childhoods of Galileo and our previous heroes. Copernicus, Tycho, Kepler, never completely severed the navel-cord which had fed into them the rich, mystic sap of the Middle Ages. Galileo is a second-generation intellectual, a second-generation rebel against authority; in a nineteenth-century setting, he would have been the Socialist son of a Liberal father.

His early portraits show a ginger-haired, short-necked, beefy young man of rather coarse features, a thick nose and conceited stare. He went to the excellent Jesuit school at the Monastery of Vallombrosa, near Florence; but Galileo senior wanted him to become a merchant (which was by no means considered degrading for a patrician in Tuscany) and brought the boy home to Pisa; then, in recognition of his obvious gifts, changed his mind and at seventeen sent him to the local university to study medicine. But Vincento had five children to look after (a younger son, Michelangelo, plus three daughters), and the University fees were high; so he tried to obtain a scholarship for Galileo. Although there were no less than forty scholarships for poor students available in Pisa, Galileo failed to obtain one, and was compelled to leave the University without a degree. This is the more surprising as he had already given unmistakable proof of his brilliance: in 1582, in his second year at the University, he discovered the fact that a pendulum of a given length swings at a constant frequency, regardless of amplitude.[2] His invention of the 'pulsilogium', a kind of metronome for timing the pulse of patients, was probably made at the same time. In view of this and other proofs of the young student's mechanical genius, his early biographers explained the refusal of a scholarship by the animosity which his unorthodox anti-Aristotelian views raised. In fact, however, Galileo's early views on physics contain nothing of a revolutionary nature.[3] It is more likely that the refusal of the scholarship was due not to the unpopularity of Galileo's views, but of his person – that cold, sarcastic presumption, by which he managed to spoil his case throughout his life.

Back home he continued his studies, mostly in applied mechanics, which attracted him more and more, perfecting his dexterity in making mechanical instruments and gadgets. He invented a hydrostatic balance, wrote a treatise on it which he circulated in manuscript, and began to attract the attention of scholars. Among these was the Marchese Guidobaldo del Monte who recommended Galileo to his brother-in-law, Cardinal del Monte, who in turn recommended him to Ferdinand de Medici, the ruling Duke of Tuscany; as a result, Galileo was appointed a lecturer in mathematics at the University of Pisa, four years after that same University had refused him a scholarship. Thus at the age of twenty-five, he was launched on his academic career. Three years later, in 1592, he was appointed to the vacant Chair of Mathematics at the famous University of Padua, again through the intervention of his patron, del Monte.

Galileo remained in Padua for eighteen years, the most creative and fertile years of his life. It was here that he laid the foundations of modern dynamics, the science concerned with moving bodies. But the results of these researches he only published towards the end of his life. Up to the age of forty-six, when the *Message from the Stars* was sent into the world, Galileo had published no scientific work.[4] His growing reputation in this period, before his discoveries through the telescope, rested partly on treatises and lectures circulated in manuscript, partly on his mechanical inventions (among them the thermoscope, a forerunner of the thermometer), and the instruments which he manufactured in large numbers with skilled artisans in his own workshop. But his truly great discoveries – such as the laws of motion of falling bodies and projectiles – and his ideas on cosmology he kept strictly for himself and for his private correspondents. Among these was Johannes Kepler.

### 3. *The Church and the Copernican System*

The first contact between the two Founding Fathers took place in 1597. Kepler was then twenty-six, a professor of mathematics in Gratz; Galileo was thirty-three, a professor of mathematics in Padua. Kepler had just completed his *Cosmic Mystery* and, profiting from a friend's journey to Italy, had sent copies of it, among others, 'to a mathematician named Galileus Galileus, as he signs himself'.[5]

Galileo acknowledged the gift in the following letter:

Your book, my learned doctor, which you sent me through Paulus Amberger, I received not a few days but merely a few hours ago; since the same Paulus informed me of his impending return to Germany, I would be ungrateful indeed not to thank you at once: I accept your book the more gratefully as I regard it as proof of having been found worthy of your friendship. So far I have only perused the preface of your work, but from this I gained some notion of its intent,* and I indeed congratulate myself on having an associate in the study of Truth who is a friend of Truth. For it is a misery that so few exist who pursue the Truth and do not pervert philosophical reason. However, this is not the place to deplore the miseries of our century but to congratulate you on the ingenious arguments you found in proof of the Truth. I will only add that I promise to read your book in tranquillity, certain to find the most admirable things in it, and this I shall do the more gladly as I adopted the teaching of Copernicus many years ago, and his point of view enables me to explain many phenomena of nature which certainly remain inexplicable according to the more current hypotheses. I have written [*conscripsi*] many arguments in support of him and in refutation of the opposite view – which, however, so far I have not dared to bring into the public light, frightened by the fate of Copernicus himself, our teacher, who, though he acquired immortal fame with some, is yet to an infinite multitude of others (for such is the number of fools) an object of ridicule and derision. I would certainly dare to publish my reflections at once if more people like you existed; as they don't, I shall refrain from doing so.

There follow more polite affirmations of esteem, the signature 'Galileus Galileus', and the date: 4 August 1597.[6]

* The preface (and first chapter) proclaim Kepler's belief in the Copernican system and outline his arguments in favour of it.

The letter is important for several reasons. Firstly, it provides conclusive evidence that Galileo had become a convinced Copernican in his early years. He was thirty-three when he wrote the letter; and the phrase 'many years ago' indicates that his conversion took place in his twenties. Yet his first explicit public pronouncement in favour of the Copernican system was only made in 1613, a full sixteen years after his letter to Kepler, when Galileo was forty-nine years of age. Through all these years he not only taught, in his lectures, the old astonomy according to Ptolemy, but expressly repudiated Copernicus. In a treatise which he wrote for circulation among pupils and friends, of which a manuscript copy, dated 1606, survives,[6a] he adduced all the traditional arguments against the earth's motion: that rotation would make it disintegrate, that clouds would be left behind, etc., etc. – arguments which, if the letter is to be believed, he himself had refuted many years before.

But the letter is also interesting for other reasons. In a single breath, Galileo four times evokes Truth: friend of Truth, investigating Truth, pursuit of Truth, proof of Truth; then apparently without awareness of the paradox, he calmly announces his intention to suppress Truth. This may partly be explained by the *mores* of late Renaissance Italy ('that age without a super-ego' as a psychiatrist described it); but taking that into account, one still wonders at the motives of his secrecy.

Why, in contrast to Kepler, was he so afraid of publishing his opinions? He had, at that time, no more reason to fear religious persecution than Copernicus had. The Lutherans, not the Catholics, had been the first to attack the Copernican system – which prevented neither Rheticus nor Kepler from defending it in public. The Catholics, on the other hand, were uncommitted. In Copernicus' own day, they were favourably inclined towards him – it will be remembered how Cardinal Schoenberg and Bishop Giese had urged him to publish his book. Twenty years after its publication, the Council of Trent re-defined Church doctrine and policy in all its aspects, but it had nothing to say against the heliocentric system of the universe. Galileo himself, as we shall see, enjoyed the active support of a galaxy of Cardinals, including the future Urban VIII, and of the leading astronomers among the Jesuits. Up to the fateful year 1616, discussion of the Copernican system was not only permitted, but encouraged by them – under the one proviso, that it should be confined to the language of science, and should not impinge on the theological matters. The situation was summed up clearly in a letter from Cardinal Dini to Galileo in 1615: 'One may write freely as long as one keeps out of the sacristy.'[7] This was precisely what the disputants failed to do, and it was at this point that the conflict began. But nobody could have foreseen these developments twenty years earlier, when Galileo wrote to Kepler.

Thus legend and hindsight combined to distort the picture, and gave rise to the erroneous belief that to defend the Copernican system as a working hypothesis entailed the risk of ecclesiastical disfavour or persecution. During the first fifty years of Galileo's lifetime, no such risk existed; and the thought did not even occur to Galileo. What he feared is clearly stated in his letter: to share the fate of Copernicus, to be mocked and derided; *ridendus et explodendum* – 'laughed at and hissed off the stage' are his exact words. Like Copernicus, he was afraid of the ridicule both of the unlearned and the learned asses, but particularly of the latter: his fellow professors at Pisa and Padua, the stuffed shirts of the peripatetic school, who still considered Aristotle and Ptolemy as absolute authority. And this fear, as will be seen, was fully justified.

## 4. *Early Quarrels*

Young Kepler was delighted with Galileo's letter. On the first occasion when a traveller left Gratz for Italy, he answered in his impulsive manner:

> Gratz, 13 October 1597.
>
> Your letter, my most excellent humanist, which you wrote on 4 August, I received on 1 September; it caused me to rejoice twice: first because it meant the beginning of a friendship with an Italian; secondly, because of our agreement on the Copernican cosmography. ... I assume that if your time has permitted it, you have by now become better acquainted with my little book, and I ardently

desire to know your critical opinion of it; for it is my nature to press all to whom I write for their unvarnished opinion; and believe me, I much prefer even the most acrimonious criticism of a single enlightened man to the unreasoned applause of the common crowd.

I would have wished, however, that you, possessed of such an excellent mind, took up a different position. With your clever secretive manner you underline, by your example, the warning that one should retreat before the ignorance of the world, and should not lightly provoke the fury of the ignorant professors; in this respect you follow Plato and Pythagoras, our true teachers. But considering that in our era, at first Copernicus himself and after him a multitude of learned mathematicians have set this immense enterprise going so that the motion of the earth is no longer a novelty, it would be preferable that we help to push home by our common efforts this already moving carriage to its destination.... You could help your comrades, who labour under such iniquitous criticism, by giving them the comfort of your agreement and the protection of your authority. For not only your Italians refuse to believe that they are in motion because they do not feel it; here in Germany, too, one does not make oneself popular by holding such opinions. But there exist arguments which protect us in the face of these difficulties.... Have faith, Galilii, and come forward! If my guess is right, there are but few among the prominent mathematicians of Europe who would wish to secede from us: for such is the force of Truth. If your Italy seems less advantageous to you for publishing [your works] and if your living there is an obstacle, perhaps our Germany will allow us to do so. But enough of this. Let me know, at least privately if you do not want to do it in public, what you have discovered in support of Copernicus....

Kepler then confessed that he had no instruments, and asked Galileo whether he had a quadrant sufficiently precise to read quarter-minutes of arc; if so, would Galileo please make a series of observations to prove that the fixed stars show small seasonal displacements – which would provide direct proof of the earth's motion.

Even if we could detect no displacement at all, we would nevertheless share the laurels of having investigated a most noble problem which nobody has attacked before us. *Sat Sapienti....* Farewell, and answer me with a very long letter.[8]

Poor, naïve Kepler! It did not occur to him that Galileo

might take offence at his exhortations, and regard them as an implied reproach of cowardice. He waited in vain for an answer to his exuberant overtures. Galileo withdrew his feelers; for the next twelve years, Kepler did not hear from him.

But from time to time unpleasant rumours reached him from Italy. Among Kepler's admirers was a certain Edmund Bruce, a sentimental English traveller in Italy, amateur philosopher and science snob, who loved to rub shoulders with scholars and to spread gossip about them. In August 1602, five years after Galileo had broken off their correspondence, Bruce wrote to Kepler from Florence that Magini (the professor of astronomy at Bologna) had assured him of his love and admiration of Kepler, whereas Galileo had admitted to him, Bruce, having received Kepler's *Mysterium*, but had denied this to Magini.

I scolded Galileo for his scant praise of you, for I know for certain that he lectures on your and his own discoveries to his pupils and others. I however, act and shall always act in a manner which serves not his fame, but yours.[9]

Kepler could not be bothered to answer this busybody, but a year later – 21 August 1603 – Bruce wrote again, this time from Padua:

If you knew how often and how much I discuss you with all the savants of Italy you would consider me not only an admirer but a friend. I spoke with them of your admirable discoveries in music, of your studies of Mars, and explained to them your *Mysterium* which they all praise. They wait impatiently for your future works. ... Galileo has your book and teaches your discoveries as his own. ...[10]

This time Kepler did answer. After apologizing for the delay and declaring himself delighted with Bruce's friendship, he continued:

But there is something about which I wish to warn you. Do not form a higher opinion of me, and do not induce others to do so, than my achievements are able to justify. ... For you certainly understand that betrayed expectations lead eventually to contempt. I wish in no way to restrain Galileo from claiming, what is mine, as his own. My witnesses are the bright daylight and time.[11]

The letter ends with 'Greetings to Magini and Galileo'.

Bruce's accusations should not be taken seriously. In fact, the opposite is true: the trouble with Galileo was not that he appropriated Kepler's discoveries – but that he ignored them, as we shall see. But the episode nevertheless sheds some additional light on the relations between the two men. Though Bruce cannot be trusted on points of fact, the inimical attitude of Galileo to Kepler emerges clearly from Bruce's letters. It fits in with the fact that he broke off the correspondence, and with later events.

Kepler, on the other hand, who had good reason to be offended by Galileo's silence, could easily have been provoked by Bruce's scandal-mongering into starting one of those juicy quarrels between scholars which were the order of the day. He was suspicious and excitable enough, as his relations with Tycho have shown. But towards Galileo he always behaved in an oddly generous way. It is true that they lived in different countries and never met personally; but hatred, like gravity, is capable of action at a distance. The reason for Kepler's forbearingness was perhaps that he had no occasion to develop an inferiority complex towards Galileo.

The year after the Bruce episode, in October 1604, a bright new star appeared in the constellation Serpentarius. It caused even more excitement than Tycho's famous *nova* of 1572, because its appearance happened to coincide with a so-called great conjunction of Jupiter, Saturn, and Mars in the 'fiery triangle' – a gala performance that occurs only once in every eight hundred years. Kepler's book *De Stella Nova* (1606) was primarily concerned with its astrological significance; but he showed that the *nova*, like the previous one, must be located in the 'immutable' region of the fixed stars, and thus drove another nail into the coffin of the Aristotelian universe. The star of 1604 is still called 'Kepler's *nova*'.*

* John Donne referred to Kepler's *nova* when he wrote (*To the Countesse of Huntingdon*):

> Who vagrant transitory Comets sees,
> Wonders, because they are rare: but a new starre
> Whose motion with the firmament agrees,
> Is miracle, for there no new things are.

Galileo, too, observed the new star, but published nothing about it. He gave three lectures on the subject, of which only fragments are preserved; he, too, seems to have denied the contention of the Aristotelians that it was a meteor or some other sublunary phenomenon, but could not have gone much further, since his lectures in defence of Ptolemy were still circulated two years later.[12]

Between 1600 and 1610, Kepler published his *Optics* (1604), the *New Astronomy* (1609) and a number of minor works. In the same period, Galileo worked on his fundamental researches into free fall, the motion of projectiles, and the laws of the pendulum, but published nothing except a brochure containing instructions for the use of the so-called militiary or proportional compass. This was an invention made in Germany some fifty years earlier,[13] which Galileo had improved, as he improved a number of other gadgets that had been known for a long time. Out of this minor publication [14] developed the first of the futile and pernicious feuds which Galileo was to wage all his life.

It began when a mathematician named Balthasar Capra in Padua published, a year after Galileo, another brochure of instructions for the use of the proportional compass.[15] Galileo's *Instructions* were in Italian, Capra's in Latin; both referred to the same subject, which interested only military engineers and technicians. It is very likely that Capra had borrowed from Galileo's *Instructions* without naming him; on the other hand, Capra showed that some of Galileo's explanations were mathematically erroneous, but again without naming him. Galileo's fury knew no bounds. He published a pamphlet *Against the Calumnies and Impostures of Balthasar Capra, etc.* (Venice, 1607), in which that unfortunate man and his teacher [16] were described as 'that malevolent enemy of honour and of the whole of mankind', 'a venom-spitting basilisque', 'an educator who bred the young fruit on his poisoned soul with stinking ordure', 'a greedy vulture, swooping at the unborn young to tear its tender limbs to pieces', and so on. He also obtained from the Venetian Court the confiscation, on the grounds of plagiarism, of Capra's *Instructions.* Not even Tycho and Ursus had sunk to such fish-wife language; yet they had fought for the authorship of a system of the universe, not of a gadget for military engineers.

In his later polemical writings, Galileo's style progressed from coarse invective to satire, which was sometimes cheap, often subtle, always effective. He changed from the cudgel to the rapier, and achieved a rare mastery of it; while in the purely expository passages his lucidity earned him a prominent place in the development of Italian didactic prose. But behind the polished façade, the same passions were at work which had exploded in the affair of the proportional compass; vanity, jealousy, and self-righteousness combined into a demoniac force, which drove him to the brink of self-destruction. He was utterly devoid of any mystical, contemplative leanings, in which the bitter passions could from time to time be resolved; he was unable to transcend himself and find refuge, as Kepler did in his darkest hours, in the cosmic mystery. He did not stand astride the watershed; Galileo is wholly and frighteningly modern.

### 5. *The Impact of the Telescope*

It was the invention of the telescope which brought Kepler and Galileo, each travelling along his own orbit, to their closest conjunction. To pursue the metaphor, Kepler's orbit reminds one of the parabola of comets which appear from infinity and recede into it; Galileo's as an eccentric ellipse, closed upon itself.

The telescope was, as already mentioned, not invented by Galileo. In September 1608, a man at the annual Frankfurt fair offered a telescope for sale which had a convex and a concave lens, and magnified seven times. On 2 October 1608, the spectacle-maker Johann Lippershey of Middleburg claimed a licence for thirty years from the Estates General of the Netherlands for manufacturing telescopes with single and double lenses. In the following month, he sold several of these, for three hundred and six hundred gilders respectively, but was not granted an exclusive licence because in the meantime two other men had claimed the same invention. Two of Lippershey's instruments were sent as a gift by the Dutch Government to the King of France; and in April 1609, telescopes could be bought in spectacle-makers' shops in Paris. In the summer of 1609, Thomas Harriot in England made telescopic observations of the moon, and drew maps of the lunar surface. In the same year, several of the Dutch telescopes found their way to Italy and were copied there.

Galileo himself claimed in the *Messenger from the Stars* that he had merely read reports of the Dutch invention, and that these had stimulated him to construct an instrument on the same principle, which he succeeded in doing 'through deep study of the theory of refraction'. Whether he actually saw and handled one of the Dutch instruments brought to Italy is a question without importance, for once the principle was known, lesser minds than Galileo's could and did construct similar gadgets. On 8 August 1609, he invited the Venetian Senate to examine his spy-glass from the tower of St Marco, with spectacular success; three days later, he made a present of it to the Senate, accompanied by a letter in which he explained that the instrument, which magnified objects nine times, would prove of utmost importance in war. It made it possible to see 'sails and shipping that were so far off that it was two hours before they were seen with the naked eye, steering full-sail into the harbour',[17] thus being invaluable against invasion by sea. It was not the first and not the last time that pure reseach, that starved cur, snapped up a bone from the warlord's banquet.

The grateful Senate of Venice promptly doubled Galileo's salary to a thousand scudi per year, and made his professorship at Padua (which belonged to the Republic of Venice) a lifelong one. It did not take the local spectacle-makers long to produce telescopes of the same magnifying power, and to sell in the streets for a few scudi an article which Galileo had sold the Senate for a thousand a year – to the great amusement of all good Venetians. Galileo must have felt his reputation threatened, as in the affair of the military compass; but, fortunately, this time his passion was diverted into more creative channels. He began feverishly to improve his telescope, and to aim it at the moon and stars, which previously had attracted him but little. Within the next eight months he succeeded, in his own words: 'by sparing neither labour nor expense, in constructing for myself an instrument so superior that objects seen through it appear magnified nearly a thousand times, and more than thirty times nearer than if viewed by the natural power of sight alone.'

The quotation is from *Sidereus Nuncius*, the *Messenger from the Stars*, published in Venice in March 1610. It was Galileo's first scientific publication, and it threw his telescopic discoveries like a bomb into the arena of the learned world. It not only contained news of heavenly bodies 'which no mortal had seen before'; it was also written in a new, tersely factual style which no scholar had employed before. So new was this language that the sophisticated Imperial Ambassador in Venice described the *Star Messenger* as 'a dry discourse or an inflated boast, devoid of all philosophy'.[18] In contrast to Kepler's exuberant baroque style, some passages of the *Sidereus Nuncius* would almost qualify for the austere pages of a contemporary *Journal of Physics.*

The whole booklet has only twenty-four leaves in octavo. After the introductory passages, Galileo described his observations of the moon, which led him to conclude:

> that the surface of the moon is not perfectly smooth, free from inequalities and exactly spherical, as a large school of philosophers considers with regard to the moon and the other heavenly bodies, but that, on the contrary, it is full of irregularities, uneven, full of hollows and protuberances, just like the surface of the earth itself, which is varied everywhere by lofty montains and deep valleys.

He then turned to the fixed stars, and described how the telescope added, to the moderate numbers that can be seen by the naked eye, 'other stars, in myriads, which have never been seen before, and which surpass the old, previously known stars in number more than ten times'. Thus, for instance, to the nine stars in the belt and sword of Orion he was able to add eighty others which he discovered in their vicinity; and to the seven in the Pleiades, another thirty-six. The Milky Way dissolved before the telescope into 'a mass of innumerable stars planted together in clusters'; and the same happened when one looked at the luminous nebulae.

But the principal sensation he left to the end:

There remains the matter which seems to me to deserve to be considered the most important in this work, namely, that I should disclose and publish to the world the occasion of discovering and observing four planets, never seen from the very beginning of the world up to our own times.

The four new planets were the four moons of Jupiter, and the reason why Galileo attributed to their discovery such capital importance he explained in a somewhat veiled aside:

Moreover, we have an excellent and exceedingly clear argument to put at rest the scruples of those who can tolerate the revolution of the planets about the sun in the Copernican system, but are so disturbed by the revolution of the single moon around the earth while both of them describe an annual orbit round the sun, that they consider this theory of the universe to be impossible.

In other words, Galileo thought the main argument of the anti-Copernicans to be the impossibility of the moon's composite motion around the earth, and with the earth around the sun; and further believed that this argument would be invalidated by the composite motion of the four Jupiter moons. It was the only reference to Copernicus in the whole booklet, and it contained no explicit commitment. Moreover, it ignored the fact that in the Tychonic system *all* the planets describe a composite motion around the sun and with the sun around the earth; and that even in the more limited 'Egyptian' system at least the two inner planets do this.

Thus Galileo's observations with the telescope produced no important arguments in favour of Copernicus, nor any clear committal on his part. Besides, the discoveries announced in the *Star Messenger* were not quite as original as they pretended to be. He was neither the first, nor the only scientist, who had turned a telescope at the sky and discovered new wonders with it. Thomas Harriot made systematic telescopic observations and maps of the moon in the summer of 1609, before Galileo, but he did not publish them. Even the Emperor Rudolph had watched the moon through a telescope before he had heard of Galileo. Galileo's star maps were so inaccurate that the

Pleiades group can only be identified on them with difficulty, the Orion group not at all; and the huge dark spot under the moon's equator, surrounded by mountains, which Galileo compared to Bohemia, simply does not exist.

Yet when all this is said, and all the holes are picked in Galileo's first published text, its impact and significance still remain tremendous. Others had seen what Galileo saw, and even his priority in the discovery of the Jupiter moons is not established beyond doubt;[18a] yet he was the first to publish what he saw, and to describe it in a language which made everybody sit up. It was the cumulative effect which made the impact; the vast philosophical implications of this further prizing-open of the universe were instinctively felt by the reader, even if they were not explicitly stated. The mountains and valleys of the moon confirmed the similarity between heavenly and earthly matter, the homogeneous nature of the stuff from which the universe is built. The unsuspected number of invisible stars made an absurdity of the notion that they were created for man's pleasure, since he could only see them armed with a machine. The Jupiter moons did not prove that Copernicus was right, but they did further shake the antique belief that the earth was the centre of the world around which everything turned. It was not this or that particular detail, but the total contents of the *Messenger from the Stars* which created the dramatic effect.

The booklet aroused immediate and passionate controversy. It is curious to note that Copernicus' *Book of Revolutions* had created little stir for half a century, and Kepler's Laws even less at their time, while the *Star Messenger*, which had only an indirect bearing on the issue, caused such an outburst of emotions. The main reason was, no doubt, its immense readability. To digest Kepler's *magnum opus* required, as one of his colleagues remarked, 'nearly a lifetime'; but the *Star Messenger* could be read in an hour, and its effect was like a punch in the solar plexus on those grown up in the traditional view of the bounded universe. And that vision, though a bit shaky, still retained an immense, reassuring coherence. Even Kepler was frightened by the wild perspective opened up by Galileo's spyglass:

'The infinite is unthinkable,' he repeatedly exclaimed in anguish.

The shock-waves of Galileo's message spread immediately, as far as England. It was published in March 1610; Donne's *Ignatius* was published barely ten months later,[19] but Galileo (and Kepler) are repeatedly mentioned in it:

> I will write [quoth Lucifer] to the Bishop of Rome:
> He shall call Galileo the Florentine to him . . .

But soon, the satirical approach yielded to the metaphysical, to a full realization of the new cosmic perspective:

> Man has weav'd out a net, and this net throwne
> Upon the Heavens, and now they are his owne . . .

Milton was still an infant in 1610; he grew up with the new wonders. His awareness of the 'vast unbounded Deep' which the telescope disclosed, reflects the end of the medieval walled universe:

> Before [his] eyes in sudden view appear
> The secrets of the hoary Deep – a dark
> Illimitable ocean, without bound,
> Without dimension . . .[20]

## 6. *The Battle of the Satellites*

Such was the objective impact on the world at large of Galileo's discoveries with his 'optick tube'. But to understand the reactions of the small, academic world in his own country, we must also take into account the subjective effect of Galileo's personality. Canon Koppernigk had been a kind of invisible man throughout his life; nobody who met the disarming Kepler in the flesh or by correspondence, could seriously dislike him. But Galileo had a rare gift of provoking enmity; not the affection alternating with rage which Tycho aroused, but the cold, unrelenting hostility which genius plus arrogance minus humility creates among mediocrities.

Without this personal background, the controversy which followed the publication of the *Sidereus Nuncius* would remain incomprehensible. For the subject of the quarrel was not the *significance* of the Jupiter satellites, but their *existence* – which some of Italy's most illustrious scholars flatly denied. Galileo's main academic rival was Magini in Bologna. In the month following the publication of the *Star Messenger*, on the evenings of 24 and 25 April 1610, a memorable party was held in a house in Bologna, where Galileo was invited to demonstrate the Jupiter moons in his spy-glass. Not one among the numerous and illustrious guests declared himself convinced of their existence. Father Clavius, the leading mathematician in Rome, equally failed to see them; Cremonini, teacher of philosophy at Padua, refused even to look into the telescope; so did his colleague Libri. The latter, incidentally, died soon afterwards, providing Galileo with an opportunity to make more enemies with the much quoted sarcasm: 'Libri did not choose to see my celestial trifles while he was on earth; perhaps he will do so now he has gone to Heaven.'

These men may have been partially blinded by passion and prejudice, but they were not quite as stupid as it may seem. Galileo's telescope was the best available, but it was still a clumsy instrument without fixed mountings, and with a visual field so small that, as somebody has said, 'the marvel is not so much that he found Jupiter's moons, but that he was able to find Jupiter itself'. The tube needed skill and experience in handling, which none of the others possessed. Sometimes, a fixed star appeared in duplicate. Moreover, Galileo himself was unable to explain why and how the thing worked; and the *Sidereus Nuncius* was conspicuously silent on this essential point. Thus it was not entirely unreasonable to suspect that the blurred dots which appeared to the strained and watering eye pressed to the spectacle-sized lens, might be optical illusions in the atmosphere, or somehow produced by the mysterious gadget itself. This, in fact, was asserted in a sensational pamphlet, *Refutation of the Star Messenger*,[20a] published by Magini's assistant, a young fool called Martin Horky. The whole controversy about optical illusions, haloes, reflections from luminous clouds, and about the unreliability of testimonies, inevitably reminds one of a similar controversy three hundred years later: the flying saucers. Here, too, emotion and prejudice combined with technical difficulties against clear-cut conclusions. And here, too, it was not unreasonable for self-respecting scholars to refuse to look at the photographic 'evidence' for fear of making fools of themselves. Similar considerations may be applied to the refusal of otherwise open-minded scholars to get involved in the ambiguous phenomena of occult seances. The Jupiter moons were no less threatening to the outlook on the world of sober scholars in 1610, than, say, extra-sensory perception was in 1950.

Thus, while the poets were celebrating Galileo's discoveries which had become the talk of the world, the scholars in his own country were, with very few exceptions, hostile or sceptical. The first, and for some time the only, scholarly voice raised in public in defence of Galileo, was Johannes Kepler's.

## 7. *The Shield Bearer*

It was also the weightiest voice, for Kepler's authority as the first astronomer of Europe was uncontested – not because of his two Laws, but by virtue of his position as Imperial Mathematicus and successor to Tycho. John Donne, who had a grudging admiration for him, has summed up Kepler's reputation 'who (as himselfe testifies of himselfe) ever since Tycho Brahe's death hath received it into his care, that no new thing should be done in heaven without his knowledge'.[21]

The first news of Galileo's discovery had reached Kepler when Wackher von Wackenfels called on him on or around 15 March 1610. The weeks that followed he spent in feverish expectation of more definite news. In the first days of April, the Emperor received a copy of the *Star Messenger* which had just been published in Venice, and Kepler was graciously permitted 'to have a look and rapidly glance through it'. On 8 April, at last, he received a copy of his own from Galileo, accompanied by a request for his opinion.

Galileo had never answered Kepler's fervent request for an opinion on the *Mysterium*, and had remained equally silent on the *New Astronomy*. Nor did he bother to put his own request for Kepler's opinion on the *Star Messenger* into a personal letter. It was transmitted to Kepler verbally by the Tuscan Ambassador in Prague, Julian de Medici. Although Kepler was not in a position to verify Galileo's disputed discoveries, for he had no telescope, he took Galileo's claims on trust. He did it enthusiastically and without hesitation, publicly offering to serve in the battle as Galileo's 'squire' or 'shield bearer' – he, the Imperial Mathematicus to the recently still unknown Italian scholar. It was one of the most generous gestures in the sour annals of science.

The courier for Italy was to leave on 19 April; in the eleven days at his disposal Kepler wrote his pamphlet *Conversation with the Star Messenger* in the form of an open letter to Galileo. It was printed the next month in Prague, and a pirated Italian translation appeared shortly afterwards in Florence.

It was precisely the support that Galileo needed at that moment. The weight of Kepler's authority played an important part in turning the tide of the battle in his favour, as shown by Galileo's correspondence. He was anxious to leave Padua and to be appointed Court Mathematician to Cosimo de Medici, Grand Duke of Tuscany, in whose honour he had called Jupiter's planets 'the Medicean stars'. In his application to Vinta, the Duke's Secretary of State, Kepler's support figures prominently:

> Your Excellency, and their Highnesses through you, should know that I have received a letter – or rather an eight-page treatise – from the Imperial Mathematician, written in approbation of every detail contained in my book without the slightest doubt or contradiction of anything. And you may believe that this is the way leading men of letters in Italy would have spoken from the beginning if I had been in Germany or somewhere far away.[22]

He wrote in almost identical terms to other correspondents, among them to Matteo Carosio in Paris:

> We were prepared for it that twenty-five people would wish to refute me; but up to this moment I have seen only one statement by Kepler, the Imperial Mathematician, which confirms everything that I have written, without rejecting even an iota of it; which statement is now being reprinted in Venice, and you shall soon see it.[23]

Yet, while Galileo boasted about Kepler's letter to the Grand Duke and his correspondents, he neither thanked Kepler nor even acknowledged it.

Apart from its strategical importance in the cosmological battle, the *Conversation with the Star Messenger* is without much scientific value; it reads like a baroque arabesque, a pattern of amusing doodles around the hard core of Galileo's treatise. It starts with Kepler voicing his hope that Galileo, whose opinion matters to him more than anybody's, would comment on the *Astronomia Nova*, and thereby renew a correspondence 'laid aside twelve years ago'. He relates with gusto how he had received the first news of the discoveries from Wackher – and how he had worried whether the Jupiter moons could be fitted into the universe built around the five Pythagorean solids. But as soon as he had cast a glance at the *Star Messenger*, he realized that 'it offered a highly important and wonderful show to astronomers and philosophers, that it invited all friends of true philosophy to contemplate matters of the highest import. . . . Who could be silent in the face of such a message? Who would not feel himself overflow with the love of the Divine which is so abundantly manifested here?' Then comes his offer of support 'in the battle against the grumpy reactionaries, who reject everything that is unknown as unbelievable, and regard everything that departs from the beaten track of Aristotle as a desecration. . . . Perhaps I shall be considered reckless because I accept your claims as true without being able to add my own observations. But how could I distrust a reliable mathematician whose art of language alone demonstrates the straightness of his judgement? . . .'

Kepler had instinctively felt the ring of truth in the *Star Messenger*, and that had settled the question for him. However much he may have resented Galileo's previous behaviour, he felt committed 'to throw himself into the fray' for Truth, Copernicus, and the Five Perfect Solids. For, having finished the Promethean labours of the *New Astronomy*, he was again steeped in the mystic twilight of a Pythagorean universe built around cubes, tetrahedra, dodecahedra, and so on. They are the *leitmotif* of his dialogue with the *Star Messenger*; neither the elliptical orbits, neither the First nor the Second Law, are mentioned even once. Their discovery appeared to him merely as a tedious detour in the pursuit of his *idée fixe*.

It is a rambling treatise, written by a hurried pen which jumps from one subject to another: astrology, optics, the moon's spots, the nature of the ether, Copernicus, the habitability of other worlds, interplanetary travel:

> There will certainly be no lack of human pioneers when we have mastered the art of flight. Who would have thought that navigation across the vast ocean is less dangerous and quieter than in the narrow, threatening gulfs of the Adriatic, or the Baltic, or the British straits? Let us create vessels and sails adjusted to the heavenly ether, and there will be plenty of people unafraid of the empty wastes. In the meantime, we shall prepare, for the brave sky-travellers, maps of the celestial bodies – I shall do it for the moon, you Galileo, for Jupiter.

Living in an atmosphere saturated with malice, Professors Magini, Horky, and even Maestlin, could not believe their ears when they heard Kepler singing Galileo's praises, and tried to discover some hidden sting in the treatise. They gloated over a passage in which Kepler showed that the principle of the telescope had been outlined twenty years before by one of Galileo's countrymen, Giovanni Della Porta, and by Kepler himself in his work on optics in 1604. But since Galileo did not claim the invention of the telescope, Kepler's historical excursion could not be resented by him; moreover, Kepler emphasized that Della Porta's and his own anticipations were of a purely theoretical nature 'and cannot diminish the fame of the inventor, whoever it was. For I know what a long road it is from a theoretical concept to its practical achievement, from the mention of the antipodes in Ptolemy to Columbus' discovery of the New World, and even more from the two-lensed instruments used in this country to the instrument with which you, O Galileo, penetrated the very skies.'

In spite of this, the German envoy in Venice, George Fugger, wrote with relish that Kepler had 'torn the mask of Galileo's face',[24] and Francis Stelluti (a member of the Lincean Academy) wrote to his brother: 'According to Kepler, Galileo makes himself out to be the inventor of the instrument, but more than thirty years ago Della Porta described it in his *Natural Magic*. ... And so poor Galileo will look foolish.'[25] Horky also quoted Kepler in his much read pamphlet against Galileo, whereupon Kepler immediately informed Horky that 'since the demands of honesty have become incompatible with my friendship for you, I hereby terminate the latter',[26] and offered Galileo to publish the rebuke; but when the youngster relented, he forgave him.

These reactions indicate the extent of dislike for Galileo in his native Italy. But whatever hidden irony the scholars had imputed to Kepler's *Dissertatio*, the undeniable fact was that the Imperial Mathematicus had expressly endorsed Galileo's claims. This persuaded some of Galileo's opponents, who had previously refused to take him seriously, to look for themselves through improved telescopes which were now becoming available. The first among the converts was the leading astronomer in Rome, the Jesuit Father Clavius. In the sequel, the Jesuit scholars in Rome not only confirmed Galileo's observations, but considerably improved on them.

## 8. *The Parting of the Orbits*

Galileo's reaction to the service Kepler had rendered him was, as we saw, complete silence. The Tuscan Ambassador at the Imperial Court urgently advised him to send Kepler a telescope to enable him to verify, at least *post factum*, Galileo's discoveries which he had accepted on trust. Galileo did nothing of the sort. The telescopes which his workshop turned out he donated to various aristocratic patrons.

Four months thus went by, Horky's pamphlet was published, the controversy had reached its peak, and so far not a single astronomer of repute had publicly confirmed having seen the moons of Jupiter. Kepler's friends began to reproach him for having testified to what he himself had not seen; it was an impossible situation.[26a] On 9 August, he again wrote to Galileo:

... You have aroused in me a great desire to see your instrument so that at last I too can enjoy, like yourself, the spectacle of the skies. For among the instruments at our disposal here the best magnifies only ten times, the others hardly thrice. . . .[27]

He talked about his own observations of Mars and the moon, expressed his indignation at Horky's knavery; and then continued:

The law demands that everybody should be trusted unless the contrary is proven. And how much more is this the case when the circumstances warrant trustworthiness. In fact, we are dealing not with a philosophical but with a legal problem: did Galileo deliberately mislead the world by a hoax? ...

I do not wish to hide from you that letters have reached Prague from several Italians who deny that those planets can be seen through your telescope.

I am asking myself how it is possible that so many deny [their existence], including those who possess a telescope. . . . Therefore I ask you, my Galileo, nominate witnesses for me as soon as possible. From various letters written by you to third persons I have learnt that you do not lack such witnesses. But I am unable to name any testimony except your own. . . .[27a]

This time Galileo hurried to answer, evidently scared by the prospect of losing his most powerful ally:

Padua, 19 August 1610.

I have received both your letters, my most learned Kepler. The first, which you have already published, I shall answer in the second edition of my observations. In the meantime, I wish to thank you for being the first, and almost the only, person who completely accept my assertions, though you had no proof, thanks to your frank and noble mind.[28]

Galileo went on to tell Kepler that he could not lend him his telescope, which magnified a thousandfold, because he had given it to the Grand Duke who wished 'to exhibit it in his gallery as an eternal souvenir among his most precious treasures'. He made various excuses about the difficulty of constructing instruments of equal excellence, ending with the vague promise that he would, as soon as possible, make new ones 'and send them to my friends'. Kepler never received one.

In the next paragraph, Horky and the vulgar crowd came in

for some more abuse; 'but Jupiter defies both giants and pygmies; Jupiter stands in the sky, and the sycophants may bark as they wish'. Then he turned to Kepler's request for witnesses, but still could not name a single astronomer; 'In Pisa, Florence, Bologna, Venice, and Padua, a good many have seen [the Medicean stars] but they are all silent and hesitate.' Instead, he named his new patron, the Grand Duke, and another member of the Medici family (who could hardly be expected to deny the existence of stars named after them). He continued:

As a further witness I offer myself, who have been singled out by our University for a lifelong salary of a thousand florins, such as no mathematician has ever enjoyed, and which I would continue to receive forever even if the Jupiter moons were to deceive us and vanish.

After complaining bitterly about his colleagues 'most of whom are incapable of identifying either Jupiter or Mars, and hardly even the moon', Galileo concluded:

What is to be done? Let us laugh at the stupidity of the crowd, my Kepler.... I wish I had more time to laugh with you. How you would shout with laughter, my dearest Kepler, if you were to hear what the chief philosophers of Pisa said against me to the Grand Duke.... But the night has come and I can no longer converse with you....

This is the second, and last, letter which Galileo ever wrote to Kepler.[29] The first, it will be remembered, was written thirteen years earlier, and its theme-song had been the perversity of philosophers and the stupidity of the crowd, concluding with the wistful remark 'if only more people like Kepler existed'. Now, writing for the first time after these thirteen years, he again singled out Kepler as a unique ally to laugh with him at the foolishness of the world. But concerning the quandary into which his loyal ally had got himself, the letter was as unhelpful as could be. It contained not a word on the progress of Galileo's observations, about which Kepler was burning to hear; and it made no mention of an important new discovery which Galileo had made, and which he had communicated, about a fortnight earlier, to the Tuscan Ambassador in Prague.[30] The communication ran as follows:

SMAISMRMILMEPOETALEUMIBUNENUGTTAURIAS

This meaningless sequence of letters was an anagram made up from the words describing the new discovery. The purpose behind it was to safeguard the priority of the find without disclosing its content, lest somebody else might claim it as his own. Ever since the affair of the proportional compass, Galileo had been very anxious to ascertain the priority of his observations – even, as we shall hear, in cases where the priority was not his. But whatever his motives in general, they can hardly excuse the fact that he asked the Tuscan Ambassador to dangle the puzzle before the tantalized eyes of Kepler, whom he could not suspect of intending to steal his discovery.

Poor Kepler tried to solve the anagram, and patiently transformed it into what he himself called a 'barbaric Latin verse': '*Salve umbistineum geminatum Martia proles*' – 'Hail, burning twin, offspring of Mars.'[31] He accordingly believed that Galileo had discovered moons around Mars, too. Only three months later, on 13 November, did Galileo condescend to disclose the solution – not, of course, to Kepler, but to Rudolph, because Julian de Medici informed him that the Emperor's curiosity was aroused.

The solution was: '*Altissimum planetam tergeminum observavi*' – 'I have observed the highest planet [Saturn] in triplet form'. Galileo's telescope was not powerful enough to disclose Saturn's rings (they were only seen half a century later by Heuygens); he believed Saturn to have two small moons on opposite sides, and very close to the planet.

A month later, he sent another anagram to Julian de Medici: '*Haec immatura a me jam frustra leguntur oy*' – 'These immature things I am searching for now in vain'. Once again Kepler tried several solutions, among them: '*Macula rufa in Jove est gyratur mathem, etc*';* then wrote to Galileo in exasperation:

> I beseech you not to withhold from us the solution for long. You must see that you are dealing with honest Germans ... consider what embarrassment your silence causes me.[32]

Galileo disclosed his secret a month later – again not directly

* 'There is a red spot in Jupiter which rotates mathematically.'

to Kepler, but to Julian de Medici: '*Cynthiae figuras aemulatur amorum*' – 'The mother of love [Venus] emulates the shapes of Cynthia [the moon].' Galileo had discovered that Venus, like the moon, showed phases – from sickle to full disc and back – a proof that she revolved around the sun. He also considered this as proof of the Copernican system – which it was not, for it equally fitted the Egyptian or the Tychonic system.

In the meantime, Kepler's dearest wish: to see for himself the new marvels, was at last fulfilled. One of Kepler's patrons, the Elector Ernest of Cologne, Duke of Bavaria, was among the select few whom Galileo had honoured with the gift of a telescope. In the summer of 1610, Ernest was in Prague on affairs of state, and for a short period lent his telescope to the Imperial Mathematicus. Thus from 3 August to 9 September, Kepler was able to watch the Jupiter moons with his own eyes. The result was another short pamphlet, *Observation-Report on Jupiter's Four Wandering Satellites*,[33] in which Kepler confirmed, this time from first-hand experience, Galileo's discoveries. The treatise was immediately reprinted in Florence, and was the first public testimony by independent, direct observation, of the existence of the Jupiter moons. It was also the first appearance in history of the term 'satellite' which Kepler had coined in a previous letter to Galileo.[34]

At this point the personal contact between Galileo and Kepler ends. For a second time Galileo broke off their correspondence. In the subsequent months, Kepler wrote several more letters, which Galileo left unanswered, or answered indirectly by messages via the Tuscan Ambassador. Galileo wrote to Kepler only once during this whole period of the 'meeting of their orbits': the letter of 19 August 1610, which I have quoted. In his works he rarely mentions Kepler's name, and mostly with intent to refute him. Kepler's three Laws, his discoveries in optics, and the Keplerian telescope, are ignored by Galileo, who firmly defended to the end of his life circles and epicycles as the only conceivable form of heavenly motion.

# 9 Chaos and Harmony

## 1. Dioptrice

We must, for the time being, let Galileo recede into the background, and complete the story of Kepler's life and work.

Galileo had transformed the Dutch spy-glass from a toy into an instrument of science, but he had nothing to say in explanation of why and how it worked. It was Kepler who did this. In August and September 1610, while he enjoyed the use of the telescope borrowed from Duke Ernest of Cologne, he wrote within a few weeks a theoretical treatise in which he founded a new science and coined a name for it: dioptrics – the science of refraction by lenses. His *Dioptrice* [1] is a classic of a strikingly un-Keplerian kind, consisting of a hundred and forty-one austere 'definitions', 'axioms', 'problems', and 'propositions' without any arabesque, ornament, or mystic flights of thought.[2] Though he did not find the precise formulation of the law of refraction, he was able to develop his system of geometrical and instrumental optics, and to deduce from it the principles of the so-called Astronomical, or Keplerian Telescope.

In his previous book on optics, published in 1604, Kepler had shown that the intensity of light diminishes with the square of distance; he had explained the principle of the *camera obscura*, the forerunner of the photographic camera, and the manner in which the spectacles for the short- and long-sighted worked. Spectacles had been in use since antiquity, but there existed no precise theory for them. Nor, if it comes to that, did a satisfactory explanation exist for the process of sight – the refraction of the incoming light by the lenses in the eye, and the projection of a reversed image on to the retina – until Kepler's first book on optics. He had modestly called it 'a Supplement to Vitellio'.[3] This Vitellio, a thirteenth-century scholar, had written a compendium of optics mainly based on Ptolemy and Alhazen, and this was the most up-to-date work on the subject till Kepler's advent. One must constantly bear in mind this lack of continuity in the development of science, the immense, dark lowlands extending between the peaks of antiquity and the watershed, to see the achievements of Kepler and Galileo in true perspective.

The *Dioptrice* is Kepler's soberest work – as sober as the geometry of Euclid. He wrote it in the same year as his punch-drunk *Conversation with the Star Messenger*. It had been one of the most exciting years in Kepler's life; it was followed by the blackest and most depressing.

## 2. *Disaster*

The year 1611 brought civil war and epidemics to Prague; the abdication of his imperial patron and provider; the death of his wife and favourite child.

Men less prone to astrology would have blamed such a series of catastrophes on the evil influence of the stars; oddly enough, Kepler did not. His astrological beliefs had become too refined for that: he still believed that the constellations influenced the formation of character, and also had a kind of catalysing effect on events; but the cruder form of direct astrological causation he rejected as superstition.

This made his position at Court even more difficult. Rudolph, sliding from apathy into insanity, was now virtually a prisoner in his citadel. His cousin Leopold had raised an army and occupied part of Prague. The Bohemian Estates appealed for help to his brother Matthias, who had already dispossessed Rudolph of Austria, Hungary, and Moravia, and was preparing to take over what was left. Rudolph craved reassurance from the stars; but Kepler was too honest to provide it. In a confidential letter to one of Rudolph's intimate advisers, he explained:

> Astrology can do enormous harm to a monarch if a clever astrologer exploits his human credulity. I must watch that this should not happen to our Emperor. . . . I hold that astrology must

not only be banished from the senate, but also from the heads of all who wish to advise the Emperor in his best interests; it must be kept entirely out of his sight.[4]

He went on to say that, consulted by the Emperor's enemies, he had pretended that the stars were favourable to Rudolph and unfavourable to Matthias; but he would never say this to the Emperor himself, lest he became over-confident and neglected whatever chance there may be left to save his throne. Kepler was not above writing astrological calendars for money, but where his conscience was involved, he acted with a scrupulousness most unusual by the standards of his time.

On 23 May, Rudolph was forced to abdicate the Bohemian crown; the following January he was dead. In the meantime, Frau Barbara contracted the Hungarian fever, which was followed by attacks of epilepsy and symptoms of mental derangement. When she got better the three children went down with smallpox, which the soldiery had imported. The oldest and youngest recovered; the favourite, six-year old Friedrich, died. Then Barbara relapsed:

Numbed by the horrors committed by the soldiers, and the bloody fighting in the town; consumed by despair of the future and by an unquenchable longing for her lost darling ... in melancholy despondency, the saddest of all states of mind, she gave up the ghost.[5]

It was the first in a series of disasters which weighed down on the last twenty years of Kepler's life. To keep going, he published his correspondence with various scholars on questions of chronology in the age of Christ. Chronology had always been one of his favourite distractions; his theory that Jesus was really born in the year 4 or 5 'B.C.' is today generally accepted. Thus he was 'marking time' in two meanings of the word; for he had secured himself a new, modest job in Linz, but could not leave Prague while Rudolph was still alive.

The end came on 20 January 1612. It was also the end of the most fertile and glorious period in Kepler's life.

## 3. *Excommunication*

The new job was that of a Provincial Mathematicus in Linz, capital of Upper Austria – similar to that he had held in his youth in Gratz. He was now forty-one years old, and he stayed in Linz for fourteen years, till he was fifty-five.

It seemed a depressing come-down after the glories of Prague; but it was not quite as bad as it seemed. For one thing, Rudolph's successor had confirmed Kepler in his title as Imperial Mathematician, which he retained throughout his life. Matthias, unlike Rudolph, had little time for his court astronomer; but he wanted him to be not too far away, and Linz, in his Austrian domain, was a satisfactory solution. Kepler himself was glad to be away from the turmoil of Prague, and to receive a salary from the Austrians which at least he was sure to get. He also had influential patrons among the local aristocracy, the Starhembergs and Liechtensteins; in fact, the job had been specially created for him, carried only theoretical obligations, and left him all the leisure he needed for his work. When the Thirty Years War began with the defenestration of Prague, he could only be thankful to be removed from the focus of events. And when he was offered the succession of Magini in the Chair for Mathematics in Bologna, he wisely refused.

But nevertheless it was a come-down. 'Linz', to Austrians, remains to this day a byword for provincialism. Barbara, whose homesickness for Austria had been one of the reasons for Kepler's choice of Linz, was dead. His desolate loneliness wrung from him one of his self-analytical outcries:

> . . . My exaggerated trustingness, display of piety, a clutching at fame by means of startling projects and unusual actions, the restless search for and interpretation of causes, the spiritual anguish for grace. . . .[6]

He had nobody to talk to, nobody even to quarrel with.

This last need, however, was fulfilled after a while by the local parson, one Daniel Hitzler. He also came from Wuerttemberg, and knew all about Kepler's scandalous crypto-Calvinist deviations. On the first occasion when Kepler came for Communion, they had an argument. Kepler denied, as he had always done, the Lutheran doctrine of the ubiquity – the omnipresence in the world, not only of the spirit, but of the body of Christ; while Hitzler insisted on a written statement of conformity to the doctrine (which, later on, was dropped by Lutheran theology). Kepler refused, whereupon Hitzler refused him Communion. Kepler complained in a fervent petition to the Church Council in Wuerttemberg; the Council answered in a long, patient, and paternally chiding letter that Kepler should stick to mathematics and leave theology to the theologians. Kepler was forced to go for Communion to a parish outside Linz, whose parson was apparently more broad-minded; the Church Council, while backing Pastor Hitzler, did nothing to prevent his colleague from giving Communion to the errant sheep. Kepler kept protesting against the curtailment of his freedom of conscience, and complaining that gossips called him an atheist and a double-dealer, who was trying to curry favour with the Catholics and flirting with the Calvinists. Yet this repeated falling between three stools seemed to agree with his innermost nature:

> It hurts my heart that the three factions have miserably torn the truth to pieces between them, that I must collect the bits wherever I can find them, and put them together again. . . . I labour to reconcile the parties with each other whenever it can be done with sincerity, so that I should be able to side with all of them. . . . Behold, I am attracted either by all three parties, or at least by two of them against the third, setting my hopes on agreement; but my opponents are only attracted by one party, imagining that there must be irreconcileable division and strife. My attitude, so help me God, is a Christian one; theirs, I do not know what.[7]

It was the language of Erasmus and Tiedemann Giese, of the Golden Age of tolerance – but entirely out of place and out of date in Germany on the eve of the Thirty Years War.

Engulfed in that European disaster, Kepler had to endure an additional ordeal: a kind of ghastly private epicycle turning on the greater wheel. His old mother had been accused of witchcraft and was threatened with being burnt alive. The proceedings lasted for six years, from 1615 to 1621; compared with this, the quasi- or semi-excommunication of Kepler himself was only a minor nuisance.

## 4. *The Witch Trial*

The witch-hunting mania, which had grown in furore throughout the sixteenth century, reached its peak in the first half of the seventeenth, both in the Catholic and Protestant parts of Germany. In Weil der Stadt, Kepler's idyllic birthplace, with a population of two hundred families, thirty-eight witches were burnt between 1615 and 1629. In the neighbouring Leonberg, where Kepler's mother now lived, a place equally small, six witches were burnt in the winter of 1615 alone. It was one of the hurricanes of madness which strikes the world from time to time, and seems to be part of man's condition.

Kepler's mother was now a hideous little old woman, whose meddlesomeness and evil tongue, together with her suspect background, predestined her as a victim. She was, we remember, an inn-keeper's daughter, brought up by an aunt who was said to have perished at the stake; and her husband had been a mercenary who vanished after barely escaping the gallows. In that same year, 1615, when Leonberg was seized with witch-hysteria, Katherine had a quarrel with another old hag, her former best friend, the wife of the glazier Jacob Reinhold. This was to be her undoing. The glazier's wife accused Katherine of having given her a witches' potion which had produced a chronic illness (in fact, her ailment was caused by an abortion). It was now remembered that various burghers of Leonberg had been taken ill at various times after being offered a drink from a tin jug which Katherine always kept hospitably prepared for her visitors. The wife of Bastian Meyer had died of it, and the schoolmaster Beutelspacher was permanently paralysed. It was remembered that once upon a time Katherine had asked the sexton for the skull of her father, which she wanted to have cast in silver as a drinking goblet for her son – that court astrologer, himself an adept of the black art. She had cast an evil eye on the children of the tailor Daniel Schmidt, who had promptly died; she was known to have entered houses through locked doors, and to have ridden to death a calf, of which she offered a cutlet to her other son, Heinrich the vagrant.

Katherine's foremost enemy, the glazier's wife, had a brother, who was court barber to the Duke of Wuerttemberg. In that fateful year, 1615, the Duke's son, Prince Achilles, came to Leonberg to hunt, with the barber in his suite. The barber and the Town Provost got drunk together, and had Ma Kepler brought to the Town Hall. Here the barber put the point of his sword on the old woman's breast and asked her to cure his sister by witches' magic of the ailment she had cast on her. Katherine had the sense to refuse – otherwise she would have convicted herself; and her family now sued for libel to protect her. But the Town Provost blocked the libel suit by starting formal proceedings against Katherine for witchcraft. The incident which provided him with an opportunity to do so involved a girl of twelve, who was carrying bricks to the kiln, and on passing Ma Kepler on the road felt a sudden pain in the arm, which led to a temporary paralysis. These sudden, stabbing pains in shoulder, arm, or hip, played a great part in the trial of Katherine and others; to this day, lumbago pains and stiff necks are called in Germany *Hexenschuss* – witches' shot.

The proceedings were long, ghastly, and squalid. At various stages, Kepler's younger brother, Christoph, drillmaster of the militia of Leonberg, and his brother-in-law, the vicar, dissociated themselves from the old woman, squabbled over the cost of defence, and would apparently have been quite glad to see their mother burnt and have done with, except for the reflection it would cast on their own bourgeois respectability. Kepler had always been fated to fight without allies, and for unpopular causes. He started with a counter-attack, accusing his mother's persecutors of being inspired by the devil, and peremptorily advised the Town Council of Leonberg to watch their steps, to remember that he was his Roman Imperial Majesty's Court Mathematicus, and to send him copies of all documents relating to his mother's case. This opening blast had the desired effect of making the Town Provost, the barber, and their clique proceed more warily, and look for more evidence before applying for formal indictment. Ma Kepler obligingly provided it, by offering the Provost a silver goblet as a bribe if he consented to suppress the report on the incident of the little girl with the bricks. After that, her son, daughter, and son-in-law decided that the only solution was flight, and bundled Ma Kepler off to Johannes in Linz, where she arrived in December 1616. This done, Christoph and the vicar wrote to the ducal Chancellery that should the Provost's accusations prove justified, they would disown old Katherine, and let justice take its course.

The old woman stayed for nine months in Linz; then she got homesick and returned to live with Margaret and the vicar, stake or no stake. Kepler followed her, reading on the journey *The Dialogue on Ancient and Modern Music* by Galileo's father. He stayed in Wuerttemberg for two months, wrote petitions, and tried to obtain a hearing of the original libel suit – to no avail. He succeeded only in obtaining permission to take his mother back with him to Linz. But the stubborn old woman refused; she did not like Austria. Kepler had to return without her.

There followed a strange lull of two years – the opening years of the Thirty Years War – during which Kepler wrote more petitions and the court collected more evidence, which now filled several volumes. Finally, on the night of 7 August 1620, Ma Kepler was arrested in her son-in-law's vicarage; to avoid scandal, she was carried out of the house hidden in an oak linen-chest, and thus transported to the prison in Leonberg. She was interrogated by the Provost, denied being a witch, and was committed to a second and last interrogation, before being put to the torture.

Margaret sent another S.O.S. to Linz, and Kepler set out once again for Wuerttemberg. The immediate result of his arrival was that the Supreme Court granted Ma Kepler six weeks to prepare her defence. She was lying in chains in a room at the Town Gate, with two full-time guards – whose salary had to be paid by the defence, in addition to the extravagant quantities of firewood which they burnt. Kepler, who had built a new astronomy on a trifle of eight minutes arc, did not neglect such details in his petitions; he pointed out that one guard would be a sufficient security precaution for his chained mother, aged seventy-three, and that the cost of the firewood should be more equitably shared. He was his irrepressible, indefatigable, passionate, and precise self. The situation, from the point of view of the authorities, was summed up by a slip in the court scribe's record: 'The accused appeared in court, accompanied, alas, by her son, Johannes Kepler, mathematician.' [8]

The proceedings lasted for another year. The accusation comprised forty-nine points, plus a number of supplementary charges – for instance, that the accused had failed to shed tears when admonished with texts from Holy Scripture (this 'weeping test' was important evidence in witch-trials); to which Ma Kepler retorted angrily, she had shed so many tears in her life that she had none left.

The Act of Accusation, read in September, was answered a few weeks later by an Act of Contestation by Kepler and counsel; this was refuted by an Act of Acceptation by the prosecution in December; in May, next year, the defence submitted an Act of Exception and Defence; in August, the prosecution answered with an Act of Deduction and Confutation. The last word was the Act of Conclusion by the defence, a hundred and twenty-eight pages long, and written mostly in Kepler's own hand. After that the case was sent, by order of the Duke, to the Faculty of Law at Tuebingen – Kepler's university. The Faculty found that Katherine ought to be questioned under torture, but recommended that the procedure should be halted at the stage of *territio*, or questioning under threat of torture.

According to the procedure laid down in such cases, the old woman was led into the torture chamber, confronted with the executioner, the instruments were shown to her, and their action on the body described in detail; then she was given a last chance to confess her guilt. The terror of the place was such that a great number of victims broke down and confessed at this stage.[9] The reactions of Ma Kepler were described in the Provost's report to the Duke as follows:

Having, in the presence of three members of the Court and of the town scribe, tried friendly persuasion on the accused, and having met with contradiction and denial, I led her to the usual place of torture and showed her the executioner and his instruments and reminded her earnestly of the necessity of telling the truth, and of the great dolour and pain awaiting her. Regardless, however, of all earnest admonitions and reminders, she refused to admit and confess to witchcraft as charged, indicating that one should do with her as one liked, and that even if one artery after another were to be torn from her body, she would have nothing to confess; whereafter she fell on her knees and said a *pater noster*, and demanded that God should make a sign if she were a witch or a monster or had ever anything to do with witchcraft. She was willing to die, she said, God would reveal the truth after her death, and the injustice and violence done to her; she would leave it all to God, who would not withdraw the Holy Ghost from her, but be her support. ... Having persisted in her contradiction and denial regarding witchcraft, and having remained steadfast in this position, I led her back to her place of custody.[10]

A week later, Ma Kepler was released, after fourteen months of imprisonment. She could not return to Leonberg, though, because the populace threatened to lynch her. Six months later, she died.

It was against this background that Kepler wrote the *Harmony of the World*,[11] in which the third planetary law was given to his engaging contemporaries.

## 5. Harmonice Mundi

The work was completed in 1618, three months after the death of his daughter Katherine, and three days after the defenestration of Prague. No irony was intended by the title. Irony he only permitted himself in a footnote (to the sixth chapter of the Fifth Book), where the sounds emitted by the various planets as they hum along their orbits are discussed: 'The Earth sings Mi-Fa-Mi, so we can gather even from this that *Mis*ery and *Fa*mine reign on our habitat.'

The *Harmony of the World* is a mathematician's Song of Songs 'to the chief harmonist of creation'; it is Job's daydream of a perfect universe. If one reads the book concurrently with his letters about the witch-trial, his excommunication, the war, and the death of his child, one has the impression of being abruptly transported from one play by his Stratford contemporary to a different one. The letters seem to echo the monologue of King Lear –

> Blow, winds, and crack your cheeks! rage! blow!
> You cataracts and hurricanoes, spout
> Till you have drench'd our steeples, drown'd the cocks! ...
> And thou, all-shaking thunder, Strike flat the thick rotundity o' the world.'

But the book's motto could be:

> Here we will sit, and let the sounds of music
> Creep in our ears: soft stillness and the night
> Become the touches of sweet harmony....
> There's not the smallest orb which thou behold'st
> But in his motion like an angel sings....
> Such harmony is in immortal souls.

The *Harmony of the World* is the continuation of the *Cosmic Mystery*, and the climax of his lifelong obsession. What Kepler attempted here is, simply, to bare the ultimate secret of the universe in an all-embracing synthesis of geometry, music, astrology, astronomy, and epistemology. It was the first attempt of this kind since Plato, and it is the last to our day. After Kepler, fragmentation of experience sets in again, science is divorced from religion, religion from art, substance from form, matter from mind.

The work is divided into five books. The first two deal with the concept of harmony in mathematics; the following three with the applications of this concept to music, astrology, and astronomy, in that order.

What exactly does he mean by 'harmony'? Certain geometrical proportions which he finds reflected everywhere, the archetypes of universal order, from which the planetary laws, the harmonies of music, the drift of the weather, and the fortunes of man are derived. These geometrical ratios are the *pure* harmonies which guided God in the work of Creation; the *sensory* harmony which we perceive by listening to musical consonances is merely an echo of it. But that inborn instinct in man which makes his soul resonate to music, provides him with a clue to the nature of the mathematical harmonies which are at its source. The Pythagoreans had discovered that the octave originates in the ratio 1:2 between the length of the two vibrating strings, the fifth in the ratio of 2:3, the fourth in 3:4, and so on. But they went wrong, says Kepler, when they sought for an explanation of this marvellous fact in occult number-lore. The explanation why the ratio 3:5, for instance, gives a concord, but 3:7 a discord, must be sought not in arithmetical, but in *geometrical* considerations. Let us imagine the string, whose vibrations produce the sound, bent into a circle with its ends joined together. Now a circle can be most gratifyingly divided by inscribing into it symmetrical figures with varying numbers of sides. Thus the side of an inscribed pentagon will divide the circumference into parts which are to the whole circle as 1/5 and 4/5 respectively – both consonant chords.

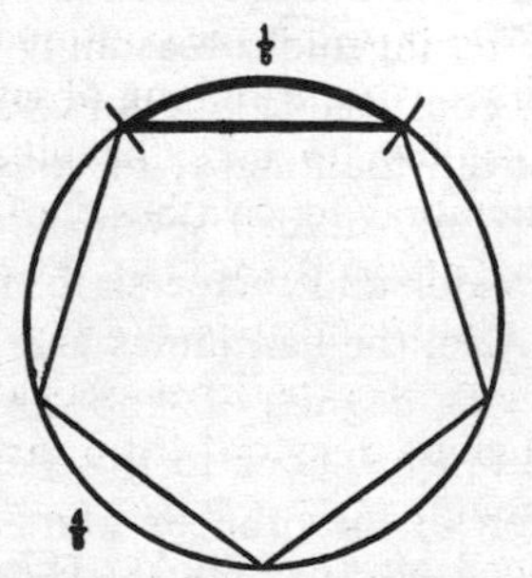

But a septagon will produce ratios of 1/7 and 6/7 – both discords. Why? The answer, according to Kepler, is: *because the pentagon can be constructed by compass and ruler, but the septagon not*. Compass and ruler are the only permissible tools in classical geometry. But geometry is the only language which enables man to understand the working of the divine mind. Therefore figures which cannot be constructed by compass and ruler – such as the septagon, the 11-, 13-, or 17-sided polygons – are somehow unclean, because they defy the intellect. They are *inscibilis*, unknowable,[12] *inefabilis*, unspeakable, *non-entia*, non-existences. 'Therein lies the reason,' Kepler explains, 'why God did not employ the septagon and the other figures of this species to embellish the world.'

Thus the pure archetypal harmonies, and their echoes, the musical consonances, are generated by dividing the circle by means of construable, regular polygons; whereas the 'unspeakable' polygons produce discordant sounds, and are useless in the scheme of the universe. To the obsession with the five perfect solids was now added the twin obsession with the perfect polygons. The former are three-dimensional bodies inscribed into the sphere, the latter are two-dimensional shapes inscribed into the circle. There is an intimate, mystical connexion between the two: the sphere, it will be remembered, is for Kepler the symbol of the Holy Trinity; the two-dimensional plane symbolizes the material world; the intersection of sphere and plane, the circle, pertains to both, and symbolizes the dual nature of man as body and spirit.

But again the facts did not tally with the scheme, and had to be explained away by ingenious reasoning. The 15-sided polygon, for instance, is construable, but does not produce a musical consonance. Moreover, the number of construable polygons is infinite, but Kepler only needed seven harmonic relations for his scale (octave, major and minor sixth, fifth, fourth, major, and minor third). Also, the harmonies had to be arranged into a hierarchy of varying degrees of 'knowability', or perfection. Kepler devoted as much labour to this fantastic enterprise as to the determination of the orbit of Mars. In the end he succeeded, to his own satisfaction, in deriving all his seven harmonies, by certain complicated rules of the game, from his perfect polygons. He had traced back the laws of music to the Supreme Geometer's mind.

In the sections which follow, Kepler applies his harmonic ratios to every subject under the sun: metaphysics and epistemology, politics, psychology, and physiognomics; architecture and poetry, meteorology and astrology. Then, in the fifth and last book, he returns to cosmology, to complete his vertiginous edifice. The universe he had built in his youth around the five perfect solids had not quite tallied with the observed facts. He now brought the two-dimensional shadow army of polygons to the rescue of the beleaguered solids. The harmonic ratios must somehow be dovetailed in between the solids to fill the gaps and to account for the irregularities.

But how could this be done? How could the harmonies be fitted into the scheme of a universe full of elliptic orbits and non-uniform motions, from which, in fact, all symmetry and harmony seemed to have departed? As usual, Kepler takes the reader into his confidence, and for his benefit recapitulates the process by which he arrived at his solution. At first, he tried to assign the harmonic ratios to the *periods of revolution* of the various planets. He drew a blank: 'We conclude that God the Creator did not wish to introduce harmonic proportions into the durations of the planetary years.'[13]

Next, he wondered whether the *sizes* or *volumes* of the various planets form a harmonic series. They do not. Thirdly, he tried to fit the greatest and smallest *solar distances* of every planet into a harmonic scale. Again no good. In the fourth place, he tried the ratios between the *extreme velocities* of each planet. Again no good. Next, the variations in the time needed by a planet to cover a *unit length* of its orbit. Still no good. Lastly, he hit on the idea of transferring the observer's position into the centre of the world, and to examine the variations in angular velocity, regardless of distance, *as seen from the sun*. And lo! it worked.

The results were even more gratifying than he had expected. Saturn, for instance, when farthest away from the sun, in its aphelion, moves at the rate of 106 seconds arc per day; when closest to the sun, and its speed is at maximum, at 135 seconds arc per day. The ratio between the two extreme velocities is 106 to 135, which only differs by two seconds from 4 : 5 – the major third. With similar, very small deviations (which were all perfectly explained away at the end), the ratio of Jupiter's slowest to its fastest motion is a minor third, Mars' the quint, and so forth. So much for each planet considered by itself. But when he compared the extreme angular velocities of *pairs* of different planets, the results were even more marvellous:

At the first glance the Sun of Harmony broke in all its clarity through the clouds.[14]

The extreme values yield in fact the intervals of the complete scale. But not enough: if we start with the outermost planet, Saturn, in the aphelion, the scale will be in the major key; if we start with Saturn in the perihelion, it will be in the minor key. Lastly, if several planets are simultaneously at the extreme points of their respective orbits, the result is a motet where Saturn and Jupiter represent the bass, Mars the tenor, Earth and Venus the contralto, Mercury the soprano. On some occasions, all six can be heard together:

The heavenly motions are nothing but a continuous song for several voices (perceived by the intellect, not by the ear); a music which, through discordant tensions, through sincopes and cadenzas, as it were (as men employ them in imitation of those natural discords), progresses towards certain pre-designed, quasi six-voiced clausuras, and thereby sets landmarks in the immeasurable flow of time. It is, therefore, no longer surprising that man, in imitation of his creator, has at last discovered the art of figured song, which was unknown to the ancients. Man wanted to reproduce the continuity of cosmic time within a short hour, by an artful symphony for several voices, to obtain a sample test of the delight of the Divine Creator in His Works, and to partake of his joy by making music in the imitation of God.[15]

The edifice was complete. Kepler finished the book on 27 May 1618, in one of the most fateful weeks of European history:

In vain does the God of War growl, snarl, roar, and try to interrupt with bombards, trumpets, and his whole tarantantaran. . . .[16] Let us despise the barbaric neighings which echo through these noble lands, and awaken our understanding and longing for the harmonies.[17]

Out of the murky abyss he soared to heights of orphic ecstasies:

The thing which dawned on me twenty-five years ago before I had yet discovered the five regular bodies between the heavenly orbits . . .; which sixteen years ago I proclaimed as the ultimate aim of all research; which caused me to devote the best years of my life to astronomical studies, to join Tycho Brahe and to choose Prague as my residence – that I have, with the aid of God, who set my enthusiasm on fire and stirred in me an irrepressible desire, who kept my life and intelligence alert, and also provided me with the remaining necessities through the generosity of two Emperors and the Estates of my land, Upper Austria – that I have now, after discharging my astronomical duties *ad satietatum*, at long last brought to light. ... Having perceived the first glimmer of dawn eighteen months ago, the light of day three months ago, but only a few days ago the plain sun of a most wonderful vision – nothing shall now hold me back. Yes, I give myself up to holy raving. I mockingly defy all mortals with this open confession: I have robbed the golden vessels of the Egyptians to make out of them a tabernacle for my God, far from the frontiers of Egypt. If you forgive me, I shall rejoice. If you are angry, I shall bear it. Behold, I have cast the dice, and I am writing a book either for my contemporaries, or for posterity. It is all the same to me. It may wait a hundred years for a reader, since God has also waited six thousand years for a witness. . . .[18]

## 6. *The Third Law*

This last quotation is from the Preface to the Fifth Book of the *Harmonice Mundi*, which contains, almost hidden among the luxuriant growth of fantasy, Kepler's Third Law of planetary motion.

It says, put into modern terms, that the squares of the periods of evolution of any two planets are as the cubes of their mean distances from the sun.[19] Here is an illustration of it. Let the earth's distance from the sun be our unit of distance, and the earth's year our unit of period. Then Saturn's distance from the sun will be a little over nine units. The cube 1 is 1; the cube of 9 is 729. The square root of 1 is 1, the square root of 729 is 27. Thus a Saturn year will be a little over twenty-seven earth years; in fact it is thirty years. Apologies for the coarse example – it is Kepler's own.[20]

Unlike his First and Second Laws, which he found by that

peculiar combination of sleepwalking intuition and wide-awake alertness for clues – a mental process on two levels, which drew mysterious benefits out of his apparent blunderings – the Third Law was the fruit of nothing but patient, dogged trying. When after endless trials, he had at last hit on the square-to-cube ratio, he of course promptly found a reason why it should be just that and none other; I have said before that Kepler's *a priori* proofs were often invented *a posteriori*.

The exact circumstances of the discovery of the Third Law were again faithfully recorded by Kepler:

> On 8 March of this present year 1618, if precise dates are wanted, [the solution] turned up in my head. But I had an unlucky hand and when I tested it by computations I rejected it as false. In the end it came back again to me on 15 May, and in a new attack conquered the darkness of my mind; it agreed so perfectly with the data which my seventeen years of labour on Tycho's observations had yielded, that I thought at first I was dreaming, or that I had committed a *petitio principi*. . . .[21]

He celebrated his new discovery, as he had celebrated his First Law, with a quotation from Virgil's *Eclogues*; in both cases Truth appears in the shape of a teasing hussy who surrenders unexpectedly to her pursuer when he has already given up hope. And in both cases also, the true solution was rejected by Kepler when it first occurred to him, and was only accepted when it crept in a second time, 'through a back-door of the mind'.

He had been searching for this Third Law, that is to say, for a correlation between a planet's *period* and its *distance*, since his youth. Without such a correlation, the universe would make no sense to him; it would be an arbitrary structure. If the sun had the power to govern the planets' motions, then that motion must *somehow* depend on their distance from the sun; but how? Kepler was the first who saw the problem – quite apart from the fact that he found the answer to it, after twenty-two years of labour. The reason why nobody before him had asked the question is that nobody had thought of cosmological problems in terms of actual physical forces. So long as cosmology remained divorced from physical causation in the mind, *the right question could not occur* in that mind. Again a parallel to the present situation imposes itself: there is, one suspects, a fragmentation in the twentieth-century mind which prevents it from asking the right questions. The offspring of a new synthesis is not a ready solution, but a healthy problem crying lustily for an answer. And *vice versa*: a one-sided philosophy – whether it be scholasticism or nineteenth-century mechanism, creates sick problems, of the sort 'What is the sex of the angels?' or 'Is man a machine?'

## 7. *The Ultimate Paradox*

The *objective* importance of the Third Law is that it provided the final clue for Newton; hidden away in it is the essence of the Law of Gravity. But its *subjective* importance to Kepler was that it furthered his chimerical quest – and nothing else. The Law makes its first appearance as 'Proposition No. 8' in a chapter characteristically called 'The Main Propositions of Astronomy which are needed for the Investigation of the Celestial Harmonies'. In the same chapter (the only one in the book which deals with astronomy proper) the First Law is merely mentioned in passing, almost shamefacedly, and the Second Law not at all. In its place Kepler once more quoted his faulty inverse ratio proposition, whose incorrectness he once knew and then forgot. Not the least achievement of Newton was to spot the Three Laws in Kepler's writings, hidden away as they were like forget-me-nots in a tropical flowerbed.

To change metaphors once more: the three Laws are the pillars on which the edifice of modern cosmology rests; but to Kepler they meant no more than bricks among other bricks for the construction of his baroque temple, designed by a moon-struck architect. He never realized their real importance. In his earliest book he had remarked that 'Copernicus did not know how rich he was'; the same remark applies to Kepler himself.

I have stressed this paradox over and again; now it is time to try to resolve it. Firstly, Kepler's obsession with a cosmos built around the Pythagorean solids and the musical harmonies, was not quite as extravagant as it seems to us. It was in keeping with the traditions of Neoplatonism, with the revival of Pythagoreanism, with the teaching of Paracelsians, Rosicrucians, astrologers, alchemists, cabbalists, and hermetists, who were still conspicuously in evidence in the early seventeenth century. When we talk of 'the age of Kepler and Galileo', we are apt to forget that they were isolated individuals, a generation ahead of the most enlightened men of their time. If the 'harmony of the world' was a fantastic dream, its symbols had been shared by a whole dreaming culture. If it was an *idée fixe*, it was derived from a collective obsession – only more elaborate and precise, enlarged on a grandiose scale, more artful and self-consistent, carried to the ultimate perfection of mathematical detail. The Keplerian cosmos is the crowning achievement of a type of cosmic architecture which began with the Babylonians and ends with Kepler himself.

The paradox, then, is not in the mystic nature of Kepler's edifice but in the modern architectural elements which it employed, in its combination of incompatible building materials. Dream-architects are not worried about imprecisions of a fraction of a decimal; they do not spend twenty years with dreary, heart-breaking computations to build their fantasy towers. Only some forms of insanity show this pedantic method in madness. In reading certain chapters of the *Harmonice*, one is indeed reminded of the explosive yet painstakingly elaborate paintings by schizophrenics, which would pass as legitimate art if painted by a savage or a child, but are judged by clinical standards if one knows that they are the work of a middle-aged chartered accountant. The Keplerian schizophrenia becomes apparent only when he is judged by the standard of his achievements in optics, as a pioneer of the differential calculus, the discoverer of the three Laws. His split mind is revealed in the manner in which he saw himself in his non-obsessional moment: as a sober 'modern' scientist, unaffected by any mystic leanings. Thus he writes about the Scottish Rosicrucian, Robert Fludd:

> It is obvious that he derives his main pleasure from unintelligible charades about the real world, whereas my purpose is, on the contrary, to draw the obscure facts of nature into the bright light of

knowledge. His method is the business of alchemists, hermetists, and Paracelsians, mine is the task of the mathematician.[22]

These words are printed in *Harmonice Mundi*, which is buzzing with astrological and Paracelsian ideas.

A second point is equally relevant to the Keplerian paradox. The main reason why he was unable to realize how rich he was – that is, to understand the significance of his own Laws – is a technical one: the inadequacy of the mathematical tools of his time. Without differential calculus and/or analytical geometry, the three Laws show no apparent connexion with each other – they are disjointed bits of information which do not make much sense. Why should God will the planets to move in ellipses? Why should their speed be governed by the area swept over by the radius vector, and not by some more obvious factor? Why should the ratio between distance and period be mixed up with cubes and squares? Once you know the inverse square law of gravity and Newton's mathematical equations, all this becomes beautifully self-evident. But without the roof which holds them together, Kepler's Laws seem to have no particular *raison d'être*. Of the first he was almost ashamed: it was a departure from the circle sacred to the ancients, sacred even to Galileo and, for different reasons, to himself. The ellipse had nothing to recommend it in the eyes of God and man; Kepler betrayed his bad conscience when he compared it to a cartload of dung which he had to bring into the system as a price for ridding it of a vaster amount of dung. The Second Law he regarded as a mere calculating device, and constantly repudiated it in favour of a faulty approximation; the Third as a necessary link in the system of harmonies, and nothing more. But then, without the notion of gravity and the method of the calculus, it *could* be nothing more.

Johannes Kepler set out to discover India and found America. It is an event repeated over and again in the quest for knowledge. But the result is indifferent to the motive. A fact, once discovered, leads an existence of its own, and enters into relations with other facts of which their discoverers have never dreamt. Apollonius of Perga discovered the laws of the useless curves which emerge when a plane intersects a cone at various angles: these curves proved, centuries later, to represent the paths followed by planets, comets, rockets, and satellites.

> One cannot escape the feeling [wrote Heinrich Herz] that these mathematical formulae have an independent existence and an intelligence of their own, that they are wiser than we are, wiser even than their discoverers, that we get more out of them than was originally put into them.

This confession of the discoverer of radio-waves sounds suspiciously like an echo of Kepler, echoing Plato, echoing Pythagoras: 'Methinks that all of nature and the graceful sky are set into symbols *in geometriam*.'

# 10 Computing a Bride

Only one circumstance, but a basic one, relieved the gloom of Kepler's later years: his second marriage, in 1613, to Susanna Reuttinger. He was forty-one, she twenty-four, the daughter of a cabinet-maker. Susanna's parents had died while she was a child; she had been brought up in the household of the Baroness Starhemberg. We do not know what position she occupied in the household, but to judge by the scandalized reactions of Kepler's correspondents, it must have been a lowly one – something between a maid and a companion.

Kepler's first marriage had been engineered by his well-wishers when he was an inexperienced and penniless young teacher. Before his second marriage, friends and go-betweens again played a prominent part – but this time Kepler had to choose between no less than eleven candidates for his hand. In a letter to an unknown nobleman, which extends to eight printed folio pages, Kepler has described in meticulous detail the process of elimination and selection that he followed. It is a curious document, and among the most revealing in his voluminous writings. It shows that he solved the problem of choosing the right wife among the eleven candidates by much the same method by which he found the orbit of Mars: he committed a series of mistakes which might have proved fatal, but cancelled out; and up to the last moment he failed to realize that he held the correct solution in his hands.

The letter is dated from Linz, 23 October 1613:[1]

Though all Christians start a wedding invitation by solemnly declaring that their marriage is due to special Divine management, I as a philosopher, would like to discourse with you, O wisest of men, in greater detail about this. Was it Divine Providence or my own moral guilt which, for two years or longer, tore me in so many different directions and made me consider the possibilities of such different unions? If it was Divine Providence, to what purpose did it use these various personalities and events? For there is nothing that I would like to investigate more thoroughly, and that I more intensely long to know, than this: can I find God, whom I can almost touch with my hands when I contemplate the universe, also in my own self? If, on the other hand, the fault was mine, in what did it consist? Cupidity, lack of judgment, or ignorance? And why, on the other hand, was there nobody among my advisers to approve of my final decision? Why am I losing their previous esteem or appear to be losing it?

What could have seemed more reasonable than that I, as a philosopher, past the peak of virility, at an age when passion is extinct, the body dried and softened by nature, should have married a widow who would look after the household, who was known to me and my first wife, and unmistakably recommended to me by her? But if so, why did nothing come of it? ...

The reasons why this first project came to nothing were, among others, that the prospective bride had two marriageable daughters, that her fortune was in the hands of a trustee; and, as an afterthought,

also a consideration of health, because, though her body was strong, it was suspect of ill-health because of her stinking breath; to this came my dubious reputation in matters of religion. In addition to this, when I met the woman after everything had been settled (I had not seen her for the last six years), there was nothing about her that pleased me. It is therefore sufficiently clear that the matter could not succeed. But why did God permit that I should be occupied with this project which was doomed to failure? Perhaps to prevent my getting involved in other perplexities while my thoughts were on this person? ... I believe that things like this happen to others too, not only once but often; but the difference is that others do not worry as much as I do, that they forget more easily and get over things quicker than I do; or that they have more self-control and are less credulous than I am. ... And now for the others.

Together with the mother, her two daughters were also offered to me – under an unfavourable omen, if an offence to probity can be interpreted as such: for the project was presented by the well-wishers of the ladies in a form which was not very proper. The ugliness of this project upset me intensely; yet I began nevertheless to inquire into the conditions. As I thus transferred my interest from widows to virgins, and continued to think of the absent one [the mother] whom, so far, I had not seen, I was captivated by the appearance and pleasant features of the one who was present [the daughter]. Her education had been, as it became sufficiently clear, more splendid than it would be useful to me. She had been brought up in luxury that was above her station, also she was not of sufficient age to run a household. I decided to submit the reasons which spoke against the marriage to the judgment of the mother, who was a wise woman and loved her daughter. But it would have been better if I had not done so, because the mother did not seem to be pleased. This was the second one, and now I come to the third.

The third was a maiden in Bohemia whom Kepler found attractive, and who took a liking to his orphaned children. He left them for a while in her care 'which was a rash act, for later on I had to fetch them back at my own expense'. She was willing to marry him, but she had, a year earlier, given her word to another man. That other man had, in the meantime, begotten a child with a prostitute, so that the maiden considered herself free; but she thought it nevertheless necessary to obtain the permission of her ex-fiancé's employer. This employer had some time ago given Kepler a letter of recommendation – and by a mysterious *non-sequitur*, Kepler states that this prevented the marriage. We are left to wonder.

The fourth he would have married gladly, in spite of her 'tall stature and athletic build', if meanwhile the fifth had not entered the scene. The fifth was Susanna, his future wife:

In comparing her to the fourth the advantage was with the latter as regards the reputation of the family, earnestness of expression, property, and dowry; but the fifth had the advantage through her love, and her promise to be modest, thrifty, diligent, and to love her step-children. . . . While I was waging my long and heavy battle with this problem, I was waiting for the visit of Frau Helmhard, wondering whether she would advise me to marry the third, who would then carry the day over the last-mentioned two. Having at last heard what this woman had to say, I began to decide in favour of the fourth, annoyed that I had to let the fifth go. As I was turning this over, and on the point of making a decision, fate intervened: the fourth got tired of my hesitations and gave her word to another suitor. Just as I had been previously annoyed about having to reject the fifth, I was now so much hurt about the loss of the fourth, that the fifth too began to lose her attraction for me. In this case, to be sure, the fault was in my feelings.

Concerning the fifth, there is also the question why, since she was destined for me, God permitted that in the course of one year, she should have six more rivals? Was there no other way for my uneasy heart to be content with its fate than by realizing the impossibility of the fulfilment of so many other desires?

And so to No. 6, who had been recommended to Kepler by his stepdaughter:

A certain nobility, and some possessions made her desirable; on the other hand, she was not old enough, and I feared the expense of a sumptuous wedding; and her noble rank in itself made her suspect of pride. In addition, I felt pity for the fifth, who had already understood what was afoot and what had been decided. This division in me between willingness and unwillingness had, on the one hand, the advantage that it excused me in the eyes of my advisers, but on the other the disadvantage that I was as pained as if I had been rejected. ... But in this case, too, Divine Providence had meant well because that woman would not have fitted in at all with my habits and household.

Now, as the fifth ruled, to my joy, alone in my heart, a fact which I also expressed to her in words, suddenly a new rival arose for her, whom I shall call No. 7 – because certain people, whom you know, suspected the humility of the fifth and recommended the noble rank of the seventh. She also had an appearance which deserved to be loved. Again I was prepared to give up the fifth, and to choose the seventh, provided it was true what they said about her. ...

But again he prevaricated; 'and what else could have been the result but a rejection, which I had quasi-provoked?'

Tongues were wagging all over Linz; to avoid more gossip and ridicule, he now turned his attention to a candidate of common origin 'who nevertheless aspired to the nobility. Though her appearance had nothing to recommend her, her mother was a most worthy person.' But she was as fickle as he was undecided, and after alternately giving him her word and retracting it on seven subsequent occasions, he again thanked Divine Providence and let her go.

His methods now became more cautious and secretive. When

he met No. 9, who, apart from a lung disease, had much to recommend her, he pretended to be in love with somebody else, hoping that the candidate's reactions might betray her feelings. Her reactions were promptly to tell Mother, who was ready to give her blessing, but Kepler mistakenly thought she had rejected him and then it was too late to put matters right.

The tenth was also of noble rank, of sufficient means and thrifty.

> But her features were most abhorrent, and her shape ugly even for a man of simple tastes. The contrast of our bodies was most conspicuous: I thin, dried-up and meagre; she, short and fat, and coming from a family distinguished by redundant obesity. She was quite unworthy to be compared with the fifth, but this did not revive love for the latter.

The eleventh and last one was again 'of noble rank, opulent, and thrifty'; but after waiting four months for an answer, Kepler was told that the maiden was not yet sufficiently grown up.

> Having thus exhausted the counsels of all my friends, I, at the last moment before my departure for Rattisbon, returned to the fifth, pledged her my word and received hers.
>
> Now you have my commentary on my remark at the beginning of this invitation. You now see how Divine Providence drove me into these perplexities that I may learn to scorn noble rank, wealth, and parentage, of which she has none, and to seek with equanimity other, simpler virtues. . . .

The letter ends with Kepler entreating his aristocratic friend to come to the wedding banquet and help him by his presence to brave the adversity of public opinion.

Susanna seemed to have justified Kepler's choice, and lived up to his expectations. There is hardly any later mention of her in his letters, and as far as Kepler's domestic life was concerned, no news is good news. She bore him seven children, of whom three died in infancy.

I have said at the beginning of this chapter, that Kepler's way of discovering the right wife for himself strangely reminds one of the method of his scientific discoveries. Perhaps, at the end of this matrimonial odyssey, this sounds less far-fetched or whimsical. There is the same characteristic split in the personality between, on the one hand, the pathetically eager, Chaplinesque figure who stumbles from one wrong hypothesis to another and from one candidate to the next – oval orbits, egg-shaped orbits, chubby-faced orbits; who proceeds by trial and error, falls into grotesque traps, analyses with pedantic seriousness each mistake and finds in each a sign of Divine Providence; one can hardly imagine a more painfully humourless performance. But on the other hand, he *did* discover his Laws and *did* make the right choice among the eleven candidates, guided by that sleepwalking intuition which made his waking errors cancel out and always asserted itself at the critical moment. Social rank and financial considerations are topmost in his waking consciousness, yet in the end he married the only candidate who had neither rank, nor money, nor family; and though he anxiously listens to everybody's advice, seems to be easily swayed and without a will of his own, he decides on the person unanimously rejected by all.

It is the same dichotomy which we observed in all his activities and attitudes. In his quarrels with Tycho and constant naggings at him, he displayed embarrassing pettiness. Yet he was curiously devoid of jealousy or lasting resentment. He was proud of his discoveries and often boasted of them (particularly of those which turned out to be worthless), but he had no proprietary feeling about them; he was quite prepared to share the copyright of the three Laws with the Junker Tengnagel and, contrary to the habits of the time, gave in all his books most generous credit to others – to Maestlin, Brahe, Gilbert, and Galileo. He even gave credit where none was due, for instance to Fabricius, whom he nearly saddled with the honour of having discovered the elliptic orbits. He freely informed his correspondents of his latest researches and naïvely expected other astronomers to part with their jealously guarded observations; when they refused, as Tycho and his heirs did, he simply pinched the material without a qualm of conscience. He had, in fact, no sense of private property concerning scientific research. Such an attitude is most unusual among scholars in our day; in Kepler's day it seemed quite insane. But it was the most endearing lunacy in his discordant, fantastic self.

# 11 The Last Years

## 1. Tubulae Rudolphinae

*Harmonice Mundi* was completed in 1618 and published the next year, when Kepler was forty-eight. His pioneering work was done; but in the remaining eleven years of his life he continued to pour out books and pamphlets – annual calendars and ephemerides, a book on comets, another on the new invention of logarithms, and two more major works: the *Epitome Astronomiae Copernicanae* and the *Rudolphine Tables.*

The title of the former is misleading. The *Epitome* is not an abstract of the Copernican system, but a textbook of the Keplerian system. The laws which originally referred to Mars only, are here extended to all planets, including the moon and the satellites of Jupiter. The epicycles are all gone, and the solar system emerges in essentially the same shape in which it appears in modern schoolbooks. It was Kepler's most voluminous work and the most important systematic exposition of astronomy since Ptolemy's *Almagest.* The fact that his discoveries are found in it once more side by side with his fantasies, does not detract from its value. It is precisely this overlapping of two universe of thought, which gives the *Epitome*, as it does to the whole of Kepler's life and work, its unique value to the history of ideas.

To realize how far ahead of his colleagues Kepler was, in spite of the residue of medievalism in his veins, one must compare the *Epitome* with other contemporary textbooks. None of them had adopted the heliocentric idea, or was to do so for a generation to come. Maestlin published a reprint of his textbook based on Ptolemy in 1624, three years after the *Epitome*; and Galileo's famous *Dialogue on the Great Systems of the World*, published another eight years later, still holds fast to cycles and epicycles as the only conceivable form of heavenly motion.

The second major work of Kepler's later years was his crowning achievement in practical astronomy: the long-awaited Rudolphine Tables, based on Tycho's lifelong labours. Their completion had been delayed for nearly thirty years by Tycho's death, the quarrel with their heirs, and the chaotic conditions created by the war – but basically by Kepler's reluctance against what one might call a Herculean donkey-work. Astronomers and navigators, calendar-makers, and horoscope-casters were impatiently waiting for the promised Tables, and angry complaints about the delay came from as far as India and the Jesuit missionaries in China. When a Venetian correspondent joined in the chorus, Kepler answered with a *cri de coeur*:

> One cannot do everything, as the saying goes. I am unable to work in an orderly manner, to stick to a time schedule and to rules. If I put out something that looks tidy, it has been worked over ten times. Often I am held up for a long time by a computing error committed in haste. But I could pour out an infinity of ideas. . . . I beseech thee, my friends, do not sentence me entirely to the treadmill of mathematical computations, and leave me time for philosophical speculations which are my only delight.[1]

At last, when he had turned the corner of fifty, he really settled down to the task at which he had only nibbled since Tycho's death. In December 1623, he triumphantly reported to an English correspondent: *video portum* – 'I can see the harbour'; and six months later to a friend: 'The Rudolphine Tables, sired by Tycho Brahe, I have carried in me for twenty-two years, as the seed is gradually developed in the mother's womb. Now I am tortured by the labours of birth.'[2]

But owing to lack of money and the chaos of the Thirty Years War, the printing took no less than four years, and consumed half of his remaining energies and life-span.

Since the Tables were to bear Rudolph's name, Kepler found it fitting that the printing should be financed by payment of the arrears due to him, amounting to 6,299 florins. He travelled to Vienna, the new seat of the Imperial Court, where he had to spend four months to obtain satisfaction. But the satisfaction was of a rather abstract nature. According to the complicated method by which the Crown's financial affairs were transacted, the Treasury transferred the debt to the three towns of Nuremberg, Memmingen, and Kempten. Kepler had to travel from town to town – partly on horseback, partly on foot because of his piles; and to beg, cajole, and threaten, until he finally obtained a total of 2,000 florins. He used them to buy the paper for the book, and decided to finance the printing out of his own pocket, 'undaunted by any fear for the future sustenance of wife and six children', and though he was forced 'to dip into the money held in trust for the children from my first marriage'. He had lost a whole year on these travels.

But this was only the beginning of his struggles; the story of the printing of the Rudolphine Tables reminds one of the Ten Egyptian Plagues. To begin with, Linz did not have an adequate printing press for such a major undertaking; so Kepler had to travel again to recruit skilled printers from other towns. When the work at last got going, the next plague descended – a familiar one this time: all Protestants in Linz were ordered either to embrace the Catholic faith, or to leave the town within six months. Kepler was again exempted, and so was his Lutheran master printer with his men; but he was requested to hand over to the authorities all books suspect of heresy. Luckily, the choice of objectionable books was left to his own judgement (which made him feel 'as if a bitch were asked to surrender one of her litter') and, thanks to the intervention of the Jesuit Father Guldin, he was able to keep them all. When the war was approaching Linz, the authorities asked Kepler's advice how to protect the books of the Provincial Library against the danger of fire; he recommended packing them tightly into wine barrels so that they could easily be rolled away from the danger spot. Incidentally, notwithstanding his excommunication (now final), Kepler kept paying visits to his beloved Tuebingen, the Lutheran stronghold, and having a jolly time with old Maestlin – all of which goes to show that the sacred cows of that bygone Age of Humanism were still held in respect during the Thirty Years War; both in Germany and Italy, as the case of Galileo will show.

The third plague was the garrisoning of Linz by Bavarian

soldiery. Soldiers were billeted everywhere, even on Kepler's printing shop. This led to a rumour which spread through the Republic of Letters, and penetrated as far as Danzig, that the soldiers had melted down Kepler's lead type to make bullets, and pulped his manuscripts for use as cartridge cases – but luckily this was not true.

Next, the Lutheran peasantry rose in bloody revolt, burnt down monasteries and castles, occupied the township of Wels, and laid siege on Linz. The siege lasted for two months, June–August 1626. There were the usual epidemics, and the populace was reduced to living on horse-flesh, but Kepler 'by the help of God and the protection of my angels' was preserved from this fate.

> You ask me [he wrote to Father Guldin] what I did with myself during the long siege. You ought to ask what one could do in the midst of the soldiery. The other houses had only a few soldiers billeted on them. Ours is on the city wall. The whole time the soldiers were on the ramparts, a whole cohort lay in our building. The ears were constantly assailed by the noise of the cannon, the nose by evil fumes, the eye by flames. All doors had to be kept open for the soldiers who, by their comings and goings, disturbed sleep at night, and work during day-time. I nevertheless considered it a great boon that the head of the Estates had given me rooms with a view over the moats and suburbs in which the fighting took place.[3]

When he did not watch the fighting, Kepler, in his unquiet study, was engaged with his old occupational therapy, the writing of a chronological work.

On 30 June, however, the peasants succeeded in setting fire to part of the town. It destroyed seventy houses, and among these was the printing shop. All the sheets that had so far been printed, went up in flame; but the angels again intervened and Kepler's manuscript escaped unscathed. This provided him with an occasion for one of his endearing understatements: 'It is a strange fate which causes these delays all the time. New incidents keep occurring which are not at all my fault.'[4]

Actually, he was not too much aggrieved by the destruction of the printing press, because he had had more than enough of Linz, and was only waiting for a pretext to move elsewhere.

He knew of a good press in Ulm, on the upper reaches of the Danube, which belonged to his Swabian homeland, and was less than fifty miles from Tuebingen – that magnetic pole which never lost its attraction. When the siege was lifted and the Emperor's consent obtained, Kepler was able, after fourteen long years, to leave Linz, which he had never liked, and which had never liked him.

But the printer at Ulm turned out a disappointment. There were quarrels from the start, and later on threats of a lawsuit. At one point, Kepler even left Ulm in a sudden huff to find a better printer – in Tuebingen, of course. He travelled on foot, because he again suffered from boils on his backside, which made riding a horse too painful. The time was February, and Kepler was fifty-six. In the village of Blaubeuren, having walked fifteen miles, he turned back and made peace with the printer (whose name was Jonas Saur, meaning sour).

Seven months later, in September 1627, the work was at long last completed. It was just in time for the annual book mart at the Frankfurt Fair. Kepler, who had bought the paper, cast some of the type, acted as printer's foreman, and paid for the whole enterprise, now travelled himself to Frankfurt, with part of the first edition of a thousand copies, to arrange for its sale. It was truly a one-man show.

The last of the Egyptian plagues he had to contend with were Tycho's heirs, who now reappeared on the scene. The Junker Tengnagel had died five years before, but George de Brahe, the misfired 'Tychonides', had continued the guerilla warfare against Kepler through all these years. He understood nothing of the contents of the work, but he objected to the fact that Kepler's preface occupied more space than his own, and to Kepler's remark that he had improved Tycho's observations, which he regarded as a slur on his father's honour. Since the work could not be published without the heirs' consent, the first two sheets, containing the dedications and prefaces, had to be reprinted twice; as a result, there exist three different versions among the surviving copies of the book.

The *Tabulae Rudolphinae* remained, for more than a century, an indispensable tool for the study of the skies – both planets and fixed stars. The bulk of the work consists of the tables and rules for predicting the positions of the planets, and of Tycho's catalogue of 777 star places, enlarged by Kepler to 1,005. There are also refraction-tables and logarithms,[5] put for the first time to astronomic uses; and a gazetteer of the towns of the world, their longitudes referred to Tycho's Greenwich – the meridian of Uraniborg on Hveen.

The frontispiece, designed by Kepler's hand, shows a Greek temple, under the columns of which five astronomers are engaged in lively dispute: an ancient Babylonian, Hipparchus, Ptolemy, Canon Koppernigk, and Tyge de Brahe. In a wall at the base of the temple, under the five immortals' feet, there is a small niche in which Kepler crouches at a rough-hewn working table, mournfully gazing at the onlooker, and to all intents like one of Snow-White's Seven Dwarfs. The table-cloth in front of him is covered with numbers, penned by a quill within reach of his hand, indicating the fact that he has no money to buy paper. Over the top of the dome-shaped roof hovers the Imperial Eagle, dropping gold ducats from its beak, a symbol of imperial largesse. Two of the ducats have landed on Kepler's table-cloth, and two more are falling through the air – a hopeful hint.

## 2. *The Tension Snaps*

The last three years of Kepler's life carry haunting echoes of the legend of the Wandering Jew. *Quis locus eligendus, vastatus an vastandus?* – 'What place shall I choose, one that is destroyed, or one that is going to be destroyed?' He had left Linz forever, and he was without a fixed domicile. Ulm was only a temporary station, for the duration of the printing. He was staying in a house that a friend had put at his disposal, and though it had been specially altered to accommodate Kepler's family, he did not have them with him. On the journey up the Danube from Linz, the river had started to freeze, and he had to continue by carriage, leaving Susanna and the children midways at Ratisbon. At least that is the explanation he gave in a letter to a correspondent; but he stayed in Ulm nearly ten months, and did not send for them.

This episode is characteristic of a certain oddness in his behaviour towards the end. It looks as if the heritage of his vagrant father and uncles was reasserting itself in his late middle age. His restlessness had found an outlet in creative achievement; when he finished the Rudolphine Tables, the tension snapped, the current was cut off, and he seemed to be freewheeling in aimless circles, driven on by an evergrowing, overriding anxiety. He was again plagued by rashes and boils; he was afraid that he would die before the printing of the Tables was finished; and the future was a waste land of famine and despair.

And yet, in spite of the war, his plight was to a large part imaginary. He had been offered the most coveted Chair in Italy, and Lord Bacon's envoy, Sir Henry Wotton, had invited him to England.* Yet he had refused:

> Am I to go overseas where Wotton invites me? I a German? I who love the firm Continent and who shrink at the idea of an island in narrow boundaries of which I feel the dangers in advance? [7]

After rejecting these tempting offers, he asked in despair his friend Bernegger in Strasburg whether he could get him a modest lectureship at that University. To attract an audience, he would be willing to cast the horoscope of every one of his hearers – because 'the threatening attitude of the Emperor, which is apparent in all his words and deeds', left him with hardly any other hope. Bernegger wrote back that his town and University would welcome Kepler with open arms if he were to honour them with his presence, and offered him unlimited personal hospitality in his spacious house with its 'very beautiful garden'. But Kepler refused 'because he could not afford the expense of the journey'; and when Bernegger tried to cheer him up with the news that a portrait of Kepler was hung on the wall of the University library: 'everybody who visits the library sees it. If only they could see you in person!', Kepler's reaction was that the portrait 'should be removed from that public place, the more so as it has hardly any likeness to myself'.[8]

* Kepler had dedicated the *Harmonice Mundi* to James I.

### 3. *Wallenstein*

The Emperor's hostility, too, existed in Kepler's imagination only. In December 1627, Kepler left Ulm for Prague – having been almost constantly on the move since the Frankfurt Fair – and was received, to his surprise, as *persona grata*. The Court had returned to Prague for the coronation of the Emperor's son as King of Bohemia. Everybody was in high spirits: Wallenstein, the new Hannibal, had expelled the Danish invaders from Prussia; he had overrun Holstein, Schleswig, and Jutland, and the enemies of the Empire were everywhere in retreat. Wallenstein himself had arrived in Prague a few weeks before Kepler; he was awarded, in addition to the Duchy of Friedland which he already held, the Duchy of Sagan in Silesia.

The Emperor's Generalissimo and his Mathematicus had crossed each other's path before. Wallenstein was addicted to astrology. Twenty years earlier, in Prague, Kepler had been requested, by a go-between, to cast the nativity of a young nobleman who wished to remain unnamed. Kepler wrote a brilliant character-analysis of the future war leader, who was then twenty-five, which testifies to his psychological insight – for Kepler had guessed the identity of his anonymous client.* Sixteen years later, he was asked, again through a middle-man, to expand the horoscope – which Wallenstein had profusely annotated on the margin – this time without the pretence of anonymity. Kepler had again obliged, but had saved his face with the usual warnings against the abuses of astrology. This second horoscope, which dates from 1624, stops at 1634 with the prophecy that March will bring 'dreadful disorders over the land': Wallenstein was murdered on 25 February of that year.†

Thus the ground was prepared for their meeting amidst the celebrations at Prague. The meeting ended, after lengthy nego-

* The name Wallenstein is written in Kepler's secret code on the original draft of the horoscope, which is still extant.

† But ten years made a round figure at which even a well-paid-for horoscope could reasonably stop.

tiations, with Kepler's appointment as Wallenstein's private mathematicus in his newly acquired Duchy of Sagan. The Emperor had no objection, and Kepler was allowed to retain his title as Imperial Mathematician for what it was worth – in terms of hard cash certainly not much; for the Crown's debts to Kepler in arrears of salary and gratuities by now amounted to 11,817 florins. The Emperor politely informed Wallenstein that he expected the latter to pay this sum – which Wallenstein, of course, never did.

The deal with Wallenstein concluded, both men left Prague in May 1628: Wallenstein to lay unsuccessful siege to Strahlsund, which was the beginning of his downfall; Kepler to visit his wife and children who were still in Ratisbon. He travelled on to Linz, to liquidate his affairs, then back to Prague, where his family joined him, and in July arrived with them in Sagan. But a considerable part of his possessions, including books and instruments needed for his work, he left behind in storage. It was the half-hearted move of an already broken man, whose behaviour became more and more erratic and devious.

Compared to Sagan, Linz had been paradise:

> I am a guest and a stranger here, almost completely unknown, and I hardly understand the dialect of the locals, who in turn consider me a barbarian. . . .[9]
>
> I feel confined by loneliness, far away from the great cities of the Empire; where letters come and go slowly, and at heavy expense. Add to this the agitations of the [counter-] reformation which, though I am not personally hit, did not leave me untouched. Sad examples are before me, or before my mind's eye, of acquaintances, friends, people of my immediate neighbourhood being ruined, and conversation with the terrorstricken is cut off by fear. . . .
>
> A little prophetess of eleven in Kottbuss, which is between here and Frankfort-on-the-Oder, threatens with the end of the world. Her age, her childish ignorance, and her enormous audiences make people believe in her.[10]

It was the same story as in Gratz and Linz: the people were compelled to become Catholics, or to leave the country. They were not even allowed to follow a Lutheran hearse to the cemetery. The privileged position that Kepler enjoyed only intensified his loneliness. He was a prisoner of constant, nagging anxieties about matters large and small:

> It seems to me that there is disaster in the air. My agent Eckebrecht in Nuremburg, who handles all my affairs, has not written for two months. ... I am worried about everything, about my account in Linz, about the distribution of the Tables, about the nautical chart for which I have given a hundred and twenty florins to my agent, about my daughter, about you, about the friends in Ulm.[11]

There was, of course, no printing press in Sagan, so he set out again on travels to procure type, machinery, and printers. This took nearly eighteen months out of the altogether two years, the last of his life, that he spent in Sagan:

> Amidst the collapse of towns, provinces, and countries of old and new generations, in the fear of barbaric raids, of the violent destruction of hearth and home, I see myself obliged, a disciple of Mars though not a youthful one, to hire printers without betraying my fear. With the help of God I shall indeed bring this work to an end, in a soldierly fashion, giving my orders with bold defiance and leaving the worry about my funeral to the morrow.[12]

## 4. *Lunar Nightmare*

When the new press was installed in December 1629, in Kepler's own lodgings, he embarked (with his assistant, Bartsch, whom Kepler had bullied into marrying his daughter Susanna) on a remunerative enterprise: the publication of ephemerides* for the years 1629–36. Ever since the *Rudolphine Tables* had come out, astronomers all over Europe were competing to publish ephemerides, and Kepler was anxious 'to join the race' as he said, on the race track that he had built. But in between, he also began the printing of an old, favourite brain-child of his: *Somnium* – a dream of a journey to the moon. He had written

* 'Ephemerides' provide detailed information about the motions of the planets for a given year, whereas 'tables' give only the broad outlines on which the calculations are based.

it some twenty years before, and from time to time had added notes to it, until these had far outgrown the original text.

The *Somnium* remained a fragment; Kepler died before he finished it, and it was only published posthumously, in 1634. It is the first work of science-fiction in the modern sense – as opposed to the conventional type of fantasy-utopias from Lucian to Campanella. Its influence on later authors of inter-planetary journeys was considerable – from John Wilkins' *Discovery of a New World* and Henry More right down to Samuel Butler, Jules Verne, and H. G. Wells.[13]

*Somnium* starts with a prelude full of autobiographical allusions. The boy Duracotus lived with his mother Fiolxhilda on Iceland 'which the ancients called Thule'.* The father had been a fisherman, who had died at the age of one hundred and fifty when the boy was only three. Fiolxhilda sold herbs in little bags of ram-skin to the seamen and conversed with demons. At fourteen, the boy curiously opened one of the little bags, whereupon his mother, in a fit of temper, sold him to a seafaring captain. The captain left him on the Isle of Hveen, where for the next five years Duracotus studied the science of astronomy under Tycho de Brahe. When he returned home, his repentant mother, as a treat, conjured up one of the friendly demons from Lavania† – the moon – in whose company selected mortals might travel to that planet.

> After completing certain ceremonies, my mother, commanding silence with her outstretched hand, sat down beside me. No sooner had we, as arranged, covered our heads with a cloth, when a hoarse, supernatural voice began to whisper, in the Icelandic language, as follows. . . .

Thus ends the prelude. The journey itself, the demon explains, is only possible during an eclipse of the moon, and must therefore be completed in four hours. The traveller is propelled by the spirits, but he is subject to the laws of physics; it is at this point that science takes over from fantasy:

* Kepler chose the name Duracotus because it sounded Scottish 'and Scotland lies on the Islandic ocean'; 'Fiolx' was the name for Iceland which he saw on an old map.

† From *Lavanah* – the Hebrew name of the moon. (Lavan=white.)

The initial shock [of acceleration] is the worst part of it, for he is thrown upward as if by an explosion of gunpowder. ... Therefore he must be dazed by opiates* beforehand; his limbs must be carefully protected so that they are not torn from him and the recoil is spread over all parts of his body. Then he will meet new difficulties: immense cold and inhibited respiration. ... When the first part of the journey is completed, it becomes easier because on such a long journey the body no doubt escapes the magnetic force of the earth and enters that of the moon, so that the latter gets the upper hand. At this point we set the travellers free and leave them to their own devices: like spiders they will stretch out and contract, and propel themselves forward by their own force – for, as the magnetic forces of the earth and moon both attract the body and hold it suspended, the effect is as if neither of them were attracting it – so that in the end its mass will by itself turn towards the moon.

In the *Astronomia Nova* Kepler had come so close to the concept of universal gravity that one had to assume the existence of a psychological blockage which made him reject it. In the passage just quoted he not only takes it for granted, but, with truly astonishing insight, postulates the existence of 'zones of zero gravity' – that nightmare of science fiction. Later on in the *Somnium* he took a further step in the same direction by assuming that there are spring tides on the moon, due to the joint attraction of sun and earth.

The journey completed, Kepler proceeds to describe conditions on the moon. A lunar day, from sunrise to sunset, lasts approximately a fortnight, and so does a moon-night – for the moon takes a month to turn once round its axis, the same time it takes to complete a revolution round the earth. As a result, it turns always the same face to the earth, which the moon-creatures call their 'volva' (from *revolvere*, to turn). This face of the moon they call the Subvolvan half, the other is the Prevolvan half. Common to both halves is that their year consists of twelve days-and-nights, and the resulting dreadful contrasts of temperature – scorching days, frozen nights. Common to both are also the queer motions of the starry sky – the sun and planets scuttle incessantly back and forth, a result

* It has recently been suggested that space-travellers should be anaesthetized during the initial acceleration.

of the moon's gyrations round the volva. This 'lunatic' astronomy – in the legitimate double-meaning of the word – which Kepler develops with his usual precision, is sheer delight; nobody before (nor since, as far as I know) had attempted such a thing. But when it comes to conditions on the moon itself, the picture becomes grim.

The Prevolvans are the worst off. Their long nights are not made tolerable by the presence of the huge volva, as on the other hemisphere, for the Prevolvans of course never see the earth. Their nights are 'bristling with ice and snow under the raging, icy winds.' The day that follows is no better: for a fortnight the sun never leaves the sky, heating the air to a temperature 'fifteen times hotter than our Africa'.

The Subvolvans are a little better off because the huge volva softens their nights by reflecting some of the light and heat [14] of the sun. The volva's surface is fifteen times that of our moon, and it stays always in the same place on the sky 'as if nailed on', but growing and waning from full-volva to new volva, as our moon. At full-volva, Africa appears as a human head severed at its shoulders; Europe, a girl in a long robe, bends down to kiss it while her long arm, stretched backwards, lures a jumping cat towards her.*

The mountains of Lavania are much higher than those on earth; so are the plants and the creatures that inhabit it. 'Growth is rapid; everything is short-lived because it develops to such an enormous bodily mass. . . . Growth and decay takes place on a single day.' The creatures are mostly like gigantic serpents. 'The Prevolvans have no fixed and safe habitations; they traverse in hordes, in a single day, the whole of their world, following the receding waters either on legs that are longer than those of our camels, or on wings, or in ships.' Some are divers, and breathe very slowly, so that they can take refuge from the scorching sun at the bottom of the deep waters. 'Those that remain on the surface are boiled by the midday sun and serve as nourishment for the approaching

* The back of the head is the Sudan, its chin Algeria; the girl's head is Spain, her open mouth at Malaga, her chin at Murcia; her arms are Italy and the British Isles, the latter luring the Scandinavian cat.

nomadic hordes. . . . Others who cannot live without breathing, retreat into caves which are supplied with water by narrow canals so that the water may gradually cool on its long way and they may drink it; but when the night approaches, they go out for prey.' Their skin is spongy and porous; but when a creature is taken unawares by the heat of the day, the skin becomes hard and scorched, and falls off in the evening. And yet they have a strange love for basking in the sun at noon – but only close to their crevices, to be able to make a swift and safe retreat. . . .

In a short appendix, the Subvolvans are allowed cities surrounded by circular walls – the craters of the moon; but Kepler is only interested in the engineering problems of their construction. The book ends with Duracotus being woken by a cloudburst from his dream – or rather, from his nightmare of prehistoric giant reptiles, of which Kepler had, of course, no knowledge whatsoever. No wonder that Henry More was inspired by the *Somnium* to a poem called *Insomnium Philosophicum.* But more amusing is Samuel Butler's paraphrase on Kepler in 'The Elephant in the Moon':

> Quoth he – Th' Inhabitants of the Moon,
> Who when the Sun shines hot at Noon,
> Do live in Cellars underground
> Of eight Miles deep and eighty around
> (In which at once they fortify
> Against the Sun an th' Enemy)
> Because their People's civiler
> Than those rude Peasants, that are found
> To live upon the upper Ground,
> Call'd Privolvans, with whom they are
> Perpetually at open War.

Although most of the *Somnium* was written much earlier, one readily understands why it was the last book on which he worked, and which he wished to see in print. All the dragons which had beset his life – from the witch Fiolxhilda and her vanished husband, down to the poor reptilian creatures in perpetual flight, shedding their diseased skin, and yet so anxious to bask under an inhuman sun – they are all there, projected into a cosmic scenery of scientific precision and rare, original beauty. All Kepler's work, and all his discoveries, were acts of catharsis; it was only fitting that the last one should end with a fantastic flourish.

## 5. *The End*

Wallenstein could not care less what Kepler did. The arrangement had been a mutual disappointment from the beginning. Unlike the aristocratic dilettantes who had patronized Tycho, Galileo, and Kepler himself in the past, General Wallenstein took no genuine interest in science. He drew a certain snob-satisfaction from having a man of European renown as his court mathematicus, but what he really wanted from Kepler was astrological advice regarding the political and military decisions he had to take. Kepler's answer to such concrete questions were always elusive – owing to his honesty, or caution, or both. Wallenstein used Kepler mainly to obtain certain exact data on the planetary motions, which he then sent on to his more willing astrologers – like the notorious Seni – as a basis for their auguries. Kepler himself rarely spoke about his personal contacts with Wallenstein. Though he once calls him 'a second Hercules',[15] his feelings were more honestly reflected in one of his last letters:

> I have returned recently from Gitschin [Wallenstein's residence], where my patron kept me in attendance for three weeks – it meant a considerable waste of time for both of us.[16]

Three months later, the pressure of Wallenstein's rivals induced the Emperor to dismiss his Generalissimo. It was only a temporary set-back in Wallenstein's dramatic career, but Kepler believed that it was the end. Once again, and now for the last time, he took to the roads.

In October, he departed from Sagan. He left his family behind, but took with him cartloads of books and documents, which were dispatched ahead to Leipzig. His son-in-law wrote later on: 'Kepler left Sagan unexpectedly, and his condition was such that his widow, his children, and friends expected to see the Last Judgment sooner than his return.'[17]

His purpose was to look out for another job, and to try to obtain some of the money owed to him by the Emperor and by the Austrian Estates. In his self-analysis, thirty-five years earlier, he had written that his constant worrying about money 'was not prompted by the desire for riches, but by fear of poverty'. This was still essentially true. He had money-deposits in various places, but he was unable to recover even the interests due to him. When he set out on that last journey across half of war-torn Europe, he took all the cash he had with him, leaving Susanna and the children penniless. Even so he had to borrow fifty florins from a merchant in Leipzig, where he stopped on the first lap of his journey.

He seems to have had one of his curious premonitions. All his life he had been in the habit of casting horoscopes for his birthdays. The horoscopes for the years preceding and following his sixtieth show merely the positions of the planets, without comment. The sixtieth, his last, is an exception; he noted on it that the positions of the planets were almost the same as at his birth.

His last letter is dated from Leipzig, 31 October, and addressed to friend Bernegger in Strasburg. He had remembered Bernegger's earlier invitation, and suddenly decided to accept it; but he seems to have forgotten it again a moment later, for in the remainder of the letter he talks of his travelling plans without any reference to it:

> Your hospitality I gladly accept. May God preserve you, and take pity on the misery of my country. In the present general insecurity one ought not to refuse any offer of shelter, however distant its location. ... Farewell to you, your wife and children. Hold fast, with me, to our only anchor, the Church, pray to God for it and for me.[18]

From Leipzig he rode on, on a miserable old horse to Nuremberg, where he visited a printer. Then on to Ratisbon, where the Diet was sitting, in full pomp, presided over by the Emperor who owed him twelve thousand florins.

He arrived in Ratisbon on 2 November. Three days later he took to bed with a fever. An eye-witness reported that 'he did not talk, but pointed his index finger now at his head, now at the sky above him'.[19] Another witness, the Lutheran preacher, Jacob Fischer, wrote in a letter to a friend:[20]

During the recent session of the Diet, our Kepler arrived in this town on an old jade (which he subsequently sold for two florins). He was only three days here when he was taken ill with a feverish ailment. At first he thought that he was suffering from *sacer ignis*, or fever pustules, and paid no attention to it. When his feverish condition worsened, they bled him, without any result. Soon his mind became clouded with ever-rising fever. He did not talk like one in possession of his faculties. Several preachers visited him and comforted him with the living waters of their sympathy.[21] In his last agony, as he gave up his ghost to God, a Protestant clergyman of Ratisbon, Sigismund Christopher Donavarus, a relative of mine, consoled him in a manly way, as behoves a servant of God. This happened on 15 November 1630. On the 19th he was buried in the cemetery of St Peter, outside the town.

The cemetery was destroyed during the Thirty Years War, and Kepler's bones were scattered; but the epitaph which he wrote for himself is preserved:

*Mensus eram coelos, nunc terrae metior umbras*
*Mens coelestis erat, corporis umbra iacet.*

I measured the skies, now the shadows I measure
Skybound was the mind, earthbound the body rests.

There is also a paragraph in one of his late letters which lingers on in memory; it is dated:

Sagan in Seliseia, in my own printing press, 6 November 1629:
When the storm rages and the state is threatened by shipwreck, we can do nothing more noble than to lower the anchor of our peaceful studies into the ground of eternity.[22]

# Part Five. The Parting of the Ways

# 1 The Burden of Proof

## 1. *Galileo's Triumph*

Once again the climate and character of this narrative must change. Personalities, intrigues, points of legal procedure will dominate the scene as we turn to the tragic conflict between the new cosmology and the Church.

Few episodes in history have given rise to a literature as voluminous as the trial of Galileo. Most of it has, unavoidably, a partisan character, ranging from crude distortion, through gentle innuendo, to attempts at impartiality thwarted by unconscious bias. Objectivity is an abstract ideal in an age which has become 'a divided house of faith and reason'; and more especially so when the episode to be treated is one of the historic causes of that division. Since it would be foolish to claim exemption from this rule, it may be as well to state my own bias before asking the reader to take on trust my brand of objectivity. Among my earliest and most vivid impressions of History was the wholesale roasting alive of heretics by the Spanish inquisition, which could hardly inspire tender feelings towards that establishment. On the other hand, I find the personality of Galileo equally unattractive, mainly on the grounds of his behaviour towards Kepler. His dealings with Urban VIII and the Holy Office can be judged in various ways, because the evidence on some vital points is based on hearsay and conjecture; but of his relationship with his German colleague, confined to a few letters, we have an unequivocal record. As a result, most of the biographers of Kepler show the same aversion towards Galileo, whereas the admirers of Galileo display towards Kepler a kind of guilty tenderness, which betrays their embarrassment.

It seems to me, then, that in so far as bias enters into the narrative, it is not based on affection for either party in this conflict, but on resentment that the conflict did occur at all.

One of the points that I have laboured in this book is the unitary source of the mystical and scientific modes of experience; and the disastrous results of their separation. It is my conviction that the conflict between Church and Galileo (or Copernicus) was not inevitable; that it was not in the nature of a fatal collision between opposite philosophies of existence, which was bound to occur sooner or later, but rather a clash of individual temperaments aggravated by unlucky coincidences. In other words, I believe the idea that Galileo's trial was a kind of Greek tragedy, a showdown between 'blind faith' and 'enlightened reason', to be naïvely erroneous. It is this conviction – or bias – that informs the following narrative.

I shall take up the thread of Galileo's life at the point where his name suddenly burst into world fame through his discovery of the Jupiter planets. The *Star Messenger* was published in March 1610; in September, he took up his new post as 'Chief Mathematician and Philosopher' to the Medicis in Florence; the following spring he spent in Rome.

The visit was a triumph. Cardinal del Monte wrote in a letter: 'If we were still living under the ancient Republic of Rome, I verily believe that there would have been a column on the Capital erected in Galileo's honour.'[1] The select *Accadèmia dei Lincei* (the lynx-eyed), presided by Prince Federico Cesi, elected him a member and gave him a banquet; it was at this banquet that the word 'telescope' was for the first time applied to the new invention.[2] Pope Paul V received him in friendly audience, and the Jesuit Roman College honoured him with various ceremonies which lasted a whole day. The chief mathematician and astronomer of the College, the venerable Father Clavius, principal author of the Gregorian Calendar reform, who at first had laughed at the *Star Messenger*, was now entirely converted; so were the other astronomers at the College, Fathers Grienberger, van Maelcote, and Lembo. They not only accepted Galileo's discoveries, but improved on his observations, particularly of Saturn and the phases of Venus. When the head of the College, the Lord Cardinal Bellarmine, asked for their official opinion on the new discoveries, they unanimously confirmed them.

This was of utmost importance. The phases of Venus, confirmed by the doyen of Jesuit astronomers, were incontrovertible proof that at least that planet revolved round the sun, that the Ptolemaic system had become untenable, and that the choice now lay between Copernicus and Brahe. The Jesuit Order was the intellectual spearhead of the Catholic Church. Jesuit astronomers everywhere in Europe – particularly Scheiner in Ingoldstadt, Lanz in Munich, Kepler's friend Guldin in Vienna, and the Roman College in a body – began to support the Tychonic system as a half-way house to the Copernican. The Copernican system itself could be freely discussed and advocated as a working hypothesis, but it was unfavourably viewed to present it as established truth, because it seemed contrary to current interpretation of scripture – unless and until definite proof could be adduced in its favour. We shall have to return more than once to this crucial point.

Within a brief period, Jesuit astronomers also confirmed the 'earthly' nature of the moon, the existence of sunspots, and the fact that comets moved in outer space, beyond the moon. This meant the abandonment of the Aristotelian doctrine of the perfect and unchangeable nature of the celestial spheres. Thus the intellectually most influential order within the Catholic Church was at that time in full retreat from Aristotle and Ptolemy, and had taken up an intermediary position regarding Copernicus. They praised and fêted Galileo, whom they knew to be a Copernican, and they kept Kepler, the foremost exponent of Copernicanism, under their protection throughout his life.

But there existed a powerful body of men whose hostility to Galileo never abated: the Aristotelians at the universities. The inertia of the human mind and its resistance to innovation are most clearly demonstrated not, as one might expect, by the ignorant mass – which is easily swayed once its imagination is caught – but by professionals with a vested interest in tradition and in the monopoly of learning. Innovation is a twofold threat to academic mediocrities: it endangers their oracular authority, and it evokes the deeper fear that their whole, laboriously constructed intellectual edifice might collapse. The academic backwoodsmen have been the curse of genius from Aristarchus to Darwin and Freud; they stretch, a solid and hostile phalanx of pedantic mediocrities, across the centuries. It was this threat – not Bishop Dantiscus or Pope Paul III – which had cowed Canon Koppernigk into lifelong silence. In Galileo's case, the phalanx resembled more a rearguard – but a rearguard still firmly entrenched in academic chairs and preachers' pulpits.

> ... There remain in opposition to my work some stern defenders of every minute argument of the Peripatetics. So far as I can see, their education consisted in being nourished from infancy on the opinion that philosophizing is and can be nothing but to make a comprehensive survey of the texts of Aristotle, that from divers passages they may quickly collect and throw together a great number of solutions to any proposed problem. They wish never to raise their eyes from those pages – as if this great book of the universe had been written to be read by nobody by Aristotle, and his eyes had been destined to see for all posterity.[3]

After his return, in the summer of 1611, from his Roman triumph to Florence, Galileo became immediately involved in several disputes. He had published a treatise on '*Things that Float on Water*' – a title that sounds harmless enough. But in this pioneer work on modern hydrostatics Galileo had embraced Archimedes' view that bodies float or sink according to their specific gravity, against the Aristotelian view that this depends on their shape. The backwoodsmen were out at once in full cry, swinging their stone axes. They were the more irate as Galileo, instead of letting the facts speak for themselves, had employed his favourite trick of anticipating the peripatetics' arguments, building them up in a mock-serious manner, and then demolishing them with glee. Their leader was a certain Lodovico delle Colombe, meaning dove; hence the name 'pigeon-league' by which Galileo and his friends called their opponents. The Aristotelians published four books in six months to refute the *Discourse on Things that Float on Water*, and the controversy went on for nearly three years. It ended in a complete rout of the attackers, both spiritual and physical.

Professors Palmerini and di Grazzia died while Galileo was preparing his riposte. Giorgio Coressio lost his chair at Pisa because he was discovered to adhere secretly to the Greek church, and went insane; the monk Francesco Sizzi, a young fanatic, who had attacked Galileo's telescopic discoveries but defended his Floating Bodies, was broken on the wheel in Paris for writing a pamphlet against the King of France.

Incidentally, the famous experiment of dropping canon balls from the leaning Tower of Pisa was carried out not by Galileo but by his opponent, the aforementioned Coressio, and not in refutation, but in confirmation of the Aristotelian view that larger bodies must fall quicker than smaller ones.[4]

## 2. *The Sunspots*

The next year (1612) brought a new controversy with more serious consequences. It concerned the sunspots.

The affair starts at Ingoldstadt in Bavaria, where Father Scheiner, a Jesuit astronomer of great repute, and his young assistant Cysat, profiting from a thick mist, turned their telescope directly at the sun. It was Cysat's turn first, who to his amazement discovered 'several black drops' on the face of the sun. He exclaimed: 'Either the sun sheddeth tears or she is blemished by spots.'[5] Then he yielded the instrument to his teacher.

After continued observations, Father Scheiner reported on his sensational discovery in several letters to Marcus Welser in Augsburg, a Maecenas of science, who also patronized Kepler. Welser had the letter promptly printed, under the pseudonym 'Apelles', as Scheiner had requested. Welser then sent the booklet to both Kepler and Galileo, for their opinions.

Kepler answered immediately. He recalled having himself observed a sunspot in 1607 'of the size of a meagre flea', which he had mistakenly assumed to be Mercury passing in front of the sun.[6] He laughed at his mistake, then quoted reports of similar observations dating back to the days of Charlemagne; then gave his opinion that the spots were a kind of dross, due to the cooling of the sun in patches.

Galileo delayed his answer for more than three months, and then claimed the priority of the discovery for himself. He alleged having observed sunspots for about eighteen months, and having shown them a year before 'to many prelates and gentlemen in Rome', but did not name any of these witnesses.

In fact, the sunspots had been discovered independently and nearly at the same time by Johannes Fabricius in Wittenberg, Thomas Harriot in Oxford, Scheiner-Cysat, and Galileo himself. Harriot seems to have been the first to observe them, but Fabricius was the first to publish, and Scheiner the second. Harriot, Fabricius, and Scheiner neither knew of the others' parallel discovery, nor did they raise any particular claim to priority. Thus Galileo's claim was untenable, firstly because Fabricius and Scheiner had been first to publish the discovery, and secondly because he could name no witnesses, or correspondents, to prove it – yet we remember how careful he was to protect his priority claims on previous occasions, by immediately sending out messages in anagram form. But Galileo had come to regard telescopic discoveries as his exclusive monopoly – as he himself stated on a later occasion:

> You cannot help it, Mr Sarsi, that it was granted to me alone to discover all the new phenomena in the sky and nothing to anybody else. This is the truth which neither malice nor envy can suppress.[7]

By his specious priority claim over the sunspots, followed by disguised attacks on Father Scheiner, Galileo had made the first enemy among the Jesuit astronomers, and started the fatal process which in the end would turn the order against him.

The whole affair was the more unfortunate, as Galileo's answer to Marcus Welser was otherwise a model of clarity and scientific method. He followed it up with two more *Letters on Sunspots*, which, the next year, were published under that title. He showed convincingly that the spots were not small planets circling the sun, as Scheiner had originally assumed, but located on, or close to, the surface of the sun itself; that they were rotating with the sun, constantly changing their shapes, and of the nature of 'vapours, or exhalations, or clouds or fumes'.[8] Thus it was proven that not only the moon, but the sun too was subject to generation and decay.

The booklet also contained Galileo's first, tentative formulation of the principle of inertia,[8a] and his first printed statement in favour of the Copernican system. Up to this date – we are now in 1613, and he is nearly fifty – he had defended Copernicus in conversations at dinner tables, but never in print. The passage in question is on the last page of the *Letters on Sunspots*; it starts with a reference to the alleged moons of Saturn and continues:

> And perhaps this planet also, no less than horned Venus, harmonizes admirably with the great Copernican system, to the universal revelation of which doctrine propitious breezes are now seen to be directed towards us, leaving little fear of clouds or crosswinds.[9]

Here it was at last, the first public commitment, though somewhat vague in form, a full quarter-century after Kepler had first sounded the Copernican trumpet in the *Mysterium*.

The book won immediate and great popular acclaim. In so far as the Church is concerned, not only was no voice raised in opposition, but Cardinals Boromeo and Barberini – the future Urban VIII – wrote letters to Galileo expressing their sincere admiration.

Not so the backwoodsmen. When Galileo's favourite pupil, the Benedictine Father Castelli (the founder of modern hydrodynamics) was called to the chair of the University of Pisa, he was expressly forbidden by the head of the University to teach the motion of the earth. The head was Arturo d'Elci, a fanatical Aristotelian and member of the 'Pigeon League', who had published one of the pamphlets against the *Things that Float on Water*.

The first serious attack against Copernicanism on *religious* grounds came also not from clerical quarters but from a layman – none other than delle Colombe, the leader of the league. His treatise *Against the Motion of the Earth* contained a number of quotations from Holy Scripture to prove that the earth was in the centre of the world. It was circulated in manuscript in 1610 or 1611, before Galileo's public committal, and did not mention Galileo's name. Galileo himself was as yet so little worried about a possible theological conflict, that he had let almost a year pass before he asked the opinion of his friend, Cardinal Conti, on the matter. The Cardinal answered that, concerning the 'immutability' of the skies, Holy Scripture seemed to favour Galileo's view rather than Aristotle's. As for Copernicus, the 'progressive' (i.e. annual) motion was admissible, but the *daily* rotation did not seem to agree with Scripture, unless it was assumed that certain passages must not be taken literally; but such an interpretation was permissible 'only in the case of the greatest necessity'.[10]

'Necessity' in this context meant once again: if and when convincing proof was shown of the reality of the earth's motion. But all this did not affect free discussion of the relative advantages of the Ptolemaic, Tychonic, or Copernican systems as mathematical hypotheses.

There the matter could have rested, and probably would have rested, but for Galileo's hypersensitivity to criticism, and his irrepressible urge to get involved in controversy. Towards the end of 1612, he was staying in the villa, near Florence, of his friend Filippo Salviati (whom he immortalized in his two great Dialogues), when some gossip reached him to the effect that a Dominican Father, Niccolo Lorini, had attacked his views in a private conversation. Galileo immediately wrote to Lorini, asking for an explanation. Lorini was a gentleman of seventy, and professor of ecclesiastical history in Florence. He wrote back:

> I have never dreamt of getting involved in such matters. . . . I am at a loss to know what grounds there can be for such a suspicion, as this thing has never occurred to me. It is indeed true that I, not with a desire to argue, but merely to avoid giving the impression of a blockhead when the discussion was started by others, did say a few words just to show I was alive. I said, as I still say, that this opinion of Ipernicus – or whatever his name is – would appear to be hostile to divine Scripture. But it is of little consequence to me, for I have other things to do. . . .[11]

The next year, 1613, brought the publication of the 'Sunspots', and general public acclaim; including, as mentioned, the future Pope's. All was sunshine. Then another piece of gossip reached Galileo, this time from Pisa. It concerned an after-dinner conversation at Duke Cosimo's table. This banal incident was the beginning of what became 'the greatest scandal in Christendom'.

### 3. *The Shifting of the Burden*

Faithful Father Castelli, now Professor of Mathematics at Pisa, the post from which Galileo had started his career, had been invited for dinner at Court. An illustrious company was present, including the Duke's mother, the Dowager Duchess Christina of Lorraine, his wife Madeleine of Austria, and several other guests, among them Dr Boscaglia, a professor of philosophy. The conversation was led by Madame Christina, who appears to have conformed to the idea of a bossy, talkative, and scatterbrained Dowager. After dinner she felt the sudden urge 'to learn all about' those Medicean planets. First she wanted to know their positions, then whether they were real of just illusions. Both Castelli and Boscaglia solemnly confirmed that they were real. Soon after that, dinner was over, and Father Castelli left.

> But I hardly come out of the place when Madame Christina's porter overtook me and told me that she wished me to return [Castelli's report to Galileo continued]. Now before I tell you what ensued, you must first know that while we were at table, Dr Boscaglia had had the ear of Madame for a while; and conceding as true all the new things you have discovered in the sky, he said that only the motion of the earth had something incredible in it, and could not take place, in particular because Holy Scripture was obviously contrary to this view.

When Castelli returned to the drawing-room, 'Madame began, after some questions about myself, to argue Holy Scripture against me. Thereupon, after having made suitable disclaimers, I commenced to play the theologian and ... carried things off like a paladine.' Everybody took the side of Castelli and Galileo, 'only Madame Christina remained against me, but from her manner I judged that she did this only to hear my answers. Professor Boscaglia never said a word.'[12]

In subsequent letters, Castelli reported that Boscaglia had once more been defeated in debate, that even the irascible Dowager had been won over, and that the subject had been dropped.

This, then, is the incident which touched off the drama.

As on that previous occasion, when Lorini had remarked on 'Ipernicus – or whatever his name is', Galileo was at once up in arms. His counter-blast to the dinner-table chirpings of the obscure Dr Boscaglia (who is never heard of again), was a kind of theological atom bomb, whose radioactive fall-out is still being felt. It took the form of a *Letter to Castelli*, enlarged a year later into a *Letter to the Grand Duchess Christina*. It was intended to be widely circulated, which indeed it was. Its purpose was to silence all theological objections to Copernicus. Its result was the precise opposite: it became the principal cause of the prohibition of Copernicus, and of Galileo's downfall.

As a work of polemical literature the *Letter* is a masterpiece. It starts:*

> Some years ago, as Your Serene Highness well knows, I discovered in the heavens many things that had not been seen before our own age. The novelty of these things, as well as some consequences which followed from them in contradiction to the physical notions commonly held among academic philosophers, stirred up against me no small number of professors – as if I had placed these things in the sky with my own hands in order to upset nature and overturn the sciences. . . .
>
> Showing a greater fondness for their own opinions than for truth, they sought to deny and disprove the new things which, if they had cared to look for themselves, their own senses would have demonstrated to them. To this end they hurled various charges and published numerous writings filled with vain arguments, and they made the grave mistake of sprinkling these with passages taken from places in the Bible which they had failed to understand properly. . . .[13]

Galileo then developed the argument which Kepler, too, had constantly used, namely that certain statements in the Bible

* I shall follow the final version of the document, i.e. the *Letter to the Grand Duchess.*

should not be taken literally because they were couched in language 'according to the capacity of the common people who are rude and unlearned':

Hence, in expounding the Bible, if one were always to confine oneself to the unadorned grammatical meaning, one might fall into error. Not only contradictions and propositions far from true might thus be made to appear in the Bible, but even grave heresies and follies. Thus it would be necessary to assign to God feet, hands, and eyes, as well as corporeal and human affections, such as anger, repentance, hatred, and sometimes even the forgetting of things past and ignorance to come. ... For that reason it appears that nothing physical which sense-experience sets before our eyes, or which necessary demonstrations prove to us, ought to be called in question (much less condemned) upon the testimony of biblical passages which may have some different meaning beneath their words.[14]

In support of this thesis, Galileo quoted at length St Augustine as a witness – not realizing that, theologically, he was walking on extremely thin ice (*see below*, p. 450). Then comes a breathtaking passage, where one can almost hear the ice cracking under his feet:

... I question whether there is not some equivocation in failing to specify the virtues which entitle sacred theology to the title of 'queen'. It might deserve that name by reason of including everything that is learned from all the other sciences and establishing everything by better methods and with profounder learning. ... Or theology might be queen because of being occupied with a subject which excels in dignity all the subjects which compose the other sciences, and because her teachings are divulged in more sublime ways.

That the title and authority of queen belongs to theology in the first sense, I think will not be affirmed by theologians who have any skill in the other sciences. None of these, I think, will say that geometry, astronomy, music, and medicine are more excellently contained in the Bible than they are in the books of Archimedes, Ptolemy, Boethius, and Galen. Hence it seems likely that regal preeminence is given to theology in the second sense; that is, by reason of its subject and the miraculous communication, by divine revelation, of conclusions which could not be conceived by men in any other way, concerning chiefly the attainment of eternal blessedness.

Let us grant then that theology is conversant with the loftiest divine contemplation, and occupies the regal throne among the sciences by this dignity. But acquiring the highest authority in this way, if she does not descend to the lower and humbler speculations of the subordinate sciences and has no regard for them because they are not concerned with blessedness, then her professors should not arrogate to themselves the authority to decide on controversies in professions which they have neither studied nor practised. Why, this would be as if an absolute despot, being neither a physician nor an architect, but knowing himself free to command, should undertake to administer medicines and erect buildings according to his whim – at grave peril of his poor patients' lives, and the speedy collapse of his edifices. . . .[15]

While reading this superb manifesto of the freedom of thought, one tends to forgive Galileo his human failings. These, however, become only too apparent in the piece of special pleading which follows the passage I have quoted, and which was to have disastrous consequences.

After invoking Augustine's authority once more, Galileo draws a distinction between scientific propositions which are 'soundly demonstrated' (i.e. proven) and others which are 'merely stated'. If propositions of the first kind contradict the apparent meaning of passages in the Bible, then, according to theological practice, the meaning of these passages must be reinterpreted – as was done, for instance, with regard to the spherical shape of the earth. So far he has stated the attitude of the Church correctly; but he continues: 'And as to the propositions which are stated but not rigorously demonstrated, anything contrary to the Bible involved by them must be held undoubtedly false and should be proved so by every possible means.'[16]

Now this was demonstrably not the attitude of the Church. 'Propositions which are stated but not rigorously demonstrated', *such as the Copernican system itself*, were not condemned outright, if they seemed to contradict Holy Scripture; they were merely relegated to the rank of 'working hypotheses' (where they rightly belong), with an implied: 'wait and see; if you bring proof, then, but only then, we shall have to reinterpret

Scripture in the light of this necessity'. But Galileo did not want to bear the burden of proof; for the crux of the matter is, as will be seen, that he had no proof. Therefore, firstly, he conjured up an artificial black-or-white alternative, by pretending that a proposition must either be accepted or outright condemned. The purpose of this sleight of hand becomes evident from the next sentence:

Now if truly demonstrated physical conclusions need not be subordinated to biblical passages, but the latter must rather be shown not to interfere with the former, then *before a physical proposition is condemned it must be shown to be not rigorously demonstrated* – and this is to be done not by those who hold the proposition to be true, but by those who judge it to be false. This seems very reasonable and natural, for those who believe an argument to be false may much more easily find the fallacies in it than men who consider it to be true and conclusive....[17]

The burden of proof has been shifted. The crucial words are those in (my) italics. It is no longer Galileo's task to prove the Copernican system, but the theologians' task to disprove it. If they don't, their case will go by default, and Scripture must be reinterpreted.

In fact, however, there had never been any question of condemning the Copernican system as a working hypothesis. The biblical objections were only raised against the claim that it was *more* than a hypothesis, that it was rigorously proven, that it was in fact equivalent to gospel truth. The subtlety in Galileo's manoeuvre is that he does not explicitly raise the claim. He cannot do so, for he had not produced a single argument in support of it. Now we understand why he needed his black-or-white alternative as a first move: to distract attention from the true status of the Copernican system as an officially tolerated working hypothesis awaiting proof. Instead, by slipping in the ambiguous words 'physical proposition' at the beginning of the italicized passage, followed by the demand that 'it must be shown to be not rigorously demonstrated', he implied (though he did not dare to state it explicitly) that the truth of the system *was* rigorously demonstrated. It is all so subtly done that the trick is almost imperceptible to the reader and, as far as I know, has escaped the attention of students to this day. Yet it decided the strategy he was to follow in coming years.

Throughout the document Galileo completely evaded any astronomical or physical discussion of the Copernican system; he simply gave the impression that it was proven beyond doubt. If he had talked to the point, instead of around it, he would have had to admit that Copernicus' forty-odd epicycles and eccentrics were not only not proven but a physical impossibility, a geometrical device and nothing else; that the absence of an annual parallax, i.e. of any apparent shift in the position of the fixed stars, in spite of the new telescopic precision, weighed heavily against Copernicus; that the phases of Venus disproved Ptolemy, but not Herakleides or Tycho; and that all he could claim for the Copernican hypothesis was that it described certain phenomena (the retrogression) more economically than Ptolemy; as against this, the above-mentioned physical objections would have carried the day.

For it must be remembered that the system which Galileo advocated was the orthodox Copernican system, designed by the Canon himself, nearly a century before Kepler threw out the epicycles and transformed the abstruse paper-construction into a workable mechanical model. Incapable of acknowledging that any of his contemporaries had a share in the progress of astronomy, Galileo blindly and indeed suicidally ignored Kepler's work to the end, persisting in the futile attempt to bludgeon the world into accepting a Ferris wheel with forty-eight epicycles as 'rigorously demonstrated' physical reality.

What was the motive behind it? For almost fifty years of his life, he had held his tongue about Copernicus, not out of fear to be burnt at the stake, but to avoid academic unpopularity. When, carried away by sudden fame, he had at last committed himself, it became at once a matter of prestige to him. He had said that Copernicus was right, and whosoever said otherwise was belittling his authority as the foremost scholar of his time. That this was the central motivation of Galileo's fight will become increasingly evident. It does not exonerate his opponents;

but it is relevant to the problem whether the conflict was historically inevitable or not.

The final section of the *Letter to the Grand Duchess* is devoted to the miracle of Joshua. Galileo first explains that the sun's rotation around its axis is the cause of all planetary motion. 'And just as if the motion of the heart should cease in an animal, all other motions of its members would also cease, so if the rotation of the sun were to stop, the rotations of all the planets would stop too.'[18] Thus he not only assumed, with Kepler, the *annual* revolutions of the planets to be caused by the sun, but also their *daily* rotation round their axes – an *ad hoc* hypothesis with no more 'rigorous proof' than the analogy with the animal's heart. He then concludes that when Joshua cried: 'Sun, stand thou still,' the sun stopped rotating, and the earth in consequence stopped both its annual and daily motion. But Galileo, who came so close to discovering the law of inertia, knew better than anybody that if the earth suddenly stopped dead in its track, mountains and cities would collapse like match-boxes; and even the most ignorant monk, who knew nothing about impetus, knew what happened when the horses reared and the mail-coach came to a sudden halt, or when a ship ran against a rock. If the Bible was interpreted according to Ptolemy, the sudden stand-still of the sun would have no appreciable physical effect, and the miracle remained credible as miracles go; if it was interpreted according to Galileo, Joshua would have destroyed not only the Philistines, but the whole earth. That Galileo hoped to get away with this kind of painful nonsense, showed his contempt for the intelligence of his opponents.

In the *Letter to the Grand Duchess Christina* the whole tragedy of Galileo is epitomized. Passages which are classics of didactic prose, superb formulations in defence of the freedom of thought, alternate with sophistry, evasion, and plain dishonesty.

### 4. *The Denunciation*

For almost a whole year after the *Letter to Castelli* nothing dramatic happened. But the damage was done. Copies of the

*Letter* were circulating and were distorted in the process, then even more distorted by rumour. People like old Father Lorini, who, a year earlier, had not even heard the name 'Ipernicus', got the impression that some new Luther had arisen, denying the miracles of the Bible and defying the authority of the Church by means of some mathematical sophistries. Typical was the reaction of the Bishop of Fiesole, who wanted Copernicus instantly jailed, and was surprised to learn that he had been dead for seventy years.

In December (we are in 1614), there was a public scandal on a minor scale. A Dominican monk, Father Thommaso Caccini, who had previously been censored in Bologna as a rabble-rouser, preached a sermon in the church of Santa Maria Novella in Florence. Choosing as his text 'Ye men of Galilee, why stand ye gazing up into the heaven?' he attacked mathematicians in general and Copernicus in particular. Galileo promptly complained to Caccini's superiors in the ecclesiastical hierarchy. As a result, Father Luigi Maraffi, Preacher General of the Dominican Order, wrote to him a sincere apology. 'Unfortunately,' wrote Maraffi, 'I have to answer for all the idiocies that thirty or forty thousand brothers may or do actually commit.'[19] The letter illustrates the contrast in attitude between the higher dignitaries of the Church and the ignorant fanatics among the lower echelons.

At the time of Caccini's sermon, Father Lorini was on a visit to Pisa. On 31 December, Castelli reported to Galileo: 'From what I hear, Father Lorini (who is here) felt very sad that your fine priest had let himself get so far out of hand.'[20] But a few days later, Lorini was shown a copy of the *Letter to Castelli.* He was profoundly shocked, and made a copy of it. On his return to his convent – St Mark's in Florence – he discussed its contents with his fellow brethren. By now the atmosphere had become so tense, that they decided the *Letter* should be forwarded to the Holy Office. On 7 February 1615, Lorini wrote to Cardinal Sfondrati:

All our Fathers of this devout convent of St Mark are of opinion that the letter contains many propositions which appear to be suspicious or presumptuous as when it asserts that the language of Holy Scripture does not mean what it seems to mean; that in discussions about natural phenomena the last and lowest place ought to be given to the authority of the sacred text; that its commentators have very often erred in their interpretation; that the Holy Scriptures should not be mixed up with anything except matters of religion. . . . Ever mindful of our vow to be the 'black and white hounds' of the Holy Office . . . when I saw that they [the 'Galileists'] expounded the Holy Scriptures according to their private lights and in a manner different from that of the common interpretation of the Fathers of the Church; that they strove to defend an opinion which appeared to be quite contrary to the sacred text; that they spoke in slighting terms of the ancient Fathers and of St Thomas Aquinas; that they were treading underfoot the entire philosophy of Aristotle which has been of such service to Scholastic theology; and, in fine, that to show their cleverness they were airing and scattering broadcast in our steadfast Catholic city a thousand saucy and irreverent surmises; when, I say, I became aware of all this, I made up my mind to acquaint your Lordship with the state of affairs, that you in your holy zeal for the Faith may, in conjunction with your most illustrious colleagues, provide such remedies as will appear advisable. . . . I, who hold that those who call themselves Galileists are orderly men and good Christians all, but a little overwise and conceited in their opinions, declare that I am actuated by nothing in this business but zeal for the sacred cause.[21]

The letter was evidently the outcome of a collective decision of the Dominicans of St Mark. It did not mention Galileo by name, only referred to 'Galileists'. It also seems that old Father Lorini was not quite definite in his mind whether the writer of the *Letter to Castelli* was Galileo or Copernicus.[21a] But the copy of the *Letter to Castelli* which he enclosed contains two deliberate errors in transcription. Galileo wrote that there were passages in Scripture which, 'taken in the strict literal meaning, look as if they differed from the truth'. In Lorini's copy this became: '. . . which are false in the literal meaning'. Galileo wrote that Scripture sometimes 'overshadows' its own meaning; in Lorini's copy 'overshadows' became 'perverts'.

The forgery is usually blamed on Lorini. By what is known

of the character of the old man and by other internal evidence, it seems much more likely that it was committed by some other hand. As will presently be seen, it made no difference to the outcome, but this first forgery should be noted because of the suspicion of a second and more important one at a later stage.

To anyone who does not remember the reverence of the higher dignitaries of the Church for Science and Scientists, the result of Father Lorini's denunciation must appear rather startling. The *Letter to Castelli* was duly forwarded to the Consultor of the Holy Office for his opinion; who pronounced that 'such words as "false" and "perverting" sound very bad'; however, considered in the general context they were not of a nature that they could be said to deviate from Catholic doctrine; and, as for the remaining contents of the *Letter*, he had no objection. The case was dismissed.

Lorini's denunciation had fallen through, but a month later Caccini appeared in Rome, undaunted by the disavowal of his superior. He approached the Holy Office 'begging to testify concerning the errors of Galileo for the exoneration of his conscience'.

Caccini beautifully fits the satirist's image of an ignorant, officious, lying, and intriguing monk of the Renaissance. His testimony before the Inquisition was a web of hearsay, innuendo, and deliberate falsehood. He named as witnesses a Spanish priest, Father Ximenes, and a young man named Atavante. Since Ximenes was travelling abroad, he could not be called till 13 November, and Atavante was called the following day. The contradictions in their evidence convinced the inquisitors that Caccini's charges of heresy and subversion were a fabrication, and the case against Galileo was again dropped.

This was in November 1615. For the next eighteen years Galileo lived honoured and unmolested, befriended by Pope Urban VIII and an impressive array of cardinals.

But the *letters* to Castelli and to the Grand Duchess remained on the files of the Inquisition, and in the minds of the theologians. The text was so carefully worded that it could not be indicted as heresy, but the intent was unmistakable; it constituted a challenge which sooner or later had to be answered. The challenge lay in the implied claim that the Copernican system belonged to the category of 'rigorously demonstrated' physical truths to which the meaning of the Bible must be adapted; and that unless it was explicitly refuted and condemned, theological objections would become irrelevant and the case would go by default.

Three months after Galileo himself had been cleared of all charges against his person, Copernicus' book was put on the Index 'pending corrections'. It is necessary to describe in some detail the events which led to this.

## 5. *The Refusal to Compromise*

Galileo's chief opponent in the historic controversy was both a bogyman and a saint. In England he was believed to be the master-mind behind the Gunpowder Plot, 'a furious and devilish Jebusite'; for a while, wine-jugs featuring a bearded head were called Bellarmines. He was beatified in 1923 and made a saint in 1930.

At the time of the controversy, Cardinal Robert Bellarmine was seventy-three, a general of the Jesuit Order, Consultor of the Holy Office, and the most respected theologian in Christendom, whose opinion carried more spiritual authority than Pope Paul V's. He was the author of the catechism in its modern form, and co-editor of the Clementine edition of the Vulgate. But his lasting fame is that of one of the great controversialists of all time. His polemics against Lutheranism, Anglicanism, and particularist tendencies in Catholic countries such as France and the Venetian Republic, were inspired by an overriding vision: the Universal Church as a super-state. This involved the rejection not only of the Protestant heresy, but also of the new nationalist tendencies derived from the principle of absolute monarchy. The idea of the Universal Church demanded a Holy Father with a universal authority overriding that of any national ruler.

However, Bellarmine was sufficiently realistic to moderate his claims of temporal power for the Papacy. Hence, on the one hand, he had to fight that other great controversialist, James I, in a long series of tracts and counter-tracts which were the scandal and delight of Western Christendom; but he also incurred the displeasure of Paul V for not claiming absolute temporal authority for the Pope. In a later controversy between Jesuits and Dominicans on the question of predestination Bellamine again took a middle line; the point of interest to us is that the Dominicans' (as later the Jansenists'), arguments were mainly based on Augustine's, so that the African saint's opinions had become a very controversial subject. Galileo's innocent reliance on Augustine's authority shows how unwise it was for a layman to venture out into the rarified but highly charged air of theology.

As an individual, Bellarmine was the opposite of what one would expect from a formidable theologian who defied popes and kings. He was a lover of music and the arts; he had lectured on astronomy in his youth. He had a simple manner and led a simple, ascetic life, in contrast to other princes of the Church; but above all he had a 'childlike quality that was noted by all who came in contact with him'. At the time of the Galileo controversy, he was writing a devotional book called *Lament of the Dove*, which his most ferocious opponent, James I, in his later years constantly carried about his person, and described as a wonderful aid to spiritual comfort.

One of Bellamine's official functions was that of a 'Master of Controversial Questions' at the Roman College. Here he was in constant touch with the leading astronomers of the capital, Fathers Clavius and Grienberger, who had been among the first converts to Galileo's telescopic discoveries, and had acclaimed him on his first visit to Rome. Thus it can hardly be said that Galileo's opposite number in the drama was an ignorant fanatic. Bellarmine's independence of mind is further illustrated by the fact that in 1890 his *magnum opus*, the *Disputationes*, was temporarily put on the Index of forbidden books.

Sixteen years before he became involved with Galileo, Bellarmine had been one of the nine Cardinal Inquisitors who participated in the trial of Giordano Bruno, and some writers have tried to see a sinister connexion between the two events. In fact, there is none. Bruno was burnt alive, on 16 February 1600, under the most horrible circumstances on the Square of Flowers in Rome, as an impenitent apostate, who during seven years of imprisonment refused to adjure his theological heresies, and persisted in his refusal to the last moment.[22] Giordano Bruno and Michael Servetus (burned, in 1553, by the Calvinists in Geneva) seem to be the only scholars of repute who became victims of religious intolerance in the sixteenth and seventeenth centuries – not, of course, because of their scientific, but because of their religious opinions. Coleridge's remark: 'If ever a poor fanatic thrust himself into the fire, it was Michael Servetus,' applies to the irascible and tempestuous Bruno as well. His teachings of the infinity of the universe and the plurality of inhabited worlds, his pantheism and universal ethics exerted a considerable influence on subsequent generations; but he was a poet and metaphysician, not a scientific writer, and thus does not enter into this narrative.[22a]

We have followed the events of 1615, from Lorini's denunciation of Galileo's *Letter* and Caccini's denunciation of his personal activities, to the collapse of the case against him in November. The proceedings were conducted in secret, and Galileo had no part in them; but his friends in Rome knew that something was up, and kept him informed of all rumours and developments. Among his informants were Cardinal Piero Dini, Archbishop of Fermo, and Monsignor Giovanni Ciàmpoli. The letters exchanged, during 1615, between these two in Rome and Galileo in Florence, are important for the understanding of the developments which led to the prohibition of Copernicus.

On 16 February, Galileo sent a copy of his *Letter to Castelli* to Dini, with the request that it should be shown to Father Grienberger and, if possible, to Cardinal Bellarmine. In his covering letter there were the usual complaints about the hostility surrounding him. He remarked that the *Letter to Castelli* was written in haste and that he was going to improve and extend it; the extended version, as we know, became the *Letter to the Grand Duchess Christina.*

Before Dini answered, Ciàmpoli wrote, at the end of February (my italics):

Cardinal Barberini [the future Pope Urban VIII], who, as you know from experience, has always admired your worth, told me only yesterday evening that with respect to these opinions he would like greater caution in *not going beyond the arguments used by Ptolemy and Copernicus,** and finally in not exceeding the limitations of physics and mathematics. For to explain the Scriptures is claimed by theologians as their field, and if new things are brought in, even by an admirable mind, not everyone has the dispassionate faculty of taking them just as they are said. . . .[23]

A few days later, on 3 March, Dini's answer arrived (my italics):

With Bellarmine I spoke at length of the things you had written. . . . And he said that as to Copernicus, there is no question of his book being prohibited; the worst that might happen, according to him, would be the addition of some material in the margins of that book to the effect that Copernicus had introduced his theory in order to save the appearances, or some such thing – just as others had introduced epicycles without thereafter believing in their existence. And *with a similar precaution you may at any time deal with these matters.* If things are fixed according to the Copernican system, [he said], it does no appear presently that they would have any greater obstacle in the Bible than the passage '[the sun] exults as a strong man to run his course,' etc., which all expositors up to now have understood by attributing motion to the sun. And although I replied that this also could be explained as a concession to our ordinary forms of expression, I was told in answer that this was not a thing to be done in haste, just as the condemnation of any of these opinions was not to be passionately hurried. . . . I can only rejoice for you. . . .[24]

On the same day – 7 March – Prince Cesi, the President of the Linceian Academy, also wrote to Galileo. His letter contained the sensational news that a Carmelite monk from Naples, Paolo Antonio Foscarini, a Provincial of his Order, had published a book in defence of Galileo and Copernicus.[25] Foscarini was now preaching in Rome and had offered to meet

* i.e., that they are to be regarded as mathematical hypotheses only, in the sense of Osiander's preface.

all comers in public discussion. He had sent a copy of his book to Bellarmine.

On 21 March, Ciàmpoli relayed further assurances by Cardinals Bellarmine and del Monte, that Galileo had nothing to fear so long as he kept to the province of physics and mathematics, and refrained from theological interpretations of Scripture.[26] He added that there was a danger of Foscarini's book being prohibited, but only because it meddled with Holy Scripture. Ciàmpoli had also been told that several Jesuit astronomers were Copernicans, but were still holding back, that it was essential to go on working until all quietened down, and to avoid new occasions for scandal-makers.[27]

Dini also warned again, in the same vein: 'One may write freely as long as one keeps out of the sacristy.'[28]

Galileo answered these admonitions in a letter to Dini dated 23 March. His answer was to refuse any compromise on the Copernican system. Copernicus did not mean it to be understood merely as a hypothesis. It was to be accepted or rejected absolutely. He agreed that the reinterpretation of Holy Scripture in the light of Copernicus should be left to the theologians, but he cannot help it if he has been forced on theological grounds, and since Bellarmine had quoted Psalm 19 to Dini, the passage that the sun 'rejoiceth as a strong man to run his course', Galileo 'in all humility' undertook to refute Bellarmine's interpretation of the Psalm. 'The running of the course' refers to the light and heat from the sun, not to the sun itself, etc., etc.[29] Dini probably had the wisdom not to show this to the greatest theologian alive.

The next utterance came from Bellarmine himself. It was a precise and authoritative statement of his attitude, and in view of his position as Consultor of the Holy Office, Master of Controversial Questions, etc., it amounted to an unofficial definition of the Church's attitude to Copernicus. The statement was occasioned by Father Foscarini's book advocating the Copernican system, and couched in the form of a letter of acknowledgement; but it was clearly addressed to Galileo as well, whose name is expressly mentioned. The letter is dated 4 April 1615; my italics.

My Very Reverend Father,

It has been a pleasure to me to read the Italian letter and the Latin paper you sent me. I thank you for both the one and the other, and I may tell you that I found them replete with skill and learning. As you ask for my opinion, I will give it as briefly as possible because, at the moment I have very little time for writing.

First, I say it seems to me that your Reverence and Signor Galileo act prudently when you content yourselves with speaking hypothetically and not absolutely, as I have always understood that Copernicus spoke. *For to say that the assumption that the Earth moves and the Sun stands still saves all the celestial appearances better than do eccentrics and epicycles* is to speak with excellent good sense and to run no risk whatever. Such a manner of speaking suffices for a mathematician.* But to want to affirm that the Sun, in very truth, is at the centre of the universe and only rotates on its axis without travelling from east to west, and that the Earth is situated in the third sphere and revolves very swiftly around the Sun, is a very dangerous attitude and one calculated not only to arouse all Scholastic philosophers and theologians but also to injure our holy faith by contradicting the Scriptures. . . .

Second, I say that, as you know, the Council of Trent forbids the interpretation of the Scriptures in a way contrary to the common agreement of the holy Fathers. Now if your Reverence will read, not merely the Fathers, but modern commentators on Genesis, the Psalms, Ecclesiastes, and Joshua, you will discover that all agree in interpreting them literally as teaching that the Sun is in the heavens and revolves round the Earth with immense speed and that the Earth is very distant from the heavens, at the centre of the universe, and motionless. Consider, then, in your prudence, whether the Church can support that the Scriptures should be interpreted in a manner contrary to that of the holy Fathers and of all modern commentators, both Latin and Greek. . . .

Third, I say that, *if there were a real proof* that the Sun is in the centre of the universe, that the Earth is in the third sphere, and that the Sun does not go round the Earth but the Earth round the Sun, then we should have to proceed with great circumspection in explaining passages of Scripture which appear to teach the contrary,

* He evidently refers here to those epicycles which were needed in the Ptolemaic system to explain the apparent retrogression of the planets and which Copernicus dispensed with.

and we should rather have to say that we did not understand them than declare an opinion to be false which is proved to be true. But I do not think there is any such proof *since none has been shown to me*. To demonstrate that the appearances are saved by assuming the sun at the centre and the earth in the heavens is not the same thing as to demonstrate that *in fact* the sun is in the centre and the earth in the heavens. I believe that the *first demonstration may exist, but I have very grave doubts about the second*; and in case of doubt one may not abandon the Holy Scriptures as expounded by the holy Fathers. . . .[30]

The italicized passage under the first heading states clearly that it is admissible not only to expound the Copernican system, but also to say that *as a hypothesis it is superior to Ptolemy's*. This is 'to speak with excellent good sense' so long as we remain in the domain of hypothesis. Under the second heading he paraphrases the legislative decision of the Council of Trent against interpreting Scripture in ways contrary to tradition (directed, of course, not against Copernicus but Luther). Under the third heading the condition is stated which would justify an exception to this rule being made; to wit, that the new cosmology should be 'really proven' (or 'truly demonstrated'). Since no proof has been shown to him, he has 'grave doubts' whether such proof exists; and in case of doubt the request for reinterpreting the Bible must be rejected. He had consulted Grienberger, and Grienberger must have truthfully informed him that no physical proof for the earth's motion had been adduced. He may have added that the absence of stellar parallax and the nine epicycles ascribed to the earth alone, were rather in the nature of a disproof.

Bellarmine had placed the burden of proof for the Copernican system back where it belonged: on the advocates of the system. There were only two possibilities left to Galileo: either to supply the required proof, or to agree that the Copernican system should be treated, for the time being, as a working hypothesis. Bellarmine had, in a tactful way, reopened the door to this compromise in the opening sentence of his letter, where he pretended that Galileo had 'contented himself with speaking hypothetically and not absolutely', had praised his prudence, and acted as if the *Letters* to Castelli and the Grand Duchess, which were before the Inquisition, did not exist.

But Galileo was by now beyond listening to reason. For, by accepting the compromise, he would disclose to the world that he had no proof, and would be 'laughed out of court'. Therefore he must reject it. It was not enough to be allowed, and even encouraged, to teach the superiority of the Copernican over the Ptolemaic hypothesis. He must insist that the Church endorse it, or reject it, absolutely – even at the risk of the latter alternative, which Bellarmine's letter, Dini's and Ciàmpoli's warnings must have made clear to him.

But how can he motivate his rejection of compromise? How can he refuse to produce proof and at the same time demand that the matter should be treated as if proven? The solution of the dilemma was to pretend that he had the proof, but to refuse to produce it, on the grounds that his opponents were too stupid, anyway, to understand. His answer to Bellarmine was contained in a letter written at some date in May to Cardinal Dini (my italics):

To me, the surest and swiftest way to prove that the position of Copernicus is not contrary to Scripture would be to give a host of proofs that it is true and that the contrary cannot be maintained at all; thus, since no truths can contradict one another, this and the Bible must be perfectly harmonious. *But how can I do this, and not be merely wasting my time, when those Peripatetics who must be convinced show themselves incapable of following even the simplest and easiest of arguments?* ...[31]

The truly staggering thing in this passage is not its contemptuous arrogance, but the fact that while talking of 'Peripatetics' it is in fact aimed at Bellarmine; for it is on him and not on the backwoodsmen, that the decision depends, and it was Bellarmine who had challenged him to produce proof.

Earlier in the same letter to Cardinal Dini, he had written:

Eight days ago I wrote to Your Reverence in reply to yours of the second of May. My answer was very brief, because I then found myself (as now) among doctors and medicines, and much disturbed in body and mind over many things, particularly by seeing no end to these rumours set in motion against me through no fault of mine, and seemingly accepted by those higher up as if I were the originator of these things. Yet for all of me any discussion of the sacred Scripture might have lain dormant forever; no astronomer or scientist who remained within proper bounds has ever got into such things. Yet while I follow the teachings of a book accepted by the Church (sic), there come out against me philosophers quite ignorant of such teachings who tell me that they contain propositions contrary to the faith. So far as possible, I should like to show them that they are mistaken, but my mouth is stopped and I am ordered not to go into the Scriptures. This amounts to saying that Copernicus' book, accepted by the Church, contains heresies and may be preached against by anyone who pleases (sic) while it is forbidden for anyone to get into the controversy and show that it is not contrary to Scripture. . . .

Galileo's style is again so convincing that one is apt to forget the facts: that Copernicus' book was only 'accepted by the Church' with the qualifications that we know; that Caccini, who had preached against it, was reprimanded by the Preacher General of his Order; and that, according to the accepted rules of the game, the scriptural objections could not be refuted on scriptural grounds, only by the scientific proofs which Bellarmine demanded and which Galileo was unable to supply.

After the passage, that I have already quoted, about the stupidity of his opponents, Galileo went on:

Yet I should not despair of overcoming even this difficulty if I were in a place where I could use my tongue instead of my pen; and if I ever get well again so that I can come to Rome, I shall do so, in the hope of at least showing my affection for the holy Church. My urgent desire on this point is that no decision be made which is not entirely good. Such it would be to declare, under the prodding of an army of malign men who understand nothing of the subject, that Copernicus did not hold the motion of the earth to be a fact of nature, but as an astronomer merely took it to be a convenient hypothesis for explaining the appearances. . . .

'The army of malign men who understand nothing of the subject' again obviously included Bellarmine, who had written that he had always understood Copernicus to speak 'hypothetically and not absolutely'.

Perhaps the one genuine sentiment in the letter was Galileo's wish to get to Rome where he could use his 'tongue instead of his pen'. Early in December he arrived in Rome; the final phase of the battle had begun.

## 6. *The 'Secret Weapon'*

This time there was no triumphant reception at the Roman College. Father Grienberger sent word that it would be better for Galileo to bring convincing scientific proof in support of Copernicus before trying to adjust Scripture to him.[32] The Tuscan Ambassador in Rome, Guicciardini, had warned Duke Cosmo against Galileo's coming to Rome, and Bellarmine, who foresaw the consequences, had also advised against it.[33] But the Duke had given in to Galileo, and on his instructions Galileo took up quarters at the Villa Medici – then the Tuscan Embassy – 'with board for himself, a secretary, a valet, and a small mule'.[34]

I have quoted some samples of Galileo's superb technique in his written polemics. According to his contemporaries, he was even more effective when he used 'his tongue instead of his pen'. His method was to make a laughing stock of his opponent in which he invariably succeeded, whether he happened to be in the right or in the wrong. Here is one Roman witness, Monsignor Querengo, describing Galileo in action:

> We have here Signor Galileo who, in gatherings of men of curious mind, often bemuses many concerning the opinion of Copernicus, which he holds for true. . . . He discourses often amid fifteen or twenty guests who make hot assaults upon him, now in one house, now in another. But he is so well buttressed that he laughs them off; and although the novelty of his opinion leaves people unpersuaded, yet he convicts of vanity the greater part of the arguments with which his opponents try to overthrow him. Monday in particular, in the house of Federico Ghisileri, he achieved wonderful feats; and what I liked most was that, before answering the opposing reasons, he amplified them and fortified them himself with new grounds which appeared invincible, so that, in demolishing them subsequently, he made his opponents look all the more ridiculous.[35]

It was an excellent method to score a moment's triumph, and make a lifelong enemy. It did not establish his own point, but it destroyed his opponent's. Yet by the force of circumstances, these were the only tactics that he could adopt: to demonstrate the absurdity of Ptolemy's epicycles and to pass in silence over the absurdity of Copernicus' epicycles. The Tuscan Ambassador reported:

> ... He is passionately involved in this quarrel, as if it were his own business, and he does not see and sense what it would comport; so that he will be snared in it, and will get himself into danger, together with anyone who seconds him. ... For he is vehement and is all fixed and impassioned in this affair, so that it is impossible, if you have him around, to escape from his hands. And this is a business which is not a joke but may become of great consequence, and this man is here under our protection and responsibility. ...[36]

But Galileo could not be persuaded to desist. He had manoeuvred himself into a position from which he could not retreat without loss of face. He had committed himself to an opinion, and he must be proved right; the heliocentric system had become a matter of his personal prestige.

An aggravating factor in the drama was the personality of Paul V Borghese 'who abhors the liberal arts and his [Galileo's] kind of mind, and cannot stand these novelties and subtleties', as Guicciardini described him.[37] 'Those who understand something and are of curious mind, if they are wise, try to show themselves quite the opposite in order not to fall into suspicion and get into trouble themselves.'

Even Bellarmine had incurred Paul's displeasure. He and the other leading dignitaries – Cardinals Barberini, Dini and del Monte, Piccolomini and Maraffi – knew how to treat him. They were anxious to avoid committing the Church to any official decision on the Copernican system, until the astronomers were able to shed more light on it, and to preserve the *status quo* as defined in Bellarmine's letter, ignoring Galileo's 'incursion into the sacristy'. But they knew that if the Pope learnt about the scandal, a showdown would be inevitable. That was probably why Bellarmine had advised against Galileo's visit to Rome.

We come to the last episode before the blow fell. Galileo had repeatedly hinted that he had discovered a decisive physical proof of the Copernican theory, but had so far refused to disclose it. When he began to feel that arguing about the miracle of Joshua and the ludicrousness of Ptolemy was no longer of avail, and that his position was becoming impossible, he produced, as a last card, his 'conclusive physical proof'. It was his theory of the tides.

Seven years earlier, in the *Astronoma Nova*, Kepler had published his correct explanation of the tides as an effect of the moon's attraction. Galileo dismissed Kepler's theory as an astrological superstition,[38] and declared that the tides were a direct consequence of the earth's combined motions which cause the sea to move at a different speed from the land. The theory will be discussed in more detail in the following chapter, p. 471. It contradicted Galileo's own researches into motion, was a relapse into crude Aristotelian physics, and postulated that there ought to be *only one high tide a day*, precisely at noon – whereas everybody knew that there were two, and that they were shifting around the clock.[38a] The whole idea was in such glaring contradiction to fact, and so absurd as a mechanical theory – the field of Galileo's own immortal achievements – that its conception can only be explained in psychological terms. It is completely out of keeping with his intellectual stature, the method and trend of his thought; it was not a mistake but a delusion.

Armed with his new 'secret weapon' (as a modern scholar has called Galileo's theory of the tides [39]), he now decided to make a direct assault on the Pope. It seems that all of Galileo's friends who had access to the Pope – Cardinals Dini, Barberini, del Monte, etc. – refused to act as intermediaries, for the mission was finally entrusted to Cardinal Alessandro Orsini, a youth of twenty-two. Galileo wrote down for him his idea of the tides; the sequel is described as follows in Ambassador Guicciardini's report to Duke Cosmo II of Tuscany:

> Galileo has relied more on his own counsel than on that of his friends. The Lord Cardinal del Monte and myself, and also several cardinals from the Holy Office, had tried to persuade him to be

quiet and not to go on irritating this issue. If he wanted to hold this Copernican opinion, he was told, let him hold it quietly and not spend so much effort in trying to have others share it. Everyone fears that his coming here may be very prejudicial and that, instead of justifying himself and succeeding, he may end up with an affront.

As he felt people cold toward his intention, after having pestered and wearied several cardinals, he threw himself on the favour of Cardinal Orsini, and extracted to that purpose a warm recommendation from Your Highness. The Cardinal, then, last Wednesday in Consistory, I do not know with what circumspection and prudence, spoke to the Pope on behalf of said Galileo. The Pope told him it would be well if he persuaded him to give up that opinion. Thereupon Orsini replied something, urging the cause, and the Pope cut him short and told him he would refer the business to the Holy Office.

As soon as Orsini had left, His Holiness summoned Bellarmine; and, after brief discussion, they decided that the opinion was erroneous and heretical; and day before yesterday, I hear, they had a Congregation on the matter to have it declared such. Copernicus, and the other authors who wrote on this, shall be amended or corrected or prohibited; I believe that Galileo personally is not going to suffer, because he is prudent and he will feel and desire as Holy Church does. [4 March.] [40]

The Tuscan Ambassador was evidently sorely tried by his guest and ward's antics, and his report is not entirely reliable, because 'last Wednesday in Consistory' places the episode on 2 March, whereas the papal decree to summon the theologians of the Holy Office to give a formal opinion on the Copernican theory is dated 19 February. But the confusion about dates may have some trivial explanation; the fact that Orsini, armed with Galileo's 'final proof', interceded with the Pope is not disputed; and whether it was this particular incident, or something of a similar kind, which brought matters to a head is not very important.[41] Galileo had done everything in his power to provoke a showdown.

## 7. *The Decree of the Holy Office*

Thus it came about that on 23 February A.D. 1616, four days after they had been summoned, the Qualifiers (i.e. theological experts) of the Holy Office met to give their opinion on the two following propositions submitted to them:

1. The sun is the centre of the world and wholly immovable of local motion.

2. The earth is not the centre of the world nor immovable, but moves as a whole, also with a diurnal motion.

The Qualifiers unanimously declared the first proposition to be

> foolish and absurd, philosophically and formally heretical inasmuch as it expressly contradicts the doctrine of Holy Scripture in many passages, both in their literal meaning and according to the general interpretation of the Fathers and Doctors.

The second proposition was declared 'to deserve the like censure in philosophy, and as regards theological truth, to be at least erroneous in faith'.[42]

But the Qualifiers' verdict was, for the time being, overruled under pressure of the more enlightened Cardinals; it was only published a full seventeen years later. Instead of it, on 5 March, the General Congregation of the Index issued a more moderate decree, in which the fatal word 'heresy' does not appear:

> ... And whereas it has also come to the knowledge of the said Congregation that the Pythagorean doctrine – which is false and altogether opposed to the Holy Scripture – of the motion of the Earth, and the immobility of the Sun, which is also taught by Nicolaus Copernicus in *De revolutionibus orbium coelestium*, and by Diego de Zuniga [in his book] on Job, is now being spread abroad and accepted by many – as may be seen from a certain letter of a Carmelite Father, entitled *Letter of the Rev. Father Paolo Antonio Foscarini, Carmelite, on the Opinion of the Pythagoreans and of Copernicus concerning the Motion of the Earth, and the Stability of the Sun, and the New Pythagorean System of the World, at Naples, Printed by Lazzaro Scoriggio*, 1615: wherein the said Father attempts to show that the aforesaid doctrine of the immobility of the sun in the centre of the world, and of the Earth's motion, is consonant with truth and is not opposed to Holy Scripture. Therefore, in order that this opinion may not insinuate itself any further to the prejudice of Catholic truth, the Holy Congregation has decreed that the said Nicolaus Copernicus, *De revolutionibus*

*orbium*, and Diego de Zuniga, *On Job*, be suspended until they be corrected; but that the book of the Carmelite Father, Paolo Antonio Foscarini, be altogether prohibited and condemned, and that all other works likewise, in which the same is taught, be prohibited, as by this present decree it prohibits, condemns and suspends them all respectively. In witness whereof the present decree has been signed and sealed with the hands and with the seal of the most eminent and Reverend Lord Cardinal of St Cecilia, Bishop of Albano, on the fifth day of March 1616.[43]

---

The document had consequences which are still felt today. It represents, as it were, the crack in the wall which led to the falling apart of Science and Faith. It is therefore important to examine its exact meaning and intent, as distinct from its psychological effect and its historic consequences.

In the first place, it must be repeated that the Qualifiers talked of heresy, the decree did not. The Qualifiers' Opinion became known to the public only in 1633, when Galileo forced a second showdown, and the Opinion was quoted in the verdict of his trial. Even then, it remained a judicial opinion, without endorsement by Papal authority, and therefore not binding on members of the Church. Accordingly, the immobility of the earth never became an article of faith, nor the immobility of the sun a heresy.

Similar considerations of a judicial nature apply to the decree itself. It was issued by the Congregation of the Index, but not confirmed by papal declaration *ex cathedra* or by Oecumenical Council, and its contents therefore never became infallible dogma. All this was deliberate policy; it is even known that it was urged upon Paul V, who would have liked to make Copernicus a heretic and *basta*, by Cardinals Barberini and Gaetani. These points have been stressed over and again by Catholic apologists, but on the man in the street such subtleties were lost; whether it became dogma or not, the condemnation of the Copernican system as 'altogether opposed to Holy Scripture' in 1616, and as 'formally heretical' in 1633, was quite enough to have its disastrous effect.

A quite different question is how the decree affected the freedom of scientific discussion. First we must note that although Galileo is the chief culprit, his name is not mentioned in the proceedings, and his works are not put on the Index. Equally striking is the distinction made in the treatment of Copernicus' *Revolutions* and of Foscarini's book. Copernicus' book is 'suspended until it be corrected'; but Foscarini's book is 'altogether prohibited and condemned'. The reason is given in the preceding sentence in the decree: Foscarini had attempted to show that the Copernican doctrine is 'consonant with truth and not opposed to Holy Scripture', whereas Copernicus is charged with no such thing. Galileo himself commented a few days after the decree that the Church

> has gone no further than to decide that [the Copernican] opinion does not concur with the Bible. Hence they have forbidden only such books which professionally attempt to sustain it as not discordant with the Bible. ... From Copernicus' own Book ten lines will be taken from the preface addressed to Pope Paul III where the author says that his doctrine does not seem to him contrary to the Bible, and I hear that here and there a word may be removed where the earth is called a star.[44]

The *Letters on Sunspots* were the only printed work by Galileo* which contained a favourable reference to the Copernican system; but since that reference treated it merely as a hypothesis, it escaped censure.

Thus the effect of the decree on scientific discussion and research was to leave things almost exactly where they had been. Astronomers could discuss Copernicus and compute the course of the planets as if they were moving round the sun, provided that they spoke hypothetically. Galileo had refused to compromise, and the compromise had been enforced by decree. But what the decree conveyed to simple sons of the Church was that to talk of the earth's motion was a Bad Thing and contrary to faith; and what it conveyed to the sceptic was that the Church had declared war on Science.

Canon Koppernigk's book remained on the Index for exactly four years. In 1620 the 'corrections' were published and turned

* The *Letter to Castelli* and the *Letter to the Grand Duchess* had not appeared in print.

out to be of the trifling nature predicted by Galileo.* They were designed by the same Cardinal Gaetani who, together with the future Urban VIII, had carried the day against the angry Paul V. From then onward, any Catholic publisher was free to reprint the *Book of Revolutions* – but no Catholic, or Protestant, publisher felt moved to do so for another three hundred years. The surviving copies of the first edition of 1543 had become treasured collectors' pieces. The book itself had become, apart from being unreadable, a mere curiosity and completely out of date owing to Tycho's observations, Kepler's discoveries, and the revelations of the telescope. Copernicanism was a slogan, but not a defendable system of astronomy.

To sum up: the temporary suspension of Copernicus' book had no ill effects on the progress of science; but it injected a poison into the climate of our culture which is still there.

It would, of course, be naïve to believe that the Church objected to the Copernican system only, or even mainly, because it seemed to disagree with the miracle of Joshua or other scriptural passages. The Council of Trent had decreed that 'petulant minds must be restrained from interpreting Scripture against the authority of tradition in matters that pertain to faith and morals'; but the 'petulant minds' at which this was aimed were the Lutherans, and not mathematicians like Copernicus, whose book had been published two years before the Council assembled, and twenty years before it ended. The real danger of removing the earth from the centre of the universe went much deeper; it undermined the whole structure of medieval cosmology.

Bellarmine had once said in a sermon: 'Men are so like frogs. They go open-mouthed for the lure of things which do not concern them, and that wily angler, the Devil, knows how

* 'Nine sentences, by which the heliocentric system was represented as certain, had to be either omitted or changed.' [45] As Santillana remarks, 'the feeling seems to have been current in Rome that the Index was a kind of administrative misadventure that occurred sooner or later to anyone writing on serious subjects and that it was a matter of waiting until the official line changed again. Of the three theologians of the Inquisition who were the experts at Galileo's trial, two subsequently incurred prohibition – and one of them a cardinal, Oregius.' [45a]

to capture multitudes of them.'[46] The people in Rome were indeed beginning to discuss questions such as whether other planets were inhabited; and if so, could their inhabitants descend from Adam? And if the earth is a planet, it needs, like other planets, an angel to move it; but where is he? They were interpreting the messages of Science in the same fundamentalist and frogmouthed way as the theologians were interpreting Faith. But Christianity had, in the past, overcome similar crises; it had digested the rotundity of the earth and the existence of the antipodes in replacement of the tabernacular universe covered by the Upper Waters. The Christian world-view had progressed from Lactantius and Augustine to the medieval cosmos of Aquinas and Albert the Great; and beyond that, to Bishop Cusa's first intimations of infinity, to the Franciscans' post-Aristotelian physics, and the Jesuits' post-Ptolemaic astronomy.

But it had been a gradual and continuous progress. The walled-in universe, the hierarchy of the Great Chain of Being could not be given up lightly, before some equally coherent vision of the world could take its place. And that vision did, as yet, not exist; it could only take shape when the Newtonian synthesis provided a new focus for the eye. Under the circumstances, the only possible policy was one of ordered retreat; to yield positions when they became untenable – such as the immutability of the sky, disproved by *novae*, comets, and sunspots, and the earth as centre of all heavenly motions, disproved by the moons of Jupiter. In all these 'dangerous innovations' astronomers of the Jesuit Order, of which Bellarmine was the General, had played a prominent part. They had quietly abandoned Ptolemy, and progressed to the Tychonic system: the planets circle the sun, and with the sun the earth (just as the four 'Medicean stars' circle Jupiter, and with Jupiter, the sun). This is as far as both metaphysical prudence and scientific caution permitted them to go — even if some Jesuits were Copernicans at heart. The reasons for metaphysical prudence were theological; the reasons for scientific caution empirical: so long as there existed no observable stellar parallax, no apparent displacement in the position of the fixed stars caused by the earth's motion through space, that motion remained unproven. Under these circumstances, the system of the universe which seemed to agree most closely with observed fact, was the Tychonic system. It also had the advantage of a compromise; by making the sun the centre of planetary motion, it prepared the way for a complete heliocentric system, should a stellar parallax be found, or some other discovery tilt the balance in its favour. But that, as we shall see, was another compromise that Galileo rejected.

Galileo's followers, whom his brilliant ways of arguing had converted, had (with a handful of exceptions) only the haziest notions of astronomy. But Bellarmine was in constant touch with the astronomers of the Roman College. He was sufficiently open-minded to know – and to say so in his letter to Foscarini – that Christianity could be reconciled with the earth's motion, as it had been reconciled with its rotundity. But he also knew that this would be a difficult readjustment, a metaphysical re-orientation on a major scale, which must be undertaken only in the case of absolute necessity. And that necessity, so far, did not exist.

The situation is summed up in a passage by Professor Burtt, which I have already quoted in part:

> It is safe to say that even had there been no religious scruples whatever against the Copernican astronomy, sensible men all over Europe, especially the most empirically minded, would have pronounced it a wild appeal to accept the premature fruits of an uncontrolled imagination, in preference to the solid inductions, built up gradually through the ages, of men's confirmed sense experience. In the strong stress on empiricism, so characteristic of present-day philosophy, it is well to remind ourselves of this fact. Contemporary empiricists, had they lived in the sixteenth century, would have been the first to scoff out of court the new philosophy of the universe.[47]

It is not surprising then, that the decree of 5 March, however fateful its consequences proved to be, and however much dismay it caused to the Galileans, was greeted with a sigh of relief by others, and not only by the fanatics and backwoodsmen. It is reflected in a letter by Monsignor Querengo, that sharp-witted observer whom I have quoted before:

> The disputes of Signor Galileo have dissolved into alchemical smoke, since the Holy Office has declared that to maintain this opinion is to dissent manifestly from the infallible dogmas of the Church. So here we are at last, safely back on a solid Earth, and we do not have to fly with it as so many ants crawling around a balloon. . . .[48]

## 8. *The Injunction*

Galileo's name had not been mentioned in public. Immediately after the decree had been issued, he wrote nonchalantly to the Tuscan Secretary of State:

> As may be seen from the very nature of this business, I am not in the least concerned, nor would I have been involved had it not been for my enemies, as I have said before.[49]

Six days after the decree, Galileo was received by the Pope, in an audience which lasted three quarters of an hour. But while everything was done to spare him public humiliation, he had been confidentially but firmly enjoined to keep within the prescribed limits. This had happened between the session of the Qualifiers on 23 February, and the publication of the decree. On Thursday, 25 February, there is the following entry in the Inquisition file (my italics):

> Thursday, 25 February 1616. The Lord Cardinal Mellini notified the Reverend Fathers, the Assessor, and the Commissary of the Holy Office that the censure passed by the theologians upon the propositions of Galileo – to the effect that the Sun is the centre of the world and immovable from its place, and that the Earth moves, and also with a diurnal motion – had been reported; and His Holiness has directed the Lord Cardinal Bellarmine to summon before him the said Galileo and admonish him to abandon the said opinion; and *in case of his refusal to obey*, that the Commissary is to enjoin on him, before a notary and witnesses, a command to abstain altogether from teaching or defending this opinion and doctrine and even from discussing it;[50] and, if he do not acquiesce therein, that he is to be imprisoned.

One of the principal points of controversy about the trial of Galileo in 1630 hinges on the question whether the procedure envisaged 'in case of his refusal to obey', took place or not. If it did, Galileo was bound by an unconditional and absolute injunction not only not to defend, but *not even to discuss* Copernicanism. If it did not take place, the obligation placed on him could be interpreted elastically.

There exist three documents bearing on this point, and they contradict each other. One was found among the *Decreta* of the Congregation. It is the minutes of a meeting on 3 March, of which the relevant passage reads:

> The Lord Cardinal Bellarmine having reported that Galileo Galilei, mathematician, had in terms of the order of the Holy Congregation been admonished to abandon the opinion he has hitherto held, that the Sun is the centre of the spheres and immovable and that the Earth moves, and has acquiesced therein. . . .

This seems to indicate that the absolute injunction foreseen 'in case of his refusal to obey' was *not* served. The second document seems to point to the same conclusion. To counter rumours that he had been humiliated and punished, Galileo asked Bellarmine for a certificate on the proceedings that had taken place, and Bellarmine wrote as follows:

> We, Roberto Cardinal Bellarmine, having heard that it is calumniously reported that Signor Galileo Galilei has in our hand abjured and has also been punished with salutary penance, and being requested to state the truth as to this, declare that the said Galileo has not abjured, either in our hand, or the hand of any other person here in Rome, or anywhere else, so far as we know, any opinion or doctrine held by him; neither has any salutary penance been imposed on him; but that only the declaration made by the Holy Father and published by the Sacred Congregation of the Index has been notified to him, wherein it is set forth that the doctrine attributed to Copernicus, that the Earth moves around the Sun, and that the Sun is stationary in the centre of the world and does not move from east to west, is contrary to the Holy Scriptures and therefore cannot be defended or held. In witness whereof we have written and subscribed these presents with our hand this twenty-sixth day of May 1616.

There is no mention here of a formal injunction, and the operative words are that the Copernican doctrine cannot be *defended or held.*[51] There is no prohibition to discuss it.

The third document is a minute in the Vatican files which seems to contradict the previous two by alleging that Galileo was formally forbidden 'to hold, teach, or defend in any way whatsoever, verbally or in writing'[52] the Copernican doctrine. This minute, of doubtful reliability, has given rise to one of the most embittered controversies in the history of science, which has now been raging for nearly a century. It may be thought that to attribute such importance to the difference between an absolute injunction and an admonition is splitting hairs. But there is, in fact, a world of difference between the admonition not to 'hold or defend' a doctrine, and the command not to teach or discuss it 'in any way whatsoever'. In the first case, it could be discussed as before, in terms of a mathematical hypothesis; in the second case, not (*see* note 52a).

Bellarmine's certificate and the Minute of 3 March seem to indicate that Galileo was under no absolute prohibition. Nevertheless, during the next few years he had to proceed more cautiously than before.

# 2 The Trial of Galileo

## 1. *The Tides*

After the issue had been formally decided by the decree of 5 March, Galileo stayed on in Rome for another three months. 'He is of a fixed humour,' the Tuscan Ambassador reported, 'to tackle the friars head on, and to fight personalities who cannot be attacked without ruining oneself. Sooner or later you will hear in Florence that he has madly tumbled into some unsuspected abyss.'[1] In the end, the alarmed Duke ordered Galileo back to Florence.

For the next seven years he published nothing. But his obsession was devouring him. It was the more self-destructive because he could not vent it. He could mutter about 'the ignorance, malice, and impiety of my opponents who had won the day'; but he must have known, without admitting it to himself, that his defeat was really due to the fact that he had been unable to deliver the required proof.

This, I suggest, explains how the delusion about the tides could gain such power over his mind. He had improvised this secret weapon in a moment of despair; one would have expected that once he reverted to a normal frame of mind, he would have realized its fallacy and shelved it. Instead, it became an *idée fixe*, like Kepler's perfect solids. But Kepler's was a creative obsession: a mystic chimera whose pursuit bore a rich and unexpected harvest; Galileo's mania was of the sterile kind. The tides, as I shall presently try to show, were an indirect substitute for the stellar parallax which he had failed to find – a substitute not only in the psychological sense, for there exists a mathematical connexion between the two, which seems to have eluded attention so far.

Galileo's theory of the tides runs, in a highly simplified form, as follows.[2] Take a point on the earth's surface – say, Venice. It has a two-fold motion: the daily rotation round the earth's axis, and its annual revolution round the sun. At night when Venice is at N, the two motions add up; in daytime, at D, they work against each other:

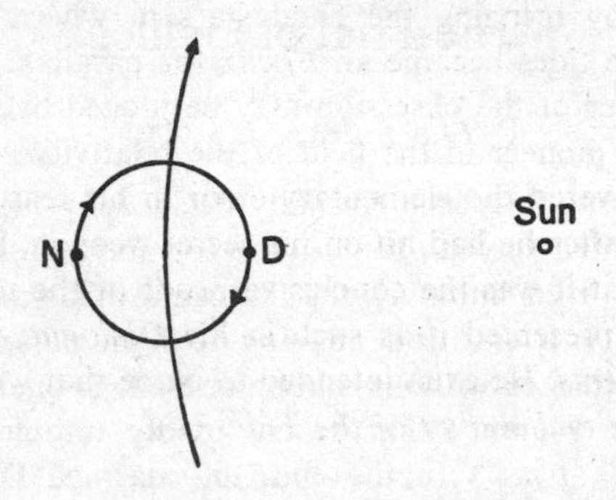


Hence Venice, and with it all the firm land, moves faster at night and slower in the daytime; as a result, the water is 'left behind' at night, and rushes ahead of the land in daytime. This causes the water to get heaped up a high tide every twenty-four hours, always around noon. The fact that there are two daily high tides at Venice instead of one, and that they wander round the clock, Galileo dismissed as due to several secondary causes, such as the shape of the sea, its depth, and so forth.

The fallacy of the argument lies in this. Motion can only be defined relative to some point of reference. If the motion is referred to the earth's axis, then any part of its surface, wet or dry, moves at uniform speed day and night, and there will be no tides. If the motion is referred to the fixed stars, then we get the periodic changes on the diagram, which are the same for land and sea, and can produce no difference in momentum between land and sea. A difference in momentum, causing the sea to 'swap over' could only arise, if the earth received a push by an external force – say, collision with another body. But both the earth's rotation and its annual revolution are inertial,[3] that is, self-perpetuating, and hence produce the same momentum in water and land; and a combination of the two motions still results in the same momentum. The fallacy in Galileo's reasoning is *that he refers the motion of the water to the earth's axis, but the motion of the land to the fixed stars.* In other words, he unconsciously smuggles in the absent parallax through the back door. No effect of the earth's annual motion relative to the fixed stars could be found. Galileo finds it in the tides, by bringing the fixed stars in where they do not belong. The tides became an *Ersatz* for parallax.

The power of the obsession may be judged by the fact that, although a pioneer in the field of the relativity of motion, he never discovered the elementary error in his reasoning; seventeen years after he had hit on his secret weapon, he still firmly believed that it was the conclusive proof of the motion of the earth, and presented it as such in his *Dialogue on the Great World Systems.* He even intended to name that work *Dialogue on the Flux and Reflux of the Tides.*

## 2. *The Comets*

The next two years he was mostly ill, but did some minor work, such as the construction of a naval telescope, and also made an attempt, which failed, to use the periods of the Jupiter moons as an aid to determine geographical longitude. It was apparently the last time he took a positive interest in astronomical research.

After two years, in 1618, he could hold back no longer, and sent his treatise on the tides to Archbishop Leopold of Austria, describing it in his covering letter as a 'poetical conceit or dream' written at the time when he believed the Copernican system to be true, and before being taught better by the decision of the authorities who were 'guided by a higher insight than any to which my humble mind can of itself attain.' He hoped, no doubt, that the treatise would be printed in Austria without formal authorization on his part, but it came to nothing.

In the same year three comets appeared in the sky. They announced the beginning both of the Thirty Years War, and of the most disastrous of the many controversies in which Galileo became involved.

It was caused by a lecture, subsequently published, by the Jesuit Father Horatio Grassi of the *Collegium Romanum.* It expressed the correct view that comets move in regular orbits like planets, at a distance far greater than the moon's. In support of this view, Grassi quoted with approval Tycho's conclusions regarding the famous comet of 1577. The treatise was a further step in the Jesuits' retreat from Aristotle, who had maintained that comets were earthly exhalations in the sublunary sphere, and a further sign of the Order's implicit endorsement of the Tychonic system.

When Galileo read the treatise, he had an outburst of fury. He covered its margins with exclamations like 'piece of asininity', 'elephantine', 'buffoon', 'evil poltroon', and 'ungrateful villain'. The ingratitude consisted in the fact that the treatise did not mention Galileo's name – whose only contribution to the theory of comets had been a casual endorsement of Tycho's views in the *Letters on Sunspots*.[4]

But now the situation had changed. The Tychonic compromise must be rejected, so that the choice should remain confined to the discredited Ptolemy and to Copernicus. Galileo abruptly reversed his own arguments: he decided that comets are not real objects at all, but optical illusions like the aurora-borealis or the mock-suns, caused by the reflection of earthly vapours, which reach up into the sky past the moon. If they were real, they ought to appear larger as they approach the earth and smaller as they recede, whereas, according to Galileo, comets appear at their full size and then vanish altogether.

Apart from wishing to prove that Tycho and Grassi understood nothing of astronomy, Galileo had another motive for denying that comets exist: their paths were so markedly elliptic, that they could not be reconciled with the circular orbits in which all real heavenly objects must move around the sun.

Galileo did not attack Grassi directly, under his own name, but let his former pupil Mario Guiducci, sign a *Discourse on Comets* – the manuscript of most of which survived, and is in Galileo's handwriting. At the end of the treatise, Grassi is reproached for failing to mention Galileo's discoveries and Father Scheiner for 'misappropriating the discoveries of others'.

Since Galileo had not signed with his own name, Grassi now replied under a transparent anagram as 'Lothario Sarsi Sigensano' (for Horatio Grassi Salonensi). He ignored Guiducci, and attacked Galileo with vehemence. He showed that Galileo claimed priority for discoveries which were not his own, and took up the challenge about the Tychonic system: since Ptolemy was refuted and Tycho rejected by Galileo, does he mean that Grassi should have endorsed Copernicus, condemned and abhorred by every good Catholic?

Grassi's pamphlet was published in 1619 under the title *The Astronomical and Philosophical Balance*. Galileo's answer was the famous *Il Saggiatore – The Assayer*, who measures things on the finer balance designed for precious metals. It took him two years to write it, and it was published in 1623, only four years after Grassi's counter-attack.

*The Assayer* was written in the form of a letter to a friend, Monsignor Cesarini, Chamberlain to the Pope. It starts with a tirade against all who tried to rob Galileo 'of the glory of his discoveries', to whose ranks he now added Marius von Gunzenhausen, the discoverer of the spiral nebula in Andromeda (the first nebula observed). It is in this context that the passage occurs which I have quoted already: 'You cannot help it, Signor Sarsi, that it was granted to me alone to discover all the new phenomena is the sky and nothing to anybody else. This is the truth which neither malice nor envy can suppress.'

*The Assayer* then sets out to demolish the reputation of Tycho by talking of his 'alleged observations', and by calling comets 'Tycho's monkey-planets'. He also explains the reason which forced him to break his previous resolve to publish no more: Galileo's enemies, having unsuccessfully tried to steal his discoveries, now try to attribute to him 'the works of others' – namely, Guiducci's tract. He indignantly denies having had any part in that tract beyond discussing the subject with Guiducci; but now he has to break his silence 'to discourage those who refuse to let sleeping dogs lie, and who stir up trouble with men that are at peace'.

The major part of the work consists of sarcastic refutations of everything Grassi had said, regardless whether the poor man had blundered – which he often had – or hit on the truth. Thus Grassi had maintained that projectiles get heated by friction with the air; Galileo answered that they got not hotter but colder: 'trying to pulverize the air is as great a waste of time as grinding water in the proverbial mortar'.[5] As it so often happens, Grassi had tried to prove a good case by a bad argument: he quoted Suidas (a tenth-century Greek lexicographer) to the effect that the Babylonians cooked eggs by whirling them swiftly through the air on a sling. This gave Galileo an opportunity of making mincemeat of his opponent in a hilarious passage which is often quoted (but mostly without mention of the context):

> If Sarsi wants me to believe with Suidas that the Babylonians cooked their eggs by whirling them in slings, I shall do so; but I must say that the cause of this effect was very different from what he suggests. To discover the true cause I reason as follows: 'If we do not achieve an effect which others formerly achieved, then it must be that in our operations we lack something that produced their success. And if there is just one single thing we lack, then that alone can be the true cause. Now we do not lack eggs, nor slings, nor sturdy fellows to whirl them; yet our eggs do not cook, but merely cool down faster if they happen to be hot. And since nothing is lacking to us except being Babylonians, then being Babylonians is the cause of the hardening of eggs, and not friction of the air.'[6]

But in between these brilliant irrelevancies and sophistries, there are again passages scattered about which have become classics of didactic literature. They concern the principles of scientific reasoning, experimental procedure, the philosopher's duty to be sceptical about authorities and principles that are being taken for granted. Above all, Galileo outlines a principle which became of outstanding importance in the history of thought: the distinction between primary qualities in nature such as the position, number, shape, and motion of bodies, and secondary qualities such as colours, odours, and tastes, which are said to exist only in the observer's consciousness.[7]

> To excite in us tastes, odours, and sounds I believe that nothing is required in external bodies except shapes, numbers, and slow or rapid movements. I think that if ears, tongues, and noses were removed, shapes and numbers and motions would remain, but not odours or tastes or sounds. The latter, I believe, are nothing more than names when separated from living beings....

Though anticipated by the Greek atomists, it is for the first time in the modern age that this distinction is made in such concise terms, the first formulation of the mechanistic view of the universe. But on most contemporary readers of *The Assayer* the significance of that passage was lost. They saw Galileo only in the role of the toreador, and the concensus was that Father Grassi had to be dragged out of the arena sprawling on his back.

Grassi was a prominent Jesuit scholar, and not at all the fool that Galileo made him out to be. He had drawn the plans for the Church of St Ignazio in Rome and had designed a submarine, based on a suggestion of Leonardo's. The treatment meted out to him, added to the equally unprovoked attacks upon Scheiner, turned these two influential members of the Jesuit Order into implacable enemies of Galileo. A third Jesuit whom he had attacked without necessity (on a question of military engineering, of all things) was Father Firenzuola, who built the fortifications of the Castle St Angelo. Twenty-five years later, Firenzuola was the Inquisition's Commissary General at Galileo's trial. The result of all this was that the Jesuits as a body turned against Galileo. Father Grienberger, who succeeded Clavius as head of the Roman College, was to remark later that 'if Galileo had not incurred the displeasure of the Company, he could have gone on writing freely about the motion of the earth to the end of his days'.[8]

The clash with the Aristotelians was inevitable. The clash with the Jesuits was not. This is not meant as an apology for the vindictiveness with which Grassi and Scheiner reacted when provoked, nor of the deplorable manner in which the Order displayed its *esprit de corps*. The point to be established is that the attitude of the *Collegium Romanum* and of the Jesuits in general changed from friendliness to hostility, not because of the Copernican views held by Galileo, but because of his personal attacks on leading authorities of the Order.

Other great scientists, including Newton, became embroiled in bitter polemics. But these were peripheral to their work, skirmishes around a solidly established position. The particular tragedy of Galileo was that his two major works were only published after his seventieth year. Up to then, his output consisted in pamphlets, tracts, manuscripts circulated privately, and oral persuasion – all of it (except the *Star Messenger*) polemical, ironically aggressive, spiced with arguments *ad hominem.* The best part of his life was spent in these skirmishes. Until the end he had no fortress in the form of a massive and solid *magnum opus* to fall back upon. The new conception of science and philosophy which he brought into the world is diffused in passages here and there among the polemics of the *Letters on Sunspots* or *The Assayer* – hidden between tangles of barbed wire, as Kepler's laws were among his harmonic labyrinths.

### 3. *Dangerous Adulation*

While he was writing *Il Saggiatore,* Galileo's loyal patron, Cosmo II, died and the formidable Dowager Christina became Regent. Bellarmine, who had been a restraining influence at the head of the Jesuit Order, died in the same year. But against these losses, fate threw into the assayer's balance the most unexpected and powerful ally: Maffeo Barberini was elected to the papacy in 1623 – just in time for Galileo to dedicate *Il Saggiatore* to him.

Maffeo Barberini was something of an anachronism: a Renaissance Pope transplanted into the age of the Thirty Years War; a man of letters who translated passages from the Bible into hexameters; cynical, vainglorious, and lusting for secular power. He conspired with Gustavus Adolphus, the Protestant heretic, against the Holy Roman Empire; and on learning of the death of Richelieu, remarked: 'If there is a God, Cardinal Richelieu will have much to answer for; if not, he has done very well.' He fortified the Castle St Angelo, and had guns cast out of the bronze ceilings of the Pantheon – which gave rise to the epigram: 'What the barbarians have not done, Barberini did.' He founded the 'Office of the Propaganda' (for missionaries), built the Barberini Palace, and was the first Pope to allow a monument to be erected to him in his lifetime. His vanity was indeed monumental, and conspicuous even in an age which had little use for the virtue of modesty. His famous statement that he 'knew better than all the Cardinals put together' was only equalled by Galileo's that he alone had discovered everything new in the sky. They both considered themselves supermen and started on a basis of mutual adulation – a type of relationship which, as a rule, comes to a bitter end.

Back in 1616, Barberini had opposed the decree of the Congregation and intervened in favour of Galileo, a fact of which he often boasted later on. In 1620, he had written an ode in honour of Galileo, with the title *Adulatio Perniciosa* – which may be translated as 'Perilous Adulation'. He even went so far as to pay homage to the memory of Copernicus – in an audience with Cardinal Hohenzollern in 1624, after he had become Pope – and added the remark that 'the Church neither had condemned, nor ever would condemn his doctrine as heretical, but only as reckless'.[9]

When Urban was installed, there began a kind of second honeymoon between the repository of Faith and the foremost representative of Science in Italy. Renuncini, a brother of Cardinal Dini, wrote to Galileo:

> I swear to you that nothing pleased his Holiness so much as the mention of your name. After I had been speaking of you for some time, I told him that you, esteemed Sir, had an ardent desire to come and kiss his toe, if his Holiness would permit it, to which the Pope replied that it would give him great pleasure, if it were not inconvenient to you . . . for great men like you must spare themselves, that they may live as long as possible.[10]

Galileo was ill, so he could only journey to Rome in the spring of the next year. He had six long audiences with Urban in the course of six weeks. The Pope showered favours on him – a pension for Galileo's son, a precious painting, a gold and silver medal. He also provided him with a glowing testimonial, addressed to the new Grand Duke, extolling the virtues and piety 'of this great man, whose fame shines in the heavens, and goes on earth far and wide'.

What exactly was said during those six audiences, has been another subject-matter of conjecture and controversy. Only a few points have been established with certainty: first, that in spite of Galileo's attempts at persuasion, Urban refused to revoke the decree of 1616; secondly, that Galileo's impression derived from the six long audiences was that he could write pretty well anything he pleased in support of Copernicus so long as he avoided theological arguments, and stuck to speaking *ex hypothesi.* Thirdly, Urban himself made a suggestion how to get around the difficulty of arguing in favour of the Copernican system without asserting it to be true. The suggestion was this: assuming that a hypothesis explains satisfactorily certain phenomena, this does not necessarily mean that it is true, for God is all-powerful and may have produced the said phenomena by some entirely different means which are not understood by the human mind. This suggestion of Urban's, on which he laid great store, played a crucial part in the sequel.

Thus encouraged, and in the full sunshine of papal favour, Galileo, who was now past sixty, felt the road at last free to embark on his great apologia of Copernicus, which, as we know, he intended to call *Dialogue on the Flux and Reflux of the Tides.* It took him nevertheless four years to write it[10a]; for nearly three, from 1626 to 1629, he seems to have laid it aside, under various excuses and against the urging of his friends. He probably felt that the favour of princes is as shortlived as the tide itself, and that his powerful enemies were working against him. One may also suspect that he was hampered by a recurrent psychological blockage, a repressed doubt in the soundness of his 'conclusive proof'.

But once again he could not retreat. In January 1630, the *Dialogue* was completed.

## 4. Dialogue on the Great World Systems

The *Dialogue* is carried on by three characters. Salviati, the brilliant savant, is Galileo's mouthpiece; Sagredo, an intelligent amateur, plays second fiddle to him under the guise of neutrality, and Simplicio, the good-humoured simpleton, defender of Aristotle and Ptolemy, fulfils the role of the clown who is kicked in the pants. Salviati and Sagredo had been friends of Galileo, and were now both dead; Simplicio, Galileo claimed, derived his name from Simplicius, the sixth-century commentator on Aristotle, but the double meaning is evident. It is Simplicio who, after being shown up as an ass over and again, trots out at the very end Pope Urban's argument as coming 'from a most eminent and learned person, and before whom one must fall silent'; whereupon the other two declare themselves silenced by 'this admirable and angelic doctrine', and decide 'to go and enjoy an hour of refreshment in the gondola that awaits us'. And thus the *Dialogue* ends with what can only be described as a rude noise at the Pope – with the consequences that one may expect.

The *Dialogue* is divided into four days. The first is devoted to the refutation of the Aristotelian view of the cosmos in general. Passages of witty journalism alternate with others which suddenly ascend to an aloof and majestic vision, and the language assumes breath-taking beauty. In attacking the Platonic dualism of earthly corruption – heavenly perfection, Sagredo explains:

> I cannot without great wonder, nay more, disbelief, hear it being attributed to natural bodies as a great honour and perfection that they are impassible, immutable, inalterable, etc.: as, conversely, I hear it esteemed a great imperfection to be alterable, generable, and mutable. It is my opinion that the Earth is very noble and admirable by reason of the many and different alterations, mutations, and generations which incessantly occur in it. And if, without being subject to any alteration, it had been all one vast heap of sand, or a mass of jade, or if, since the time of the deluge, the waters freezing which covered it, it had continued an immense globe of crystal, wherein nothing had ever grown, altered, or changed, I should have esteemed it a wretched lump of no benefit to the Universe, a mass of idleness, and in a word superfluous, exactly as if it had never been in Nature. The difference for me would be the same as between a living and a dead creature. I say the same concerning the Moon, Jupiter, and all the other globes of the Universe. The more I delve into the consideration of the vanity of popular discourses, the more empty and simple I find them. What greater folly can be imagined than to call gems, silver, and gold noble, and earth and dirt base? For do not these persons consider that, if there were as great a

scarcity of earth as there is of jewels and precious metals, there would be no king who would not gladly give a heap of diamonds and rubies and many ingots of gold to purchase only so much earth as would suffice to plant a jessamine in a little pot or to set a tangerine in it, that he might see it sprout, grow up, and bring forth such goodly leaves, fragrant flowers, and delicate fruit?

It is scarcity and plenty that make things esteemed and despised by the vulgar, who will say that here is a most beautiful diamond, for it resembles a clear water, and yet would not part with it for ten tons of water. These men who so extol incorruptibility, inalterability, and so on, speak thus, I believe, out of the great desire they have to live long and for fear of death, not considering that, if men had been immortal, they would not have had to come into the world. These people deserve to meet with a Medusa's head that would transform them into statues of diamond and jade, that so they might become more perfect than they are.[11]

The battle for and against Copernicus is really joined on the second day, on which the objections against the earth's motion are refuted in terms of terrestrial physics. The central part of the argument concerns the relativity of motion. The classic objections had all been variations on the same theme: that if the earth rotated, everything not firmly attached to it would be left behind – cannon balls, falling stones, birds, clouds, and so forth. In his refutation, Galileo comes very near to a correct theory of impetus, and to Newton's First Law. He shows that a stone dropped from the top of a moving ship's mast would not be left behind because the stone shares the ship's momentum; and by analogy, that a stone dropped from a tower, or a cannon ball in flight, shares the earth's momentum.

But he could not entirely break loose from the Aristotelian dogma about circular motion. He postulates that if a body is left to itself, it will continue, under its initial momentum, to move, not in a straight line, but in a circular orbit through eternity. The reason for this Galileo explains in the opening part of the first day, and repeats over and again:

... straight motion being by nature infinite (because a straight line is infinite and indeterminate), it is impossible that anything should have by nature the principle of moving in a straight line; or, in other words, towards a place where it is impossible to arrive, there being no finite end. For nature, as Aristotle well says himself, never undertakes to do that which cannot be done, nor endeavours to move whither it is impossible to arrive.[12]

This belief contradicts Galileo's intimate knowledge of centrifugal forces, the tendency of an object moving in a circle to fly off at a tangent in a straight line. On the second day another classic objection against the earth's rotation, that bodies not attached to the earth would fly off into space, is admitted by Galileo as valid in theory, but negligible in practice, because the centrifugal force is so much smaller than the earth's attraction.[13] He thus asserts in one passage that a stone lying in a field has a natural tendency to persist in its circular motion, and in another that it has a natural tendency to fly off in a straight line. Similarly, he believed that freely falling bodies describe a circular path.[14] Thus even this most determined opponent of Aristotelianism could not rid himself of the old circular obsession – which partly explains Galileo's rejection of Kepler's Laws.

The second day ends, on Galileo's own admission, in a stalemate. He has refuted the objection that on a rotating earth detached bodies would be left behind, etc.; but he has not proved that the earth does rotate. On either hypothesis, whether she moves or stands still, stones would fall and birds would fly as they do.

The third day is concerned with the astronomical arguments for and against Copernicus, and here Galileo is downright dishonest. He first shows that the Copernican system is superior to the Ptolemaic by the familiar arguments from the Jupiter moons and the phases of Venus. He then explains that to 'save' the planets' apparent stations and retrogressions, Ptolemy had to introduce 'very great epicycles' which Copernicus was able to dispense 'with one single motion of the earth'. But he breathes not a word about the fact that Copernicus, too, needs a whole workshop full of epicycles; he keeps silent about the eccentricity of the orbits, the various oscillations and librations, the fact that the sun is neither in the centre of the motions, nor lies in their plane; in a word, he deliberately evades the real problems of astronomy which had started Tycho and Kepler on their quest. The planets all move in perfect circles at uniform linear velocities around the sun (which, for instance, would make Saturn's period twenty-four years instead of thirty).[15] All problems appear solved 'with admirable facility'; for 'in the Ptolemaic hypothesis there are the diseases, and in the Copernican their cure'.[16]

It is true that Galileo was writing for a lay audience, and in Italian; his account however, was not a simplification but a distortion of the facts, not popular science, but misleading propaganda. Even his latest, admiring biographer is prompted to the remark:

A drastic simplification of Copernicus may have seemed to him an easier didactic device. This is, at least, the charitable hypothesis. But the problem remains of how Galileo could commit the capital error, against which he had warned others so many times, of constructing theories in defiance of the best results of observation.[17]

Even so, the arguments are again inconclusive, for all that Salviati succeeds in proving against Simplicio is that the heliocentric system saves the phenomena more elegantly than the geocentric, but not that it is true. Moreover, he keeps silent about the fact that the Tychonic system fits the phenomena equally well.

To break the stalemate, the famous theory of the tides is brought out on the fourth day. But before that, at the end of the third, a new and unexpected argument makes its appearance. It is derived from the sunspots, and is introduced with a flourish:

Hearken then to this great and new wonder. The first discoverer of the solar spots, as also of all other celestial novelties, was our Lincean academician, and he discovered them *anno* 1610. ...[18]

The 'Lincean academician' is the expression by which the Galileo in the *Dialogue* refers to himself.

After thus reasserting his spurious claim, he proceeds to claim another discovery of Scheiner's: that the sun, and with it the spots, rotate on an axis which is tilted against the plane of the ecliptic. As a result of this, the spots also travel round the sun in 'tilted' circles (as seen from the earth); and their curve changes according to the earth's position – just as the curvature of a tilted spinning top changes to the eye as we walk around it. Ergo, Galileo concludes, the changing curves along which the sunspots travel prove, in a manner 'so solid and rational as never before', that the earth moves round the sun.[19]

At this point poor Simplicius turns into a relativist and correctly observes that the curves of the spots would look just the same whether the sun travelled round the earth or the earth round the sun. Salviati proceeds to demolish this objection: if we assume that the sun travels round the earth, the spots will look the same only if we also assume that the sun's axis always remains parallel to itself; and this he finds *'very hard and almost impossible to believe'*.[20] Simplicius, intimidated, pipes down; Sagredo exclaims 'that amongst all the ingenious subtleties I ever heard, I have never met with anything of greater admiration to my intellect or that has more absolutely captivated my judgement'.[21]

One simply gapes. Salviati wins his case by pretending that it was virtually impossible for one heavenly body to travel round another while its axis remains parallel to itself. Yet that is, of course, what the earth does while travelling round the sun: its axis remains parallel to itself at a constant tilt of 23½ degrees. If it was impossible to believe that the sun could move thus, then it was equally impossible that the earth should move thus. Yet in a later section Galileo discusses at great length the reasons *why* the earth moves thus, and explains that the preservation of the fixed tilt of its axis 'is far from having any repugnance or difficulty in it'.[22]

The changing faces of the sunspot-paths were as obvious a consequence of the tilt in the sun's axis as the changing seasons are a consequence of the tilt in the earth's axis. It was as simple as that. But the two pages in which Galileo argues the point against Simplicius[23] are among the most obscure and incomprehensible in the book. He employs his usual tactics of refuting his opponent's thesis without proving his own; in this case not by sarcasm, but by confusing the issue.

There can be no doubt that Galileo's theory of the tides was based on unconscious self-deception; but in the light of the above there can also be little doubt that the sunspot argument was a deliberate attempt to confuse and mislead. To represent the constant tilt of a rotating body as a new and inconceivable hypothesis, when every student since Pythagoras knew that this was the reason why summer followed winter; to obscure this simple issue by the novelty of curving sunspots, while making the complexities of Copernicus appear deceptively simple, was part of a deliberate strategy, based on Galileo's contempt for the intelligence of his contemporaries. We have seen that scholars have always been prone to manias and obsessions, and inclined to cheat about details; but impostures like Galileo's are rare in the annals of science.

The fourth and last day of the *Dialogue* is taken up almost entirely by the theory of the tides, which is elaborated in more detail. The annual variations in the tides are explained by the tilt of the earth's axis, the monthly variations by monthly changes in orbital velocity.[24] Kepler's explanation of the tides by the moon's attraction is rejected with the remark that 'despite his open and penetrating mind' he has 'lent his ear and his assent to the moon's dominion over the waters, to occult properties [gravity] and such-like little fancies'.[24a]

Another surprising thing about the *Dialogue* is that Galileo not only misrepresented the Copernican system as a beautifully simple affair, but seems to have been himself unaware of its complexities. He had never taken much interest in the tiresome details of planetary theory, and there was no real reason for him to plod through the technical chapters in the *Revolutions* from cover to cover. If he had done so, he could not have believed that all planets move with the same angular velocity, nor attributed the idea to Copernicus that the moon either shines in her own light or is transparent to the light of the sun.[25] About the insoluble difficulties of the Copernican system we learn only by an aside:

. . . how each planet governs itself in its particular revolutions, and how precisely the structure of its circle is framed, which is what is commonly called the Theory of the Planets, we cannot yet undoubtedly resolve. Mars, that has so much puzzled our modern astronomers, is a proof of this.[26]

This was written some twenty years after Kepler's determination of the Martian orbit laid a new foundation for planetary theory.* The truth is that after his sensational discoveries in 1610, Galileo neglected both observational research and astronomic theory in favour of his propaganda crusade. By the time he wrote the *Dialogue* he had lost touch with new developments in that field, and forgotten even what Copernicus had said.

### 5. *The* Imprimatur

The manuscript was completed in January 1630.

Galileo intended to supervise the printing of the book in Rome, but could not go at once. His friends assured him that there would be no difficulty, and that all was sunshine. Faithful Father Castelli, who now lived in Rome, wrote that Urban VIII had assured Campanella in an audience that 'if it had depended on him, the prohibition of 1616 would never have been passed'.[27] Another member of the old guard, Monsignor Ciàmpoli, who was now Papal Secretary, wrote that at the Vatican 'they were longing for Galileo more than for any beloved damsel'.[28]

He arrived in Rome at the beginning of May, and was received in a long audience by Urban VIII. The Pope again confirmed that there was no objection against discussing the merits of the Copernican system, provided that it was treated strictly as a hypothesis. He objected, however, to the intended title, *Dialogue on the Flux and Reflux of the Tides*, which laid too much emphasis on physical proof, and suggested that it should be called a *Dialogue on the Great World Systems* instead. He was, of course, far too busy to read the book himself, and left this task to the censors.

The function of Chief Censor and Licenser was exercised by the 'Master of the Palace', Father Niccolo Riccardi. He was

* It will be remembered that the *New Astronomy's* sub-title is *Studies on Mars.*

another Florentine, a member of the Castelli-Ciàmpoli clique, and accordingly devoted to Galileo, though he believed that the Ptolemaic and Copernican systems were mere highbrow pastimes since the ultimate truth was that the stars were moved by angels. But this did not prevent him from admiring the ingenuity of men like Galileo who were plotting the courses of these angelic gymnastics. Because of his enormous girth, Riccardi had been called by the King of Spain *Il Padre Mostro*, the Father Monster, and all his friends referred to him by this affectionate nickname. By a perverse twist of history it was this pure-hearted and lovable man who, by his bungling, became the chief cause of the tragedy.

The Father Monster read through the manuscript of the *Dialogue*, and came to the conclusion that it was above his head. He knew that His Holiness had approved of the idea of the book, had showered favours on Galileo, and encouraged him to go ahead. But he also sensed, though unable to follow the argument in detail, that the book was thinly disguised propaganda for Copernicus, and contradicted in spirit and letter the decree of 1616. To escape the dilemma he instructed his assistant, Father Visconti, to work through the text and to make suitable alterations.

Visconti was equally unfit for the task. He made some minor corrections which aimed at making the pro-Copernican arguments appear more 'hypothetical', then handed the text back to his superior.

Riccardi felt even more helpless than before. He played for time, and finally decided that he must shoulder his responsibilities and revise the text himself. But he now met with the concerted pressure of Galileo and his allies: to wit, the Papal Secretary Ciàmpoli, who indirectly represented the will of His Holiness, and the new Tuscan Ambassador, Niccolini, who was married to the Father Monster's favourite cousin, Caterina.

The result of this pressure was that Riccardi agreed to make an unusual deal: to save time, he granted the *imprimatur* for the book in advance, on condition that he would revise it himself, and then pass on each revised sheet to the printer. He was to be assisted in his task by the universally respected President of the Lincean Academy, Prince Cesi.

As soon as this agreement was concluded, Galileo returned to Florence to escape the heat of Rome, on the understanding that he would be back in the autumn. But soon after his departure, Prince Cesi died. Another few weeks later, the plague broke out, and the strict quarantine made communications between Rome and Florence difficult. This provided a welcome opportunity for Galileo to wriggle out of the conditions under which the *imprimatur* had been granted: he demanded that the book be printed in Florence out of Riccardi's control. The devoted Castelli again played a fateful part in this manoeuvre, by feeding Galileo's suspicions with dark hints about 'most weighty reasons which he did not wish to commit to paper'[29] – just as he had done, years ago, by the exaggerated importance he had given to the dinner talk with the Grand Duchess Christina.

Riccardi at first flatly refused to grant permission for printing the book in Florence without revising it; he demanded that Galileo should send the manuscript to Rome for this purpose. Galileo answered that quarantine regulations made the safe dispatch of the manuscript impossible, and insisted that the final revision should be done by a Florentine censor. He enlisted the support of the Grand Duke (to whom Riccardi, as a Florentine, owed allegiance). The Tuscan Ambassador, Niccolini, and the Papal Secretary, Ciàmpoli, also renewed the pressure. The Father Monster was a constant guest at the Niccolinis'; in the end it was his beautiful Cousin Caterina who made him yield, over a bottle of Chianti at her dinner table. He agreed that the work should be revised and printed in Florence, except for the preface and concluding paragraphs which must be submitted to himself.

The revision was to be done by the Florentine Inquisitor, Father Clemente Egidii. But this was not to Galileo's taste, who proposed Father Stefani instead of Egidii. Riccardi again agreed. Evidently Father Stefani was entirely under Galileo's sway, for he was 'moved to tears at many passages by the humility and reverend obedience' of the book. Stefani made a few corrections, for form's sake, and the printing began early in 1631. Riccardi, who had dark forebodings, still tried to play for time by withholding the preface and concluding sections. Once again the Niccolinis' help was enlisted. They managed to wrench the revised preface and conclusion from their cousin, though he only consented 'dragged by the hair', as Niccolini himself described it. And thus, in February 1632, the first printed copies of the *Dialogue* came from the press.

It took only a few weeks for Urban and the Holy Office to discover that they had been outwitted. By August, the book was confiscated, and in October Galileo was summoned to appear before the Inquisition in Rome. He succeeded in delaying his journey on grounds of ill health and other pretexts for four months; but in February '33, he had to go. He took up quarters at the Tuscan Embassy, as before; but for another three months nothing happened. It was on 12 April that his first interrogation at the Holy Office took place.

There is little doubt that the decision to instigate proceedings was Urban VIII's, who felt that Galileo had played a confidence trick on him. There is equally little doubt that the Jesuits used their influence to have the book banned, and to turn the Pope against its author. Apart from solidarity with Fathers Grassi and Scheiner, they were probably moved by the consideration that Galileo's rejection of the Tychonic compromise would hamper the gradual evolution of the Church towards the new cosmology, and that his all-or-nothing gamble, based on the spurious arguments about the sunspots and the tides, might play into the hands of the reactionary forces inside the Church, upsetting their careful cosmic strategy.

But it did not require much Jesuit cunning to turn Urban's perilous adulation into the fury of the betrayed lover. Not only had Galileo gone, in letter and spirit, against the agreement to treat Copernicus strictly as a hypothesis, not only had he obtained the *imprimatur* by methods resembling sharp practice, but Urban's favourite argument was only mentioned briefly at the very end of the book, and put into the mouth of the simpleton who on any other point was invariably proved wrong. Urban even suspected that Simplicius was intended as a caricature of his own person. This, of course, was untrue; but Urban's suspicion persisted long after his fury had abated:

> I hear from Rome [Galileo wrote three years after his trial] that his Eminence Cardinal Antonio Barberini and the French Ambassador have seen his Holiness and tried to convince him that I never had the least idea of perpetrating so sacrilegious an act as to make game of his Holiness, as my malicious foes have persuaded him, and which was the primary cause of all my troubles.[30]

If corroboration were needed, it is to be found in Niccolini's reports. They stress that Urban 'was so incensed that he treated this affair as a personal one',[31] and quote Urban's 'bitter remark' that Galileo had deceived him.

## 6. *The Trial*

The proceedings against Galileo began with the appointment of a special commission to investigate the whole affair. The Commission's findings were that Galileo had transgressed orders in deviating from the hypothetical treatment of Copernicus, and maintaining absolutely the earth's motion; that he had erroneously ascribed the phenomena of the tides to it; and thirdly, that he had been deceitfully silent about the command laid upon him by the Holy Office in 1616 'to relinquish altogether the said opinion ... nor henceforth to hold, teach, or defend it in any way whatsoever, verbally or in writing'. This third point referred to the controversial minute about the serving of an absolute injunction (*see above*, p. 468 f.) which the Commission had discovered in the archives.

The Commission did not recommend any specific steps to be taken against Galileo; as for his book, the contents were indicted on eight counts, but the Commission suggested that all these matters could be corrected if the book were thought to be of value. The report was then handed over for further action to the Inquisition, which issued its summons in October 1632, and first interrogated Galileo on 12 April of the following year.

According to the basic rule of inquisitorial procedure, the charges were not communicated to the accused; he was, on the contrary, asked whether he knew or guessed on what grounds he had been summoned.* Galileo said that he believed it was on account of his latest book. The Commissary, Firenzuola, then questioned him in detail about the events of 1616. Galileo stated that he had been told by the Lord Cardinal Bellarmine that 'the opinion of Copernicus, if adopted absolutely, was contrary to Holy Scripture and must neither be held nor defended, but that it could be taken and used hypothetically'. He affirmed that he had 'not in any way disobeyed this command, that is, had not by any means held or defended the said opinion'. The Inquisitor then read to him the alleged absolute injunction of 1616 that Galileo must 'neither hold, defend, nor teach that opinion *in any way whatsoever*'. Galileo did not directly deny the absolute injunction, but said that he could not remember the words 'not to teach' and 'in any way'; he referred to Bellarmine's certificate which did not contain these words. The Inquisitor then went over the whole story of the negotiations concerning the *imprimatur*. He asked whether when he applied for permission to print the *Dialogue*, Galileo had informed Father Riccardi about the command that had been issued to him. Galileo answered that he did not think it necessary to do so, 'for I have neither maintained nor defended in that book the opinion that the earth moves and that the sun is stationary, but have rather demonstrated the opposite of the Copernican opinion, and shown that the arguments of Copernicus are weak and not conclusive'.[32]

With that the first hearing ended.

Five days after the hearing, three experts of the Inquisition, who had been appointed to examine the contents of the book,

* This has also become the standard procedure in trials by the Soviet State Police. The 'inquisitorial' character of OGPU methods is more than a figure of political jargon. The absolute secrecy enjoined on the accused regarding the proceedings and even the fact that he is under investigation; the absence of lawyers for the defence, and the assumption that he is guilty unless proved innocent; the methods of psychological pressure, the alternation between threats and paternal reassurance, and above all, the metaphysical axiom of the 'union of wills' between Church and penitent are only the more salient features which the OGPU copied after thorough study of the Inquisition's methods and procedure.

handed in their reports which, by the consensus of historians, were accurate and fair. By a long list of quotations they proved beyond doubt that Galileo had not only discussed the Copernican view as a hypothesis, but that he had taught, defended, and held it, and that he had called those who did not share it 'mental pygmies', 'dumb idiots', and 'hardly deserving to be called human beings'.

To pretend, in the teeth of the evidence of the printed pages of his book, that it said the opposite of what it did, was suicidal folly. Yet Galileo had had several months' respite in which to prepare his defence. The explanation can only be sought in the quasi-pathological contempt which Galileo felt for his contemporaries. The pretence that the *Dialogue* was written in refutation of Copernicus was so patently dishonest that his case would have been lost in any court.

The next, unexpected turn of events is best described in the words of one of the principal personalities in the drama, the Commissary of the Inquisition, Firenzuola. In a letter to Urban's brother, Cardinal Francesco Barberini, who was one of the judges at the trial, he reported:[33]

In compliance with the commands of His Holiness, I yesterday informed the Most Eminent Lords of the Holy Congregation of Galileo's case, the position of which I briefly reported. Their Eminences approved of what has been done thus far and took into consideration, on the other hand, various difficulties with regard to the manner of pursuing the case and of bringing it to an end. More especially as Galileo has in his examination denied what is plainly evident from the book written by him, since in consequence of this denial there would result the necessity for greater rigour of procedure and less regard to the other considerations belonging to this business. Finally, I suggested a course, namely, that the Holy Congregation should grant me permission to treat extra-judicially with Galileo, in order to render him sensible of his error and bring him, if he recognizes it, to a confession of the same. This proposal appeared at first too bold, not much hope being entertained of accomplishing this object by merely adopting the method of argument with him; but, upon my indicating the grounds upon which I made the suggestion, permission was granted me. That no time might be lost, I entered into discourse with Galileo yesterday afternoon, and after many and many arguments and rejoinders had passed between us, by God's grace, I attained my object, for I brought him to a full sense of his error, so that he clearly recognized that he had erred and had gone too far in his book.* And to all this he gave expression in words of much feeling, like one who experienced great consolation in the recognition of his error, and he was also willing to confess it judicially. He requested, however, a little time in order to consider the form in which he might most fittingly make the confession, which, as far as its substance is concerned, will, I hope, follow in the manner indicated.

I have thought it my duty at once to acquaint your Eminence with this matter, having communicated it to no one else; for I trust that His Holiness and your Eminence will be satisfied that in this way the affair is being brought to such a point that it may soon be settled without difficulty. The court will maintain its reputation; it will be possible to deal leniently with the culprit; and, whatever the decision arrived at, he will recognize the favour shown him, with all the other consequences of satisfaction herein desired. Today I think of examining him in order to obtain the said confession; and having, as I hope, received it, it will only remain to me further to question him with regard to his intention and to receive his defence plea; that done, he might have his house assigned to him as a prison, as hinted to me by your Eminence, to whom I offer my most humble reverence.

Your Eminence's most humble and most obedient servant,

FRA VINC°. DA FIRENZUOLA

Rome, 28 April 1633.

The letter speaks for itself: the tradition of the sacred cows was still alive, in spite of everything.

Two days after the interview on 30 April, Galileo was called for examination a second time, and was asked whether he had anything to say. He made the following statement:

In the course of some days' continuous and attentive reflection on the interrogations put to me on the twelfth of the present month, and in particular as to whether, sixteen years ago, an injunction was intimated to me by order of the Holy Office, forbidding me to

* I was amused and pleased to find that Santillana comments on the Commissary's private surprise visit to the accused Galileo: 'It was Ivanov coming to Rubashov.'

hold, defend, or teach 'in any manner' the opinion that had just been condemned – of the motion of the Earth and the stability of the Sun – it occurred to me to re-peruse my printed *Dialogue*, which for three years I had not seen, in order carefully to note whether, contrary to my most sincere intention, there had, by inadvertence, fallen from my pen anything from which a reader, or the authorities, might infer not only some taint of disobedience on my part, but also other particulars which might induce the belief that I had contravened the orders of the Holy Church.

Being, by the kind permission of the authorities, at liberty to send about my servant, I succceeded in procuring a copy of my book, and, having procured it, I applied myself with the utmost diligence to its perusal and to a most minute consideration thereof. And, as owing to my not having seen it for so long, it presented itself to me, as it were, like a new writing and by another author, I freely confess that in several places it seemed to me set forth in such a form that a reader ignorant of my real purpose might have had reason to suppose that the arguments brought on the false side, and which it was my intention to confute, were so expressed as to be calculated rather to compel conviction by their cogency than to be easy of solution.

Two arguments there are in particular – the one taken from the solar spots, the other from the ebb and flow of the tide – which, in truth, come to the ear of the reader with far greater show of force and power than ought to have been imparted to them by one who regarded them as inconclusive, and who intended to refute them, as indeed I truly and sincerely held and do hold them to be inconclusive and admitting of refutation. And, as an excuse to myself for having fallen into an error so foreign to my intention, not contenting myself entirely with saying that when a man recites the arguments of the opposite side with the object of refuting them, he should, especially if writing in the form of dialogue, state these in their strictest form and should not cloak them to the disadvantage of his opponent – not contenting myself, I say, with this excuse, I resort to that of the natural complacency which every man feels with regard to his own subtleties and in showing himself more skilful than the generality of men in devising, even in favour of false propositions, ingenious and plausible arguments. With all this, although with Cicero '*avidior sim gloriæ quam sat est*', if I had now to set forth the same reasonings, without doubt I should so weaken them that they should not be able to make an apparent show of that force of which they are really and essentially devoid. My error, then, has been – and I confess it – one of vainglorious ambition and of pure ignorance and inadvertence.

This is what it occurs to me to say with reference to this particular and which suggested itself to me during the re-perusal of my book.[34]

When he had finished this statement, the hearing was closed; but Galileo, after being dismissed, returned and volunteered the following supplementary statement:

And in confirmation of my assertion that I have not held and do not hold as true the opinion which has been condemned, of the motion of the Earth and stability of the Sun – if there shall be granted to me, as I desire, means and time to make a clearer demonstration thereof, I am ready to do so; and there is a most favourable opportunity for this, seeing that in the work already published the interlocutors agree to meet again after a certain time to discuss several distinct problems of Nature not connected with the matter discoursed of at their meetings. As this affords me an opportunity of adding one or two other 'days', I promise to resume the arguments already brought in favour of the said opinion, which is false and has been condemned, and to confute them in such most effectual manner as by the blessing of God may be supplied to me. I pray, therefore, this holy Tribunal to aid me in this good resolution and to enable me to put it in effect.[35]

I have criticized Galileo freely, but I do not feel at liberty to criticize the change in his behaviour before the Inquisition. He was seventy, and he was afraid. That his fears were exaggerated, and that his self-immolatory offer (which the Inquisitors discreetly allowed to drop as if it had never been made) was quite unnecessary, is beside the point. His panic was due to psychological causes: it was the unavoidable reaction of one who thought himself capable of outwitting all and making a fool of the Pope himself, on suddenly discovering that he has been 'found out'. His belief in himself as a superman was shattered, his self-esteem punctured and deflated. He returned to the Tuscan Embassy, in Niccolini's words 'more dead than alive'. From then on he was a broken man.

He was called again ten days later, on 10 May, to a purely formal hearing, at which he handed in his written defence.[36]

In the first part he argued – 'in order to demonstrate the purity of my intention, ever foreign to the practice of dissimulation or deceit in any operation I engage in' – that he was unaware of a specific and absolute injunction in 1616, and made out a convincing case for this. The main point of his defence was that

those faults which are seen scattered throughout my book, have not been artfully introduced with any concealed or other than sincere intention, but have only inadvertently fallen from my pen, owing to a vainglorious ambition and complacency in desiring to appear more subtle than the generality of popular writers, as indeed in another deposition I have confessed; which fault I shall be ready to correct with all possible industry whenever I may be commanded or permitted by Their Most Eminent Lordships.

He concludes on a tone of a humble personal appeal:

Lastly, it remains for me to beg you to take into consideration my pitiable state of bodily indisposition, to which, at the age of seventy years, I have been reduced by ten months of constant mental anxiety and the fatigue of a long and toilsome journey at the most inclement season – together with the loss of the greater part of the years to which, from my previous condition of health, I had the prospect. I am persuaded and encouraged to do so by the faith I have in the clemency and goodness of the most Eminent Lords, my judges; with the hope that they may be pleased, in answer to my prayer, to remit what may appear to their entire justice the rightful addition that is still lacking to such sufferings to make up an adequate punishment for my crimes, out of consideration for my declining age, which, too, humbly commends itself to them. And I would equally commend to their consideration my honour and reputation, against the calumnies of ill-wishers, whose persistence in detracting from my good name may be inferred from the necessity which constrained me to procure from the Lord Cardinal Bellarmine the attestation which accompanies this.

The remainder of the trial was now expected to be a mere formality. Throughout the proceedings Galileo had been treated with great consideration and courtesy. Against all precedent he was not confined to the dungeons of the Inquisition, but was allowed to stay as the Tuscan Ambassador's guest at the Villa Medici, until after his first examination. Then he had to surrender formally to the Inquisition, but instead of being put into a cell, he was assigned a five-roomed flat in the Holy Office itself, overlooking St Peter's and the Vatican gardens, with his own personal valet and Niccolini's major domo to look after his food and wine. Here he stayed from 12 April to the third examination on 10 May. Then, before his trial was concluded, he was allowed to return to the Tuscan Embassy – a procedure quite unheard of, not only in the annals of the Inquisition but of any other judiciary. Contrary to legend, Galileo never spent a day of his life in a prison cell.

The sentence did not come until six weeks later. On 16 June, the following decision was entered into the acts:

> ... Sanctissimus decreed that said Galileo is to be interrogated as to his intention [in writing the *Dialogue*] under the threat of torture; and if he kept firm he is to be called upon to abjure before a plenary assembly of the Congregation of the Holy Office, and is to be condemned to imprisonment at the pleasure of the Holy Congregation, and ordered not to treat further, in whatever manner, either in words or in writing, of the mobility of the Earth and the stability of the Sun; otherwise he will incur the penalties of relapse. The book entitled *Dialogo di Galileo Galilei Linceo* is to be prohibited. Furthermore, that these things may be known by all, he ordered that copies of the sentence shall be sent to all Apostolic Nuncios, to all Inquisitors against heretical pravity, and especially the Inquisitor in Florence, who shall read the sentence in full assembly and in the presence of most of those who profess the mathematical art.[37]

Two days after this decision was taken, the Pope received Niccolini in audience, hinted at the sentence to come, and added:

> However, after the publication of the sentence we shall see you again, and shall consult together so that he may suffer as little distress as possible, since matters cannot be let pass without some demonstration against his person.

Another three days later Galileo was convoked for his third and last examination. After he had taken the oath, he was questioned about his real conviction concerning the two cosmological systems. He answered that before the decree of 1616 he had considered that either Ptolemy or Copernicus might be true in nature, 'but after the said decision, assured of the wisdom of the authorities, I ceased to have any doubt; and I held, as I still hold, as most true and indisputable the opinion of Ptolemy, that is to say, the stability of the Earth'.[38]

He was then told that by the manner in which the subject was treated in the *Dialogue*, and the fact alone that he had written the said book, he was presumed to have held the Copernican opinion, and was asked a second time to state the truth freely. He answered that he had written the book to confer a common benefit by setting forth the arguments for both sides, and repeated again 'I do not now hold the condemned opinion, and have not held it since the decision of the authorities.'[39]

He was admonished a third time that on the contents of the book he was presumed to hold with Copernicus, or at least to have done so at the time he wrote it, and that therefore 'unless he made up his mind to confess the truth, recourse would be had against him to the appropriate remedies of the law.' Galileo answered: 'I do not hold, and have not held, this opinion of Copernicus since the command was intimated to me that I must abandon it; for the rest I am here in your hands – do with me what you please.' When he was for a last time bidden to speak the truth, under threat of torture, Galileo repeated, 'I am here to obey and I have not held this opinion since the decision was pronounced, as I have stated.'[40]

If it had been the Inquisition's intention to break Galileo, this obviously was the moment to confront him with the copious extracts from his book – which were in the files in front of the judge – to quote to him what he had said about the sub-human morons and pygmies who were opposing Copernicus, and to convict him of perjury. Instead, immediately following Galileo's last answer, the minutes of the trial say:

'And as nothing further could be done in execution of the decree, his signature was obtained to his deposition and he was sent back.'[41]

Both the judges and the defendant knew that he was lying;

both the judges and he knew that the threat of torture (*territio verbalis*)* was merely a ritual formula, which could not be carried out; and that the hearing was a pure formality. Galileo was led back to his five-room apartment, and on the next day the sentence was read out to him.† It was signed by only seven of the ten judges. Among the three who abstained was Cardinal Francesco Barberini, Urban's brother. The *Dialogue* was prohibited; Galileo was to abjure the Copernican opinion, was sentenced to 'formal prison during the Holy Office's pleasure'; and for three years to come, was to repeat once a week the seven penitential psalms. He was then presented with the formula of abjuration,‡ which he read out. And that was the end of it.

The 'formal prison' took the form of a sojourn at the Grand Duke's villa at Trinita del Monte, followed by a sojourn in the palace of Archbishop Piccolomini in Siena, where, according to a French visitor, Galileo worked 'in an apartment covered in silk and mostly richly furnished'.[44] Then he returned to his farm at Arcetri, and later to his house in Florence, where he spent the remaining years of his life. The recital of the penitential psalms was delegated, with ecclesiastical consent, to his daughter, Sister Marie Celeste, a Carmelite nun.[44]

From the purely legal point of view the sentence was certainly a miscarriage of justice. If one works through the maze of verbiage, it appears that he was found guilty on two counts: firstly, of having contravened both Bellarmine's admonition, and the alleged formal injunction of 1616, and having 'artfully and cunningly extorted the licence to print by not notifying the censor of the command imposed upon him'; and secondly, of having rendered himself 'vehemently suspect of heresy, namely, of having believed and held the doctrine which is contrary to sacred Scripture that the sun is the centre of the world'. Concerning the first count, no more need be said about the dubious character of the document referring to the alleged absolute in-

* As opposed to *territio realis* where the instruments of torture are shown to the accused, as in the case of Kepler's mother.

† See the full text, Note 42.

‡ See the full text, Note 43.

junction; as for the second, the sun-centred universe had never been officially declared a heresy, since neither the opinion of the Qualifiers, nor the decree of the Congregation of 1616, had been confirmed by infallible pronouncement *ex cathedra* or by Ecumenic Council. Had not Urban himself said that the Copernican opinion 'was not heretical but merely reckless'?

On the other hand, the judgement hushes up the incriminating contents of the book by stating that Galileo had represented the Copernican system as merely 'probable' – which is a whale of an understatement. It also hushes up the fact that Galileo had been lying and perjuring himself before his judges by pretending that he had written the book in refutation of Copernicus, that he had 'neither maintained nor defended the opinion that the earth moves', and so forth. The gist of the matter is that Galileo could not be legally convicted without completely destroying him – which was not the intention of the Pope or the Holy Office. Instead, they resorted to a legally shaky concoction. The intention was, clearly, to treat the famous scholar with consideration and leniency, but at the same time to hurt his pride, to prove that not even a Galileo was allowed to mock Jesuits, Dominicans, Pope, and Holy Office; and lastly, to prove that, in spite of his pose as a fearless crusader, he was not of the stuff of which martyrs are made.

The only real penalty inflicted on Galileo was that he had to abjure his conviction. On the other hand, up to the age of fifty Galileo had been hiding that conviction, and at his trial he had twice offered to add a chapter to the *Dialogue* refuting Copernicus. To recant in the Basilica of the Convent of Minerva, when everybody understood that this was an enforced ceremony, was certainly much less dishonourable for a scholar than to publish a scientific work contrary to his convictions. One of the paradoxa of this perverse story is that the Inquisition had thus in fact saved Galileo's honour in the eyes of posterity – no doubt unintentionally.

Shortly after the conclusion of the trial, a copy of the prohibited *Dialogue* was smuggled out to Kepler's old friend, the faithful Bernegger in Strasburg, who arranged for a Latin translation; it was published in 1635 and circulated widely in

Europe. A year later, Bernegger also arranged for Italian and Latin versions of the *Letter to the Grand Duchess Christina* to be published in Strasburg.

Galileo himself spent the year following the trial in writing the book on which his true and immortal fame rests: the *Dialogues concerning Two New Sciences.* At long last, in his seventies, he rediscovered his real vocation: the science of dynamics. He had abandoned it a quarter century before, when he embarked on his propaganda crusade for the heliocentric astronomy of which he had only a sketchy knowledge. The crusade had ended in a fiasco; and out of the shambles modern physics was born.

The book was completed in 1636, when Galileo was seventy-two. As he could not hope for an *imprimatur* in Italy, the manuscript was smuggled out to Leyden and published by the Elzevirs; but it could also have been printed in Vienna where it was licensed, probably with Imperial consent, by the Jesuit Father Paulus.

In the following year his right eye was blinded by an inflammation, and by the end of the year both eyes were lost.

> Alas, [he wrote to his friend Diodati] your friend and servant Galileo has been for the last month hopelessly blind; so that this heaven, this earth, this universe, which I, by marvellous discoveries and clear demonstrations, have enlarged a hundred thousand times beyond the belief of the wise men of bygone ages, henceforward for me is shrunk into such small space as is filled by my own bodily sensations.[46]

Yet he kept on dictating additional chapters to the *Two New Sciences*, and received a stream of distinguished visitors; among them Milton in 1638.

He died at the age of seventy-eight, in 1642, the year Newton was born, surrounded by friends and pupils – Castelli, Toricelli, Viviani.

His bones, unlike Kepler's, were not scattered into the wind; they rest in the Pantheon of the Florentines, the Church of Santa Croce, next to the remains of Michelangelo and Machiavelli. His epitaph was written for him by posterity: *eppur si* *muove* – the famous words which he never uttered at his trial. When his friends wanted to erect a monument over his grave, Urban told the Tuscan Ambassador that this would be a bad example for the world, since the dead man 'had altogether given rise to the greatest scandal throughout Christendom'. That was the end of the 'perilous adulation', and the end of one of the most disastrous episodes in the history of ideas; for it was Galileo's ill-conceived crusade which had discredited the heliocentric system and precipitated the divorce of science from faith.*

* An unexpected confirmation of the view expressed in the foregoing sections came my way belatedly, when the text was already in page proof, and can therefore be mentioned only briefly. It is a well known fact that the Jesuit missionaries in China in the sixteenth and seventeenth centuries owed their influence at the court in Peking in the first place to their services as astronomers; but I was surprised to discover that the type of astronomy they taught, from the end of the seventeenth century onward, was the *Copernican* system of the world; and that the rapid spreading through China and Japan, of the doctrine of the earth's motion was thus primarily due to the Society of Jesus, working under the supervision of the Sacred Congregation of the Propaganda in Rome. See B. Szczesniak, *The Penetration of the Copernican Theory into Feudal Japan*, Journal of the Royal Asiatic Society, 1944, parts I and II; and C. R. Boxer, *Jan Compagnie in Japan*, The Hague, 1936, p. 52 *seq.*

# 3 The Newtonian Synthesis

## 1. *'Tis all in Pieces*

On the opening pages of this book, 2,300 years earlier in this story, I compared the intellectual situation of Greece in the sixth pre-Christian century to an orchestra which is tuning up, each player absorbed in his own instrument, while waiting for the entrance of the conductor. In the seventeenth Christian century, the second heroic age of science, the situation repeated itself. The conductor who pulled the orchestra together and made a new harmony out of the caterwauling discords was Isaac Newton, born on Christmas Day, 1642, eleven months after Galileo had died.

It is appropriate that this survey of man's ideas about the universe should end with Newton for, in spite of more than two centuries that have passed since his death, our vision of the world is by and large still Newtonian. Einstein's correction of Newton's formula of gravity is so small that for the time being it only concerns the specialist. The two most important branches of modern physics, relativity and quantum mechanics, have not so far been integrated into a new universal synthesis; and the cosmological implications of Einstein's theory are still fluid and controversial. Until a new maestro emerges, or perhaps until space travel provides new observational data on our cosmic environment, the blueprint of the universe remains essentially the one that Newton drew for us, in spite of all disturbing rumours about the curvature of space, the relativity of time, and the runaway nebulae. There, after the long voyage from the Babylonian star gods, the Greek crystal spheres, the medieval walled universe, our imagination has temporarily come to rest.

During the last quarter-millennium of unprecedented human change, Newton has enjoyed an influence and authority only comparable to that of Aristotle in the two previous millennia. If one had to sum up the history of scientific ideas about the universe in a single sentence, one could only say that up to the seventeenth century our vision was Aristotelian, after that Newtonian. Copernicus and Tycho, Kepler and Galileo, Gilbert and Descartes lived in the no-man's-land between the two – on a kind of table-land between two wide plains; they remind one of stormy mountain streams, whose confluence finally gave rise to the broad, majestic river of Newtonian thought.

Unfortunately, we know very little about the intimate working of Newton's mind and the method by which he achieved his monumental synthesis. I shall not go into his life; any attempted contribution to the vast literature on Newton would be a separate undertaking. Instead, I shall briefly describe the scattered cosmological jigsaw puzzle as it presented itself to the young Newton; how he succeeded in perceiving that the odd disjointed bits were pieces of a single puzzle, and how he managed to put them together, we do not know. What he achieved was rather like an explosion in reverse. When a projectile blows up, its shiny, smooth, symmetrical body is shattered into jagged, irregular fragments. Newton found fragments and made them fly together into a simple, seamless, compact body, so simple that it appears as self-evident, so compact that any grammar-schoolboy can handle it.

The following, then, were the parts of the puzzle which confronted Newton in the 1660s, thirty years after Kepler's, twenty years after Galileo's death. The key pieces were Kepler's laws of the motion of heavenly bodies, and Galileo's laws of the motions of bodies on earth. But the two fragments did not fit together (any more than relativity and quantum mechanics do today). The forces which drove the planets in the Keplerian model did not stand up to the physicist's scrutiny. And *vice versa*, Galileo's laws of falling bodies and projectiles had no apparent bearing on the motions of planets or comets. According to Kepler, planets moved in ellipses, according to Galileo in circles. According to Kepler, they were driven along by 'spokes' of a force issuing from the rotating sun; according to Galileo, they were not driven at all because circular motion was self-perpetuating. According to Kepler, the laziness or inertia of the planets made them tend to lag behind; according to Galileo, the very principle of inertia made them persist in going round in circles. ''Twas all in pieces all cohesion gone.'

The confusion was made even worse by the last of the pre-Newtonian giants, Descartes. According to him, inertia made bodies persist not in circular but in straight motion. This was the most bewildering view of all, for heavenly bodies might move in circles or ellipses, but they certainly did not move along straight lines. Descartes therefore assumed that the planets were whirled round by vortices in an all-pervading ether – an elaboration of Kepler's rotating, sweeping brooms.[1]

There was then complete disagreement (*a*) on the nature of the force which drives the planets round and keeps them in their orbits, and (*b*) on the question what a body in the vastness of space would do with itself if it were left alone, that is, without external agents acting on it. These questions were inextricably mixed up with the problem of what 'weight' really meant, with the mysterious phenomenon of magnetism, and with the perplexities of the emergent concepts of physical 'forces' and 'energies'.

## 2. *What is 'Weight'?*

The telescope had shown that the moon had a rugged surface much like the earth, and that the sun was apt to break out in spots; this led to a growing conviction that heavenly bodies were of earthly nature and would tend to behave in the same manner as things behave on earth. Now the most conspicuous quality in which all bodies on earth shared was weight – the tendency to press or fall downward (unless forced upward by the pressure of heavier substances). In the old philosophy, this was satisfactorily explained by the fact that every earthly object tended to move towards the centre of the world or away from it – whereas objects in the sky obeyed different laws. In the new philosophy, this dualism was denied, and the position of the earth in the centre of the world was equally denied. But while it played havoc with the old commonsense beliefs, the new philosophy provided no answers to the problems which it raised. If the moon, the planets, and comets were of the same nature as bodies on earth, then they too must have 'weight'; but what exactly does 'the weight' of a planet mean, what does it press against or where does it tend to fall? And if the reason why a stone falls to earth is not the earth's position in the centre of the universe, then just why does the stone fall?

One may note in passing that some of our logical positivists transferred into the seventeenth century would have dismissed the question what a planet 'weighs' as meaningless with an airy wave of the hand; and if their attitude had prevailed, the scientific revolution would not have taken place. As it happened, the leaders of the movement tried to wriggle out from between the horns of the dilemma, each after his own fashion, without much regard for semantic purity. Copernicus tentatively suggested that objects on the sun and moon had weight as bodies on earth, and that 'weight' meant the tendency of all matter to arrange itself in spherical shape around a centre. Galileo believed that 'weight' was an absolute quality of all terrestrial matter, which did not require a cause and was in fact indistinguishable from its inertia; whereas in heavenly bodies 'weight' became somehow identical with their persistence in moving along a circular path. Kepler was the first to explain 'weight' as the *mutual attraction* between two bodies; he had even postulated that two bodies in space, exposed to no other influence, would approach each other and would meet at an intermediary point, so that the distances covered by each would be in inverse ratio to their masses, and he correctly attributed the tides to the attraction of the sun and moon; yet, as we saw, at the decisive moment he shrank back from the fantastic notion of a gravitational *anima mundi*.

## 3. *The Magnetic Confusion*

The confusion was further increased by William Gilbert's sensational theory that the earth was a giant loadstone, which induced Kepler to identify the sun's action on the planets as a 'magnetic' force. It was quite natural, and indeed logical, that this confusion between magnetism and gravity should arise, for the loadstone was the only concrete and tangible demonstration of the mysterious tendency of matter to join matter under the influence of a 'force' which acted at a distance without contact or intermediaries. Hence the magnet became the archetype of action-at-a-distance and paved the way for universal gravity. Without Dr Gilbert, man would have been much less prepared to exchange the homely and traditional view that 'weight' meant the natural tendency of bodies to fall towards the centre for the adventurous notion that it meant the grappling of bodies at each other across empty space. Magnetism demonstrated that this grappling by ghostly fingers was a fact, that iron filings rushed to a magnet as by secret command, as stones rushed to the earth; and for about half a century the two phenomena were either identified, or regarded as Siamese twins. Moreover, the word 'magnetism' was used in a broader, metaphorical sense; it had a profoundly appealing ambiguity as another Janus-faced agency which pertained both to the world of the spirit and of matter. On the one hand, the magnet sent out its energy, as exact science demanded, 'without error . . . quick, definite, constant, directive, motive, imperant, harmonious'; on the other hand, it was something animate and living, it 'imitates a soul', nay, it was the very 'soul of the earth', its 'instinct of self-preservation'. 'The magnetic effluvium of the earth reaches out like an arm clasping round the attracted body and drawing it to itself ' This arm 'must needs be light and spiritual so as to enter the iron', but at the same time it must also be material – a thin and rare ether.[2]

One may again note in passing, that this Janus-faced quality is equally present, though expressed in less poetic language, in the contemporary theories of matter as both a corpuscle and a wave, according to which face it presents. Magnetism, gravity, and action-at-a-distance have not lost an iota of their baffling mystery since Gilbert.

Kepler was not the only victim of this inevitable confusion; Galileo too believed that Gilbert had provided the explanation why the earth's axis always points in the same direction in space – the axis was simply a kind of magnetic needle. Even Robert Boyle, the father of modern chemistry and one of the principal influences on Newton, thought that gravity may be due to 'magnetic vapours' issuing from the earth.

Only the most implacably sceptical and logical brain among them all, that of Descartes, repudiated magnetism, gravity, and any form of action-at-a-distance. Descartes took matters a decisive step forward by letting bodies persist in their motion, not in a Galilean circle, but in a straight line.* At the same time, however, he took an equally important step backward by explaining magnetism and gravity as whirlpools in the ether. It is a measure of Newton's daring that even Descartes, who promised to reconstruct the whole universe from matter and extension alone, who invented the most beautiful tool of mathematical reasoning, analytical geometry, who was more ruthless in his methods of thought than any of his predecessors – that even Descartes, this Robespierre of the scientific revolution, rejected attraction-at-a-distance at the price of filling all space with monstrous eddies and vortices. Like Kepler who hit on the concept of gravity, then kicked it away, like Galileo who rejected even the moon's influence on the tides, Descartes' wide-open mind boggled in horror at the idea of ghost arms clutching through the void – as unprejudiced intelligence was indeed bound to do, until 'universal gravity' or 'electromagnetic field' became verbal fetishes which hypnotized it into quiescence, disguising the fact that they are metaphysical concepts dressed in the mathematical language of physics.

## 4. *Enter Gravity*

These, then, were the pieces of the chaotically scattered jigsaw puzzle confronting Newton. Contradictory theories regarding the behaviour of objects in space in the absence of interfering forces; contradictory theories about the forces which make planets revolve; confusing fragments of information about inertia and momentum, weight and free fall, gravity and magnetism; doubts about the location of the centre of the universe and whether it had a centre; and overshadowing it all, the

* Newton's First Law of Motion was in fact formulated by Descartes.

question where the God of the Scriptures fitted into the picture.

There had been some vague conjectures in the right direction, but unsupported by precise argument. The French mathematician, Giles Peron de Roberval, for instance, had suggested in the year following Galileo's death that all matter in the universe was drawn together, and that the moon would fall on the earth if the ether did not act as a supporting cushion between them. Giovanni Borelli, who occupied Galileo's erstwhile chair in Pisa, took up an ancient Greek suggestion that the moon behaved 'like a stone in a sling' whose flying force prevented it from falling to the earth. But he contradicted himself by believing, with Kepler, that the moon needed to be pushed round in a circle by an invisible broom – that is, that the moon had no impetus of its own; then why should it try to fly away?

Newton was twenty-four when, in 1666, he found the key to the solution; but then his interest turned to other matters, and it was only twenty years later that he completed the synthesis. It is, alas, impossible to reconstruct his struggle on the rungs of Jacob's ladder with the angel who guards the secrets of the cosmos – as we have been able to do in Kepler's case; for Newton was not communicative about the genesis of his discoveries, and the scant information he provides sounds like rationalizations after the fact. Besides, part of the thinking was done collectively by the circle round the Royal Society – Hooke, Halley, Christopher Wren – and influenced by kindred minds like Huygens' in Holland; so that it is impossible to know precisely which intermediary step was first taken by whom.

It is equally impossible to discover when and under what precise circumstances the cornerstone of the theory was laid – the Law of Gravity, which states that the force of attraction is proportionate to the attracting masses, and diminishes with the square of the distance. It had been suggested, but without concrete proof, as far back as 1645 by Boulliau. Perhaps it was derived by analogy from the diffusion of light which, as Kepler knew, also diminishes in intensity with the square of distance. Another suggestion is that it was deduced from Kepler's Third Law; Newton himself says that he found the formula by calculating the force required to counterbalance the moon's centrifugal force – but it does not sound entirely convincing.

The details are obscure, but the grand outline is dazzlingly clear. With true sleepwalker's assurance, Newton avoided the booby-traps strewn over the field: magnetism, circular inertia, Galileo's tides, Kepler's sweeping-brooms, Descartes' vortices – and at the same time knowingly walked into what looked like the deadliest trap of all: action-at-a-distance, ubiquitous, pervading the entire universe like the presence of the Holy Ghost. The enormity of this step can be vividly illustrated by the fact that a steel cable of a thickness equalling the diameter of the earth would not be strong enough to hold the earth in its orbit. Yet the gravitational force which holds the earth in its orbit is transmitted from the sun across ninety-three million miles of space without any material medium to carry that force.[2a] The paradox is further illustrated by Newton's own words, which I have quoted before, but which bear repeating:

> It is inconceivable, that inanimate brute matter should, without the mediation of something else, which is not material, operate upon, and affect other matter without mutual contact. . . . And this is one reason, why I desired you would not ascribe innate gravity to me. That gravity should be innate, inherent, and essential to matter, so that one body may act upon another, at a distance through a vacuum, without the mediation of anything else, by and through which their action and force may be conveyed from one to another, is to me so great an absurdity, that I believe no man who has in philosophical matters a competent faculty of thinking, can ever fall into it. Gravity must be caused by an agent acting constantly according to certain laws; but whether this agent be material or immaterial, I have left to the consideration of my readers.

The 'agent' to which he refers is the interstellar ether, which was supposed somehow to transmit the force of gravity. But how this is done remained unexplained; and whether the ether was something material or not, remained an open question – not only in the reader's, but evidently also in Newton's mind. He sometimes called it a medium, but on other occasions used the term 'spirit'. Thus the ambiguity which we noted in Kepler's use of the term 'force' as a half animistic, half mechanistic concept, is equally present (though less explicity stated) in Newton's concept of gravity.

Another appalling difficulty of this concept was that a universe filled with gravity ought to collapse, i.e., all the fixed stars should rush together and meet in a kind of final, cosmic super-explosion.* The difficulty was indeed unsurmountable, and Newton found no other solution than to assign to God the function of counteracting gravity and keeping the stars in their places:

> And though the matter were divided at first into several systems, and every system by a divine power constituted like ours; yet would the outside systems descend towards the middlemost; so that this frame of things could not always subsist without a divine power to conserve it. . . .[3]

It is only by bringing into the open the inherent contradictions, and the metaphysical implications of Newtonian gravity, that one is able to realize the enormous courage – or sleepwalker's assurance – that was needed to use it as the basic concept of cosmology. In one of the most reckless and sweeping generalizations in the history of thought, Newton filled the entire space of the universe with interlocking forces of attraction, issuing from all particles of matter and acting on all particles of matter, across the boundless abysses of darkness.

But in itself this replacement of the *anima mundi* by a *gravitatio mundi* would have remained a crank idea or a poet's cosmic dream; the crucial achievement was to express it in precise mathematical terms, and to demonstrate that the theory fitted the observed behaviour of the cosmic machinery – the moon's motion round the earth and the planet's motions round the sun.

## 5. *The Final Synthesis*

His first step was to do in imagination what history had failed to achieve: to bring Kepler and Galileo together. More pre-

* The reason why this does not happen is in the enormous distances and relative velocities of the stars, galaxies, and nebulae, of which Newton was unaware.

cisely: to join one half of Kepler to one half of Galileo, and to discard each redundant half.

The meeting place was the moon. Young Jeremiah Horrocks – the English prodigy who died at twenty-one – had applied Kepler's Laws to the orbit of the moon. This provided Newton with one half of the synthesis. The second half he found in Galileo's Laws of the motion of projectiles in the immediate vicinity of the earth. *Newton identified the Keplerian orbit of the moon with the Galilean orbit of a projectile*, which was constantly falling downward towards the earth, but was unable to reach it, owing to its fast forward motion. In his *System of the World*, his process of reasoning is described as follows:

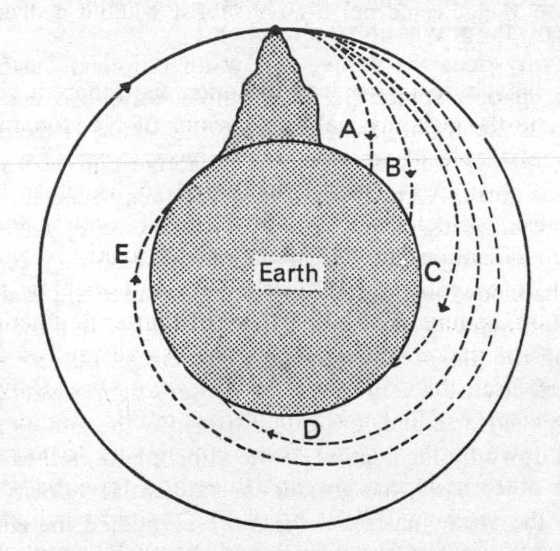


If a projectile is fired from the top of a mountain, it will be deflected from its straight path by the earth's attraction. According to the initial velocity imparted to it, it will follow the curves A, B, C, D, or E; and if the initial velocity exceeds a certain critical value the projectile will describe a circle or an ellipse 'and return to the mountain from which it was projected'. Moreover, according to Kepler's Second Law, 'its velocity when it returns to the mountain will be no less than it was at first: and retaining the same velocity, it will describe the same curve over and over by the same law ... and go on revolving through the heavens just as the planets do in their orbs'. In other words, Newton, by thought-experiment, created an artificial satellite nearly three hundred years before technology was able to implement it.

Thus the basic idea of Newton's celestial mechanics is the interaction of two forces: the force of gravity, which pulls the planet towards the sun, and the centrifugal force, which counteracts it. The usual way to demonstrate the idea is to whirl around a stone at the end of a string. The force which keeps the string taut is the stone's centrifugal force; the cohesion of the string which holds the stone captive in its orbit represents the gravitational attraction.

But why does the planet follow an elliptical, instead of a circular path? To put it in a simple way, because when I whirl a stone round, the length of the string is fixed, and it won't stretch – whereas the sun's attractive pull varies according to distance. Accordingly, the stone goes round in a perfect circle, whereas the planet would go round in a perfect circle only if its tangential velocity, and the resulting centrifugal force, happened just exactly to counterbalance the sun's attraction. If its velocity is either greater or smaller than the required amount, the planet will move not in a circle, but in an ellipse. If its velocity were very small, the planet's orbit would intersect the sun's surface, and it would crash into the sun as meteors, slowed down by the friction of the atmosphere, fall to to earth. On the other hand, the greater the tangential velocity in relation to the attracting force, the more elongated the ellipse will be; until one end of it will be stretched open toward infinity, as it were, and the ellipse will change into a parabola – the assumed path of certain comets which come from the depth of space, are deflected from their course by the sun, but not sufficiently to be captured, and recede again into infinity.

Why planets should move in ellipses is easy to show in mathematical terms; leaving mathematics aside, one may visualize the mechanism as a tug-of-war between gravity and centrifugal force. If the string to which the revolving stone is attached be made of elastic material, one can imagine it alternately expanding and contracting, thus making the stone's orbit an oval*. Or one can visualize the process as follows: as the planet approaches the sun, its speed increases. It shoots past the sun, but as it does so, the clutching hand of gravity swings it round – as a running child grabbing at a maypole is swung around it – so that it now continues in the opposite direction. If its velocity on the approach-run had been exactly the amount required to prevent it from falling into the sun, it would continue in a circle. But as it was slightly greater, the receding run will carry it into an elongated path, which the planet pursues at slackening speed in the teeth of the sun's attraction, as it were, gradually curving inward; until, after passing the aphelion, the curve again approaches the sun and the whole cycle starts again.

The 'eccentricity' of the ellipse is the amount by which it deviates from the circle. The eccentricities of the planets are small, owing to the common origin of the solar system which make their tangential velocities almost precisely balance gravity.

But all this was as yet merely conjecture; and the days of purely speculative hypotheses were past. It was wild conjecture to postulate that the moon was constantly 'falling' towards the earth, like a projectile, or like the famous apple in the garden at Woolsthorpe – in other words, that the earth's attraction reached as far as the moon, the sun's attraction as far as the planets, and that interstellar space was indeed 'filled' or 'charged' with gravity. To transform a wild guess into scientific theory, Newton had to provide rigorous mathematical proof.

This means he had to calculate: (*a*) the centrifugal force of the moon;[4] (*b*) the gravitational force which the earth was supposed to exert on the moon; and (*c*) he had to show that the interaction of these two forces produced a theoretical orbit which agreed with the moon's observed orbit.

In order to carry out this operation, he must first of all know

* The analogy between elastic resistance and gravitational force is, of course, quite wrong, but it may help to get the 'feel' of the elliptic orbit.

at what rate the earth's gravity diminished with distance. The apple fell from the tree at a known acceleration of approximately ten yards added speed per second; but what would be the acceleration of the distant moon towards the earth? In other words, he had to discover the Law of Gravity – that the force diminishes with the square of distance. In the second place, he had to know the exact value of the moon's distance. Thirdly, he had to decide whether it was legitimate to treat two huge globes like the earth and the moon in an abstract manner, as if their whole mass were concentrated in a single central point. Lastly, to reduce the mathematical difficulties, the moon's orbit had to be treated as if it were a circle instead of an ellipse.

As a result of all these difficulties, Newton's first calculations only agreed 'pretty nearly' with the facts; and that was not good enough. For nearly twenty years he dropped the whole issue.

During these twenty years, Jean Picard's expedition to Cayenne produced much improved data on the earth's diameter and its distance from the moon; Newton himself developed his own brand of infinitesimal calculus, the indispensable mathematical tool for attacking the problem; and the Halley-Hooke-Wren trio kept fitting together further bits of the puzzle. The orchestra had now reached the stage where whole groups of instruments could be discerned running through certain passages; only the rap of the conductor's baton was needed to make everything fall into place.

In 1686, goaded on by Halley, Newton arrived at his ultimate synthesis. He computed the force of the earth's attraction on the moon, and showed that this, combined with the moon's own centrifugal force, satisfied the moon's observed motions. Next he computed the sun's attraction on the planets, and demonstrated that the orbit produced by a force of attraction which diminished with the square of distance was a Keplerian ellipse with the sun in one focus; and conversely, that an elliptic orbit required a gravitational force obeying the inverse square ratio. Kepler's Third Law relating the duration of the planets' periods to their mean distances from the sun became a cornerstone of the system; and the Second Law – equal areas being swept out in equal times – was now shown to hold for any central orbit. Comets were shown to move either in very elongated ellipses or in parabolas, receding into the infinity of space. Newton further proved that any object above the earth's surface behaved as if the whole mass of the earth were concentrated in its centre; which made it possible to treat all heavenly bodies as if they were mathematical points. Lastly, all observable motion in the universe was reduced to four basic laws: the Law of Inertia; the Law of Acceleration under an impressed force; the Law of Reciprocal Action and Reaction; and the Law of Gravity.

The miracle was completed; the fragments had all flown together in this reversed explosion and were fused into a smooth, compact, innocent-looking body; and had Donne still been alive, he could have reversed his lament into a triumphant: ' 'Tis all in one piece, all coherence now.'

The motions of sun, moon, and the five vagabond stars had been the main problem of cosmology since the days of the Babylonians. Now that they were all shown to follow the same simple laws, the solar system was recognized as an integrated unit. The rapid progress of astronomy and astro-physics soon led to the further realization that this unit was merely a subdivision of a larger one: our galaxy of millions of stars of roughly the same nature as our sun, some of them, no doubt, also surrounded by planets; and that our galaxy again was merely one among other galaxies and nebulae in various stages of their evolution, yet all governed by the same universal set of laws.

But these developments no longer concern us. With the publication of Newton's *Principia* in 1687 A.D., cosmology became a disciplined science; and at this point our narrative of man's changing vision of the universe must end. The wild dance of shadows thrown by the stars on the wall of Plato's cave was settling into a decorous and sedate Victorian waltz. All mysteries seemed to have been banished from the universe, and divinity reduced to the part of a constitutional monarch, who is kept in existence for reasons of decorum, but without real necessity and without influence on the course of affairs.

It remains to discuss some implications of the story.

## *Chronological Table to Parts Four and Five*

| | TYCHO DE BRAHE | GALILEO | KEPLER |
|---|---|---|---|
| A.D. 1546 | Born on 14 Dec., at Knudstrup. | | |
| ,, 1559 to 1572 | Studies at Copenhagen, Germany and Switzerland. | A.D. 1564 Born on 15 Feb., at Pisa. | |
| | | | A.D. 1571 Born on 16 May, at Weil-der-Stadt. |
| ,, 1572 | Appearance of 'Tycho's Nova'. | | |
| ,, 1576 | Receives the island of Hveen. | | Left in care of grand-parents. |
| 1581 | Works at Hveen. | Matriculates at University of Pisa. | 'Put to hard work in the country.' |
| ,, 1584 | | | Enters theological Seminary. |
| ,, 1589 | | Appointed Lecturer on Mathematics, University of Pisa. | Matriculates at Tuebingen University. |
| ,, 1592 | | Professor of Mathematics, Padua University. | |
| ,, 1593 | | | Teacher of Mathematics, Provincial School in Gratz. |
| ,, 1597 | Leaves Hveen. | Writes pro-Copernican letter to Kepler. | Publishes *Mysterium* Cosmographicum. |
| ,, 1599 | Appointed Imperial Mathematicus to Rudolph II. | | Exiled from Gratz; school closed down. |
| ,, 1600–1 | Collaboration Tycho-Kepler. | | Kepler at Benatek and Prague. |
| 1601 | Dies on 13 Oct., Prague. | | Appointed Tycho's successor. |

| | | GALILEO | KEPLER |
|---|---|---|---|
| „ | 1609 | | Publishes *Astronomia Nova* (First and Second Law). |
| „ | 1610 | Telescopic discoveries. *The Star Messenger*. Appointed 'Chief Mathematician and Philosopher' at the Court of Cosmo II de Medici. | *Conversation with the Star Messenger.* |
| „ | 1611 | Triumphal visit to Rome. | *Dioptrice.* |
| „ | 1612 | Writes *Letters on Sunspots.* | Death of Rudolph; departure for Linz. Excommunication. |
| „ | 1613 | Writes *Letter to Castelli.* | |
| „ | 1614 | Caccini preaches against Galileans. | |
| „ | 1615 | Lorini denounces Galileans. Galileo in Rome. Theory of the tides. | Proceedings against mother start. |
| „ | 1616 | Copernicus' book banned 'until corrected'. Galileo instructed to abandon it. | |
| „ | 1618 | Start of dispute on comets. | Outbreak of Thirty Years War. |
| „ | 1619 | | *Harmonice Mundi* published (Third Law). |
| „ | 1620 | Copernicus' book, with minor corrections, again permissible reading. | Mother arrested. |
| „ | 1621 | | Mother acquitted; dies. Publication of *Epitome* completed. |
| „ | 1623 | Barberini becomes Urban VIII. *Il Saggiatore* published. | |
| „ | 1625 | Starts writing *Dialogue.* | Printing of Rudolphine Tables begun. |
| „ | 1626 | | Siege of Linz. Destruction of printing press. Departure for Ulm. |
| „ | 1627 | | Printing of Tables completed. |
| „ | 1628 | | Erratic travels. Obtains post with Wallenstein at Sagan. |

| | GALILEO | KEPLER |
|---|---|---|
| A.D. 1630 | *Dialogue* completed. Negotiations about the *imprimatur*. | Work on *Somnium*. Last journey to Ratisbon. Dies on 15 Nov. |
| „ 1632 | *Dialogue* published and banned. Galileo ordered to Rome. | |
| „ 1633 | Trial of Galileo. | |
| „ 1637 | Goes blind in both eyes. | |
| „ 1638 | *Two New Sciences* published in Leyden. | |
| „ 1642 | Dies at Arcetri, on 8 Jan. | |

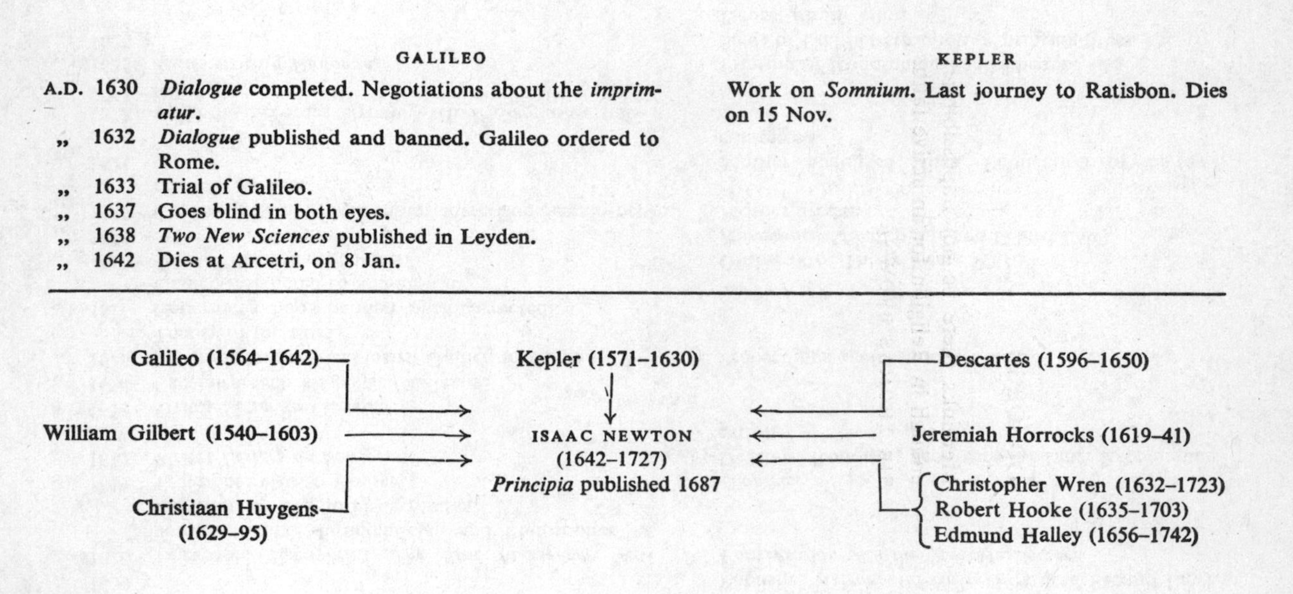

# Epilogue

Methinks there be not impossibilities enough in Religion for an active faith.
SIR THOMAS BROWNE

## 1. *The Pitfalls of Mental Evolution*

We are in the habit of visualizing man's political and social history as a wild zigzag which alternates between progress and disaster, but the history of science as a steady, cumulative process, represented by a continuously rising curve, where each epoch adds some new item of knowledge to the legacy of the past, making the temple of science grow brick by brick to ever greater height. Or alternately, we think in terms of 'organic' growth from the magic-ridden, myth-addicted infancy of civilization through various stages of adolescence, to detached, rational maturity.

In fact, we have seen that this progress was neither 'continuous' nor 'organic'. The philosophy of nature evolved by occasional leaps and bounds alternating with delusional pursuits, *culs-de-sac*, regressions, periods of blindness, and amnesia. The great discoveries which determined its course were sometimes the unexpected by-products of a chase after quite different hares. At other times, the process of discovery consisted merely in the cleaning away of the rubbish that blocked the path, or in the rearranging of existing items of knowledge in a different pattern. The mad clockwork of epicycles was kept going for two thousand years; and Europe knew less geometry in the fifteenth century than in Archimedes' time.

If progress had been continuous and organic, all that we know, for instance, about the theory of numbers, or analytical geometry, should have been discovered within a few generations after Euclid. For this development did not depend on technological advances or the taming of nature: the whole corpus of mathematics is potentially there in the ten billion neurons of the computing machine inside the human skull. Yet the brain is supposed to have remained anatomically stable for something like a hundred thousand years. The jerky and basically irrational progress of knowledge is probably related to the fact that evolution had endowed *homo sapiens* with an organ which he was unable to put to proper use. Neurologists have estimated that even at the present stage we are only using two or three per cent of the potentialities of its built-in 'circuits'. The history of discovery is, from this point of view, one of random penetrations into the uncharted Arabias in the convolutions of the human brain.

This is a very curious paradox indeed. The senses and organs of all species evolve (*via* mutation and selection as we suppose), according to adaptive needs; and novelties in anatomical structure are by and large determined by those needs. Nature meets its customers' requirements by providing longer necks to graze off the top of trees, harder hooves and teeth to cope with the coarse grass of the drying steppes; by shrinking the smell-brain and enlarging the visual cortex of birds, arboreals, and bipeds as they slowly raise their heads above ground. But it is entirely unprecedented that nature should endow a species with an extremely complex luxury organ far exceeding its actual and immediate needs, which the species will take millennia to learn to put to proper use – if it ever does. Evolution is supposed to cater for adaptative demands; in this case the goods delivered anticipated the demand by a time-stretch of geological magnitude. The habits and learning potentialities of all species are fixed within the narrow limits which the structure of its nervous system and organs permits; those of *homo sapiens* seem unlimited precisely because the possible uses of that evolutionary novelty in his skull were quite out of proportion with the demands of his natural environment.

Since evolutionary genetics is unable to account for the fact that a biologically more or less stable race should mentally evolve from cave-dwellers to spacemen, we can only conclude that the term 'mental evolution' is more than a metaphor; and that it refers to a process in which some factors operate to which as yet we have not got a clue. All we know is that mental evolution cannot be understood either as a cumulative, linear process, or as a case of 'organic growth' comparable to the maturing of the individual; and that it would perhaps be better to consider it in the light of biological evolution, of which it is a continuation.

It would indeed seem more expedient to treat the history of thought in terms borrowed from biology (even if they can yield no more than analogies) than in terms of an arithmetical progression. 'Intellectual progress' has, as it were, linear associations – a continuous curve, a steadily rising water level; whereas 'evolution' is known to be a wasteful, fumbling process characterized by sudden mutations of unknown cause, by the slow grinding of selection, and by the dead-ends of over-specialization and rigid inadaptability. 'Progress' can by definition never go wrong; evolution constantly does; and so does the evolution of ideas, including those of 'exact science'. New ideas are thrown up spontaneously like mutations; the vast majority of them are useless crank theories, the equivalent of biological freaks without survival-value. There is a constant struggle for survival between competing theories in every branch of the history of thought. The process of 'natural selection', too, has its equivalent in mental evolution: among the multitude of new concepts which emerge only those survive which are well adapted to the period's intellectual *milieu*. A new theoretical concept will live or die according to whether it can come to terms with this environment; its survival value depends on its capacity to yield results. When we call ideas 'fertile' or 'sterile', we are unconsciously guided by biological analogy. The struggle between the Ptolemaic, Tychonic, and Copernican systems, or between the Cartesian and Newtonian views of gravity, was decided by those criteria. Moreover, we find in the history of ideas mutations which do not seem to correspond to any obvious need, and at first sight appear as mere playful whimsies – such as Appollonius' work on conic sections, or the non-Euclidian geometries, whose practical value became apparent only later. Conversely, there are organs which have lost their purpose and are yet carried over as an evolutionary legacy: modern science is full of appendices and rudimentary monkey-tails.

There occur in biological evolution periods of crisis and transition when there is a rapid, almost explosive branching out in all directions, often resulting in a radical change in the dominant trend of development. The same kind of thing seems to have happened in the evolution of thought at critical periods like the sixth century B.C. or the seventeenth A.D. After these stages of 'adaptative radiations', when the species is plastic and malleable, there usually follow periods of stabilization and specialization along the new lines – which again often lead into dead ends of rigid over-specialization. When we look back at the grotesque decline of Aristotelian scholasticism, or the blinkered single-mindedness of Ptolemaic astronomy, we are reminded of the fate of those 'orthodox' marsupials, like the koala, who changed from tree-climbers into tree-clingers. Their hands and feet turned into hooks, their fingers no longer served to pluck fruit and explore objects but degenerated into curved claws with the sole purpose of fixing the animal to the bark of the tree to which it hangs on for dear life.

To quote a last analogy, we find 'faulty linkages' in evolution which remind one of certain ideological *mésalliances.* The central nerve chain of an invertebrate such as the lobster runs beneath its alimentary canal, whereas the main portion of its rudimentary brain is placed above it, in its forehead. In other words, the lobster's gullet, from mouth to stomach, has to pass through the midst of its brain ganglia. If its brain were to expand – and expand it must if the lobster is to grow in wisdom – its gullet would be squeezed and it would starve. In spiders and scorpions something like this did actually happen: their brain mass has so compressed their alimentary tube that only fluid food can pass through: they had to become blood-suckers. *Mutatis mutandis*, something on these lines happened when the stranglehold of Neoplatonism prevented man from taking in any solid empirical food for thought, and forced him to feed throughout the Dark Ages on a liquid diet of other-worldliness. And did not the stranglehold of mechanistic materialism in the nineteenth century produce the opposite effect, spiritual starvation? In the first case, religion had entered into a misalliance with a nature-rejecting ideology; in the second, science became allied to an arid philosophy. Or again, the stranglehold of the dogma of uniform motion in perfect circles turned the Copernican system into a kind of crustacean ideology. The analogies may seem far-fetched, which indeed they are, but all they are meant to demonstrate is the fact that such faulty linkages of a self-defeating nature occur in the realms of both biological and mental evolution.

## 2. *Separations and Reintegrations*

The process of evolution may be described as differentiation of structure and integration of function. The more differentiated and specialized the parts, the more elaborate coordination is needed to create a well-balanced whole. The ultimate criterion of the value of a functional whole is the degree of its internal harmony or integratedness, whether the 'functional whole' is a biological species or a civilization or an individual. A whole is defined by the pattern of relations between its parts, not by the sum of its parts; and a civilization is not defined by the sum of its science, technology, art, and social organization, but by the total pattern which they form, and the degree of harmonious integration in that pattern. A physician has recently said that 'the organism in its totality is as essential to an explanation of its elements as its elements are to an explanation of the organism'. This is as true when we talk about the supra-renal gland as it is when we talk of the elements of a culture – Byzantine art, or medieval cosmology, or utilitarian ethics.

Conversely, a diseased state of organism, a society, or culture, is characterized by a weakening of the integrative controls, and the tendency of its parts to behave in an independent and self-assertive manner, ignoring the superior interest of the whole, or trying to impose their own laws on it. Such states of imbalance may be caused either by the weakening of the co-ordinating powers of the whole through growth beyond a critical limit, senescence, and so forth; or by excessive stimulation of an organ or part; or its cutting off from communication with the integrative centre. The isolation of the organ from central control leads according to circumstances to its hyperactivity or degeneration. In the realm of the mind, the 'splitting off' of thoughts and emotions, of some aspect of the personality, leads to similar results. The term schizophrenia is directly derived from this splitting-off process; 'repressed' and 'autonomous' complexes point in the same direction. In the obsessional neuroses, in the 'fixed ideas' and 'fixed behaviour patterns', we see parts of the personality dissociating themselves from the whole.

In a society or culture the degree of integration between its parts, or fields of endeavour, is equally decisive. But here the diagnosis of dis-integrative symptoms is vastly more difficult and always controversial, because there exists no criterion of normality. I believe, nevertheless, that the story outlined in this book will be recognized as a story of the splitting-off, and subsequent isolated development, of various branches of knowledge and endeavour – sky-geometry, terrestrial physics, Platonic and scholastic theology – each leading to rigid orthodoxies, one-sided specializations, collective obsessions, whose mutual incompatibility was reflected in the symptoms of double-think and 'controlled schizophrenia'. But it is also a story of unexpected reconciliations and new syntheses emerging from apparently hopeless fragmentation. Can we derive some positive hints from the conditions under which these apparently spontaneous cures occur?

### 3. *Some Patterns of Discovery*

In the first place, a new synthesis never results from a mere adding together of two fully developed branches in biological or mental evolution. Each new departure, each reintegration of what has become separated, involves the breaking down of the rigid, ossified patterns of behaviour and thought. Copernicus failed to do so; he tried to mate the heliocentric tradition with orthodox Aristotelian doctrine, and failed. Newton succeeded because orthodox astronomy had already been broken up by Kepler and orthodox physics by Galileo; reading a new pattern into the shambles, he united them in a new conceptual frame. Similarly, chemistry and physics could only become united after physics had renounced the dogma of the indivisibility and impermeability of the atom, thus destroying its own classic concept of matter, and chemistry had renounced its doctrine of ultimate immutable elements. A new evolutionary departure is only possible after a certain amount of de-differentiation, a cracking and thawing of the frozen structures resulting from isolated, over-specialized development.

Most geniuses responsible for the major mutations in the history of thought seem to have certain features in common; on the one hand scepticism, often carried to the point of iconoclasm, in their attitude towards traditional ideas, axioms, and dogmas, towards everything that is taken for granted; on the other hand, an open-mindedness that verges on naïve credulity towards new concepts which seem to hold out some promise to their instinctive gropings. Out of this combination results that crucial capacity of perceiving a familiar object, situation, problem, or collection of data, in a sudden new light or new context: of seeing a branch not as part of a tree, but as a potential weapon or tool; of associating the fall of an apple not with its ripeness, but with the motion of the moon. The discoverer perceives relational patterns of functional analogies where nobody saw them before, as the poet perceives the image of a camel in a drifting cloud.

This act of wrenching away an object or concept from its habitual associative context and seeing it in a new context is, as I have tried to show, an essential part of the creative process.[1] It is an act both of destruction and of creation, for it demands the breaking up of a mental habit, the melting down, with the blow-lamp of Cartesian doubt, of the frozen structure of accepted theory, to enable the new fusion to take place. This perhaps explains the strange combination of scepticism and credulity in the creative genius.[2] Every creative act – in science, art, or religion – involves a regression to a more primitive level, a new innocence of perception liberated from the cataract of accepted beliefs. It is a process of *reculer pour mieux sauter,* of disintegration preceding the new synthesis, comparable to the dark night of the soul through which the mystic must pass.

Another pre-condition for basic discoveries to occur, and to be accepted, is what one might call the 'ripeness' of the age.

It is an elusive quality, for the 'ripeness' of a science for a decisive change is not determined by the situation in that particular science alone, but by the general climate of the age. It was the philosophical climate of Greece after the Macedonian conquest that nipped in the bud Aristarchus' heliocentric concept of the universe; and astronomy went on happily with its impossible epicycles, because that was the type of science that the medieval climate favoured.

Moreover, it *worked*. This ossified discipline, split off from reality, was capable of predicting eclipses and conjunctions with considerable precision, and of providing tables which were by and large adequate to the demand. On the other hand, the seventeenth century's 'ripeness' for Newton, or the twentieth's for Einstein and Freud, was caused by a general mood of transition and awareness of crisis, which embraced the whole human spectrum of activities, social organization, religious beliefs, art, science, fashions.

The symptom that a particular branch of science or art is ripe for a change is a feeling of frustration and malaise, not necessarily caused by any acute crisis in that specific branch – which might be doing quite well in its traditional terms of reference – but by a feeling that the whole tradition is somehow out of step, cut off from the mainstream, that the traditional criteria have become meaningless, divorced from living reality, isolated from the integral whole. This is the point where the specialist's *hubris* yields to philosophical soul-searching, to the painful reappraisal of his basic axioms and of the meaning of terms which he had taken for granted; in a word, to the thaw of dogma. This is the situation which provides genius with the opportunity for his creative plunge under the broken surface.

## 4. *Mystic and Savant*

The most disturbing aspect of this story of separations and reintegrations, the one to which I have been constantly harping back, concerns the mystic and the savant.

At the beginning of this long journey, I quoted Plutarch's

comment on the Pythagoreans: 'The contemplation of the eternal is the aim of philosophy, as the contemplation of the mysteries is the aim of religion.' For Pythagoras as for Kepler, the two kinds of contemplation were twins; for them philosophy and religion were motivated by the same longing: to catch glimpses of eternity through the window of time. The mystic and the savant jointly satisfied the dual urge of allaying the self's cosmic anxiety and of transcending its limitations; its dual need for protection and liberation. They provided reassurance by explanation, by reducing threatening, incomprehensible events to principles familiar to experience: lightning and thunder to temperamental outbursts of man-like gods, eclipses to the greed of moon-eating pigs; they asserted that there was rhyme and reason, a hidden law and order behind the seemingly arbitrary and chaotic flux, even behind the death of a child and the eruption of a volcano. They jointly satisfied man's basic need and voiced his basic intuition, that the universe is meaningful, ordered, rational, and governed by some form of justice, even if its laws are not transparent.

Apart from reassuring the conscious mind by investing the universe with meaning and value, religion acted in a more direct manner on the unconscious, pre-rational layers of the self, providing it with intuitive techniques to transcend its limitations in time and space by a mystical short-circuit, as it were. The same duality of approach – the rational and the intuitive – characterizes, as we saw, the scientific quest. It is therefore a perverse mistake to identify the religious need solely with intuition and emotion, science solely with the logical and the rational. Prophets and discoverers, painters and poets, all share this amphibial quality of living both on the contoured drylands and in the boundless ocean. In the history of the race as of the individual, both branches of the cosmic quest originate in the same source. The priests were the first astronomers; the medicine-men were both prophets and physicians; the techniques of hunting, fishing, sowing, and reaping were imbued with religious magic and ritual. There was division of labour and diversity of method in the symbols and techniques, but unity of motive and purpose.

The first separation, insofar as our knowledge of history goes, occurred between Olympian religion and Ionian philosophy. The Ionians' polite atheism reflected the degeneration of the State Religion into an elaborate and specialized ritual, its loss of cosmic consciousness. The Pythagorean synthesis was made possible by the loosening up of that rigid theological structure through the mystic revival which Orphism brought in its wake. A similar situation occurred in the sixteenth century, when the religious crisis shook up medieval theology, and enabled Kepler to build his new model of the universe *ad majorem gloriam Dei* – that short-lived Neo-Pythagorean union of mystic inspiration and empirical fact.

Throughout the Dark Ages, the monasteries had been oases of learning in a desert of ignorance, and the monks the guardians of the dried-up wells. There was dearth, but no quarrel between theology and philosophy: both agreed that vulgar Nature was no worthy object of knowledge. It was an age of double-think, of a culture divided from reality, but the partition was not between theologian and scientist, because the latter did not exist.

The later medieval cosmology of the Great Chain of Being was a highly integrated one. It is true that the 'Venus riding on the third epicycle' of the *Divine Comedy* could not be represented by a mechanical model; but here again the dividing wall stood not between religious and natural philosophy, but between mathematics and physics, physics and astronomy, as the Aristotelian doctrine demanded. It is also true that the Church was partly responsible for this state of affairs because she had allied herself with Aristotle, as she had before with Plato; but it was not an absolute alliance, as the example of the Franciscans and the Ockhamist schools prove.

There is no need to recapitulate the reinstatement by Aquinas of the Light of Reason as an active partner of the Light of Grace; nor the leading part played in the revival of learning by Dominicans and Franciscans, ecclesiastics like Bishops Oresme, Cusa, or Giese; nor to dwell again on the joint impact of the recovery of the Greek texts of the Septuagint and Euclid. The reformation of religion and the renaissance of science were related processes of breaking up petrified patterns of development, and going back to their sources to discover where things had gone wrong. Erasmus and Reuchlin, Luther and Melanchton, went back to Greek and Hebrew texts as Copernicus and his successors went back to Pythagoras and Archimedes – prompted by the same urge of *reculer pour mieux sauter*, of regaining a unifying vision lost by doctrinaire over-specialization. Throughout the golden age of humanism, and even the gunpowder age of the Counter-reformation, the scientists remained the sacred cows of cardinals and popes, from Paul III to Urban VIII; at the same time the Roman College and the Jesuit Order took over the lead in mathematics and astronomy.

The first open conflict between Church and Science was the Galileo scandal. I have tried to show that unless one believes in the dogma of historic inevitability – this form of fatalism in reverse gear – one must regard it as a scandal which could have been avoided; and it is not difficult to imagine the Catholic Church adopting, after a Tychonic transition, the Copernican cosmology some two hundred years earlier than she eventually did. The Galileo affair was an isolated, and in fact quite untypical, episode in the history of the relations between science and theology, almost as untypical as the Dayton monkey-trial was. But its dramatic circumstances, magnified out of all proportion, created a popular belief that science stood for freedom, the Church for oppression of thought. That is only true in a limited sense for a limited period of transition. Some historians, for instance, wish to make us believe that the decline of science in Italy was due to the 'terror' caused by the trial of Galileo. But the next generation saw the rise of Toricelli, Cavallieri, Borelli, whose contributions to science were more substantial than those of any generation before or during Galileo's lifetime; the shift of the centre of scientific activity to England and France and the gradual decline of Italian science, as of Italian painting, was due to different historical causes. Never since the Thirty Years War has the Church oppressed freedom of thought and expression to an extent comparable to the terror based on the 'scientific' ideologies of Nazi Germany or Soviet Russia.

The contemporary divorce between faith and reason is not the result of a contest for power or for intellectual monopoly, but of a progressive estrangement without hostility or drama, and therefore all the more deadly. This becomes evident if we shift our attention from Italy to the Protestant countries of Europe, and to France. Kepler, Descartes, Barrow, Leibniz, Gilbert, Boyle, and Newton himself, the generation of pioneers contemporary with and succeeding Galileo, were all deeply and genuinely religious thinkers. But their image of the godhead had undergone a subtle and gradual change. It had been freed from its rigid scholastic frame, it had receded beyond the dualism of Plato to the mystic, Pythagorean inspiration of God the chief mathematician. The pioneers of the new cosmology, from Kepler to Newton and beyond, based their search into nature on the mystic conviction that there must exist laws behind the confusing phenomena; that the world was a completely rational, ordered, harmonic creation. In the words of a modern historian, the

> aspiration to demonstrate that the universe ran like a piece of clock-work ... was itself initially a religious aspiration. It was felt that there would be something defective in Creation itself – something not quite worthy of God – unless the whole system of the universe could be shown to be interlocking, so that it carried the pattern of reasonableness and orderliness. Kepler, inaugurating the scientist's quest for a mechanistic universe in the seventeenth century, is significant here – his mysticism, his music of the spheres, his rational deity demand a system which has the beauty of a piece of mathematics.[3]

Instead of asking for specific miracles as proof of God's existence, Kepler discovered the supreme miracle in the harmony of the spheres.

## 5. *The Fatal Estrangement*

And yet this new Pythagorean unity lasted only a short time, and was followed by a new estrangement which seems to us more irrevocable than any before. The first signs of this estrangement appear already in Kepler's own writings.

What else can the human mind hold besides numbers and magnitudes? These alone we apprehend correctly, and if piety permits to say so, our comprehension is in this case of the same kind as God's, at least insofar as we are able to understand it in this mortal life.[4]

Geometry is unique and eternal, a reflection of the mind of God. That men are able to participate in it is one of the reasons why man is an image of God.[5]

Therefore I chance to think that all Nature and the graceful sky are symbolized in the art of geometry. ... Now as God the maker play'd He taught the game to Nature whom He created in His image; taught her the self-same game which He played to her.[6]

All this was wholly admirable and unexceptionable from the theologian's point of view. But in Kepler's later writings, a new note becomes discernible. We hear that 'geometry provided the Creator with a model for the decoration of the whole world'.[7] that geometry somehow preceded the Creation of the world, and that 'quantities are the archetypes of the world'.[8]

There is a subtle shift of emphasis here, which conveys the impression that God copied the universe from geometrical archetypes which co-existed with Him from eternity, and that in the act of Creation He was somehow bound by blueprints. Paracelsus expressed the same idea in a less delicate manner: 'God can make an ass with three tails, but not a triangle with four sides.'[9]

For Galileo, too, 'the book of nature is written in the mathematical language ... without its help it is impossible to comprehend a single word of it'.[10] But Galileo's 'chief mathematician' is called 'Nature', not God, and his references to the latter sound like lip-service. Galileo takes the hyperstatization of mathematics a decisive step further by reducing all nature to 'size, figure, number, and slow or rapid motion', and by relegating into the limbo of 'subjective' or 'secondary' qualities everything that cannot be reduced to these elements – including, by implication, ethical values, and the phenomena of the mind.

The division of the world into 'primary' and 'secondary' qualities was completed by Descartes. He further reduced primary qualities to 'extension' and 'motion', which form the 'realm of extension' – *res extensa* – and he lumped together everything else in the *res cognitans*, the realm of the mind, housed in a somewhat niggardly manner in the tiny pituitary gland. Animals, for Descartes, are robot mechanisms, and so is the human body; and the universe (with the exception of a few million pea-sized pituitary glands) was now so completely mechanized that he could claim 'give me matter and motion and I will construct the world'. And yet Descartes, too, was a profoundly religious thinker, who deduced his law of the immutability of the total amount of motion in the universe* from the immutability of God. But since, given matter and motion, he would have created the same universe governed by the same laws, was the deduction from God's mind really necessary? The answer is contained in Bertrand Russell's aphorism on Descartes: 'No God, no geometry; but geometry is delicious, therefore God must exist.'

As for Newton, who was a better scientist and hence a more muddled metaphysician than Galileo or Descartes, he assigned to God a two-fold function as Creator of the universal clockwork, and as its Supervisor for maintenance and repair. He believed that the placing of all planetary orbits into a single plane and in such orderly manner, and the fact that there was only a single sun in the system sufficient to provide the rest with light and heat, instead of having several suns or no sun at all, were proof that Creation was the work of an 'intelligent agent ... not blind or fortuitous, but very well skilled in mechanics and geometry'.[11] He further believed that under the pressure of gravity the universe would collapse 'without a divine power to support it'; [12] and moreover, that the small irregularities in the planetary motions would accumulate and throw the whole system out of gear if God did not from time to time set it right.

Newton was a crank theologian like Kepler, and like Kepler addicted to chronology; he dated the Creation from 4004 B.C., after Bishop Usher, and held that the tenth horn of the fourth beast of the Apocalypse represented the Roman Church. He

* The forerunner of the law of conservation of energy.

desperately tried to find a niche for God somewhere between the wheels of the mechanical clockwork – as Jeans and others later tried to find it in Heisenberg's principle of indeterminacy. But, as we have seen, such mechanical addings together of two fully grown specialized disciplines never work. The Kant-Laplace theory of the origin of the solar system showed that its orderly arrangement could be explained on purely physical grounds, without recourse to divine intelligence; and God's alleged duties as a maintenance engineer were derided as absurd by Newton's own contemporaries, foremost among them Leibnitz:

> According to their [Newton and his followers'] doctrine, God Almighty wants to wind up his watch from time to time, otherwise it would cease to move. He had not, it seems, sufficient foresight to make it a perpetual motion. Nay, the machine of God's making is so imperfect according to these gentlemen, that he is obliged to clean it now and then by an extraordinary concourse, and even to mend it as a clockwork-maker mends his work. ... And I hold that when God works miracles, He does not do it in order to supply the wants of Nature, but those of grace. Whoever thinks otherwise must needs have a very mean notion of the wisdom and power of God.[13]

In a word, atheists were the exception among the pioneers of the scientific revolution. They were all devout men who did not want to banish deity from their universe, but could find no place for it – just as, quite literally, they were unable to reserve sites for Paradise and Hell. The Chief Mathematician became redundant, a polite fiction gradually absorbed into the tissues of natural law. The mechanical universe could accommodate no transcendental factor. Theology and physics parted ways not in anger, but in sorrow, not because of Signor Galileo, but because they became bored with and had nothing more to say to each other.

The divorce led to consequences which are familiar to us from similar occasions in the past. Cut off from what was once called the philosophy of nature and now exact science, theology continued in its own specialized, doctrinal line. The age of Benedictine, Franciscan, Thomist, Jesuit leadership in research was past. To the inquiring intellect, the established churches became venerable anachronisms – though still capable of giving sporadic uplift to a diminishing number of individuals at the price of splitting his mind into incompatible halves. Whitehead's admirable summing up of the situation in 1926 is even truer today, a generation after he wrote it:

> There have been reactions and revivals. But on the whole, during many generations, there has been a gradual decay of religious influence in European civilization. Each revival touches a lower peak than its predecessor, and each period of slackness a lower depth. The average curve marks a steady fall in religious tone. ... Religion is tending to degenerate into a decent formula wherewith to embellish a comfortable life.
>
> ... For over two centuries religion has been on the defensive, and on a weak defensive. The period has been one of unprecedented intellectual progress. In this way a series of novel situations have been produced for thought. Each such occasion has found the religious thinkers unprepared. Something, which has been proclaimed to be vital, has finally, after struggle, distress, and anathema, been modified and otherwise interpreted. The next generation of religious apologists then congratulates the religious world on the deeper insight which has been gained. The result of the continued repetition of this undignified retreat, during many generations, has at last almost entirely destroyed the intellectual authority of religious thinkers. Consider this contrast: when Darwin or Einstein proclaim theories which modify our ideas, it is a triumph for science. We do not go about saying that there is another defeat for science, because its old ideas have been abandoned. We know that another step of scientific insight has been gained.
>
> Religion will not regain its old power until it can face change in the same spirit as does science. Its principles may be eternal, but the expression of those principles requires continual development. ...
>
> The religious controversies of the sixteenth and seventeenth centuries put theologians into a most unfortunate state of mind. They were always attacking and defending. They pictured themselves as the garrison of a fort surrounded by hostile forces. All such pictures express half-truths. That is why they are so popular. But they are dangerous. This particular picture fostered a pugnacious party spirit which really expresses an ultimate lack of faith. They dared not

modify, because they shirked the task of disengaging their spiritual message from the associations of a particular imagery. . . .

. . . We have to know what we mean by religion. The churches, in their presentation of their answers to this query, have put forward aspects of religion which are expressed in terms either suited to the emotional reactions of bygone times or directed to excite modern emotional interests of non-religious character. . . .

Religion is the vision of something which stands beyond, behind, and within, the passing flux of immediate things; something which is real, and yet waiting to be realized; something which is a remote possibility, and yet the greatest of present facts; something that gives meaning to all that passes, and yet eludes apprehension; something whose possession is the final good, and yet is beyond all reach; something which is the ultimate ideal, and the hopeless quest.[14]

## 6. *The Vanishing Act*

To the other divorced party, science, the parting of the ways seemed at the beginning to be an unmitigated boon. Freed from mystical ballast, science could sail ahead at breathtaking speed to its conquest of new lands beyond every dream. Within two centuries it transformed the mental outlook of *homo sapiens* and transformed the face of his planet. But the price paid was proportionate: it carried the species to the brink of physical self-destruction, and into an equally unprecedented spiritual impasse. Sailing without ballast, reality gradually dissolved between the physicist's hands; matter itself evaporated from the materialist's universe.

This uncanny vanishing act began, as we saw, with Galileo and Descartes. In that famous passage in *The Assayer* (see p. 476), Galileo banished the qualities which are the very essence of the sensual world – colour and sound, heat, odour, and taste – from the realm of physics to that of subjective illusion. Descartes carried the process one step further by paring down the reality of the external world to particles whose only quality was extension in space and motion in space and time. At first this revolutionary approach to nature looked so promising that Descartes believed he would be able to complete the whole edifice of the new physics by himself. His less sanguine contemporaries thought that it might take as much as two generations to wrest its last secret from nature. 'The particular phenomena of the arts and sciences are in reality but a handful,' said Francis Bacon. 'The invention of all causes and sciences would be the labour of but a few years.'[15]

But in the two centuries that followed, the vanishing act continued. Each of the 'ultimate' and 'irreducible' primary qualities of the world of physics proved in its turn to be an illusion. The hard atoms of matter went up in fireworks; the concepts of substance, force, of effects determined by causes, and ultimately the very framework of space and time turned out to be as illusory as the 'tastes, odours, and colours' which Galileo had treated so contemptuously. Each advance in physical theory, with its rich technological harvest, was bought by a loss in intelligibility. These losses on the intellectual balance sheet, however, were much less in evidence than the spectacular gains; they were light-heartedly accepted as passing clouds which the next advance would dissolve. The seriousness of the impasse became only apparent in the second quarter of our century, and then only to the more philosophically-minded among scientists, who had retained a certain immunity against what one might call the new scholasticism of theoretical physics.

Compared to the modern physicist's picture of the world, the Ptolemaic universe of epicycles and crystal spheres was a model of sanity. The chair on which I sit seems a hard fact, but I know that I sit on a nearly perfect vacuum. The wood of the chair consists of fibres, which consist of molecules, which consist of atoms, which are miniature solar systems with a central nucleus and electrons for planets. It all sounds very pretty, but it is the dimensions that matter. The space which an electron occupies is only one fifty-thousandth in diameter of its distance from the nucleus; the rest of the atomic interior is empty. If the nucleus were enlarged to the size of a dried pea, the nearest electron would circle around it at a distance of about a hundred and seventy-five yards. A room with a few specks of dust floating in the air is overcrowded compared to the emptiness which I call a chair and on which my fundaments rest.

But it is doubtful whether it is permissible to say that the electron 'occupies space' at all. Atoms have the capacity of swallowing energy and of spitting out energy – in the form of light rays, for instance. When a hydrogen atom, the simplest of all, with a single electron-planet, swallows energy, the planet jumps from its orbit to a larger orbit – say, from the orbit of Earth to the orbit of Mars; when it emits energy, it jumps back again into the smaller orbit. But these jumps are performed by the planet without it passing through the space that separates the two orbits. It somehow de-materializes in orbit A and re-materializes in orbit B. Moreover, since the amount of 'action' performed by the hydrogen electron while going once round its orbit is the indivisibly smallest quantum of action (Planck's basic constant 'h'), it is meaningless to ask at what precise point of its orbit the electron is at a given moment of time. It is equally everywhere.[15a]

The list of these paradoxa could be continued indefinitely; in fact the new quantum-mechanics consist of nothing but paradoxa, for it has become an accepted truism among physicists that the sub-atomic structure of any object, including the chair I sit on, cannot be fitted into a framework of space and time. Words like 'substance' or 'matter' have become void of meaning, or invested with simultaneous contradictory meanings. Thus beams of electrons, which are supposedly elementary particles of matter, behave in one type of experiment like little pellets, but in another type of experiment they behave like waves; conversely, rays of light behave sometimes like waves and at other times like bullets. Consequently, the ultimate constituents of matter are both substance and non-substance, lumps and waves. But waves in, on, of what? A wave is movement, undulation; but what is it that moves and undulates, producing my chair? It is nothing the mind can conceive of, not even empty space, for each electron requires a three-dimensional space for itself, two electrons need six dimensions, three electrons nine dimensions, to co-exist. In some sense these waves are real: we can photograph the famous dart-board pattern they produce when they pass through a diffraction grate; yet they are like the grin of the Cheshire cat.

For ought we know [says Bertrand Russell] at atom may consist entirely of the radiations which come out of it. It is useless to argue that radiations cannot come out of nothing. ... The idea that there is a little hard lump there, which *is* the electron or proton, is an illegitimate intrusion of commonsense notions derived from touch. ... 'Matter' is a convenient formula for describing what happens where it isn't.[16]

These waves, then, on which I sit, coming out of nothing, travelling through a non-medium in multi-dimensional non-space, are the ultimate answer modern physics has to offer to man's question after the nature of reality. The waves that seem to constitute matter are interpreted by some physicists as completely immaterial 'waves of probability' marking out 'disturbed areas' where an electron is likely to 'occur'. 'They are as immaterial as the waves of depression, loyalty, suicide, and so on, that sweep over a country.'[17] From here there is only one step to calling them abstract, mental, or brain waves in the Universal Mind – without irony. Imaginative scientists of such different persuasion as Bertrand Russell on the one hand, Eddington and Jeans on the other, have indeed come very near to taking this step. Thus Eddington wrote:

The stuff of the world is mind-stuff. The mind-stuff is not spread in space and time; these are part of the cyclic scheme ultimately derived out of it. But we must presume that in some other way or aspect it can be differentiated into parts. Only here and there does it rise to the level of consciousness, but from such islands proceeds all knowledge. Besides the direct knowledge contained in each self-knowing unit, there is inferential knowledge. The latter includes our knowledge of the physical world.[18]

Jeans went even further:

The concepts which now prove to be fundamental to our understanding of nature – a space which is finite; a space which is empty, so that one point [which appears to us occupied by a material body] differs from another solely in the properties of the space itself;

four-dimensional, seven and more dimensional spaces; a space which for ever expands; a sequence of events which follows the laws of probability instead of the laws of causation – or, alternatively, a sequence of events which can only be fully and consistently described by going outside space and time, all these concepts seem to my mind to be structures of pure thought, incapable of realization in any sense which would properly be described as material.[19]

And again:

Today there is a wide measure of agreement, which on the physical side of science approaches almost to unanimity, that the stream of knowledge is heading towards a non-mechanical reality; the universe begins to look more like a great thought than like a great machine. Mind no longer appears as an accidental intruder into the realm of matter; we are beginning to suspect that we ought rather to hail it as the creator and governor of the realm of matter. . . .[20]

Thus the medieval walled-in universe with its hierarchy of matter, mind, and spirit, has been superseded by an expanding universe of curved, multi-dimensional empty space, where the stars, planets, and their populations are absorbed into the space-crinkles of the abstract continuum – a bubble blown out of 'empty space welded on to empty time'.[21]

How did this situation come about? Already in 1925, before the new quantum mechanics came into being, Whitehead wrote that 'the physical doctrine of the atom has got into a state which is strongly suggestive of the epicycles of astronomy before Copernicus'.[22] The common feature between pre-Keplerian astronomy and modern physics is that both have developed in relative isolation as 'closed systems', manipulating a set of symbols according to certain rules of the game. Both systems 'worked'; modern physics yielded nuclear energy, and Ptolemaic astronomy yielded predictions whose precision bowled over Tycho. The medieval astronomers manipulated their epicyclic symbols as modern physics manipulates Schroedinger's wave equations or Dirac's matrices, and it worked – though they knew nothing of gravity and elliptic orbits, believed in the dogma of circular motion, and had not the faintest idea *why* it worked. We are reminded of Urban VIII's famous argument which Galileo treated with scorn: that a hypothesis which works must not necessarily have anything to do with reality for there may be alternative explanations of how the Lord Almighty produces the phenomena in question. If there is a lesson in our story it is that the manipulation, according to strictly self-consistent rules, of a set of symbols representing one single aspect of the phenomena may produce correct, verifiable predictions, and yet completely ignore all other aspects whose ensemble constitutes reality:

> ... Science deals with but a partial aspect of reality, and ... there is no faintest reason for supposing that everything science ignores is less real than what it accepts. ... Why is it that science forms a closed system? Why is it that the elements of reality it ignores never come in to disturb it? The reason is that all the terms of physics are defined in terms of one another. The abstractions with which physics begins are all it ever has to do with. ...[23]

Modern physics is not really concerned with 'things' but with the mathematical relations between certain abstractions which are the residue of the vanished things. In the Aristotelian universe, quantity was merely one attribute of things, and one of the least important. Galileo's 'the book of nature is written in the language of mathematics' was regarded by his contemporaries as a paradox; today it has become unquestioned dogma. For a long time the reduction of quality to quantity – of colour, sound, radiation to vibrational frequencies – was so eminently successful that it seemed to answer all questions. But when physics approached the ultimate constituents of matter, quality took its revenge: the method of reduction to quantity still worked, but we no longer know just what it is that is being thus reduced. All we do in fact know is that we read our instruments – the number of clicks in the Geiger counter, or the position of a pointer on a dial – and interpret the signs according to the rules of the game:

> And so in its actual procedure physics studies not these inscrutable qualities [of the material world], but pointer-readings which we can observe. The readings, it is true, reflect the fluctuations of the world-qualities; but our exact knowledge is of the readings, not

of the qualities. The former have as much resemblance to the latter as a telephone number has to a subscriber.[24]

Bertrand Russell expressed this state of affairs even more succinctly:

> Physics is mathematical not because we know so much about the physical world, but because we know so little: it is only its mathematical properties that we can discover.[25]

## *7. The Conservatism of Modern Science*

There are two ways of interpreting this situation.

Either the structure of the universe is indeed of such a nature that it cannot be comprehended in terms of human space and time, human reason, and human imagination. In this case Exact Science has ceased to be the Philosophy of Nature, and no longer has much inspiration to offer to the questing human mind. In this case it would be legitimate for the scientist to withdraw into his closed system, to manipulate his purely formal symbols, and to evade questions concerning the 'real meaning' of these symbols as 'meaningless', as it has become the fashion. But if this be the case, he must accept his role as a mere technician whose task is to produce, on the one hand, better bombs and plastic fibres, and on the other, more elegant systems of epicycles to save the phenomena.

The second possibility is to regard the present crisis in physics as a temporary phenomenon, the result of a one-sided, overspecialized development like the giraffe's neck – one of those *culs-de-sac* of mental evolution which we have so often observed in the past. But if that is the case, where, on the three-centuries' journey from 'natural philosophy' to 'exact science', did the estrangement from reality begin; at what point was the new version of Plato's curse uttered: 'Thou shalt think in circles'? If we knew the answer, we would, of course, also know the remedy; and once the answer is known, it will again appear as heartbreakingly obvious as the sun's central position in the solar system. 'We are indeed a blind race,' wrote a contemporary scientist, 'and the next generation, blind to its own blindness, will be amazed at ours.'[26]

I shall quote two examples which seem to me to illustrate this blindness. The materialist philosophy in which the average modern scientist was reared has retained its dogmatic power over his mind, though matter itself has evaporated; and he reacts to phenomena which do not fit into it much in the same manner as his scholastic forebears reacted to the suggestion that new stars might appear in the immutable eighth sphere. Thus for the last thirty years, an impressive body of evidence has been assembled under strict laboratory conditions which suggests that the mind might perceive stimuli emanating from persons or objects without the intermediary of the sensory organs; and that in controlled experiments, these phenomena occur with a statistical frequency which invites scientific investigation. Yet academic science reacts to the phenomena of 'extra-sensory perception' much as the Pigeon League reacted to the Medicean Stars; and, it seems to me, for no better reason. If we have to accept that an electron can jump from one orbit into the other without traversing the space between them, why are we bound to reject out of hand the possibility that a signal of a nature no more puzzling than Schroedinger's electron-waves should be emitted and received without sensory intervention? If modern cosmology has a single comprehensive lesson it is that the basic events in the physical world cannot be represented in three-dimensional space and time. Yet the modern version of scholasticism denies additional dimensions to the mind, or brain, which it readily accords to the particles of a piece of lead. I am not playing on the word 'dimension' as a mechanical analogy – after the manner of the 'fourth dimension' of occult quacks. I am merely saying that since the space-time framework, the concepts of matter and causality as understood both by classical physics and by commonsense experience, have been abandoned by modern physics, there seems to be no justification in refusing to investigate empirical phenomena because they do not fit into that already abandoned philosophy.

A second example of the *hubris* of contemporary science is the rigorous banishment of the word 'purpose' from its vocabulary. This is probably an aftermath of the reaction against the animism of Aristotelian physics, where stones accelerated their fall because of their impatience to get home, and against a teleological world-view in which the purpose of the stars was to serve as chronometers for man's profit. From Galileo onward, 'final causes' (or 'finality' for short) were relegated into the realm of superstition, and mechanical causality reigned supreme. In the mechanical universe of indivisible hard little atoms, causality worked by impact, as on a billiard table; events were caused by the mechanical push of the past, not by any 'pull' of the future. That is the reason why gravity and other forms of action-at-a-distance did not fit into the picture and were regarded with suspicion; why ethers and vortices had to be invented to replace the occult pull by a mechanical push. The mechanistic universe gradually disintegrated, but the mechanistic notion of causality survived until Heisenberg's indeterminacy principle proved its untenability. Today we know that on the sub-atomic level the fate of an electron or a whole atom is not determined by its past. But this discovery has not led to any basically new departure in the philosophy of nature, only to a state of bewildered embarrassment, a further retreat of physics into a language of even more abstract symbolism. Yet if causality has broken down and events are not rigidly governed by the pushes and pressures of the past, may they not be influenced in some manner by the 'pull' of the future – which is a manner of saying that 'purpose' may be a concrete physical factor in the evolution of the universe, both on the organic and unorganic levels. In the relativistic cosmos, gravitation is a result of the curvatures and creases in space which continually tend to straighten themselves out – which, as Whittaker remarked,[27] 'is a statement so completely teleological that it would certainly have delighted the hearts of the schoolmen.' If time is treated in modern physics as a dimension almost on a par with the dimensions of space, why should we *a priori* exclude the possibility that we are pulled as well as pushed along its axis? The future has, after all, as much or as little reality as the past, and there is nothing logically inconceivable in introducing, as a working hypothesis, an element of finality, supplementary to the element of causality, into our equations.

It betrays a great lack of imagination to believe that the concept of 'purpose' must necessarily be associated with some anthropomorphic deity.

These are matters of speculation and possibly quite beside the point; but we have learnt from the past that impasses in evolution can only be overcome by some new departure in an unexpected direction. Whenever a branch of knowledge became isolated from the mainstream, its frozen surface had to crack up and thaw before it could be reunited with living reality.

## 8. *From Hierarchy to Continuum*

As a result of their divorce, neither faith nor science is able to satisfy man's intellectual cravings. In the divided house, both inhabitants lead a thwarted existence.

Post-Galilean science claimed to be a substitute for, or the legitimate successor of, religion; thus its failure to provide the basic answers produced not only intellectual frustration but spiritual starvation. A summary recapitulation of European men's view of the world before and after the scientific revolution may help to put the situation into sharper relief. Taking the year 1600 as our dividing line or watershed, we find indeed virtually all rivers of thought and currents of feeling flow into opposite directions. The 'pre-scientific' European lived in a closed universe with firm boundaries in space and time a few million miles in diameter and a few thousand years of duration. Space as such did not exist as an abstract concept, merely as an attribute of material bodies – their length, width, and depth; hence empty space was unthinkable, a contradiction in terms, and infinite space even more so. Time, similarly was simply the duration of an event. Nobody in his senses would have said that things move *through* or *in* space or time – how can a thing move in or through an attribute of itself, how can the concrete move through the abstract?

In this safely bounded world of comfortable dimensions, a well-ordered drama was taking its pre-ordained course. The stage remained static from beginning to end: there was no change in the species of animals and plants, no change in the nature, social order, and mentality of man. There was neither progress nor decline within the natural and spiritual hierarchy. The total body of possible knowledge was as limited as the universe itself; everything that could be known about the Creator and his creation had been revealed in Holy Scripture and the writings of the ancient sages. There existed no sharp boundaries between the natural and the supernatural: matter was imbued with animal spirits, natural law was interpenetrated with divine purpose; there was no event without a final cause. Transcendental justice and moral values were inseparable from the natural order; no single event or fact was ethically neutral; no plant or metal, no insect or angel exempt from moral judgement; no phenomenon was outside the hierarchy of values. Every suffering had its reward, every disaster its meaning; the plot of the drama had a simple outline, a clear beginning and end.

This briefly, was our forebear's view of the world less than fifteen generations ago. Then, roughly within the five generations from Canon Koppernigk to Isaac Newton, *homo sapiens* underwent the most decisive change in his history:

> The gloriously romantic universe of Dante and Milton, that set no bounds to the imagination of man as it played over space and time, had now been swept away. Space was identified with the realm of geometry, time with the continuity of number. The world that people had thought themselves living in – a world rich with colour and sound, redolent with fragrance, filled with gladness, love, and beauty, speaking everywhere of purposive harmony and creative ideals – was crowded now into minute corners in the brains of scattered organic beings. The really important world outside was a world hard, cold, colourless, silent, and dead; a world of quantity, a world of mathematically computable motions in mechanical regularity. The world of qualities as immediately perceived by man became just a curious and quite minor effect of that infinite machine beyond.[28]

The *uomo universale* of the Renaissance, who was artist and craftsman, philosopher and inventor, humanist and scientist, astronomer and monk, all in one, split up into his component parts. Art lost its mythical, science its mystical inspiration; man became again deaf to the harmony of the spheres. The Philosophy of Nature became ethically neutral, and 'blind' became the favourite adjective for the working of natural law. The space-spirit hierarchy was replaced by the space-time continuum.

As a result, man's destiny was no longer determined from 'above' by a super-human wisdom and will, but from 'below' by the sub-human agencies of glands, genes, atoms, or waves of probability. This shift of the locus of destiny was decisive. So long as destiny had operated from a level of the hierarchy higher than man's own, it had not only shaped his fate, but also guided his conscience and imbued his world with meaning and value. The new masters of destiny were placed lower in the scale than the being they controlled; they could determine his fate, but could provide him with no moral guidance, no values and meaning. A puppet of the Gods is a tragic figure, a puppet suspended on his chromosomes is merely grotesque.

Before the shift, the various religions had provided man with explanations of a kind which gave to everything that happened to him meaning in the wider sense of transcendental causality and transcendantal justice. But the explanations of the new philosophy were devoid of meaning in this wider sense. The answers of the past had been varied, contradictory, primitive, superstitious, or whatever one likes to call them, but they had been firm, definite, imperative. They satisfied, at least for a given time and culture, man's need for reassurance and protection in an unfathomably cruel world, for some guidance in his perplexities. The new answers, to quote William James, 'made it impossible to find in the driftings of the cosmic atoms, whether they work on the universal or on the particular scale, anything but a kind of aimless weather, doing and undoing, achieving no proper history, leaving no result.' In a word, the old explanations, with all their arbitrariness and patchiness, answered the question after 'the meaning of life' whereas the new explanations, with all their precision, made the question of meaning itself meaningless. As man's science grew more abstract, his art became more esoteric, and his pleasures more chemical. In the end he was left with nothing but 'an abstract heaven over a naked rock'.

Man entered upon a spiritual ice age; the established Churches could no longer provide more than Eskimo huts where their shivering flock huddled together, while the campfires of rival ideologies drew the masses in wild stampedes across the ice.[29]

## 9. *The Ultimate Decision*

Coincident with this progressive spiritual dessication, the post-Renaissance centuries brought an unprecedented rise in both constructive and destructive power. The operative word here is 'unprecedented'. All comparisons with past epochs break down before the fact that our species has acquired the means to annihilate itself, and make the earth uninhabitable; and that in the foreseeable future, it will be within its power to turn this planet into a *nova*, a rival sun in the solar system. Every age had its Cassandras, and one tends to draw comfort from the fact that mankind has, after all, managed to survive regardless of their pessimistic prophecies. But such analogies are no longer valid, for no past age, however convulsed, had the actual means of committing racial suicide and interfering with the order of the solar system.

The basic novelty of our age is the combination of this sudden, unique increase in physical power with an equally unprecedented spiritual ebb-tide. To appreciate this novelty one must abandon the limited perspective of European history, and think in terms of the history of the species. Elsewhere I have suggested that the process which led to our present predicament could be represented by two curves similar to temperature charts, one showing the growing physical power of the race, the other its spiritual insight, moral awareness, charity, and related values. For something of the order of several hundred thousand years, from Cro-Magnon man to about 5000 B.C., the first curve would depart little from the horizontal. With the invention of the pulley, the lever, and other simple mechanical devices, the muscular strength of man would appear amplified, say, five-fold; after that the curve would again remain horizontal for five or six thousand years. But in the course of the last two hundred years – a stretch less than one-thousandth of the total on the chart – the curve would, for the first time in the history of the species, suddenly rise in leaps and bounds; and in the last fifty years – about one-hundred-thousandth of the total – the curve rises so steeply that it now points almost vertically upward. A single example will illustrate this: after the First World War, less than a generation before Hiroshima, statisticians reckoned that on an average ten thousand rifle bullets, or ten artillery shells, were needed to kill one enemy soldier.

Compared to the first, the second curve will show a very slow rise during the nearly flat pre-historic miles; then it will undulate with indecisive ups and downs through civilized history; finally, on the last dramatic fraction of the chart where the power curve shoots upward like a cobra stabbing at the sky, the spiritual curve goes into a steep decline.

The diagram may be oversimplified, but it is certainly not overdramatized. To draw it true to scale, one would have to use paper about a hundred yards long, but even so the relevant portion would occupy only an inch. We would be obliged to use units of time at first of a hundred thousand, then of a thousand years, while, as we approach the present, the vertical rise of the power-curve is greater in a single year than it was in ten thousand earlier ones.

Thus within the foreseeable future, man will either destroy himself or take off for the stars. It is doubtful whether reasoned argument will play any significant part in the ultimate decision, but if it does, a clearer insight into the evolution of ideas which led to the present predicament may be of some value. The muddle of inspiration and delusion, of visionary insight and dogmatic blindness, of millennial obsessions and disciplined double-think, which this narrative has tried to retrace, may serve as a cautionary tale against the *hubris* of science – or rather of the philosophical outlook based on it. The dials on our laboratory panels are turning into another version of the shadows in the cave. Our hypnotic enslavement to the numerical aspects of reality has dulled our perception of non-quantitative moral values; the resultant end-justifies-the-means ethics may be a major factor in our undoing. Conversely, the example of Plato's obsession with perfect spheres, of Aristotle's arrow propelled by the surrounding air, the forty-eight epicycles of Canon Koppernigk and his moral cowardice, Tycho's mania of grandeur, Kepler's sun-spokes, Galileo's confidence tricks, and Descartes' pituitary soul, may have some sobering effect on the worshippers of the new Baal, lording it over the moral vacuum with his electronic brain.

*March 1955–May 1958*

# Acknowledgements

The publishers wish to thank the following for permission to quote from various works: Sheed & Ward, London (*The Confessions of St Augustine*, translated by F. J. Sheed); the University of Chicago Press (*Dialogue on the Great World Systems*, by Professor Georgio de Santillana, *c.* 1953, by the University of Chicago, and *The Crime of Galileo*, also by Santillana, *c.* 1955, by the University of Chicago); Edward Arnold (Publishers) Ltd, London (*The Waning of the Middle Ages*, by J. Huizinga); Columbia University Press, New York (*Three Copernican Treatises*, translated by Professor E. Rosen); The Johns Hopkins Press, Baltimore (*From the Closed World to the Infinite Universe*, by Professor Alexandre Koyré); Doubleday & Co., Inc., New York (*Discoveries and Opinions of Galileo*, translated by Stillman Drake, *c.* 1957, by Stillman Drake); Cambridge University Press (*Science and the Modern World*, by A. N. Whitehead); Wm Collins, Sons & Co. Ltd and The Macmillan Company, New York (*The Trail of the Dinosaur*, by Arthur Koestler).

# Selected Bibliography


ARMITAGE, A., *Copernicus the Founder of Modern Astronomy* (London, 1938).

DE BRAHE, TYCHO, *Opera Omnia* (Copenhagen, 1913–29).

DE BRAHE, TYCHO, see also Dreyer, J. L. E.

BRUNO, GIORDANO, *On the Infinite Universe and Worlds* – see Singer, D. W.

BURNET, J., *Early Greek Philosophy* (London, 1908).

BURNET, J., *Greek Philosophy*, Part I, *Thales to Plato* (London, 1914).

BURTT, E. A., *The Metaphysical Foundations of Modern Physical Science* (London, 1924).

BUTTERFIELD, H., *The Origins of Modern Science* (London, 1949).

CASPAR, M., *Johannes Kepler* (Stuttgart, 1948).

CASPAR, M., and V. DYCK, W., *Johannes Kepler in seinen Briefen* (Muenchen and Berlin, 1930).

COOPER, L., *Aristotle, Galileo and the Tower of Pisa* (Ithaca, 1935).

COPERNICUS, NICOLAS, *On the Revolutions of the Heavenly Spheres* (trans. Wallis C. G., Chicago, 1952).

COPERNICUS, NICOLAS, *Commentariolus* (trans. Rosen E. – *Three Copernican Treatises*, Columbia, 1939).

COPERNICUS, NICOLAS, *Letter Against Werner* (trans. Rosen E. – *Three Copernican Treatises*, Columbia, 1939).

COPERNICUS, NICOLAS, see also Prowe, L.

CORNFORD, F. M., *From Religion to Philosophy* (London, 1912).

CUSANUS, NICOLAS, *On Learned Ignorance* (trans. Fr G. Heron, Yale, 1954).

DELAMBRE, J. B. J., *Histoire de l'astronomie moderne* (Paris, 1821).

DELATTE, A., *Études sur la litterature pythagoricienne* (Paris, 1915).

DINGLE, H., *The Scientific Adventure* (London, 1952).

DRAKE, ST., *Discoveries and Opinions of Galileo* (New York, 1957).

DREYER, J. L. E., *History of the Planetary Systems from Thales to Kepler* (Cambridge, 1906).

DREYER, J. L. E., *Tyco Brahe* (Edinburgh, 1890).

DUHEM, P., *Le Système du monde – Histoire des doctrines cosmologiques de Plato à Copernic* (Paris, 1913–17).

DUHEM, P., *Études sur Leonard de Vinci* (Paris, 1906–13).

EDDINGTON, SIR ARTHUR, *The Philosophy of Physical Science* (London, 1939).

DE L'ÉPINOIS, *Les Pièces du procés de Galilée* (Rome, Paris, 1877).

FARRINGTON, B., *Greek Science* (London, 1953).

GALILEO, GALILEI, *Opere* (Ediz. Naz., Florence, 1929–39).

GALILEO, GALILEI, *Opere* (ed. F. Flora, 1953).

GALILEO, GALILEI, *Dialogue on the Great World Systems* – see de Santillana.

GALILEO, GALILEI, *Dialogue Concerning Two New Sciences* (trans. Crew, H., Evanston, Ill., 1950).

GALILEO, GALILEI, *The Star Messenger, The Assayer*, etc. – see Drake, St.

V. GEBLER, K., *Galileo Galilei and the Roman Curia* (London, 1879).

GILBERT, W., *On the Loadstone and Magnetic Bodies* (trans. Mottelay, New York, 1893).

GRISAR, H., *Galileistudien* (Regensburg, New York, and Cincinnati, 1882).

HEATH, TH. L., *Greek Astronomy* (London, 1932).

HEATH, TH. L., *The Copernicus of Antiquity* (London, 1920).

HUIZINGA, J., *The Waning of the Middle Ages* (London, 1955).

JEANS, SIR JAMES, *The Mysterious Universe* (Cambridge, 1937).

JEANS, SIR JAMES, *The Growth of Physical Science* (Cambridge, 1947).

KEPLER, JOHANNES, *Opera Omnia*, ed. Ch. Frisch (Frankfurt and Erlangen, 1858–71).

KEPLER, JOHANNES, *Gesammelte Werke*, ed. Caspar and v. Dyck (Munich, 1938– ).

KEPLER, JOHANNES, see Caspar, M.

KOESTLER, A., *Insight and Outlook* (London and New York, 1949).

KOYRÉ, A., *Études Galiléennes* (Paris, 1939–40).

KOYRÉ, A., *From the Closed World to the Infinite Universe* (Baltimore, 1957).

LOVEJOY, A. O., *The Great Chain of Being* (Cambridge, Mass., 1936).

NEWTON, SIR ISAAC, *Opera Omnia* (London, 1779–85).

NEWTON, SIR ISAAC, *The Mathematical Principles of Natural Philosophy* (trans. Motte, London, 1803).

NICOLSON, M., *Science and Imagination* (Oxford, 1956).

PACHTER, H. M., *Magic into Science* (New York, 1951).

PLEDGE, H. T., *Science since 1500* (London, 1939).

PROWE, L., *Nicolaus Copernicus* (Berlin, 1883–4).

PTOLEMY, CLAUDIUS, *The Almagest* (trans. Taliaferro, R. C., Chicago, 1952).

REICKE, E., *Der Gelehrte, Monographien zur deutschen Kulturgeschichte* (Leipzig, 1900).

REUSCH, F. H., *Der Process Galilei's und die Jesuiten* (Bonn, 1879).

RHETICUS, JOACHIM, *Narratio Prima* (trans. Rosen, E. – *Three Copernican Treatises*).

ROSEN, E., *Three Copernican Treatises* (Columbia, 1939).

ROSEN, E., *The Naming of the Telescope* (New York, 1947).

RUDNICKI J., *Nicolas Copernicus* (*Mikolaj Kopernik*) (London, 1943).

DE SANTILLANA, G. (ed.), *Galileo Galilei Dialogue on the Great World Systems* (Chicago, 1953).

DE SANTILLANA, G., *The Crime of Galileo* (Chicago, 1955).

SARTON, G., *The History of Science and the New Humanism* (Cambridge, Mass., 1937).

SELTMAN, CH., Pythagoras (*History Today*, London, August, September, 1956).

SHERWOOD TAYLOR, F., *Science Past and Present* (London, 1949).

SHERWOOD TAYLOR, F., *Galileo and the Freedom of Thought* (London, 1938).

SINGER, C., *A Short History of Science to the Nineteenth Century* (Oxford, 1941).

SINGER, D. W., *Giordano Bruno, His Life and Thought with Annotated Translation of His Work, On the Infinite Universe and Worlds* (New York, 1950).

STIMSON, D., *The Gradual Acceptance of the Copernican Theory of the Universe* (New York, 1917).

SULLIVAN, J. W. N., *The Limitations of Science* (New York, 1949).

TILLYARD, E. M. W., *The Elizabethan World Picture* (London, 1943).

WHITEHEAD, A. N., *Science and the Modern World* (Cambridge, 1953).

WHITTAKER, SIR EDMUND, *Space and Spirit* (London, 1946).

WOLF, A., *A History of Science, Technology and Philosophy in the Sixteenth and Seventeenth Centuries* (London, 1935).

ZINNER, E., *Enstehung und Ausbreitung der Copernicanischen Lehre* (Erlangen, 1943).

# Notes

## *Preface*

1 *A Study of History, Abridgement of Vols. I–VI* by D. C. Somervell (Oxford, 1947). In the complete ten-volume edition there are three brief references to Copernicus, two to Galileo, three to Newton, none to Kepler. All references are by way of asides.

2 Cf. *Insight and Outlook, An Inquiry into the Common Foundations of Science, Art and Social Ethics* (London and New York, 1949).

## *Part One. The Heroic Age*

### 1: DAWN

1 *Ency. Brit.* 1955 ed., II–582c.

2 ibid., II–582d.

3 F. Sherwood Taylor, *Science Past and Present* (London, 1949), p. 13.

'From the beginning of the reign of Nabonassar, 747 B.C.,' Ptolemy reported some 900 years later, 'we possess the ancient observations continued practically to the present day.' (Th. L. Heath, *Greek Astronomy* (London, 1932), p. xiv f.)

The Babylonian observations, incorporated by Hipparchus and Ptolemy into the main body of Greek data, were still an indispensable aid to Copernicus.

4 Plato, *Thaetetus*, 174 A., quoted by Heath, op. cit., p. 1.

5 Compressed from the *Fragments*, quoted i.a. by John Burnet, *Early Greek Philosophy* (London, 1908), p. 126 *seq.*

6 ibid., p. 29.

### 2: THE HARMONY OF THE SPHERES

1 Cf. John Burnet, *Greek Philosophy Part I Thales to Plato* (London, 1914), pp. 42, 54.

2 Aristoxenus of Tarentum, *Elements of Harmony*, quoted by Burnet, op. cit., p. 41. Aristoxenus, a fourth cent. peripatetic, studied under the Pythagoreans and Aristotle.

For a critical evaluation of the sources on Pythagoras see Burnet's *Early Greek Philosophy*, p. 91 *seq.*; and A. Delatte, *Études sur la litterature pythagoricienne* (Paris, 1915). For the astronomy of the Pythagoreans, J. L. E. Dreyer, *History of the Planetary Systems from Thales to Kepler* (Cambridge, 1906) and Pierre Duhem, *Le Système du Monde – Histoire des Doctrines Cosmologiques de Plato à Copernic*, vol. I (Paris, 1913).

3 The discovery of the sphericity of the earth is variously attributed to Pythagoras and/or Parmenides.

4 *Hist. nat.*, II, p. 84, quoted by Dreyer, op. cit., p. 179.

5 *The Merchant of Venice*, V.i.

6 Euripides, *The Bacchae*, a new translation by Philip Vellacott (London, 1954).

7 Burnet, *Early Greek Phil.*, p. 88.

8 Quoted by B. Farrington, *Greek Science* (London, 1953), p. 45.

9 F. M. Cornford, *From Religion to Philosophy* (London, 1912), p. 198.

10 Book III, ch. 13. Quoted by Ch. Seltman, 'Pythagoras', in *History Today*, August 1956.

11 Quoted by T. Danzig, *Number, The Language of Science* (London, 1942), p. 101.

12 Farrington, op. cit., p. 43.

### 3: THE EARTH ADRIFT

1 *Hist.* IV, 25, 42; quoted by Dreyer, op. cit., p. 39.

2 Duhem (op. cit., p. 17) is inclined to believe that the counter-earth was always in opposition to the earth, on the other side of the central fire. But in this view (deduced from an ambiguous passage in Pseudo-Plutarch) the antichton would have no practical function. If the earth were to complete a revolution in twenty-four hours round the central, her angular velocity would become prohibitive unless the central fire were quite close; in which case the counter-earth seems to be really needed to prevent her from going up in smoke.

3 Number-lore was indeed the Achilles heel of the Pythagoreans; but lest we become too smug about antique superstitions, what about 'Bode's Law'? In 1772, Johannes Daniel Titius of Wittenberg announced that he had discovered a simple (but quite arbitrary) numerical law, according to which the relative distances of all planets from the sun can be expressed by the series 0, 3, 6, 12, 24, etc., by adding 4 to each term. The result is the

series 4, 7, 10, 16, 28, 52, 100, 196. This corresponded surprisingly closely to the relative distances of the seven planets known in A.D. 1800; but the eighth planet, with a distance 28 did not exist. Accordingly, in that year, a party of six German astronomers set out to look for the missing planet. They found the planetoid Ceres; since then over five hundred planetoids have been discovered in the neighbourhood, presumed to be the fragments of a former full-sized planet in the predicted place. But to the question, *why* that arbitrary number-sequence should so closely correspond to fact, no satisfactory answer has so far been found.

| | *Bode's law* | *Observed distance* |
|---|---|---|
| Mercury .. .. .. .. | 4 | 3·9 |
| Venus .. .. .. .. | 7 | 7·2 |
| Earth .. .. .. .. | 10 | 10 |
| Mars .. .. .. .. | 16 | 15·2 |
| ? .. .. .. .. | 28 | ? |
| Jupiter .. .. .. .. | 52 | 52 |
| Saturn .. .. .. .. | 100 | 95 |
| Uranus .. .. .. .. | 196 | 192 |

The table reminds one curiously of Mendeleyev's periodical table, before the discovery of isotopes.

4 The explanation is Schiaparelli's. Cf. Duhem, op. cit., I, 12.

5 To whom the hypothesis of the earth's rotation on its axis is due, we do not know. Two Pythagoreans are mentioned as responsible for it: Hyketas (some sources call him Niketas) and Ekphantus, both supposedly from Syracuse; but they remain shadows, and we do not know even their dates. Cf. Dreyer, p. 49 *seq.*; and Duhem, I, p. 21 *seq.*

6 The precession of the equinoxes was not discovered, or at least not seriously considered, until Hipparchus, who flourished *c.* 125 B.C.

7 As Venus' angular velocity exceeds that of the earth, she will, when in opposition move clockwise, in conjunction anti-clockwise, as seen from the earth.

8 Yet according to Saidas, when Plato left for Sicily, he left the Academy in Herakleides' charge. Ency. Brit. XI–454d.

9 Schiaparelli, Paul Tannery, and Pierre Duhem; see Duhem, op. cit., I, p. 410. But there exists no evidence in support of this hypothesis. The 'Tychonic' system would have been a logical stepping-stone from Herakleides to Aristarchus; but if somebody advocated it, it should have left some trace. It seems more prob-

able, as Dreyer argues (pp. 145 ff.) that Aristarchus performed a kind of mental jump from Fig. B. to Fig. D.

10 Dreyer's translation, op. cit., p. 137.

11 *De facie in orbe lunae*, ch. 6. Quoted by Heath, *Greek Astronomy*, p. 169.

12 Except for a single Babylonian astronomer named Seleukos, who lived a whole century after Aristarchus and developed a theory of the tides based on the earth's rotation.

13 Heath, *The Copernicus of Antiquity* (London, 1920), p. 38.

## 4: THE FAILURE OF NERVE

1 Quoted by Farrington, op. cit., p. 81.

2 *The Republic of Plato*, transl. Thomas Taylor, Book VII.

3 Loc. cit.

4 G. B. Grundy's article on 'Greece', *Ency. Brit.*, x–780c.

5 Bertrand Russell, *Unpopular Essays* (London, 1950), p. 16.

6 *Politics*, quoted by K. R. Popper, *The Open Society and its Enemies* (London 1945), vol. II, p. 2.

7 *Metaphysics*, quoted by Farrington, op. cit., p. 131.

8 *Timaeus*, 90, 91.

9 Spenser, *The Faerie Queene*.

10 *Phaedo*, quoted by Bertrand Russell, *A History of Western Philosophy* (London, 1946), p. 159.

11 There has been an unending controversy about a single word, *ειλομένην* or *ιλλομένην* in a phrase in *Timaeus*, 40B, which reads, in Dreyer's translation: 'But the earth, our nourisher, *packed* round the axis that extends through the universe, He formed as the guardian and artificer of night and day, the first and most ancient of the gods that have been generated within the universe' (op. cit., p. 71 f.) Burnet reads instead of 'packed': 'going to and fro' or 'backwards and forwards' (*Greek Philosophy*, p. 348); Professor A. E. Taylor (quoted by Heath, *Greek Astronomy*, p. xli) suggests that the phrase must be taken as meaning that the earth is 'sliding up and down the axis of the universe', and that Plato was merely quoting a Pythagorean theory (which he evidently got all muddled up), without subscribing to it. Apart from this nebulous sentence, Plato nowhere alludes to any motion of the earth. Plutarch, in discussing the system of Philolaus with its central fire, reports that 'These ideas are said to have been entertained by Plato also in his old age; for he too thought that the earth was in a subordinate position,

and that the centre of the universe was occupied by some nobler body.' (Plutarch's *Life of Numa*, ch. 11, quoted by Dreyer, p. 82.) Though it is possible that the ageing Plato toyed with the idea of the 'central fire' from a quasi-mythological point of view, he again nowhere alludes to it in his writings.

12 *Timaeus*, 33B–34B, quoted by Heath, op. cit., p. 49 f.

13 Farrington, op. cit., p. 56.

14 For a concise summing up of Aristotle's and Plato's different attitudes to Change, see Popper, op. cit., vol. II, pp. 4–6, and particularly Note 11, p. 271 f.

## 5: THE DIVORCE FROM REALITY

1 Eudoxus' is the first serious attempt to put astronomy on an exact geometrical basis. His model could lay no claim on representing physical reality, but for sheer geometrical elegance it is unrivalled in pre-Keplerian astronomy, and superior to Ptolemy's. It worked, briefly as follows: The outermost ($S_4$) of the four spheres which form a planet's 'nest' reproduced the apparent diurnal rotation; the Axis ($A_3$) of $S_3$ was perpendicular to the ecliptic, so that its equator turned in the plane of the ecliptic, in the outer planet's zodiacal period, and the inner planets in one year. The two innermost spheres served to account for the movement in latitude, and for the stations and retrogressions. $S_2$ had its poles on the equator of $S_3$, i.e. on the zodiacal circle; $S_2$ rotated in the synodic period of the planet. $S_1$ rotated in the same period but in the opposite direction; and $A_1$ was inclined to $A_2$ at an angle different for each planet. The planet sat on the equator of $S_1$. The combined rotations of $S_1$ and $S_2$ made the planet describe a lemniscate (i.e. an elongated 'figure of eight') lying along the zodiac. For details see Dreyer, op. cit., ch. 4; and Duhem, op. cit., I, pp. 111–23.

2 Nevertheless the theories of Eudoxus and his followers fail to save the phenomena, and not only those which were first noticed at a later date, but even those which were known before, and actually accepted by the authors themselves. . . . I refer to the fact that the planets appear at times to be near to us and at times to have receded. This is indeed obvious to our eyes in the case of some of them; for the star called after Aphrodite and also the star of Ares seem, in the middle of the retrogradations, to be many times as large, so much so that the star of Aphrodite actually makes bodies cast shadows on moonless nights. The moon also, even in the perception of our eye, is clearly not always at the same distance from us, because it does not

always seem to be of the same size under the same conditions as to medium. The same fact is, moreover, confirmed if we observe the moon by means of an instrument; for it is at one time a disk of eleven fingerbreadths, and again at another time a disk of twelve fingerbreadths, which when placed at the same distance from the observer hide the moon (exactly) so that his eye does not see it.

(Simplicius on *De Caelo*, quoted by Heath, op. cit., p. 68 f.)

3 It is perhaps significant that Ptolemy, alone among famous astronomers, was also a famous map-maker. The re-discovery of his *Geography*, which was translated into Latin in 1410, marked the beginning of scientific geography in Europe. Copernicus and Kepler, who were also entrusted with map-making, considered it a tedious task to be evaded. Even Hipparchus and Tycho, the greatest star-map-makers, avoided earthly geography. But it was Hipparchus who outlined the *principles* of mathematical map-making by regular projection, which Ptolemy adopted. Both the epicyclic universe and the Geography of Ptolemy are painstaking executions of Hipparchus' original designs.

4 From *Al-majisty*, an Arabic corruption of the Greek *Megisty Syntaxis*.

5 Dreyer, op. cit., p. 175

6 ibid., p. 184. The distance of the sun could not be calculated, even approximately, before the invention of the telescope: Ptolemy gave 610 earth diameters (true value 11,500); but Copernicus again knew no better: his estimate was 571 earth diameters (Dreyer, pp. 185 and 339). As for the fixed stars, Ptolemy knew that their distance was enormous compared to the solar system; he says that compared to the sphere of the stars 'the earth is like a point'.

7 Except, of course, the ellipticity of the orbits; but see below, note 15.

8 Quoted by Ernst Zinner, *Entstehung und Ausbreitung der Copernicanischen Lehre* (Erlangen, 1943), p. 49.

9 loc. cit.

10 op. cit., p. 52 f.

11 ibid., p. 50.

12 loc. cit.

13 *De facie orbe lunae*, ch. 6, quoted by Heath, op. cit., p. 169

14 The Ionian philosophers had been suspected of atheism and brought astronomy into somewhat ill repute; but that had been centuries before, and even then they had not come to harm.

Plutarch reports in the Life of Nicias, the sixth-century Greek general, that he was afraid of eclipses, that the people were equally superstitious, and that

in those days there was no tolerance for the natural philosophers or 'babblers about things in heaven' as they were called. They were charged with explaining away the divine and substituting for it irrational causes, blind forces, and the sway of necessity. So Protagoras was banished, Anaxagoras was gaoled and it was all that Pericles could do to get him out; and Socrates, though he had nothing to do at all in the matter, was put to death for being a philosopher. It was only much later, through the brilliant repute of Plato, that the reproach was removed from astronomical studies and access to them opened up for all. This was on account of the respect in which his life was held and because he made natural laws subordinate to the authority of divine principles.

(Quoted by Farrington, op. cit., p. 98 f.)

Now neither Socrates, nor Protagoras, had anything to do with astronomy, and the only instance of persecution throughout antiquity is the imprisonment of Anaxagoras, in the sixth century B.C., though, according to another source, he was merely fined and temporarily exiled; he died at seventy-two.

In the light of this one can hardly agree with Duhem's comment:

The obstacles by which, in the seventeenth century, the Protestant, and later the Catholic church, obstructed the progress of the Copernican doctrine, can only convey to us a feeble idea of the charges of impiety incurred, in pagan antiquity, by the mortal who dared to shake the perpetual immobility of the Hearth of the Divinity (*sic*), and to put those incorruptible and divine beings, the stars on the same footing as the earth, humble domain of generation and decay.

(op. cit., I, p. 425)

The only support for this statement is again Plutarch's anecdotical remark about Cleanthes. It should be noted that in Duhem's version Aristotle's metaphysics is treated as if it had become the pagan equivalent of Christian dogma; at the same time Aristotle himself becomes a heretic, for he, too, had laid hands on 'the Hearth of Divinity'. The reasons for this slip-up, and for the inflated importance given to the Cleanthes story, become evident when Duhem proceeds to quote with approval Paul Tannery (whose religious convictions he shares), to the effect that though Galileo was mistakenly condemned by the

Inquisition, 'he would probably have incurred much more serious dangers if he had to fight against the star-worshipping superstitions of antiquity'. Owing to Duhem's authority, the Cleanthes legend has found its way into most of the popular histories of science (as a twin to the equally apocryphal '*eppur si muove*'); and is quoted in support of the view (which was certainly not Duhem's intention) that an innate and irreconcilable hostility always existed, and must always exist, between religion, in any form, and science. A notable exception is Dreyer (cf. op. cit., p. 148), who comments simply that in Aristarchus' days 'the time was long past when a philosopher might be judicially called to account for proposing startling astronomical theories', and that 'the accusation of "impiety", if it was really brought forward, can hardly have done the theory much harm.'

15 One other attempted explanation must be briefly discussed. Dreyer sees the reason for the abandonment of the heliocentric system in the rise of observational astronomy in Alexandria. Aristarchus could explain the retrograde motions of the planets and their change of brilliancy, but not the anomalies arising from the ellipticity of their orbits; and 'the hopelessness of trying to account for them by the beautifully simple idea of Aristarchus must have given the death blow to his system' (p. 148). Duhem's explanation is the same (pp. 425–6). But this seems rather to beg the question, for the so-called 'second anomaly' could just as well be saved by epicycles in the heliocentric as in the geocentric system; and this is indeed what Copernicus did. In other words, either system could serve as a starting point for building a 'Ferris wheel'; but with Aristarchus as a starting point the task would have been incomparably simpler, because the 'first anomaly' was already eliminated. On second thoughts, Dreyer seems to have realized this, for he subsequently (p. 201 f.) says:

> To the modern mind, accustomed to the heliocentric idea, it is difficult to understand why it did not occur to a mathematician like Ptolemy to deprive all the outer planets of their epicycles, which were nothing but reproductions of the earth's annual orbit transferred to each of these planets, and also to deprive Mercury and Venus of their deferents and place the centres of their epicycles in the sun, as Herakleides had done. It is in fact possible to reproduce Ptolemy's values of the ratio of the radii of epicycle and deferent from the semi-axis major of each planet expressed in units of that of the earth. . . . Obviously the heliocentric idea of Aristarchus might just as well have

sprung out of the epicyclic theory as from that of movable eccentrics . . .

He further points out that the Ptolemaic system failed to save the phenomena even more drastically than the Aristarchian in the case of the moon, whose apparent diameter ought to vary, according to Ptolemy, to an extent contradicted by the simplest observation (p. 201).

16 *Almagest III*, ch. 2, quoted by Duhem, p. 487.
17 ibid., II, quoted by Zinner, p. 35.
18 In a later and shorter work, *Hypotheses concerning the Planets*, Ptolemy made a half-hearted attempt to give his system a semblance of physical reality by representing each epicycle by a sphere or disk, gliding between a convex and a concave spherical surface, ball-bearing fashion. But the attempt defeated itself. Cf. Duhem, II, pp. 86–99.
19 Quoted by Dreyer, p. 168.
20 *Almagest, I.*
21 Cf. Zinner, op. cit., p. 48.
22 Joh. Kepler. Letter to D. Fabricius, 4.7.1603, *Gesammelte Werke*, vol. XIV, p. 409 *seq.*
23 Quoted by R. H. Wilenski, *Modern French Painters* (London, 1940), p. 202.
24 ibid., p. 221.

## *Part Two. Dark Interlude*

### 1: THE RECTANGULAR UNIVERSE

1 Edmund Whittaker, *Space and Spirit* (London, 1946), p. 11.
2 *The Confessions of St Augustine*, transl F. J. Sheed (London, 1944), p. 111.
3 ibid., p. 113.
4 ibid., p. v f.
5 Dr Th. A. Lacey on 'Augustine', *Ency. Brit.* II–685c.
6 ibid., II–684a.
7 Christopher Dawson, quoted in Preface to *The Confessions*, p. v.
8 *The City of God*, quoted by Russell, *A History of Western Philosophy*, p. 381.
9 ibid., VIII, 5.
10 Whittaker, op. cit., p. 12.
11 *The Confessions*, p. 197 f.
12 Quoted by Russell, op. cit., p. 362.

13 Dreyer, op. cit., p. 210.
14 ibid., p. 211.
15 ibid., p. 213.
16 ibid., p. 212; Duhem II, p. 488 f.
17 Dreyer, p. 211.

## 2: THE WALLED-IN UNIVERSE

1 *Comment in Somnium Scipionis,* I, 14, 15. Quoted by A. O. Lovejoy, *The Great Chain of Being* (Cambridge, Mass., 1936), p. 63.
2 The *Primum Mobile* no longer was an unmoved mover, since Hipparchus discovered the precession of the equinoxes. Its task was now to account for that motion, whose slowness – one revolution in 26,000 years – was explained by its desire to share in the perfect rest of the adjoining tenth sphere, the Empyrean.
3 Dante, *Convito*, ii. 6; quoted by Dreyer, p. 237.
4 *De animalibus historia* viii, i, 588b; quoted by Lovejoy, op. cit., p. 56.
5 *Summa contra gentiles*, II, 68.
6 Lovejoy, op. cit., 102.
7 *Essays*, II, 2.
8 *The Faerie Queene.*
9 *An Essay on Man.*
10 *History of the World*, quoted by E. M. W. Tillyard, *The Elizabethan World Picture* (London, 1943), p. 9.
11 Olivier de la Marche, *L'État de la Maison du Duc Charles de Burgogne*, quoted by J. Huizinga, *The Waning of the Middle Ages* (London, 1955), p. 42 f.
12 H. Zinsser, *Rats, Lice and History* (1937), quoted by Popper, II, p. 23.
13 *Troilus and Cressida.*
14 Cf. Duhem, op. cit., III, pp. 47–52.
15 There exist, i.a., two manuscripts written in the name of the Venerable Bede, but clearly long after his death, expounding the Herakleidian system. The first is known as the 'Pseudo-Bede' and dates from the ninth century or later; the second is now attributed to William of Conches, a Norman, who lived in the twelfth century. Cf. Dreyer, pp. 227–30; Duhem III, p. 76 *seq.*
16 Duhem III, p. 110.
17 The earliest preserved Portolano maps date from the thirteenth century, but show a long-established tradition, whereas the circu-

lar Hereford map (*c.* 1280), and the 'T and O' maps of the fifteenth century show that 'theoretical' and 'practical' maps of the world must have overlapped for several centuries.

18 Huizinga, op. cit., p. 68.

19 ibid., pp. 45, 50.

### 3: THE UNIVERSE OF THE SCHOOLMEN

1 The *Categories* and the *De Interpretatione.*

2 Whitehead, *Science and the Modern World* (Cambridge, 1953), p. 15.

3 *De Caelo*; *De Generatione et Corruptione*, quoted by Whittaker, op. cit., p. 27.

4 There were of course notable exceptions: Bacon, the Francisan school, and the Parisian school of the fourteenth century. But the anti-Aristotelian physics of Ockham, Buridan, and Oresme bore no immediate fruit; Copernicus and Kepler, for instance, knew nothing of their revolutionary theory of impetus (though Leonardo did); and their triumph over Aristotle came only three centuries later, through Galileo – who never acknowledged his indebtedness to them.

5 Because a thing cannot be both in act and in potency at the same time and in the same respect. But 'potency' and 'act', as applied to a moving body, are meaningless terms. For a simple exposition of the Aristotelian-Ockhamist controversy on motion see Whittaker, op. cit., appendix.

6 H. Butterfield, *The Origins of Modern Science* (London, 1949), p. 14.

7 See above note 4. But even in antiquity, this blindness was not a total one. Thus Plutarch argues in *On the Face of the Moon* that the moon is of earthy, solid stuff, and that in spite of its weight it does not fall down on the earth, because:

> . . . The moon has a security against falling in her very motion and the swing of her revolution, just as objects put in slings are prevented from falling by the circular whirl; *for everything is carried along by the motion natural to it if it is not deflected by anything else.* Thus the moon is not carried down by her weight because her natural tendency is frustrated by the revolution.
>
> (Heath, op. cit., p. 170; my italics.)

The translation is by Heath, who comments: 'This is practically Newton's first Law of Motion.' (Heath, op. cit., p. 170.) It it curious that this passage has aroused so little comment. The context makes it clear that Plutarch did not hit on the idea of momentum by a lucky chance, but that he had the 'feel' of it, as it were. So, of course, had every spear-thrower (and his victim).

8 Butterfield, op. cit., p. 7.

9 *Morias Enkognion*, Basilieae, 1780, p. 218 *seq.*

10 Gilbert Murray, *Five Stages of Greek Religion* (London, 1935), p. 144

11 *Science and the Modern World*, p. 7.

## *Part Three. The Timid Canon Copernicus*

The standard biography of Copernicus is still Leopold Prowe's *Nicolaus Copernicus* (Berlin, 1883–4).

The most important recent work about the Copernican theory, its origins and repercussions, is Ernst Zinner's *Entstehung und Ausbreitung der Copernicanischen Lehre* (*Sitzungsberichte der Physikalisch-medizinischen Sozietaet zu Erlangen*, 74 *Band*, Erlangen, 1943).

Concise summaries of the Copernican system are given in Angus Armitage's *Copernicus the Founder of Modern Astronomy* (London, 1938), and in Dreyer, op. cit.

Prowe's work was published in two volumes, the first consisting of two parts separately numbered. References to Volume I are accordingly marked Prowe, I, 1, and I, 2. The first volume contains the biography, the second, documents in Latin, Greek and medieval German. All references to Prowe Vol. II refer to the Latin originals.

### 1: THE LIFE OF COPERNICUS

1 The surname is variously spelt in documents as Coppernic, Koppernieck, Koppernik, Koppernigk, Kopperlingk, Cupernick, and Kupernick. The most usual is Koppernigk (the spelling which Prowe also adopted). He himself signed his name on different occasions as Copernic, Coppernig, Coppernik, Cophernic, and in later years mostly Copernicus.

2 *De revolutionibus orbium coelestium, Libri VI* (Nuremberg, 1543). In the text the work will be referred to either as *Book of the Revolutions*, or *Revolutions* for short.

3 *De revolutionibus*, Lib. V. Cap. 30.

4 ibid., Lib. IV, Cap. 7.

5 *Wagner's Staats Lexikon* (1862) vol. II, describes Frauenburg 'as a small town on the Vistula'.

5a Prowe, I, 2, p. 4 n.

6 There is only one other case known of a similar geographical displacement of Frauenburg by one of its citizens;; Tiedmann Giese in 1536 dated a letter to Erasmus of Rotterdam 'From the shores of the Vistula' – and Canon Giese was Canon Koppernigk's closest friend. Cf. Prowe, I, 2, p. 4.

7 Rheticus, *Ephemerides novae* (Leipzig, 1550), p. 6, quoted by Prowe, I, 2, p. 58.

8 Prowe, I, 2, p. 314.

9 Prowe, I, 1, p. 111.

10 Thus, for instance, in 1943, the 'Copernicus Quater-centenary Celebration Committee, London' published a monograph *Nicolas Copernicus* (*Mikolaj Kopernik*) by Dr Jozef Rudnicki, which, in describing Copernicus' studies in Italy, omits the fact that he appears on the register of the German *natio* in Bologna and then says about Copernicus' next university, Padua: '... the Polish "nation" was one of the largest in the university. There, according to the historian of Padua, N. C. Papadopoli, "Copernicus devoted himself to the study of philosophy and medicine for four years, as is known from the entries in the register of Polish students." '

Now Copernicus would have been quite capable of joining the German *natio* in Bologna and the Polish *natio* in Padua, but the fact is that documentary evidence exists for the former, but none for the latter, and that the Papadopoli quoted as a source has been exposed as a fraud by his Italian compatriots who had no stake in the Polish-German dispute (Cf. Prowe, I, 1, p. 297). The feud is also waged in the field of spelling; thus Rudnicki turns the Tyrolean Georg Joachim von Lauchen into a Slav, by translating his Latin *nom de plume* Rheticus into Retyk (p. 9). It must be noted, though, that the booklet was written during the war. See also notes 28 and 89.

11 Carlo Malagola, *Della Vita e delle Opere di Antonio Urceo detto Codro* (Bologna, 1878).

12 Rheticus, *Narratio Prima*, trans. Edward Rosen, *Three Copernican Treatises* (Columbia, 1939), p. 111.

13 Prowe, I, 1, p. 266.

14 Prowe, I, 1, p. 89.

15 Zach's *Monatliche Korrespondenz*, vol. II, p. 285, quoted by Prowe loc cit.

16 Prowe, I, 2, p. 313.

17 Prowe, I, 1, p. 359.

18 See note 33.
19 *Ency. Brit.* xx–696d.
20 Bernhardy, *Grundriss der Grieschischen Litteratur,* I, p. 583, quoted by Prowe, I, 1, p. 393.
21 Prowe, II, pp. 124–7, which also contain the Greek original.
22 Prowe, II, p. 51.
23 *Ency. Brit.* IX–732b, 13th edition, 1926. (All other references are to the 1955 edition.)
24 Quoted by Prowe, I, 1, p. 402.
25 H. R. Trevor-Roper, 'Desiderius Erasmus' (*Encounter*, London, May 1955).
26 See below note 20 to ch. 2.
27 Known as the *Letter against Werner*. See p. 203.
28 The treatise was originally written in German and submitted to the Prussian Diet of 1522, then rewritten in Latin for the *Landtag* of 1528. Its aim was to remedy the debasement of the Prussian coinage (which had been aggravated by the war), by means of state monopolies in the minting of coins, the control of the amount of coins in circulation, and of the quantity of base metal in the alloy. It is sometimes claimed that Copernicus anticipated Gresham's Law that 'bad money drives out good'; in fact the principle seems to have been first enunciated two centuries earlier by Nicolas of Oresme, whose economic teachings formed the basis of the monetary reform of Charles V.

Both versions of Copernicus' treatise are printed in Prowe, II, pp. 21–9, and analysed in Prowe, I, 1, pp. 139–52 and 193–201.

It is amazing to note that even this subject has been dragged into the Polish-German controversy. Thus Rudnicki (op. cit., p. 24), in spite of Prowe's exhaustive treatment of the matter, flatly states, 'it is noteworthy that the Germans pass over Copernicus' economic treatises in silence', and considers the treatise as further proof that Copernicus' 'outlook is Polish to the core' (p. 26) because he suggested that the new coins of Polish Prussia should bear as crest the royal crown of Poland; but he omits to mention the fact that the treatise itself was written in German.

On the other hand, Zinner omits to mention that one of the first teachers of Copernicus was apparently a master with the indubitably Polish name Mikolaj Vodka – who later latinized his name into Abstemius. ... Cf. L. A. Birkenmajer, *Mikolaj Wodka z Kwidzyna zwany Abstemius lekarz i astronom polski XV – go stulecia* (Thorn, 1926). See also notes 10 and 89.
29 Quoted by Prowe, I, 2, p. 177.

30 *Flosculorum Lutheranorum de fide et operibus ανδηλογιχον* (Cracow), quoted by Prowe, I, 2, p. 172.

31 Compare below p. 156 *seq.*, the equally complicated compromise formula for the publication of Rheticus' *Narratio prima.*

32 The date of the *Commentariolus* is uncertain, but internal evidence points to the years 1510–14. See Zinner, op. cit., p. 185, and A. Koyré, *Nicolas Copernic Des Revolutions des Orbes Célestes* (Paris, 1934), p. 140.

33 *Nicolai Copernici de hypothesibus motuum coelestium a se constitutis commentariolus.* I have translated 'Commentariolus' as 'Brief outline'. Hand-written copies of it were still circulating among scholars towards the end of the century. Then the treatise vanished from sight until two copies were found independently in 1878 and 1881 in Vienna and Stockholm. The full text was first published by Prowe, together with a German translation of the opening section. A complete English translation was publisted by Edward Rosen (op. cit.).

34 That is to say, the planet's angular velocity is not uniform relative to the centre of its epicycle; it is only uniform relative to another point, the *punctum equans* situated on the axis major of its orbit. See below p. 205 f. *seq.*

35 *Epistolae diuersorum philosophorum, oratorum, rhetorum sex et viginti* (Padua, 1499).

36 *Bessarionis Cardinalis Niceni et Patriarchae Constantinopolitani in calumniatorem Platonis libri quatuor* (Padua, 1503).

37 Quoted by Prowe, II, p. 132–7.

38 *De revolutionibus*, prefatory matter.

39 Quoted by Prowe, I, 2, p. 274.

40 There is talk of a new astrologer who wants to prove that the earth moves and goes round instead of the sky, the sun, and moon, just as if somebody were moving in a carriage or ship might hold that he was sitting still and at rest while the earth and the trees walked and moved. But that is how things are nowadays: when a man wishes to be clever he must needs invent something special, and the way he does it must needs be the best! The fool wants to turn the whole art of astronomy upside-down. However, as Holy Scripture tells us, so did Joshua bid the sun to stand still and not the earth.

(Luther's *Tischreden*, ed. Walch, p. 2260, quoted by Prowe, I, 2, p. 232.)

41 Quoted by Prowe, I, 2, p. 233.

42 Georg Joachim Rheticus, *Narratio prima – Encomium Borussiae*

(Danzig, 1540), trans. Rosen, op. cit., p. 191. In subsequent quotations from the *Narratio prima* I have followed E. Rosen's translation except for some minor paraphrasings.

43 op. cit., trans. Rosen, pp. 192–5.
44 cf. above, p. 144, and note 31.
45 op. cit., trans. Rosen, p. 186 f.
46 ibid., p. 187.
47 ibid., p. 126.
48 ibid., p. 131.
49 Johannes Kepler, *Gesammelte Werke*, vol. III (Munich, 1937).
50 Kepler to Longomontanus, spring 1605, *Gesemmelte Werke*, vol. XV (Munich, 1591), p. 134 *seq.*
51 Rheticus, op. cit., p. 163 f.
52 ibid., p. 188.
53 ibid., trans. Rosen, p. 195.
54 See below, note 13 to ch. 2.
55 Canon Alexander Sculteti, about whom more below. Not to be confused with Bernard Sculteti, see above pp. 133, 149.
56 *De Lateribus et Angulis Triangulorum* (Wittenberg, 1542).
57 Zinner, op. cit., p. 243.
58 ibid., p. 244.
59 The full text of Osiander's preface is as follows (trans. Rosen, op. cit., p. 24 f.):

TO THE READER
CONCERNING THE HYPOTHESES OF THIS WORK

Since the novelty of the hypotheses of this work has already been widely reported, I have no doubt that some learned men have taken serious offence because the book declares that the earth moves, and that the sun is at rest in the centre of the universe; these men undoubtedly believe that the liberal arts, established long ago upon a correct basis, should not be thrown into confusion. But if they are willing to examine the matter closely, they will find that the author of this work has done nothing blameworthy. For it is the duty of an astrologer to compose the history of the celestial motions through careful and skilful observation. Then turning to the causes of these motions or hypotheses about them, he must conceive and devise, since he cannot in any way attain to the true causes, such hypotheses as, being assumed, enable the motions to be calculated correctly from the principles of geometry, for the future as well as for the past. The present author has performed both these duties excellently. For these hypotheses need not be true nor even probable; if they provide a calculus consistent with the observations, that alone is sufficient. Perhaps there is someone who is so ignorant of geometry and optics that he regards the epicycle of Venus as probable, or thinks that it is the reason why Venus sometimes precedes and sometimes follows the sun by forty degrees and even more. Is there anyone who is not aware that from this assumption it necessarily follows that the diameter of the planet in perigree should appear more than four times, and the body of the planet more than sixteen times, as great as in the apogee, a result contradicted by the experience of every age? In this study there are other no less important absurdities, which there is no need to set forth at the moment. For it is quite clear that the causes of the apparent unequal motions are completely and simply unknown to this art. And any causes are devised by the imagination, as indeed very many are, they are not put forward to convince anyone that they are true, but merely to provide a correct basis for calculation. Now when from time to time there are offered for one and the same motion different hypotheses (as eccentricity and an epicycle for the sun's motion), the astronomer will accept above all others the one which is the easiest to grasp. The philosopher will perhaps rather seek the semblance of the truth. But neither of them will understand or state anything certain, unless it has been divinely revealed to him. Let us therefore permit these new hypotheses to become known together with the ancient hypotheses, which are no more probable; let us do so especially because the new hypotheses are admirable and also simple, and bring with them a huge treasure of very skilful observations. So far as hypotheses are concerned, let no one expect anything certain from astronomy, which cannot furnish it, lest he accept as the truth ideas conceived for another purpose, and depart from this study a greater fool than when he entered it. Farewell.

60 Copernicus' letter, dated 1 July 1540, is lost.

61 Osiander's answer was dated 20 April 1541. It is quoted in Kepler's *Apologia Tychonis contra Ursum*, published in Kepler's *Opera Omnia*, ed. Frisch, I, pp. 236–76.

62 Same source, loc. cit.

63 ibid.

63a *De Revolutionibus,* Dedication to Paul III.

64 Johannes Kepler, *Astronomia Nova*, Prefatory matter, *Gesammelte Werke,* vol. III. I have followed Rosen's English translation of the passage.

65 Johannes Praetorius to Herwart von Hohenburg. The letter was first published by Zinner, op. cit., p. 454.

66 ibid., p. 453.

67 ibid., p. 424.

67a It is equally suspect that Kepler, having seen the whole Osiander–Copernicus correspondence, quotes Osiander's letters to Copernicus and Rheticus verbatim, but summarizes the much more important reply from Copernicus to Osiander in a single phrase about Copernicus' 'stoical firmness of mind'. The *Astronomia Nova* attempts to put the Copernican system on a physical basis, and Kepler could not admit that Copernicus had any doubts concerning its physical reality or was prepared to compromise on that question.

68 Private communication, 5 August 1955.

69 Careful reading of Osiander's preface will show that his charges of 'improbability' and 'absurdity' are directed against geometrical details of the Copernican system, but not against the basic concept of the earth's motion. On this central idea he shared Copernicus' beliefs, as shown by his letters to Copernicus and Rheticus, and by his devotion to the project. His stressing of the formal or fictitious nature of the system was partly prompted by diplomacy, but partly by a genuine disbelief in the reality of the epicyclic machinery. Copernicus' attitude was essentially the same; the long and heated controversy on this point is mostly based on a failure to distinguish between the heliocentric idea and the epicyclic detail of the system. Concerning the former, the text of the dedication to Paul III is alone sufficient proof that Copernicus was convinced of its physical truth. Concerning the latter, a series of passages in the text show that he regarded epicycles and eccenters as not more than computing devices. Hence Copernicus was neither a 'realist' (to use Duhem's terms) nor a 'fictionalist', but a realist regarding the immobility of the sun and the fixed stars, and a fictionalist regarding the motions of the planets. The fictionalist attitude is particularly evident in the treatment of the rectilinear oscillatory motions of all planets in latitude, of Mercury in longitude and of the earth's axis, which could not be represented by any model with even a remote resemblance of reality.

For a brief and sensible discussion of the issue, with a list of some relevant passages in the *Revolutions*, see Armitage, op. cit., pp. 84–7.

70 The only protest on record came from the loyal Giese, who saw the printed book only after Copernicus' death. Copernicus died in May 1543, when Giese was away in Cracow to attend the marriage of the King of Poland. He returned to Prussia in July, and found two copies of the *Revolutions*, which Rheticus had sent him from Nuremberg with a personal dedication. Only now did Giese see Osiander's preface, and he considered it a desecration of the memory of his dead friend. He wrote (on 26 July) to

Rheticus, blaming both Osiander and the printer Petreius, and suggesting that the opening pages of the book should be reprinted, Osiander's preface eliminated, and Rheticus' biography of Copernicus, as well as this theological defence of the Copernican system, inserted in its place. He also asked Rheticus to intervene with the Nuremberg City Fathers (to whom Giese had written directly) that they should force Petreius to comply. Rheticus did as he was asked, but the Corporation of Nuremberg, after investigating the matter, resolved on 20 August: 'To forward to Bishop Tiedeman at Culm the written answer to his letter by Johan Petrius (after eliminating its harshness and toning it down), with the comment that in view of the contents of the answer, no action could be taken against him.'

(Cf. Prowe, I, 2, 535 *seq.*, and Zinner, p. 255 f.)

Petrius' answer is lost, but it is evident that he made out a good case against Giese's accusation that he had acted against the author's wishes. It is equally evident that if Copernicus had given his explicit or silent consent to the compromise suggested by Osiander, he would have kept the matter from Giese who, in the light of their past arguments, would be certain to disapprove.

71 Zinner, op. cit., p. 246.

72 Prowe, II, pp. 419–21.

73 *Orationes de Astronomia Geographica et Physica* (Nuremberg, 1542), reprinted in Prowe, II, pp. 382–6.

74 *Ency. Brit.* XVIII–162c.

75 Prowe, I, 2, p. 334.

76 loc. cit.

77 The full Latin texts are published by Prowe, II, pp. 157–68.

78 Prowe, II, p. 157.

79 Prowe, II, pp. 158–9.

80 Prowe, I, 2, p. 325.

81 Oxford, 1934. *The Oxford Glossary of Later Latin* (1949) has 'Soldier's concubine'.

82 See note 55.

83 Prowe, I, 2, p. 364.

84 ibid., p. 360.

85 Zinner, op. cit., p. 222 f.

86 Prowe, I, 2, p. 366 f.

87 Shortly after he took possession of the Ermland see, in January 1538, Dantiscus procured a canonry for one of his favourites. This was the future Cardinal Stanislaw Hosius (1504–79), the moving spirit of the Counter-reformation in Poland, the man

who introduced the Jesuit Order into Prussia, and had a decisive part in bringing the semi-autonomous parts of Prussia under Polish and Catholic rule. He was variously known as 'the Hammer of the Heretics' and 'the Death of Luther'; the Polish Queen described him as a person who united the innocence of a dove with the cunning of a snake. He was the symbol of the new age of fanaticism and godly massacre, which followed the age of humanism and tolerance, of Erasmus and Melanchton. Dantiscus, the friend of Melanchton, was a child of that earlier age, and himself never became a fanatic; but as an experienced diplomat he knew the forces that were active in Europe, and he was aware that the Prussian border province which he ruled must either become Protestant and German, or Catholic and Polish. Not only his religious and national allegiance, but his whole philosophy made him opt for the continuity and traditions of the Roman Church, and the civilizing influence of Poland in the golden age of the Jagellones. Accordingly, when he accepted the vacant Bishopric of Kulm, his efforts were already aimed at Ermland; for Kulm, which belonged to 'Royal' Prussia, was safe for Poland, whereas Ermland was the strategical and political key to the whole of East Prussia, the former domain of the Teutonic Knights. The Bishop of Ermland enjoyed *de facto* the status of a ruling Prince; he had great influence in the Prussian Diet, over which he presided, and his Chapter fulfilled the functions of government and administration.

By arranging a canonry for Hosius, Dantiscus introduced a kind of Trojan horse into the Chapter. A bare few months later, Hosius was nominated as a candidate for a precentorship which had become vacant. The Chapter, jealous of its quasi-autonomous status towards the Polish Crown, blocked this move by electing to the post another member: Alexander Sculteti. In spite of considerable pressure by Dantiscus, Sculteti refused to yield. This was the beginning of a long and bitter struggle, apparently between two individuals, Hosius and Sculteti, in fact between the Polish Crown on the one hand, and certain powers at the Papal Court on the other, who backed Sculteti in an attempt to thwart Polish ambitions and to keep Ermland under the direct influence of Rome. Though Sculteti did have several children by his housekeeper, the accusations brought against him of leading an improper life, and holding heretical views, must be seen against this political background. In 1540, he was expelled from the Chapter and banished by royal edict from all territories under Polish sovereignty. The next six or seven years, Sculteti lived in Rome, fighting various legal actions, which ended with his vindication by the Papal Court. The Ermland Chapter, however, under Polish pressure, refused to acknowledge this, with the result that all its members in residence at Frauenburg were excommunicated. The whole complicated intrigue ended with the victory of Hosius, who, in 1551, became Bishop of Ermland, and made it safe for the Polish Crown.

88 Prowe, I, 2, p. 361.

89 Thus the otherwise so scholarly and reliable Zinner explains Dantiscus' demand that Copernicus should part with his housekeeper, by Dantiscus' 'hatred and urge to oppress his intellectual superior and to deprive him of the leisure required to complete his work. Dantiscus achieved his aim. The work was never completed.' (p. 224).

In describing the relations between Copernicus and Dantiscus, Zinner omits to mention that Dantiscus sent Copernicus a contribution (see below) to be included in the *Revolutions*. He refers to Dantiscus' contribution only as an aside, in a different context (p. 239). Zinner's bias against Dantiscus again seems to have a political motivation. He describes him as a careerist (p. 224) who 'entered the service of the Polish King and supported Polish claims against his own country, Prussia' (p. 221). He also repeats the legend according to which Copernicus refused to obey Dantiscus' 'order' to break off relations with Sculteti, and declared that 'he had a higher regard for Sculteti than for the other Canons'. In the light of Copernicus' letters to Dantiscus this seems hard to believe. The source of this version is a Polish author named Szulc, quoted by Prowe (I 2, p. 361). Prowe, however, points out in a footnote that Szulc does not give his source for the alleged statement by Copernicus 'though otherwise he always does so'. Prowe himself is scrupulously fair to Dantiscus and displays a detached attitude to the nationalist controversy.

90 Prowe, II, p. 168.

91 Prowe, II, p. 418 f.

92 Prowe, I, 2, p. 554.

93 Zinner, op. cit., p. 244.

94 ibid., p. 245.

95 Quoted by Zinner, p. 466.

95a ibid., p. 259.

96 loc. cit.

97 In his letter to Rheticus of 26 July 1543 (see note 70), Giese says

that the 'elegant' biography of Copernicus which Rheticus had written requires only the addition of the facts about the master's death to be complete. In the same letter he also refers to the treatise which Rheticus had written to prove that the doctrine of the motion of the earth does not contradict the Holy Scripture.

98 Zinner, op. cit., p. 259.
99 ibid., p. 261.
100 loc. cit.
101 ibid., p. 262.
102 Prowe, II, p. 389, and Ency Brit. XIX–246d. Zinner (p. 262) gives the date of Rheticus' death as 1574.
103 Prowe, I, 2, p. 387 f.

2: THE SYSTEM OF COPERNICUS

1 The first complete English translation was published in 1952 in the 'Great Books of the Western World' series (vol. 16, Chicago, trans. Charles Glenn Wallis).
2 *Ency. Brit.* II–584a.
3 Zinner, op. cit., pp. 273–8.
4 H. Dingle, *The Scientific Adventure* (London, 1951), p. 74.
5 London, 1932, p. 26.
6 London, 1949, pp. 26–7.
7 London, 1939, p. 38.
8 Oxford, 1941, p. 182.
9

| *Earth* | *De revolutionibus* | *Commentariolus* |
|---|---|---|
| Diurnal rotation .. .. .. | 1 | 1 |
| Motions in longitude .. .. | 3 | 1 |
| Conic motion of earth's axis to account for its fixed direction in space* and for precession .. | 1 | 1 |
| Two rectilineal oscillations to account for (imaginary) fluctuations in the rate of precession and in the value of the obliquity†; resolved into 2 circular motions each .. .. | 4 | |
| | —— 9 | —— 3 |

* Copernicus thought with the ancients of the earth's axis as quasi-mechanically attached to the orbital ring (on the analogy of the moon turning always the same face to the earth) and therefore had to introduce a special motion to keep the axis parallel to itself in space.

† See below, p. 204 f.

| | | | | |
|---|---|---|---|---|
| *Moon* | | | | |
| Motions in longitude .. .. | 3 | | 3 | |
| Motion in latitude .. .. .. | 1 | | 1 | |
| | | 4 | | 4 |
| *Three Outer Planets* | | | | |
| Motions in longitude 3 × 3 = .. | 9 | | 9 | |
| Oscillations in latitude resolved into 2 circular motions apiece, 3 × 2 = | 6 | | 6 | |
| | | 15 | | 15 |
| *Venus* | | | | |
| Motions in longitude .. .. | 3 | | 3 | |
| 3 oscillatory motions in latitude resolved into 6 circular motions .. | 6 | | 2 | |
| | | 9 | | 5 |
| *Mercury* | | | | |
| Motions in longitude (including one oscillatory motion) .. .. | 5 | | 5 | |
| Motions in latitude (as Venus) .. | 6 | | 2 | |
| | | 11 | | 7 |
| | | 48 | | 34 |

The count refers to circles in general, that is to say, eccenters, epicycles, deferents, and cycloids to account for rectilineal oscillations.

Apart from the erroneous reference to 34 epicycles, I have nowhere seen a count made of the number of circles in *De revolutionibus.*

Incidentally, as Zinner has pointed out (op. cit., p. 187) even the famous count at the end of the *Commentariolus* is wrong as Copernicus forgot to account for the precession, the motions of the aphelia and the lunar nodes. Taking these into account, the *Commentariolus* uses thirty-eight not thirty-four circles.

10 This was pointed out by A. Koyré, *Nicolas Copernic Des Revolutions des Orbes Célestes* (Paris, 1934), p. 18 n.

11 Peurbach, *Epitomae.* In his *Theoricae,* which is a simplified popular exposure of the system, Peurbach only gives twenty-seven epicycles. Quoted by Professor Koyré (in a private communication to the author, 20 December 1957).

12 The reasons why Copernicus had to increase the number of his circles are:

(a) to compensate for the abolition of Ptolemy's equants;

(b) to account for the imaginary fluctuation in the rate of precession and of the value of the obliquity;

(c) to account for the constant angle of the earth's axis;

(d) because he insisted on resolving rectilinear oscillations into circular motion – which Ptolemy, who was less of a purist, did not bother to do.

This made a total of twenty-one additional epicycles as against a gain of thirteen (five from the annual and eight from the diurnal motion of the earth).

13 The *editio princeps* and the three subsequent editions (Nuremberg, Basle, Amsterdam, and Warsaw) were based not on Copernicus' manuscript but on the copy made of it by Rheticus, which differed from the manuscript in numerous details. Copernicus' original manuscript was only discovered in the 1830's in the library of Count Nostitz in Prague. Nevertheless the Warsaw edition of 1854 still followed the earlier ones, and only the Torun edition of 1837 took account of the discovery of the original.

14 Butterfield, op. cit., p. 30.

15 *De revolutionibus*, Lib. I, Cap. 9.

16 ibid., Lib. I, Cap. 8.

17 H. M. Pachter, *Magic into Science* (New York, 1951), pp. 26, 30.

18 *Letter against Werner*, Prowe, II, p. 176 f. An English translation was published by Rosen, op. cit.

19 Rheticus, *Ephemerides Novae* (Leipzig, 1950), quoted by Prowe, II, p. 391.

20 The last observation of his own (an eclipse of Venus by the moon) which he used in the *Revolutions* was made in March 1529. The book went to the printers in 1542. During the intervening thirteen years, Copernicus continued to make observations and jotted down twenty-two results, but he did not use these in the *Revolutions.*

This enables us to determine the date of the completion of the manuscript with reasonable certainty. It must have been completed after 1529, since the Venus observation just mentioned is entered in the body of the text. It is unlikely that it was completed later than 1532, since observations made in that year are not entered into the text, but inserted on a separate leaf.

He continued to make corrections and alterations in subsequent years, but these were of a minor character.

The statement in the dedication to Paul III that he withheld his work for 'four times nine years' cannot be taken literally. (It is actually an allusion to Horace's *Epistle ad Pisonus*). He evidently brought the heliocentric idea back from Italy when he returned to Ermland in 1506 – which happens to be almost precisely four times nine years before he published the *Revolutions*; the details of the system must have been gradually taking shape in his mind between that earlier date and 1529. He was then in his middle fifties, and after that made no serious attempt to revise his theory.

21 *De revolutionibus*, Lib. III, Cap. 1–4. Misled by these data, Copernicus erroneously concluded that the rate of precession of the equinoxes was non-uniform, and sought to account for its imaginary fluctuations, and the equally imaginary fluctuations of the obliquity of the ecliptic, by two independent oscillatory motions of the earth's axis.

22 *De revolutionibus*, Lib. III, Cap. 4.

23 *Commentariolus*, trans. Rosen, p. 57.

24 ibid., p. 57 f. He gives the same reason in the dedicatory preface to the *Revolutions*. Ptolemy's system, he explains there, agrees fairly well with the phenomena, but it violates 'the first principle of uniformity of motion'. Rheticus too, in the *Narratio prima*, keeps harping on the same subject: 'You see that here in the case of the moon we are liberated from an equant by the assumption of this theory. . . . My teacher dispenses with equants for the other planets as well. . . .' (trans. Rosen, p. 135.) '. . . My teacher saw that only on this [i.e. Copernicus'] theory could all the circles in the universe be satisfactorily made to evolve uniformly and regularly about their own centres, and not about other centres – an essential property of circular motion.' (ibid., p. 137.) [Non-uniform circular motion around a centre is] 'a relation which nature abhors' (ibid., p. 166).

25 *De revolutionibus*, Dedication to Pope Paul III.

26 ibid., Lib. I, Cap. 5.

27 The pseudo-Plutarch's *De placiti philosophorum*, from which Copernicus quoted the passage about Philolaus, Herakleides, etc., says, a few pages earlier (II, 24 quoted by Armitage, p. 88):

> Aristarchus places the sun among the fixed stars, and holds that the earth revolves round the sun.

In Copernicus' version on the manuscript of the *Revolutions*, this became transformed into:

Philolaus perceived the mobility of the earth, and some say that Aristarchus of Samos was of the same opinion. (Prowe II, p. 129.)

But even this watered-down tribute is crossed out in the manuscript. The *name* of Aristarchus actually occurs three times in the *Revolutions* (in Book III, chapters 2, 6, and 13). but these passages merely refer to his observations on the obliquity of the ecliptic and the length of the tropical year. The fact that Aristarchus fathered the heliocentric idea on which Copernicus built his system is nowhere mentioned.

Apart from the brief reference in the pseudo-Plutarch, Copernicus knew about Aristarchus' theory from the classic passage in Archimedes' *Sand-reckoner* (see above, Part One, ch. III, 3), which Regiomontanus, too, had specially marked (cf. Zinner, p. 178).

28 Averroes, *Commentary on Aristotle's Metaphysics*, quoted by Rosen, op. cit., p. 194 f.
29 *De docta ignorantia* (Basle, 1514).
30 op. cit., II, 11, 12, quoted by Armitage, p. 89 f.
31 ibid., p. 102 *seq.*, quoted by Koyré, *From the Closed World to the Infinite Universe* (Baltimore, 1957), p. 14 f.
32 ibid., p. 105 *seq.*, quoted by Koyré, pp. 20, 22.
33 loc. cit.
34 Zinner, op. cit., p. 97.
35 ibid., p. 100.
36 ibid., p. 97.
36a Cf. Prowe, I, 2, p. 480 ff.
37 Zinner, op. cit, p. 133.
38 ibid., p. 132.
39 loc. cit.
40 ibid., p. 135. The daily rotation leaves the apparent movements of the firmament unaltered; the annual revolution should produce a small stellar parallax.
41 There is no direct evidence that Copernicus knew Calcagnini, but they were contemporaries at the small University of Ferrara, and Professor Antonius Leutus, who on 31 May 1503, handed Copernicus the insignia of his doctor's title, was the godfather of Calcagnini.
42 Butterfield, op. cit., p. 29.
43 The semi-diameter of the earth was known to be approx. 4,000

miles and Copernicus believed the earth's distance from the sun to be approx. 1,200 semi-diameters (*De revolutionibus*, Lib. IV, Cap. 21). Hence the diameter of the earth's orbit was believed to be 9·6 million miles.

44 The annual parallax was only proved in 1838 by Bessel.

45 *De revolutionibus*, Lib I, Cap. 10.

46 Burtt, op. cit., p. 25.

47 *De revolutionibus*, Lib. I, Cap. 8.

## *Part Four. The Watershed*

*Joannis Kepleri Astronomi Opera Omnia*, ed. Ch. Frisch, 8 vols., Frankofurti et Erlangae, 1858–71.

A modern collected edition of Kepler's work and correspondence *Johannes Kepler, Gesammelte Werke*, ed. W. v. Dyck ✝ and Max Caspar, in collaboration with Franz Hammer, was begun in 1938. Up to date (1963, volumes I to VII, IX, XIII to XVIII are available. The texts are in the original Latin and medieval German.

The only serious modern work of biography is Max Caspar's *Johannes Kepler* (Stuttgart, 1948).

*Abbreviations*

O.O. – *Opera Omnia.*

*G.W.* – Gesammelte Werke.

Ca. – Caspar's Biography.

### I: THE YOUNG KEPLER

1 O.O., vol. VIII, p. 670 *seq.*, henceforth referred to as 'Horoscope'.

2 In 1945, a French unit was advancing on the town and started shelling it in the mistaken belief that the retreating German army had left a rearguard between its walls. At the critical moment a French officer – whose name was given to me as Colonel de Chastigny – arrived at the scene, identified it as Kepler's birthplace, stopped the firing and saved Weil from destruction.

3 'One of my ancestors, Heinrich, and his brother, Friedrich, were knighted ... in 1430, by the Emperor [Sigismond] on the bridge over the Tiber in Rome.' (Letter from Kepler to Vincento Bianchi, 17 February 1619; G.W., vol. XVII, p. 321). The Patent of Nobility is still extant, but the two Keplers knighted in 1430 were called Friedrich and Konrad, not Friedrich and Heinrich.

4 'Horoscope.'

5 ibid.
6 ibid.
6a Kretschmer, *The Psychology of Men of Genius,* trans. R. B. Cattell (London, 1931).
7 i.e., placed very close to the sun.
8 O.O., vol. V, p. 476 *seq.*; henceforth referred to as 'Memoir'.
9 'Memoir.' Cf. also letter to Herwart von Hohenburg. 9/10 April 1599, G.W. vol. XIII, pp. 305 ff.
10 'Horoscope.'
11 *Johannes Kepler in seinen Briefen,* ed. Caspar and v. Dyck (Munich and Berlin, 1930), vol. I, p. 36.
12 'Memoir.'
13 G.W., vol. XIII, p. 19 f.
14 *Tertius interveniens,* G.W., vol. IV, p. 145 *seq.*
15 *De Stella nova in pede Serpentarii,* G.W., vol. I, p. 147 *seq.*
16 *Tertius interveniens.*
17 *De stella nova,* Cap. 28.
18 *Antwort auf Röslini Diskurs,* G.W., vol. IV, p. 99 *seq.*
19 Ca., 108.
20 *Tertius interveniens.*
21 'Memoir.'
21a *Antwort auf Röslini Diskurs,* p. 127.
22 *Tertius interveniens.*
23 To Herwart, G.W., vol. XIII, pp. 305 ff.

## 2: THE 'COSMIC MYSTERY'

1 *Mysterium Cosmographicum* (G.W., vol. I), *Preface to the Reader.*
2 ibid., loc. cit.
3 loc. sit.
4 Particularly striking is Kepler's advanced relativistic position in the first chapter of the *Mysterium.* For 'metaphysical and physical' reasons, he says, the sun must be in the centre of the world, but this is not necessary for a formally correct description of the facts. Concerning the Ptolemaic and Copernican views of the apparent motion of the fixed stars, he says: 'It is sufficient that both should say (what both really say) that this phenomenon is derived from a contrasting motion between earth and sky.' Regarding the annual revolution, he says that the universe of Tycho (where five planets revolve round the sun and the sun revolves round the earth) is pragmatically as legitimate as the

Copernican. 'Indeed, the proposition "the sun rests in the centre" is too narrow, goes too far. It is sufficient to postulate more generally: "the sun in the centre of the five planets".'

5 In England, the significance of Copernicus was recognized earlier than on the Continent, mainly thanks to two works: firstly Thomas Digges' *A Perfit Description of the Caelestiall Orbes according to the most auncient doctrine of the Pythagorians, latelye reuiued by Copernicus and by Geometricall Demonstrations approued*, which he added, in 1576, to a new edition of his father Leonard Digges' *Prognostication euerlasting*; and secondly, Giordano Bruno's *La cena de le ceneri*, which Bruno wrote during his English sojourn, and which was first published by Carlewood in London in 1584.

6 Cap. 13.

7 By inscribing Mercury's sphere not into the faces of the octahedron, as it ought to be done, but into the square formed by the four median edges. Cap. 13, Note 4.

8 Cap. 15.

9 Cap. 18.

10 ibid., Note 8.

11 Cap. 20.

12 ibid., Notes 2 and 3.

13 The law resulting from this first attempt was: $R_1 : R_2 = P_1 : \frac{P_1 + P_2}{2}$ where $P_1$, $P_2$ are the period, $R_1$, $R_2$ the mean solar distances of two planets. The correct law (Kepler's 'Third Law') is, of course: $R_1 : R_2 = P_1^{\frac{2}{3}} : P_2^{\frac{2}{3}}$.

14 Cap. 21.

15 ibid., Note 7.

16 Ca., p. 78.

17 *Mysterium Cosmographicum*, Dedication of the 2nd Edition.

18 *Astronomia Nova*, summary of Cap. 45.

19 Letter to Maestlin, 3 October 1595. G.W., vol. XIII, p. 33 ff.

20 *Tertius Interveniens*.

21 *Harmonice Mundi*, Lib. IV, Cap. I. G.W., Vol. VI.

22 *Mysterium Cosmographicum*, Cap. XXI, Notes 8 and 11.

23 It is curious to note that no authority writing on Kepler seems to have noticed this stubborn omission of the word 'ellipse'; perhaps because historians of science recoil from the irrationality of their heroes, as Kepler himself recoiled from the apparent irrationality of the elliptic orbits which he discovered.

24 Burtt, *The Metaphysical Foundations of Modern Physical Science* (London, 1932, rev. ed.), p. 203. Burtt is a notable exception from the attitude referred to in the previous note.
25 *Tertius Interveniens.*
26 *Mysterium Cosmographicum*, Preface to the Reader.
26a ibid, Note 8.

3: GROWING PAINS

1 Letter to Friedrich, Duke of Wuerttemberg, 27 February 1596. G.W., vol. XIII, pp. 50 ff.
2 G.W., vol. XIII, pp. 162 ff.
3 Letter to Maestlin, 11 June 1598. G.W., vol. XIII, p. 218 *seq.*
4 'Horoscope.' Cf. also Letter to Maestlin, 10 February 1597, G.W., vol. XIII, p. 104 *seq.*
5 Letter to Maestlin, 9 April 1597, G.W., vol. XIII, p. 113 *seq.*
6 Letter to Herwart, 9/10 April 1599, G.W., vol. XIII, p. 305 *seq.*
7 Letter to an anonymous woman, *c.* 1612, G.W., vol. XVII, p. 39 *seq.*
8 ibid.
9 Letter to Maestlin, 15 March 1598, G.W., vol. XIII, p. 185.
10 E. Reicke, *Der Gelehrte, Monographien zur deutschen Kulturgeschichte*, vol. VII (Leipzig, 1900), p. 120.
11 G.W., vol. XIII, p. 84 f.
12 G.W., vol. XIII, p. 207.
13 Letter to Herwart, 16 December 1598, G.W., vol. XIII, p. 264 *seq.*
14 From the failure of his efforts, Kepler concluded that the parallax of the Polar Star must be smaller than 8′, 'because my instrument does not allow me to measure angles smaller than this. Hence the semi-diameter of the earth's orbit must be smaller than 1/500 of the semi-diameter of the sphere of the fixed stars.'

(Letter to Herwart, G.W., vol. XIII, p. 267 f.)

Copernicus assumed the mean distance of the earth from the sun=1142 earth radii (*De revolutionibus*, Lib. IV, Cap. 21). In round figures, the radius of the earth's orbit thus amounts to $1200 \times 4000 = 4{\cdot}8$ million miles; and the minimum radius of the universe to $4{\cdot}8 \times 500 = 2400$ million miles. Later on, however, in the *Epitome*, he enlarged the radius of the universe to sixty million earth radii, i.e. $24{\cdot}10^{10}$ miles. He arrived at this figure by assuming that the orbital radius of Saturn was the geometrical mean between the radius of the sun and the radius of the sphere of the fixed stars; and the radius of the sun to be fifteen times the radius of the earth. (*Epitome*, IV, 1, O.O., VI, p. 332).

15 Letter to Herwart, 16 December 1598, loc. cit. Kepler himself never accepted infinity. He believed that the fixed stars were all placed at almost exactly the same distance from the sun, so that their 'sphere' (which, of course, he did not regard as real) was only 'two German miles' in thickness. (*Epitome*, IV, 1, O.O., VI, p. 334).

16 Letter to Maestlin, 16 February 1599, G.W., vol. XIII, p. 289 *seq.*

17 ibid.

18 For a profound analysis of the subjective elements in Newton's cosmology, see Burtt, op. cit.

19 Kepler's discoveries were not of the kind that 'lie in the air'; the three laws were the result of a devious *tour de force*, and represent a rather exceptional 'one-man show', as it were. Even Galileo failed to see their point.

20 Letter to Maestlin, 8 December 1598, G.W., vol. XIII, p. 249 *seq.*

21 12 September 1597, G.W., vol. XIII, p. 131 *seq.*

22 Letter to Herwart, 16 December 1598, G.W., vol. XIII, p. 264 *seq.*

23 Letter to Maestlin, 29 August 1599, G.W., vol. XIV, p. 43 *seq.*

24 Letter to Maestlin, 22 November 1599, G.W., vol. XIV, p. 86 *seq.*

25 Maestlin to Kepler, 25 January 1600, G.W., vol. XIV, p. 105 *seq.*

## 4: TYCHO DE BRAHE

1 J. L. E. Dreyer, *Tycho Brahe* (Edinburgh, 1890), p. 27. Dreyer's is the moderin standard biography of Tycho. He also edited Tycho's *Opera Omnia.*

2 loc. cit.

3 op. cit., p. 14.

4 To be precise, he used two threads, passing through two pairs of stars, and intersecting in the Nova.

5 op. cit., p. 86 f.

6 *An Itinerary written by Fynes Morison*, etc. (London, 1617), fol., p. 60, quoted by Dreyer, p. 89.

7 Dreyer, op. cit., p. 105.

8 ibid., p. 262 n.

8a His other, principal achievements were: improved approximations of the sun's and moon's orbits; discovery of the 'moon's equation' (independently from Kepler); demolition of the

Copernican belief in a periodic inequality in the precession of the equinoxes.

9 ibid, p. 261

10 ibid., p. 249 f.

11 ibid., p. 279.

12 *Nicolai Raimari Ursi Dithmarsi Fundamentum astronomicum* (Strasburg, 1588).

13 The only differences between the system of Ursus and the Tychonic system were that in the former the daily rotation was attributed to the earth, in the latter to the fixed stars; and that different orbits were attribued to Mars.

14 To Ursus, 15 November 1595, G.W., vol. XIII, p. 48 f.

15 *Nicolai Raimari Ursi Dithmarsi de astronomicis Hypothesibus,* etc. (Prague, 1597).

16 To Tycho, 13 December 1597, G.W., vol. XIII, p. 154.

17 Tycho to Kepler, 1 April 1598, G.W., vol. XIII, p. 197 *seq.*

18 21 April 1598, G.W., vol. XIII, p. 204 f.

19 19 February 1599, G.W., vol. XIII, p. 286 f.

20 The passage runs:

> Some doctor stopped on his return journey from Italy in Gratz and showed me a book of his [Ursus'] which I hurriedly read in the three days for which I was permitted to keep it. I found in it . . . certain golden rules which, as I remembered, Maestlin had frequently used in Tuebingen, and also the science of the *sine* and of the computation of triangles – subjects which, though generally known, were new to me . . . for afterwards I found in Euclid and Regiomontanus most of what I had ascribed to Ursus.

21 G.W., vol. XIV, p. 89 *seq.*

## 5: TYCHO AND KEPLER

1 Dreyer, op. cit., p. 279.

2 Letter to Herwart, 12 July 1600, G.W., vol. XIV, p. 128 *seq.*

3 Ca., p. 117.

4 To Herwart, 12 July 1600, G.W., vol. XIV, p. 128 *seq.*

5 Ca., p. 119.

6 Tycho to Jessenius, 8 April 1600, G.W., vol. XIV, p. 112 *seq.*

7 April 1600, G.W., vol. XIV, p. 114 *seq.*

7a He had signed, however, a written undertaking to keep all information he obtained from Tycho 'in highest secrecy', i.e. he could publish nothing without Tycho's consent.

8 9 September 1600, G.W., vol. XIV, p. 150 *seq.*

9 9 October 1600, G.W., vol. XIV, p. 115 *seq.*
10 28 August 1600, G.W., vol. XIV, p. 145 *seq.*
11 F. Morison, op. cit.
12 Dreyer, op. cit., p. 386 f.
13 Quoted by Kepler in *Astronomia Nova,* I, Cap. 6.

## 6: THE GIVING OF THE LAWS

*1. ASTRONOMIA NOVA ΑΙΤΙΟΛΟΓΗΤΟΣ, sev PHYSICA COELESTIS, tradita commentariis DE MOTIBUS STELLÆ MARTIS, Ex observationibus G. V. TYCHONIS BRAHE.*

1a *Astronomia Nova,* G.W., vol. III, Preamble to the Table of Contents.
2 ibid., II, Cap. 7.
3 ibid., Dedication.
4 'It is inconceivable that a non-material force should be present in a non-body and should move through space and time,' ibid., I, Cap. 2.
5 ibid, II, Cap. 14.
6 ibid., II, Cap. 14.
7 At a later stage, however, he reverted to the Ptolemaic position.
8 Altogether, Tycho had observed ten oppositions, and Kepler himself two (1602 and 1604). The Tychonic data which he used were those for 1587, 1591–3–5.
8a Letter to Herwart, 12 July 1600, G.W., vol. XIV, p. 132 f.
9 *Astronomia Nova,* II, Cap. 18.
10 ibid., II, Cap. 19.
11 *Science and the Modern World* (Cambridge, 1953 [reprint], p. 3.
12 *Astronomia Nova,* II, Cap. 20, III, Cap. 24.
14 ibid., III, Cap. 22.
14 loc. cit.
15 The observer on Mars went into action each time Mars returned to a given position in its orbit, i.e. had the same heliocentric longitude. Since the sidereal period of Mars was known, the times when this occurred could be determined, and the different positions which the earth occupied at these times could also be determined. The method yielded a series of triangles Mars–Sun–Earth: $MSE_1$, $MSE_2$, $MSE_3$, where the angles at S and E were known (from Tycho's data and/or from Kepler's previously established method of approximation). These yielded the ratios $SE_1/SM$, $SE_2/SM$, $SE_3/SM$; and it was now a simple problem

in geometry to determine the earth's orbit (still assumed to be circular), its eccentricity, and the position of the *punctum equans*. The same method enabled him later on to determine the relative distances Mars–Sun for any observed geocentric longitude of Mars.

16 At the beginning of III, Cap. 33.
17 Table of Contents, summary of Cap. 32.
18 'At other places [not in the vicinity of aphelion and perihelion] there is a very small deviation. . . .' The passage implies that the deviation is negligible. This is true of the earth's orbit, because of its small eccentricity, but not at all true of Mars, with its large eccentricity.
19 That Descartes derived his theory of vortices from Kepler is probable, but unproven.
20 *Astronomia Nova*, III, Cap. 40.
21 loc. cit.
22 loc. cit.
23 To sum up, the three incorrect assumptions are: (a) that the planet's velocity varies in reverse ratio with its distance from the sun; (b) the circular orbit; (c) that the sum of eccentric radii vectors equals the area. The erroneous *physical* hypotheses played only an indirect part in the process.
24 Letter to Longomontanus, 1605, G.W., vol. XV, p. 134 *seq.*
25 *Astronomia Nova*, IV, Cap. 45.
26 loc sit.
27 Letter to D. Fabricius, 18 December 1604, G.W., vol. XV, p. 78 *seq.*
28 Letter to D. Fabricius, 4 July 1603, G.W., vol. XIV, p. 409 *seq.*
29 Letter to D. Fabricius, 18 December 1604.
30 *Astronomia Nova*, IV, Cap. 55.
31 ibid., IV, Cap. 56.
32 ibid., IV, Cap. 58.
33 1605; G.W., vol. XV, p. 134 *seq.*
34 *Mysterium Cosmographicum*, Cap. 18.
35 Cf. *Insight and Outlook* (London and New York, 1949).
35a *Astronomia Nova*, Introduction.
36 Delambre, *Histoire de l'Astronomie Moderne* (Paris, 1891), vol. I, p. 394.
37 Third Letter to Bentley, *Opera Omnia* (London, 1779–85), IV, 380. Quoted by Burtt, op. cit., p. 265 f.
38 Thus, for instance, in Galileo's *Dialogue on the Great World Systems*, it is Simplicius, the naïve Aristotelian, who says: 'The

cause [why bodies fall] is manifest, and everyone knows that it is gravity'; but he is promptly rebuked with: 'You are out, Simplicius; you say that everyone knows that it is *called* gravity, and I do not question you about the name but about the essence of the thing. Of this you know not a tittle more than you know the essence of the mover of the stars in gyration.' (Salusbury trans., ed. Santillana (Chicago, 1953), p. 250.

39 10 February 1605; G.W., vol. XV, p. 145 *seq.*

40 *Astronomia Nova,* III, Cap. 33.

41 ibid., III, Cap. 38.

42 ibid., I, Cap. 6.

43 Max Caspar's introduction to his German translation of the *Astronomia Nova* (Munich and Berlin, 1929), p. 54.

7: KEPLER DEPRESSED

1 Letter to Heydon, October 1605, G.W., vol. XV, p. 231 *seq.*

2 Letter to D. Fabricius, 1 October 1602, G.W., vol. XIV, p. 263 *seq.*

2a Letter to D. Fabricius, February 1604, G.W., vol. XV, p. 17 *seq.*

3 Greetings to the reader! I had intended to address thee, reader, with a longer preface. Yet the mass of political affairs which keep me more than usually busy these days, and the hasty departure of our Kepler, who intends to leave for Frankfurt within the hour, only left me a moment's time to write. But I thought nevertheless that I ought to address a few words to thee, lest ye should become confused by the liberties which Kepler takes in deviating from Brahe in some of his expositions, particularly those of a physical nature. Such liberties can be found in all philosophers since the world existed; and it in no way affects the work of the Rudolphine tables. [This refers to the planetary tables dedicated to Rudolph which Tengnagel had promised to produce, and never did.] You will be able to see from this work that it has been built on the foundations of Brahe ... and that the entire material (I mean the observations) was collected by Brahe. In the meantime, consider Kepler's excellent work ... as a prelude to the Tables and to the Observations to follow which, for the reasons explained, can only be published slowly. Pray with me to the Almighty and all-wise Lord for the rapid progress of this much desired work and for happier days.

Franz Gransneb Tengnagel,
In Campp. Counsellor of his Imperial Majesty.

4 G.W. vol. XV, p. 131 *seq.*

5 D. Fabricius to Kepler, 20 January 1607, G.W., vol. XV, p. 376 *seq.*

6 30 October 1607, G.W., vol. XVI, p. 71.
7 The writer is the Danzig astronomer P. Crueger, quoted in W. v. Dyck and M. Caspar *Nova Kepleriana* 4, Abhandlungen der Bayrischen Ak. d. Wiss, XXXI, p. 105 *seq.*
8 loc. cit.
9 *Astronomiae Pars Optica*, Dedication to Rudolph II, G.W., vol II.
10 Letter to Besold, 18 June 1607, G.W., vol. XV, p. 492.
10a Letter to Herwart, 10 December 1604, G.W., vol. XV, p. 68 f.
11 Letter to Herwart, 24 November 1607, G.W., vol. XVI, p. 78 *seq.*
12 Letter to D. Fabricius, 11 October 1605, G.W., vol. XV, p. 240 *seq.*
13 *Dissertatio cum Nuncio Sidero*, G.W., vol. IV, p. 281 *seq.*
14 There has been some controversy on the question whether the title meant 'messenger' or 'message' – cf. Stillman Drake, *Discoveries and Opinions of Galileo* (New York, 1957), p. 19. Stillman Drake translates the title as *The Starry Messenger*; de Santillana (see below), as *Sidereal Message* (*Dialogue*) or *Starry Message* (*The Crime of Galileo*). I propose to use *Messenger from the Stars*, oI *Star Messenger* for short.

## 8: KEPLER AND GALILEO

1 F. Sherwood Taylor, *Galileo and the Freedom of Thought* (London, 1938), p. 1.
2 This is strictly true for small angles only, but sufficient for practical purposes of time-measurement. The correct law for the pendulum was discovered by Huygens.

The candelabra still shown at the Cathedral of Pisa, whose oscillations are alleged to have given Galileo his idea, was only installed several years after the discovery.
3 His manuscript treatise *De Motu*, written about 1590, and privately circulated, certainly deviates from Aristotelian physics, but by subscribing to the entirely respectable theory of impetus which had been taught by the Paris school in the fifteenth century and by several of Galileo's predecessors and contemporaries. Cf. A. Koyré, *Études Galileennes* (Paris, 1939).
4 About his technical treatise on the proportional compass, see below.
5 Letter to Maestlin, September 1597, G.W., vol. XIII, p. 140 *seq.*
6 G.W., vol. XIII, p. 130 f.
6a *Trattato della Sfera, Opere, Ristampa della Ediz. Nazionale*

(Florence, 1929–39), vol. II, pp. 203–55. Henceforth 'Opere' refers to this edition, except when marked 'Ed. F. Flora', which refers to the handier selection of works and letters in one volume, published in 1953.

7 Quoted by Sherwood Taylor, op. cit., p. 85.

8 G.W., vol. XIII, p. 144 *seq.*

9 G.W., vol. XIV, p. 256.

10 ibid., p. 441.

11 ibid., p. 144 f.

12 It is surprising to read that Prof. Charles Singer attributes the discovery that the nova of 1604 had no parallax to Galileo, and moreover, passing in silence over Tycho's classic book on the nova of 1572, writes:

New stars when previously noticed had been considered to belong to the lower and less perfect regions near the earth. Galileo had thus attacked the incorruptible and interchangeable heavens and had delivered a blow to the Aristotelian scheme, wellnigh as serious as the experiment on the tower of Pisa (sic).

(Ch. Singer, A *Short History of Science to the Nineteenth Century* (Oxford, 1941), p. 206.

Since that experiment is also legendary, Prof. Singer's comparison contains an ironic truth; but this triple misstatement is characteristic of the power of the Galileo myth over some eminent historians of science. Prof. Singer also seems to believe that Galileo invented the telescope (op. cit., 217), that in Tycho's system 'the sun revolves round the earth in twenty-four hours carrying all the planets with it' (ibid., p. 183), that Kepler's Third Law was 'enunciated in the *Epitome Astronomiae*' (ibid., p. 205), etc.

13 Cf. Zinner, op. cit., p. 514.

14 *Le Operazioni delle Compasso Geometrico e Militare* (Padua, 1606); *Opere*, II, pp. 362–405.

15 *Usus et Fabrica Circiui Cuiusdam Proporziones* (Padua, 1607), *Opere*, II, pp. 425–511.

16 Capra's teacher was the distinguished astronomer Simon Marius (1573–1624), discoverer of the Andromeda Nebula, with whom Galileo later became involved in another priority quarrel. See below, p. 468.

17 Letter to B. Landucci, quoted by Gebler, *Galileo Galilei and the Roman Curia* (London, 1879), p. 19.

18 George Fugger (a member of the famous banker's family) in a letter to Kepler, 16 April 1610, G.W., vol. XVI, p. 302.

18a Cf. Zinner, op. cit., p. 345 f.

19 This refers to the first, Latin edition.

20 *Paradise Lost*, book ii, l. 890.

20a *Peregrinatio contra Nuncium Sydereum* (Mantua, 1610).

21 *Ignatius his Conclave.*

22 *Opere*, ed. F. Flora (Milan–Naples, 1953), p. 887 *seq.*

23 ibid., p. 894 *seq.*

24 28 May 1610, G.W., vol. XVI, p. 314.

25 Quoted by E. Rosen, *The Naming of the Telescope* (New York, 1947).

26 Letter to Horky, 9 August 1610, G.W., vol. XVI, p. 323.

26a Poor Kepler is unable to stem the feeling against Your Excellency, for Magini has written three letters, which were confirmed by 24 learned men from Bologna, to give effect that they had been present when you tried to demonstrate your discoveries . . . but failed to see what you pretended to show them.

(M. Hasdale to Galileo, 15 and 28 April 1610, G.W. Vol. XVI, pp. 300 f., 308.)

27 G.W., vol. XVI, p. 319 *seq.*

27a It was probably this letter which led Prof. de Santillana to the erroneous statement: 'It took even Kepler, always generous and open-minded, a whole five months before rallying to the cause of the telescope. . . . His first *Dissertatio cum Nuncio sidereo*, of April 1610, is full of reservations.' (*Dialogue on the Great World Systems* (Chicago 1937, p. 98 n.) Kepler's reservations referred, as we saw, to the priority of the invention of the telescope, not to Galileo's discoveries with it.

28 G.W., vol. XVI, p. 327 *seq.*

29 Except for a short note of introduction to Kepler, which Galileo gave a traveller seventeen years later, in 1627, *Opere*, XIII, p. 347 f.

30 Gebler, op. cit., p. 24.

31 At least, that seems to be the meaning. The word '*umbistineum*' does not exist and may either be derived from '*ambustus*', burnt up, or '*umbo*' = boss, projection.

32 9 January 1611, G.W., vol. XVI, p. 356 *seq.*

33 *Narratio de Observatis a se quatuor Iovis sattelitibus erronibus.*

34 25 October 1610, G.W., vol. XVI, p. 431.

## 9: CHAOS AND HARMONY

1 The book should really be called 'Dioptrics and Catoptrics', for it deals with both refraction and reflection.

2 Except for the Preface.
3 *Ad Vitellionem Paralipomena, quibus Astronomiae Pars Optica.*
4 3 April 1611, G.W., vol. XVI, p. 373 *seq.*
5 Dedication of the *Eclogae Chronicae*, 13 April 1612, quoted in *Johannes Kepler in seinen Briefen*, vol. I, p. 391 *seq.*
6 Ca., p. 243.
7 Ca., p. 252 f.
8 Ca., p. 300.
9 Galileo, as we shall see, was submitted to the much milder form of *territio verbalis*, without actually being led into the torture chamber.
10 Quoted in *Johannes Kepler in seinen Briefen*, vol. II, p. 183 f.
11 *Harmonices Mundi Libri V* (Linz, 1619). The work is sometimes erroneously referred to as 'Harmonices', as if the 's' stood for the plural, whereas it stands, of course, for the genitive.
12 Kepler's translation of the word is *unwissbar.*
13 *Harmonice Mundi*, Book V, Cap. 4.
14 loc. cit.
15 ibid, Cap. 7.
16 Dedication of the *Ephemerides* for 1620 to Lord Napier.
17 ibid.
18 *Harmonice Mundi*, Introduction to Book V.
19 '*Sed res est certissima exactissimaque, quod proportio, quae est inter binorum quorumconque planetarum tempora periodica, sit praecise sesquialtera proportionis mediarum distantiarum, id est orbium ipsorum.*' (ibid., V, Cap. 3, Proposition No. 8.)
20 loc. cit.
21 loc. cit.
22 ibid., appendix to Book V.

## 10: COMPUTING A BRIDE

1 G.W., vol. XVII, p. 79 *seq.* The following is a compressed version.

## 11: THE LAST YEARS

1 To Bianchi, 17 February 1619, G.W.; vol. XVII, p. 321 *seq.*
2 To Bernegger, 20 May 1624, *Johannes Kepler in seinen Briefen*, II, p. 205 f.
3 1 October, 1626. ibid., II, p. 222 ff.
4 loc. cit.

5 Kepler became acquainted with Napier's logarithms in 1617: 'A Scottish baron has appeared on the scene (his name I have forgotten) who has done an excellent thing by transforming all multiplication and division into addition and subtraction....' ibid., II, p. 101.) Since Napier did not at first explain the principle behind it, the thing looked like black magic and was received with scepticism. Old Maestlin remarked: 'It is not fitting for a professor of mathematics to manifest childish joy just because reckoning is made easier.' (Ca. p. 368).

6 To Bernegger, 20 May 1624, see note 2.

7 Ca., p. 302.

8 To Bernegger, 6 April 1627, *Johannes Kepler in seinen Briefen*, 11, p. 236 *seq.*

9 To Bernegger, 22 July 1629, ibid., p. 292.

10 To Bernegger, 2 March 1629, ibid., p. 284 f.

11 To Bernegger, 29 April 1629, ibid., p. 286 f.

12 To Ph. Mueller, 27 October 1629, ibid., p. 297.

13 Cf. Marjorie Nicolson's essay, *Kepler, the Somnium, and John Donne*, in her *Science and Imagination* (Oxford, 1956).

14 Kepler added the following note to this passage:

> We can feel the warmth of the moonlight with the help of an apparatus. For if one gathers the rays of the full moon in a concave parabolical or spherical mirror, then one feels in its focus, where the rays meet, a warm breath, as it were. I noticed that in Linz when I was engaged in other experiments with mirrors, without thinking of the warmth; I involuntarily turned around to see whether somebody was breathing on my hand.

As Ludwig Gunther, who edited and translated the *Somnium* (Kepler's *Traum vom Mond* (Leipzig, 1898), pointed out, this passage establishes Kepler's priority in discovering that the moon reflects not only the light, but also some of the heat of the sun – a fact which was by no means obvious and which (according to Gunther, p. 131) was only established by C. V. Boyse in the 1890s. The ancient believed that the sunlight lost all its warmth when reflected by the moon (cf. Plutarch's *On the Face in the Moon Disc*).

15 Letter to Bartsch, 6 November 1629, *Johannes Kepler in seinen Briefen*, II, p. 303.

16 To Ph. Mueller, 22 April 1630, ibid., p. 316.

17 Bartsch to Ph. Mueller, 3 January 1631, ibid. II, p. 329.

18 ibid. II, p. 325.

19 Ca., p. 431.
20 Quoted by S. Lansius to anon., 24 January 1631, *Johannes Kepler in seinen Briefen*, II, p. 333.
21 From which expression one might infer that they refused him the last sacraments.
22 To Bartsch, *Johannes Kepler in seinen Briefen*, II, p. 308.

## *Part Five. The Parting of the Ways*

### 1: THE BURDEN OF PROOF

1 To Cosmo II, 3 May 1611, quoted by Gebler, op. cit. p. 36.
2 The term was devised by a member of the Lynxes, Demisiani, and announced at the banquet on 14 April 1611. See E. Rosen, *The Naming of the Telescope* (New York, 1947).
3 *Letters on Sunspots*, Third Letter, 1612, trans. Stillman Drake, op. cit., p. 126 f.
4 About this hilarious chapter of science-mythology, see Lane Cooper, *Aristotle, Galileo and the Tower of Pisa* (Ithaca, 1935).
5 Zinner, op. cit., p. 346.
6 This episode was a typically Keplerian comedy of errors. On 28 May 1607, Kepler had observed the sun through a kind of improvised *camera obscura.* It consisted in narrow gaps between the shingles on the roof of his house in Prague. The gaps let the rain into the attic, but each gap deputized for the aperture of a (lensless) camera; by holding a sheet of paper under the slit, Kepler obtained a projected image of the sun. On that particular day he observed on the projected disc of the sun 'a small, almost black dot, approximately like a meagre flea'. When he moved the paper away from the gap, thus enlarging the disc to the size of his palm, the spot grew to the size of 'a small mouse'. Kepler was convinced that the spot was the shadow of Mercury, and that he was observing a transit of that planet across the disc of the sun. He raced up the Hradshin to the Emperor's palace, and conveyed the news by way of a flunkey to Rudolph; raced down again, induced several people to convince themselves of the existence of the black dot and to sign documents testifying to it, and in 1609 published a treatise on the event: *Mercurius in Sole.*
7 *Il Saggiatore*, quoted by Zinner, p. 362.
8 *Letters on Sunspots*, trans. Stillman Drake, op. cit., p. 100.
8a ibid., p. 113 f.
9 ibid., p. 144.

10 Conti to Galileo, 7 July 1612. Quoted by G. de Santillana, *The Crime of Galileo* (Chicago, 1955), p. 27 f.

11 *Opere*, XI, 427, quoted by Stillman Drake, p. 146 f. A number of historians (including, recently, Professor de Santillana) have tried to lend this incident more weight by stating that Lorini had preached a public sermon against Galileo. But had he done so ('on All Souls' Day', as Santillana says), it would be fantastic to assume that he could have denied the fact in writing; besides, Galileo himself says that the incident occurred 'in private discussion' (*Opere*, V, 291, quoted by Drake, p. 147 n.).

12 *Opere*, XI, p. 605 f.; quoted by Drake, p. 151 f.

13 Trans. Drake, op. cit., p. 175.

14 ibid., pp. 181–3.

15 ibid., p. 192 f.

16 ibid., p. 194.

17 ibid., p. 194 f.

18 ibid., p. 213.

19 10 January 1615. Quoted by Gebler, op. cit., p. 52.

20 *Opere*, XII, p. 123. Quoted by Drake, p. 115.

21 Trans. Santillana, op. cit., p. 45 f.

21a Gebler, op. cit., p. 53.

22 He maintained, i.a., 'that Christ was not God, merely an unusually skilled magician ... and that the Devil will be saved'. (Catholic Encyclopaedia on Bruno.)

22a It is surprising to note how indifferently scholars reacted to Bruno's martyrdom, at any rate in Germany. This is illustrated by Kepler's voluminous correspondence, in which every subject under the sky is discussed, but Bruno is hardly mentioned. One of Kepler's favourite pen-friends during his Prague period was the physician Brengger in Kaltbeuren, a man of great erudition and a wide range of interests. In a letter dated 1 September 1607, Brengger mentioned in passing the theory of 'Jordano Bruno of Nola' about the plurality of worlds. This was nearly eight years after Bruno had been executed, but Brengger was evidently unaware of the fact. Kepler answered (on 30 November 1607) that 'not only the unfortunate Bruno, who was roasted on coals in Rome, but my venerated Tycho too believed that the stars are inhabited'. (He actually committed one of his dreadful puns: '*... infelix ille Prunus prunis tostus Romae*'.) In his next letter (7 March 1608) Brengger wrote: 'You say that Jordano Bruno was roasted on coals, from which I gather that he was burnt,' and inquired why this had been done: 'I pity the man.' Kepler answered (on 5 April): 'That Bruno was burnt in Rome I learnt from Master Wackher; he suffered his fate steadfastly. He had asserted the vanity of all religions and had substituted circles and points for God.' Brengger concluded that Bruno must have been insane and wondered where his fortitude had come from if he denied God (25 May 1608). This, then, was the comment of two contemporary scholars on the burning alive of Giordano Bruno (G.W., vol. XVI, pp. 39, 116, 142, 166.)

23 *Opere*, XII, pp. 145–7, quoted by Drake, p. 158.
24 *Opere*, XII, p. 151, quoted by Drake, p. 159.
25 *Lettera del R.P. Maestro Paolo Antonio Foscarini, Carmelitano, sopra l'opinione de i Pittagorici e del Copernico dèlla mobilita della Terra e stabilita del Sole, e il nuove Sisteme del Mondo* (Naples, 1615).
26 Gebler, op. cit., p. 61.
27 Santillana, op. cit., p. 91.
28 Sherwood Taylor, op. cit., p. 85.
29 *Opere* (ed. F. Flora), pp. 999–1007.
30 *Opere*, XII, p. 171 f. Trans. Drake, pp. 162–4, and Santillana, pp. 98–100.
31 *Opere*, XII, pp. .183–5. Trans. Drake, pp. 165–7.
32 Santillana, op. cit., p. 118.
33 Drake, p. 170.
34 Santillana, op. cit., p. 110.
35 Letter to Cardinal Allessandro d'Este, 20 January 1616, trans. Santillana, p. 112 f.
36 ibid., p. 117.
37 ibid., p. 116.
38 *Dialogue of the Great World Systems*, Salusbury trans., ed. Santillana (Chicago, 1953), p. 469; henceforth referred to as *Dialogue*. The Italian title *Dialogo ... sopra i due Massime Sistemi del Mondo* expressly mentions *two* great world systems, the Ptolemaic and the Copernican; but since I have followed Santillana's edition of the Salusbury translation I must refer to it by the title which the editor has given it.
38a He explained this as due to secondary causes operating in inland seas such as the Mediterranean and the Adriatic. See below, pp. 472, 486.
39 H. Butterfield, op. cit., p. 63.
40 Trans. Santillana, op. cit., p. 119.
41 Some of Galileo's biographers are anxious to give the impression that the decree of 5 March was not caused by Galileo's persistent

provocations, but the result of a coldly planned inquisitorial campaign to stifle the voice of science. To prove this, they maintain that the convocation of the Qualifiers was not an *ad hoc* decision, provoked either by Orsini's *démarche* with the Pope or by Galileo's general behaviour in Rome, but that it was the conclusion of continued inquisitorial proceedings, starting with the denunciation by Lorini and Caccini, or even earlier. The 'even earlier' refers to a meeting of the Congregation of the Holy Office back in 1611 at which Bellarmine introduced 'a small item on the agenda': 'Find out whether, in the proceedings against Dr Cesare Cremonini, there is a mention of Galileo, Professor of Philosophy and Mathematics.' Cremonini was an Aristotelian enemy of Galileo's at the University of Padua; he was never brought to trial. The entry dates from the days of Galileo's triumphant visit in Rome, and the matter is never again mentioned in the file. Then there is nothing in the file for five years, until Lorini's charges against the 'Letter to Castelli', which are dismissed, Caccini's testimony in February, and the testimonies of Ximenes and Atavante in November, which brought the proceedings to a close.

But Caccini had mentioned the 'Letters on Sunspots', and on 25 November there is a note in the file referring to an Instruction by the Congregation: 'See the "Letters on Sunspots" by the said Galile.' Then nothing until 23 February next year, when the Qualifiers are convoked to pronounce on the two propositions submitted to them, but without mention of either the 'Sunspots' or Galileo's name. Nevertheless, the abovementioned entry of 25 November is construed as indicating that the proceedings had never been dropped, merely delayed, and that the calling in of the Qualifiers was the final and inevitable result of 'historic fatality'.

The fact is that the Qualifiers were *not* asked to see or censure the 'Letters on Sunspots'; that whoever looked at the book must have seen at once that it contained a single and unobjectionable reference to the Copernican system as a hypothesis; and that the matter was dismissed as the Cremonini and Caccini and Lorini denunciations had been dismissed before.

The absence of any preconceived plan is also illustrated by Bellarmine's letter to Foscarini, and by the clumsy wording of the second question to the Qualifiers, that the earth moves 'according to the whole of itself, also with a diurnal motion' (*ma si move secondo sè tutta, etiam di moto diurno*). Santillana has shown (op. cit., p. 139) that the words, which really make no sense, were picked out of Caccini's garbled version of Copernicanism. If the convocation of the Qualifiers had been planned beforehand, and not an *ad hoc* measure ordered by an irate Pope, the Inquisitor in charge of formulating the questions could surely have prepared something more precise than the phrase he picked from a hurried perusal of the file.

Of the two most recent serious works on Galileo, Stillman Drake maintains that it was Orsini's urging the Pope to rule in favour of Galileo's views that resulted in their prohibition (op. cit., p. 152 n.), whereas Santillana opines that the Orsini story was deliberately 'leaked' by the Inquisition to the Tuscan Ambassador to deceive him, 'whereas the decision had already been taken in secret session many days before. In this way the informers were shielded; things were made to look as though only Galileo's impatience and indiscretion had goaded the long-suffering authorities to action; and with Guiccardini's cooperation the best way had been found to discredit Galileo with the Grand Duke' (op. cit., p. 120). But the reference to 'shielding informers' makes no sense in the context, and the intention to discredit Galileo with the Grand Duke is hardly compatible with the fact that a week after issuing the decree, Pope Paul V received Galileo in gracious audience, and Bellarmine issued a certificate of honour to him. The showdown provoked by Galileo had become unavoidable; once it was over, soothing honours were paid to the Grand Duke's Mathematician.

42 Trans. Santillana, p. 121.
43 ibid., p. 123.
44 To Picchena, 6 March 1616, quoted by Drake, p. 218 f.
45 *Cath. Ency.*, article on 'Galileo'.
45a Santillana, op. cit., p. 90 n.
46 ibid., p. 88.
47 Burtt, op. cit., p. 25.
48 Santillana, op. cit., p. 124.
49 To Picchena, 6 March 1616.
60 *Ut omnino abstineat . . . docere aut defendere sue de ea tractare* (L'Épinois, *Les Pièces du Procès de Galilée* [Rome, Paris, 1877, p. 40]).
51 *Non si possa difendere, ne tenere* (ibid., pp. 72, 75).
52 *Quovis modo teneat, doceat, aut defendat, verbo aut scriptis.* (ibid., p. 40 f.)
52a The latest contribution to the controversy is Santillana's *The*

*Crime of Galileo*, which I have quoted frequently and to which my indebtedness is evident. It is all the more regrettable that on this crucial issue he omits to mention some relevant facts, which to a large extent vitiates his conclusions on the Galileo trial. On p. 128 he says about the controversial minute of 26 February that 'it was a very Catholic historian, but a distinguished one, Professor Franz Reusch, who in the 1870s drew attention' to certain suspicions concerning the form in which the minute of 26 February was written. On p. 131, n., he repeats: 'We have said earlier, and we must emphasize it here, that the first Catholic historian, to our knowledge, to have found that there is something strange about the document is Professor Reusch.' Actually, the first suspicion on the document was cast not by Reusch, but by Emil Wohlwill in *Der Inquisitionsprocess des Galileo Galilei*, published in 1870. This could be regarded as a minor *lapsus* (though the whole Galileo controversy echoes with the name of Wohlwill, who started this particular hare); but since Santillana professes such respect for Reusch it is incomprehensible why he omits to say that it was in fact Reusch who, notwithstanding his initial suspicions about the document, adduced some important arguments in favour of its authenticity. The principal argument of Wohlwill and his followers (Gebler, Cantor, Scartazzini, and others) against the authenticity of the minute had turned on three words: '*successive ac incontinenti*'. The minute said that after Bellarmine had admonished Galileo to abandon his Copernican opinions, *successive ac incontinenti* the Commissary of the Inquisition 'commanded and enjoined' on Galileo the absolute injunction. But, the argument runs, the Holy Office had decreed that the absolute injunction should only be served *in case of Galileo's refusal to submit*, and the words *successive ac incontinenti* indicate that the injunction was served immediately after the admonition without giving Galileo an opportunity to refuse; in other words, that the procedure described by the minute of 26 February contradicted the procedure ordered by the decree of the previous day.

Against this argument Reusch proved that the words '*successive ac incontinenti*' meant in the Vatican usage of the time not 'immediately afterwards' or 'without pause', but simply 'in the sequel' or 'later on'.* The passage is impossible to miss as it is

* F. H. Reusch, *Der Process Galilei's und die Jesuten* (Bonn, 1879), p. 136 f.

specially marked in the list of contents of Reusch's book (p. ix) and it once and for all settled this particular argument. H. Grisar, a Jesuit, dotted the i's by proving that the expression in question was even used to refer to events several days apart.* Yet Santillana (p. 26), ignoring all this (in the same chapter in which he twice quotes Reusch), translates the words '*successive ac inconinenti*' by 'immediately thereafter'.

The auxiliary arguments about the form of the minute, the absence of the notary's signature, etc., which have also been exhaustively dealt with by Reusch and others, are listed by Santillana as if he were unaware of the long and complicated controversy on the subject. He fails to mention that the minute of the meeting of 25 February and the minute of the procedure of 26 February were written by the hand of the same notary. Not the least omission is Santillana's failure to point out that the terms of the injunction as served according to the minute of 26 February, were actually less harsh than those foreseen in the meeting of 25 February. On 25 February, the Holy Office had ordered that in case of Galileo's refusal to obey he should be commanded 'to abstain altogether from teaching or defending this opinion or doctrine and even from discussing it.' But the injunction according to the minute of 26 February, only forbade him to 'hold, teach, or defend in any way whatsoever verbally or in writing' the Corpernican doctrine; the words 'and even from discussing it' are omitted in the minute of 26 February. If that minute had been a fabrication aimed at framing Galileo, why did the fabricator omit precisely those words which would have provided a cast-iron reason for convicting him? It was this last point which convinced Reusch that the charge of a fabrication was logically untenable (op. cit., pp. 144–5).

What are we to conclude? (a) The possibility of a technical forgery has been eliminated by careful analysis of the paper and ink (cf. Gebler, op. cit., pp. 90, 334 *seq.*). (b) The possibility of a *mala fide* fabrication which the notary wrote down on the instructions of some high-placed enemy or enemies of Galileo in the Holy Office is logically untenable on the grounds just explained, and a number of other reasons. (c) Yet certain discrepancies between the minutes of the decision of 25 February, the procedure on 26 February, and Bellarmine's certificate re-

* H. Grisar, S. J. *Galileistudien, Regensburg* (New York and Cincinnati, 1882), pp. 50–1.

main. The fact that the notary did not record Galileo's refusal to acquiesce in Bellarmine's admonition is one; but the shortness and summary nature of the minute (twenty lines in all in L'Épionois' *Pièces du Procès*) might explain this; besides, Galileo may not have formally refused to obey, but merely argued, as was his wont. The watering down of the text of the injunction, and the face-saving testimonial which Bellarmine gave Galileo at his request might perhaps be explained, again with Reusch, by Bellarmine's diplomacy, who, on the one hand, wanted to put an end to the Galilean agitation and, on the other hand, wished to spare his and Duke Cosmo's feelings. This at least seems to be the most plausible assumption, particularly if we remember Bellarmine's letter to Foscarini in which he praised Galileo for acting 'prudently' by treating Copernicus merely as a working hypothesis, when Bellarmine knew the opposite to be the case. But certainty will only become possible when the complete Vatican file is at last made accessible to scholars.

## 2: THE TRIAL OF GALILEO

1 Santillana, op. cit., p. 136.
2 *Dialogue on the Great World Systems*, p. 425 *seq.*
3 Apart from gravity, of course, which does not enter Galileo's picture.
4 Second letter to Mark Welser, trans. Drake, p. 118 f.
5 Trans. Drake, p. 266.
6 ibid., p. 272.
7 ibid., p. 276 f.
8 Santillana, p. 233.
9 ibid., p. 162 f.
10 Gebler, op. cit., p. 115.
10a Some parts of the *Dialogue* were actually written as far back as 1610.
11 *Dialogue*, p. 68 f.
12 ibid., p. 24.
13 ibid., p. 200 *seq.*
14 ibid., p. 178 *seq.*
15 This is not expressly stated, but clearly implied in pp. 458–60.
16 ibid., p. 350.
17 Santillana, in a footnote to *Dialogue*, p. 349.
18 ibid., p. 354.

19 ibid., p. 357.
20 ibid., p. 364.
21 ibid., p. 365.
22 ibid., p. 407.
23 ibid., pp. 362–4.
24 Owing to the moon's revolutions round the earth, the centre of gravity of these two bodies travels in a smaller or larger orbit, and by analogy with an isochronous pendulum, its velocity must also vary. *Dialogue*, pp. 458–60. By the same analogy the tangential velocity of all planets ought to be the same (see above, note 15).
24a ibid., p. 469. The word (which Salusbury translates by 'trifles') is *fanciullezze*.
25 ibid., p. 342 f.
26 ibid., p. 462
27 Santillana, op. cit., p. 183.
28 loc. cit.
29 ibid., p. 184.
30 Gebler, op. cit., p. 161.
31 ibid., p. 183.
32 Santillana, p. 241.
33 ibid., p. 252 ff.
34 ibid., p. 255 f.
35 ibid., p. 256.
36 ibid., pp. 258–60.
37 ibid., p. 292 f.
38 ibid., p. 302.
39 ibid., p. 303.
40 loc. cit.
41 loc. cit.
42 Whereas you, Galileo, son of the late Vincenzo Galilei, Florentine, aged seventy years, were in the year 1615 denounced to this Holy Office for holding as true the false doctrine taught by some that the Sun is the centre of the world and immovable and that the Earth moves, and also with a diurnal motion; for having disciples to whom you taught the same doctrine; for holding correspondence with certain mathematicians of Germany concerning the same; for having printed certain letters, entitled 'On the Sunspots', wherein you developed the same doctrine as true; and for replying to the objections from the Holy Scriptures, which from time to time were urged against it, by glossing the said Scriptures according to your own meaning: and whereas there was thereupon produced the copy of a document in the form of a letter, purporting to be written by you to one formerly

your disciple, and in this divers propositions are set forth, following the position of Copernicus, which are contrary to the true sense and authority of Holy Scripture:

This Holy Tribunal being therefore of intention to proceed against the disorder and mischief thence resulting, which went on increasing to the prejudice of the Holy Faith, by command of His Holiness and of the Most Eminent Lords Cardinals of this supreme and universal Inquisition, the two propositions of the stability of the Sun and the motion of the Earth were by the theological Qualifiers qualified as follows:

The proposition that the Sun is the centre of the world and does not move from its place is absurd and false philosophically and formally heretical, because it is expressly contrary to the Holy Scripture.

The proposition that the Earth is not the centre of the world and immovable but that it moves, and also with a diurnal motion, is equally absurd and false philosophically and theologically considered at least erroneous in faith.

But whereas it was desired at that time to deal leniently with you, it was decreed at the Holy Congregation held before His Holiness on 25 February 1616, that his Eminence the Lord Cardinal Bellarmine should order you to abandon altogether the said false doctrine and, in the event of your refusal, that an injunction should be imposed upon you by the Commissary of the Holy Office to give up the said doctrine and not teach it to others, not to defend it, nor even discuss it; and failing your acquiescence in this injunction, that you should be imprisoned. And in execution of this decree, on the following day, at the Palace, and in the presence of his Eminence, the said Lord Cardinal Bellarmine, after being gently admonished by the said Lord Cardinal, the command was enjoined upon you by the Father Commissary of the Holy Office of that time, before a notary and witnesses, that you were altogether to abandon the said false opinion and not in future to hold or defend or teach it in any way whatsoever, neither verbally nor in writing; and, upon your promising to obey, you were dismissed.

And, in order that a doctrine so pernicious might be wholly rooted out and not insinuate itself further to the grave prejudice of Catholic truth, a decree was issued by the Holy Congregation of the Index prohibiting the books which treat of this doctrine and declaring the doctrine itself to be false and wholly contrary to the sacred and divine Scripture.

And whereas a book appeared here recently, printed last year at Florence, the title of which shows that you were the author, this title being: 'Dialogue of Galileo Galilei on the Great World Systems'; and whereas the Holy Congregation was afterwards informed that through the publication of the said book the false opinion of the motion of the Earth and the stability of the Sun was daily gaining ground, the said book was taken into careful consideration, and in it there was discovered a patent violation of the aforesaid injunction that had been imposed upon you, for in this book you have defended the said opinion previously condemned and to your face declared to be so, although in the said book you strive by various devices to produce the impression that you leave it undecided, and in express terms as probable: which, however, is a most grievous error, as an opinion can in no wise be probable which has been declared and defined to be contrary to divine Scripture.

Therefore by our order you were cited before this Holy Office, where, being examined upon your oath, you acknowledged the book to be written and published by you. You confessed that you began to write the said book about ten or twelve years ago, after the command had been imposed upon you as above; that you requested license to print it without, however, intimating to those who granted you this licence that you had been commanded not to hold, defend, or teach the doctrine in question in any way whatever.

You likewise confessed that the writing of the said book is in many places drawn up in such a form that the reader might fancy that the arguments brought forward on the false side are calculated by their cogency to compel conviction rather than to be easy of refutation, excusing yourself for having fallen into an error, as you alleged, so foreign to your intention, by the fact that you had written in dialogue and by the natural complacency that every man feels in regard to his own subtleties and in showing himself more clever than the generality of men in devising, even on behalf of false propositions, ingenious and plausible arguments.

And, a suitable term having been assigned to you to prepare your defense, you produced a certificate in the handwriting of his Eminence the Lord Cardinal Bellarmine, procured by you, as you asserted, in order to defend yourself against the calumnies of your enemies, who charged that you had abjured and had been punished by the Holy Office, in which certificate it is declared that you had not abjured and had not been punished but only that the declaration made by His Holiness and published by the Holy Congregation of the Index had been announced to you, wherein it is declared that the doctrine of the motion of the Earth and the stability of the Sun is contrary to the Holy Scriptures and therefore cannot be defended or held. And, as in this certificate there is no mention of the two articles of the injunction, namely, the order not 'to teach' and 'in any way', you represented that we ought to believe that in the course of fourteen or sixteen years you had lost all memory of them and that this was why you said nothing of the injunction when you requested permission to print your book. And all this you urged not by way of excuse for your error but that it might be set down to a vainglorious ambition rather than to malice. But this certificate produced by you in your defence has only aggravated your delinquency, since, although it is there stated that said opinion is contrary to Holy Scripture, you have nevertheless dared to discuss and defend it and to argue its probability; nor does the licence artfully and cunningly extorted by you avail you anything, since you did not notify the command imposed upon you.

And whereas it appeared to us that you had not stated the full truth with regard to your intention, we thought it necessary to subject you to a rigorous examination at which (without prejudice, however, to the matters confessed by you and set forth as above with regard to your said intention) you answered like a good Catholic. Therefore, having seen and maturely considered the merits of this your case, together with your confessions and excuses abovementioned, and all that ought justly to be seen and considered, we have arrived at the underwritten final sentence against you:

Invoking, therefore, the most holy name of our Lord Jesus Christ and of His most glorious Mother, ever Virgin Mary, by this our final sentence, which sitting in judgment, with the counsel and advice of the Reverend Masters of sacred theology and Doctors of both Laws, our assessors, we deliver in these writings, in the cause and causes at present before us between the Magnificent Carlo Sinceri, Doctor of both Laws, Proctor Fiscal of this Holy Office, of the one part, and you Galileo Galilei, the defendant, here present, examined, tried, and confessed as shown above, of the other part –

We say, pronounce, sentence, and declare that you, the said Galileo, by reason of the matters adduced in trial, and by you confessed as above, have rendered yourself in the judgment of this Holy Office vehemently suspected of heresy, namely, of having believed and held the doctrine – which is false and contrary to the sacred and divine Scriptures – that the Sun is the centre of the world and does not move from east to west and that the Earth moves and is not the centre of the world; and that an opinion may be held and defended as probable after it has been declared and defined to be contrary to the Holy Scripture; and that consequently you have incurred all the censures and penalties imposed and promulgated in the sacred canons and other constitutions, general and particular, against such delinquents. From which we are content that you be absolved, provided that, first, with a sincere heart and unfeigned faith, you abjure, curse, and detest before us the aforesaid errors and heresies and every other error and heresy contrary to the Catholic and Apostolic Roman Church in the form to be prescribed by us for you.

And, in order that this your grave and pernicious error and transgression may not remain altogether unpunished and that you may be more cautious in the future and an example to others that they may abstain from similar delinquencies, we ordain that the book of the 'Dialogue of Galileo Galilei' be prohibited by public edict.

We condemn you to the formal prison of this Holy Office during our pleasure, and by way of salutary penance we enjoin that for three years to come you repeat once a week the seven penitential Psalms. Reserving to ourselves liberty to moderate, commute, or take off, in whole or in part, the aforesaid penalties and penance.

And so we say, pronounce, sentence, declare, ordain, and reserve in this and in any other better way and form which we can and may rightfully employ. (ibid., pp. 306–10.)

43 I, Galileo, son of the late Vincenzo Galilei, Florentine, aged seventy years, arraigned personally before this tribunal and kneeling before you, Most Eminent and Reverend Lord Cardinals Inquisitors-General against heretical pravity throughout the entire Christian commonwealth having before my eyes and touching with my hands the Holy Gospels, swear that I have always believed, do believe, and by God's help will in the future believe all that is held, preached, and taught by the Holy Catholic and Apostolic Church. But, whereas – after an injunction had been judicially intimated to me by this Holy Office to the effect that I must altogether abandon the false opinion that the Sun is the centre of the world and immovable and that the Earth is not the centre of the world and moves and that I must not hold, defend, or teach in any way whatsoever, verbally or in writing, the said false doctrine, and after it had been notified to me that the said doctrine was contrary to Holy Scripture – I wrote and printed a book in which I discuss this new doctrine already condemned and adduce arguments of great cogency in its favour without presenting any solution of these, I have been pronounced by the Holy Office to be vehemently suspected of heresy, that is to say, of having held and believed that the Sun is the centre of the world and immovable and that the Earth is not the centre and moves:

Therefore, desiring to remove from the minds of your Eminences, and of all faithful Christians, this vehement suspicion justly conceived against me, with sincere heart and unfeigned faith I abjure, curse, and detest the aforesaid errors and heresies and generally every other error, heresy, and sect whatsoever contrary to the Holy Church, and I swear that in future I will never again say or assert, verbally or in writing, anything that might furnish occasion for a similar suspicion regarding me; but, should I know any heretic or person suspected of heresy, I will denounce him to this Holy Office or to the Inquisitor or Ordinary of the place where I may be. Further, I swear and promise to fulfil and observe in their integrity all penances that have been, or that shall be, imposed upon me by this Holy Office. And, in the event of my contravening (which God forbid!) any of these my promises and oaths, I submit myself to all the pains and penalties imposed and promulgated in the sacred canons and other constitutions, general and particular, against such delinquents. So help me God and these His Holy Gospels, which I touch with my hands. (ibid., p. 312)

44 ibid., p. 325.

45 During his Padua days, Galileo had lived with a Venetian woman, Marina Gamba, who bore him two daughters and a son. He parted from her when he moved to the Court of the Medicis in Florence.

46 *Opere*, XVII, p. 247.

3: THE NEWTONIAN SYNTHESIS

1 There is, however, no direct evidence that Descartes derived his vortices from Kepler.

2 William Gilbert. *On the Loadstone and Magnetic Bodies*, trans. Mottelay (New York, 1893), quoted by Burtt, op. cit., p. 157 f.

2a This illustration is from D. Bohm *Causality and Chance in Modern Physics* (London, 1957), p. 43 f.

3 Third Letter to Bentley, *Opere*, IV.

4 The formula for the centrifugal force had been found by Huygens, in his *Horologium Oscillatorium* (1673).

## *Epilogue*

1 See *Insight and Outlook* (London and New York, 1949).

2 Cf. i.a. Ernest Jones, *The Nature of Genius*, British Medical Journal, 4 August 1956.

3 H. Butterfield, op. cit., p. 105.

4 To Herwart, 9–10 April 1599.

5 Ca., p. 105 f.

6 *Tertius Interveniens.*

7 Ca., p. 314.

8 ibid., p. 320.

9 Quoted by Pachter, op. cit., p. 225.

10 *Il Saggiatore*, *Opere*, VI, p. 232.

11 First Letter to Bentley, *Opere*, IV.

12 Third Letter to Bentley, ibid.

13 Quoted by Burtt, p. 289.

14 op. cit., pp. 233–8.

15 Quoted by Butterfield, p. 90.

15a The Bohr theory to which this refers, was the last which, in spite of its paradoxa, provided a kind of imaginable model of the atom. It has now been abandoned in favour of a purely mathematical treatment, which banishes from atomic physics the very idea of a 'model', with the sternness of the Second Commandment ('Thou shalt not make unto thee any graven image').

16 *An Outline of Philosophy*, pp. 163 and 165.

17 J. W. N. Sullivan, *The Limitations of Science* (New York, 1949), p. 68.

18 Quoted by Sullivan, p. 146.

19 *The Mysterious Universe* (Cambridge, 1937), p. 122 f.

20 ibid., p. 137.

21 ibid., p. 100.

22 op. cit., p. 164.

23 Sullivan, op. cit., p. 147.

24 Eddington, *The Domain of Physical Science*, quoted by Sullivan, p. 141.

25 *An Outline of Philosophy*, p. 163.

26 L. L. Whyte, *Accent on Form* (London, 1955), p. 33.

27 *Space and Spirit* (London, 1946), p. 103.

28 Burtt, op. cit., p. 236 f.

29 *The Trail of the Dinosaur* (London and New York, 1955), p. 245 seq. I have borrowed several other passages from that essay, without quotation marks.

# Index